国家卫生和计划生育委员会"十三五"英文版规划教材
全国高等学校教材

供临床医学专业及来华留学生（MBBS）双语教学用

Molecular Biology

医学分子生物学

Jordanka Zlatanova, Kensal E. van Holde

主　编　吕社民
Chief Editor　Shemin Lyu

副主编　陈　娟　　侯　琳
Vice Chief Editor　Juan Chen　Lin Hou

人民卫生出版社
·北　京·

版权所有，侵权必究！

Molecular Biology: structure and dynamics of genomes and proteomes 1st Edition / by Jordanka Zlatanova, Kensal E. van Holde / ISBN:978-0-8153-4504-6

Copyright © 2016 by Graland Science, Taylor & Francis Group, LLC

Authorized translation from English language edition published by Graland Science, part of Taylor & Francis Group LLC; All rights reserved; 本书原版由 Taylor & Francis 出版集团旗下，Graland Science 出版公司出版，并经其授权出版. 版权所有，侵权必究！

People's Medical Publishing House' is authorized to publish and distribute exclusively the Bilingual (China Simplified Characters and English) language edition. This edition is authorized for sale throughout Mainland of China. No part of the publication may be reproduced or distributed by any means, or stored in a database or retrieval system, without the prior written permission of the publisher. 本书中英文双语版授权由人民卫生出版社有限公司独家出版并仅限在中国大陆地区销售。未经出版者书面许可，不得以任何方式复制或发行本书的任何部分。

Copies of this book sold without a Taylor & Francis sticker on the cover are unauthorized and illegal. 本书封面贴有 Taylor & Francis 公司防伪标签，无标签者不得销售。

图书在版编目（CIP）数据

医学分子生物学：英文 /（美）乔丹卡·兹拉坦诺瓦（Jordanka Zlatanova），（美）肯索尔·E. 范霍德（Kensal E. van Holde）著；吕社民主编. —北京：人民卫生出版社，2022.1

全国高等学校临床医学专业第一轮英文版规划教材

ISBN 978-7-117-32810-4

Ⅰ.①医… Ⅱ.①乔…②肯…③吕… Ⅲ.①医学－分子生物学－医学院校－教材－英文 Ⅳ.①Q7

中国版本图书馆 CIP 数据核字（2022）第 006873 号

| 人卫智网 | www.ipmph.com | 医学教育、学术、考试、健康，购书智慧智能综合服务平台 |
| 人卫官网 | www.pmph.com | 人卫官方资讯发布平台 |

图字：01-2018-4644 号

医学分子生物学
Yixue Fenzi Shengwuxue

主　　编：吕社民
出版发行：人民卫生出版社（中继线 010-59780011）
地　　址：北京市朝阳区潘家园南里 19 号
邮　　编：100021
E - mail：pmph @ pmph.com
购书热线：010-59787592　010-59787584　010-65264830
印　　刷：人卫印务（北京）有限公司
经　　销：新华书店
开　　本：850×1168　1/16　　印张：30
字　　数：887 千字
版　　次：2022 年 1 月第 1 版
印　　次：2022 年 3 月第 1 次印刷
标准书号：ISBN 978-7-117-32810-4
定　　价：128.00 元

打击盗版举报电话：010-59787491　E-mail：WQ @ pmph.com
质量问题联系电话：010-59787234　E-mail：zhiliang @ pmph.com

编者（按姓氏笔画排序）

马　宁 Ning Ma	哈尔滨医科大学 Harbin Medical University
王海河 Haihe Wang	中山大学中山医学院 Sun Yat-sen University Zhongshan School of Medicine
王梁华 Lianghua Wang	海军军医大学 Naval Medical University
吕社民 Shemin Lyu	西安交通大学医学部 Xi'an Jiaotong University Health Science Center
关一夫 Yifu Guan	中国医科大学 China Medical University
李冬民 Dongmin Li	西安交通大学医学部 Xi'an Jiaotong University Health Science Center
李冠武 Guanwu Li	汕头大学医学院 Shantou University Medical College
杨　洋 Yang Yang	北京大学医学部 Peking University Health Science Center
张素梅 Sumei Zhang	安徽医科大学 Anhui Medical University
陈　舌 She Chen	复旦大学上海医学院 Fudan University Shanghai Medical College
陈　娟 Juan Chen	华中科技大学同济医学院 Huazhong University of Science and Technology Tongji Medical College
郑　芳 Fang Zheng	西安交通大学医学部 Xi'an Jiaotong University Health Science Center
侯　琳 Lin Hou	青岛大学青岛医学院 Qingdao University Qingdao Medical College
袁　栎 Li Yuan	南京医科大学 Nanjing Medical University
蔡　蓉 Rong Cai	上海交通大学医学院 Shanghai Jiao Tong University School of Medicine

秘　书　郑　芳　Fang Zheng（兼）

全国高等学校临床医学专业第一轮英文版规划教材·编写说明

1995年,我国首次招收全英文授课医学留学生,到2015年,接收临床医学专业MBBS (Bachelor of Medicine & Bachelor of Surgery)留学生的院校达到了40余家,MBBS院校数量、规模不断扩张;同时,医学院校在临床医学专业五年制、长学制教学中陆续开展不同规模和范围的双语或全英文授课,使得对一套符合我国教学实际、成体系、高质量英文教材的需求日益增长。

为了满足教学需求,进一步落实教育部《关于加强高等学校本科教学工作提高教学质量的若干意见(教高[2001]4号)》和《来华留学生医学本科教育(英文授课)质量控制标准暂行规定(教外来[2007]39号)》等相关文件的要求,规范和提高我国高等医学院校临床医学专业五年制、长学制和来华留学生(MBBS)双语教学及全英文教学的质量,推进医学双语教学和留学生教育的健康有序发展,完善和规范临床医学专业英文版教材的体系,人民卫生出版社在充分调研的基础上,于2015年召开了全国高等学校临床医学专业英文版规划教材的编写论证会,经过会上及会后的反复论证,最终确定组织编写一套全国规划的、适合我国高等医学院校教学实际的临床医学专业英文版教材,并计划作为2017年春季和秋季教材在全国出版发行。

本套英文版教材的编写结合国家卫生和计划生育委员会、教育部的总体要求,坚持"三基、五性、三特定"的原则,组织全国各大医学院校、教学医院的专家编写,主要特点如下:

1. 教材编写应教学之需启动,在全国范围进行了广泛、深入调研和论证,借鉴国内外医学人才培养模式和教材建设经验,对主要读者对象、编写模式、编写科目、编者遴选条件等进行了科学设计。

2. 坚持"三基、五性、三特定"和"多级论证"的教材编写原则,组织全国各大医学院校及教学医院有丰富英语教学经验的专家一起编写,以保证高质量出版。

3. 为保证英语表达的准确性和规范性,大部分教材以国外英文原版教科书为蓝本,根据我国教学大纲和人民卫生出版社临床医学专业第八轮规划教材主要内容进行改编,充分体现科学性、权威性、适用性和实用性。

4. 教材内部各环节合理设置,根据读者对象的特点,在英文原版教材的基础上结合需要,增加本章小结、关键术语(英中对照)、思考题、推荐阅读等模块,促进学生自主学习。

本套临床医学专业英文版规划教材共38种,均为国家卫生和计划生育委员会"十三五"规划教材,计划于2017年全部出版发行。

English Textbook for Clinical Medicine · Introduction

In 1995, China recruited overseas medical students of full English teaching for the first time. Up to 2015, more than 40 institutions enrolled overseas MBBS (Bachelor of Medicine & Bachelor of Surgery) students. The number of MBBS institutions and overseas students are continuously increasing. At the meantime, medical colleges' application for bilingual or full English teaching in different size and range in five-year and long-term professional clinical medicine teaching results to increasingly demand for a set of practical, systematic and high-qualified English teaching material.

In order to meet the teaching needs and to implement the regulations of relevant documents issued by Ministry of Education including "Some Suggestions to Strengthen the Undergraduate Teaching and to Improve the Teaching Quality" and "Interim Provisions on Quality Control Standards of International Medical Undergraduate Education (English teaching)", as well as to standardize and improve the quality of the bilingual teaching and English teaching of the five-year, long-term and international students (MBBS) of clinical medicine in China's higher medical colleges so as to promote the healthy and orderly development of medical bilingual teaching and international students education and to improve and standardize the system of English clinical medicine textbooks, after full investigation, People's Medical Publishing House (PMPH) held the writing discussion meeting of English textbook for clinical medicine department of national colleges and universities in 2015. After the repeated demonstration in and after the meeting, PMPH ultimately determined to organize the compilation of a set of national planning English textbooks which are suitable for China's actual clinical medicine teaching of medical colleges and universities. This set will be published as spring and autumn textbooks of 2017.

This set of English textbooks meets the overall requirements of the Ministry of Education and National Health and Family Planning Commission, the editorial committee includes the experts from major medical colleges and universities as well as teaching hospitals, the main features are as follows:

1. Textbooks compilation is started to meet the teaching needs, extensive and deep research and demonstration are conducted across the country, the main target readers, the model and subject of compilation and selection conditions of authors are scientifically designed in accordance with the reference of domestic and foreign medical personnel training model and experience in teaching materials.

2. Adhere to the teaching materials compiling principles of "three foundations, five characteristics, and three specialties" and "multi-level demonstration", the organization of English teaching experts with rich experience from major medical schools and teaching hospitals ensures the high quality of publication.

3. In order to ensure the accuracy and standardization of English expression, most of the textbooks are modeled on original English textbooks, and adapted based on national syllabus and main content of the eighth round of clinical medicine textbooks which were published by PMPH, fully reflecting the scientificity, authority, applicability and practicality.

4. All aspects of teaching materials are arranged reasonably, based on original textbooks, the chapter summary, key terms (English and Chinese), review questions, and recommended readings are added to promote students' independent learning in accordance with teaching needs and the characteristics of the target readers.

This set of English textbooks for clinical medicine includes 38 species which are among "13th Five-Year" planning textbooks of National Health and Family Planning Commission, and will be all published in 2017.

全国高等学校临床医学专业第一轮英文版规划教材·教材目录

	教材名称		主审	主编	
1	人体解剖学	Human Anatomy		刘学政	
2	生理学	Physiology		闫剑群	
3	医学免疫学	Medical Immunology		储以微	
4	生物化学	Biochemistry		张晓伟	
5	组织学与胚胎学	Histology and Embryology		李 和	
6	医学微生物学	Medical Microbiology		郭晓奎	
7	病理学	Pathology		陈 杰	
8	医学分子生物学	Molecular Biology		吕社民	
9	医学遗传学	Medical Genetics		傅松滨	
10	医学细胞生物学	Medical Cell Biology		刘 佳	
11	病理生理学	Pathophysiology		王建枝	
12	药理学	Pharmacology		杨宝峰	
13	临床药理学	Clinical Pharmacology		李 俊	
14	人体寄生虫学	Human Parasitology		李学荣	
15	流行病学	Epidemiology		沈洪兵	
16	医学统计学	Medical Statistics		郝元涛	
17	核医学	Nuclear Medicine		黄 钢	李 方
18	医学影像学	Medical Imaging		申宝忠	龚启勇
19	临床诊断学	Clinical Diagnostics		万学红	
20	实验诊断学	Laboratory Diagnostics		胡翊群	王 琳
21	内科学	Internal Medicine		文富强	汪道文
22	外科学	Surgery		陈孝平	田 伟
23	妇产科学	Obstetrics and Gynecology	郎景和	狄 文	曹云霞
24	儿科学	Pediatrics		黄国英	罗小平
25	神经病学	Neurology		张黎明	
26	精神病学	Psychiatry		赵靖平	
27	传染病学	Infectious Diseases		高志良	任 红

全国高等学校临床医学专业第一轮英文版规划教材·教材目录

	教材名称		主审	主编
28	皮肤性病学	Dermatovenereology	陈洪铎	高兴华
29	肿瘤学	Oncology		石远凯
30	眼科学	Ophthalmology		杨培增　刘奕志
31	康复医学	Rehabilitation Medicine		虞乐华
32	医学心理学	Medical Psychology		赵旭东
33	耳鼻咽喉头颈外科学	Otorhinolaryngology-Head and Neck Surgery		孔维佳
34	急诊医学	Emergency Medicine		陈玉国
35	法医学	Forensic Medicine		赵　虎
36	全球健康学	Global Health		吴群红
37	中医学	Chinese Medicine		王新华
38	医学汉语	Medical Chinese		李　骢

DEDICATION

We dedicate this work to the memory of Dr. E. Morton Bradbury, a physicist who taught us how the most sophisticated physical techniques could be applied to solve problems in biology.

ACKNOWLEDGMENTS

The authors and publisher of *Molecular Biology: Structure and Dynamics of Genomes and Proteomes* specially acknowledge and thank Kristopher J. Koudelka (Point Loma Nazarene University) for creating the tutorials. The contributions of the following people are also gratefully acknowledged:

Ivan Dimitrov (The University of Texas Southwestern Medical Center) for Figure 6.18, 6.19, 6.20; Aleksandra Kuzmanov (University of Wyoming) for the gene therapy section of Chapter 3; Jean Marc Victor (Université Pierre-et-Marie-Curie) and Thomas Bishop (Louisiana Tech University) for help with Chapter 5. We also thank the numerous researchers that provided useful resources in terms of figures and expert advice.

The following scientists and instructors provided valuable commentary as readers, reviewers, and advisors during the development of the book:

Steven Ackerman (University of Massachusetts Boston); Paul Babitzke (The Pennsylvania State University); Aaron Cassill (The University of Texas at San Antonio); Scott Cooper (University of Wisconsin-La Crosse); Raymond Deshaies (California Institute of Technology); Martin Edwards (Newcastle University [UK]); Yiwen Fang (Loyola Marymount University); Errol C. Friedberg (The University of Texas Southwestern Medical Center); Fátima Gebauer Hernández (Centre for Genomic Regulation [Spain]); Paul D. Gollnick (State University of New York at Buffalo); Paul Gooley (The University of Melbourne [Australia]); Leslie A. Gregg-Jolly (Grinnell College); Andrew W. Grimson (Cornell University); David Hess (Santa Clara University); Walter E. Hill (The University of Montana); Peter L. Jones (University of Massachusetts Medical School); Nemat O. Keyhani (University of Florida); Raida Wajih Khalil (Philadelphia University [Jordan]); Hannah Klein (New York University); Kristopher J. Koudelka (Point Loma Nazarene University); Stephen Kowalczykowski (University of California, Davis); Krzysztof Kuczera (The University of Kansas); Gary R. Kunkel (Texas A&M University); Richard LeBaron (The University of Texas at San Antonio); Boris Lenhard (Imperial College London [UK]); Diego Loayza (Hunter College of The City University of New York); William F. Marzluff (The University of North Carolina at Chapel Hill); Mitch McVey (Tufts University); Marcel Mechali (National Centre for Scientific Research [France]); Corinne A. Michels (Queens College, City University of New York); Peter B. Moore (Yale University); Daniel Moriarty (Siena College); Greg Odorizzi (University of Colorado Boulder); Wilma K. Olson (Rutgers, The State University of New Jersey); Wade H. Powell (Kenyon College); Susan A. Rotenberg (Queens College, City University of New York); Wilma Saffran (Queens College, City University of New York); Michael J. Smerdon (Washington State University); Kathryn Leigh Stoeber (Anglia Ruskin University [UK]); Francesca Storici (Georgia Institute of Technology); Andrew Arthur Travers (University of Cambridge [UK]); Edward N. Trifonov (University of Haifa [Israel]); Peter H. von Hippel (University of Oregon); Hengbin Wang (The University of Alabama at Birmingham); Carol Wilusz (Colorado State University); Xuewu Zhang (The University of Texas Southwestern Medical Center); Zhaolan Zhou (University of Pennsylvania).

LIST OF AUTHORS

1. Jordanka Zlatanova
Professor Emeritus
Department of Molecular Biology at the University of Wyoming
She earned her Ph.D. and D.Sc. degrees in Cellular and Molecular Biology from the Bulgarian Academy of Sciences.

2. Kensal E. van Holde
Distinguished Professor
Emeritus in the Department of Biochemistry and Biophysics at Oregon State University
He earned his Ph.D. in Physical Chemistry at the University of Wisconsin-Madison.

ADAPTATION PREFACE

Molecular biology is a discipline to study the phenomena, principles, and essence of life at the molecular level. Life molecules called biomacromolecules include nucleic acids, proteins, polysaccharides, and lipids, all of which are the objectives that traditional biochemists work with. This indicates no difference between biochemistry and molecular biology, which in many aspects is correct since their contexts are overlapped and both study life in the molecular life. Of course, many biochemists believe that molecular biology is the later stage of biochemistry development since the seminal finding of the double helix model of DNA. New generation biologists including the discoverers of the critical findings in studying the gene think themselves different from the conventional biochemists. Molecular biology is defined as studying the gene, a segment of DNA expressing functional products, such as RNA and proteins. In addition, the knowledge and technology of molecular biology have been incorporated into the diverse disciplines of medicine, which promotes and develops a new branch of medicine called molecular medicine. Molecular medicine means understanding etiology, pathogenesis, development and prognosis of the disease, and diagnosing, treating, and preventing the disease by using molecular biology. So as medical students they must have been familiar with molecular biology not just for reading the paper, doing experiments also for using proper molecular diagnosis and treatment. The course and textbook of molecular biology suitable to the students, Chinese or international, training under the Chinese medical education system are fundamental and necessary.

There are a series of eminent English textbooks on molecular biology. Although they are prevalent and have received excellent feedback from the teachers and students, under comprehensive consideration we have chosen the textbook titled Molecular Biology: Structure and Dynamics of Genomes and Proteomes written by Drs. Jordanka Zlatanova and Kensal E. van Holde, edited by Summers Scholl, and published by Garland Science as our original model. According to the contract of copyright transition between the two companies, we rearranged the chapters and remolded each chapter's context including deleting and adding some contexts. We added especially two new chapters that are: Chapter 15 Disease Gene, Chapter 16 Gene Diagnosis and Therapy to meet the needs of Chinese medical students. So, the revised textbook includes in total 16 chapters, and each chapter contains the main context: Key concepts, Key words (in English and Chinese), Questions, and References.

Under the help of the People's Medical Publishing House we recruited 15 professors from 13 universities and performed editor committer. As a professor in molecular biology, all the members have long-term experience in teaching molecular biology and are good at using English. We have had several meetings to discuss the outline and have reviewed the related chapters with each other. During this writing period, the COVID-19 has been pandemic in the world. All the members overcame different difficulties, took efforts, and concentrated on the writing. Finally, they submitted good-quality chapters following the schedule. Without their professional contribution, this work could never be finished. So, I should express my great gratitude to all the professors who actively participated in this work.

We also indebted the company to have agreed on us to use their products as the model and original authors who have written the excellent textbook. Dr. Fang Zheng, the secretary of the editor committer, has made efforts and given lot of time to communicate with all the members. It is she who has made the members from different universities became a harmonious and effective working group. We appreciated her fantastic work for completing the task, and her English ability and high effectiveness have given us deep expression.

Finally, we should say that this is our first time to had edited an academic book in English. Another is that English is not our mother tongue per se, although almost all the members have some experience studying in western countries. There are a few errors in the book we guess. We hope all the readers would tell us if you find it and we will correct in the next printing or edition.

Shemin Lyu, M.D., Ph.D.

PREFACE

Molecular biology is the study of biology taken to the molecular level. It reveals the essential principles behind the transmission and expression of genetic information in terms of DNA, RNA, and proteins.

Having been long engaged in collaborative research, we realized that the field of molecular biology has undergone great changes in recent years, becoming a highly structure-based science. Whether it is the elegant, dynamic structure of the ribosome, and how that structure explains the details of protein synthesis, or the complex organization of the human genome and its transcription, structural information is the key to understanding molecular mechanisms. For over a decade, one of us has taught an advanced molecular biology course, both in the classroom and online, and simply could not find a suitable text. A new textbook with contemporary coverage was needed.

Molecular Biology: Structure and Dynamics of Genomes and Proteomes elucidates the exquisite relationship between molecular structure and dynamics and the transmission and expression of genetic information. New techniques, ranging from single-molecule methods to whole-genome sequencing, have deepened our understanding of molecular processes and widened the scope of our vision of their interrelationships. These important advances present a new paradigm and must be considered in context of the dramatic history of the field—one of the most remarkable in modern sciences. An important example is our frequent reference to the ENCODE project and its implications. At the other end of the scale are the numerous illustrations of molecular interactions by cryo-electron microscopy and single-molecule dynamics.

From a pedagogical point of view, molecular biology is becoming more dependent on visual representations, and our goal was to create a dynamic and engaging illustration program. Accordingly, nearly 700 illustrations comprise a substantial portion of the book. We feel that legends should fully complement the figures so that figure and legend together can be essentially self-standing. Therefore, we have provided comprehensive explanatory detail within the figures themselves. The text is written primarily for students at the upper undergraduate to first-year graduate level. The book can be adapted for students with or without an introductory course in biochemistry. A number of features make it flexible in approach: the first four introductory chapters could be covered very quickly as a review by an advanced audience, but they provide essential background for students with less experience in the biological sciences. In a similar vein, we have organized contextual and practical material (history, techniques, general cell biology background, and medical applications) into supplemental boxes. So that this book is useful to as wide a spectrum of students as possible, certain boxes represent a deeper analysis of a topic than will be needed for every course—labeled as "A Closer Look." These insights can be bypassed for a more direct, compact course, but they are available to those instructors or students who wish to explore more deeply into the topic. An additional feature is our focus on references that cover cutting-edge techniques, like single-molecule methods, that allow true molecular insights. Every effort has been made to consult and refer to the most current work in every area.

With these devices, the text offers readable, essential coverage of the field. The book is organized as follows. We begin with two introductory chapters, first presenting the basic ideas and development of molecular biology and elements of genetics. There follow two more chapters detailing the substances under study: proteins and nucleic acids. These four chapters could be used for a brief review in an advanced course. The processes by which genetic information is expressed, regulated, and maintained compose the core of the book. These are taken in the order of transcription, translation, and replication, with each topic requiring several chapters to cover structures, mechanisms, and regulation. Special chapters in the core are devoted to recombinant DNA techniques and the structure of the genome. The book concludes with chapters on genetic recombination and DNA repair.

Finally, a special note to the student: you are entering one of the most dynamic and exciting areas of science. Enjoy it!

Jordanka Zlatanova
Kensal E. van Holde

CONTENTS

Chapter 1　From Classical Genetics to Molecular Genetics ··· 1
1.1　Introduction ··· 1
1.2　Classical genetics and the rules of trait inheritance ··· 1
  1.2.1　Gregor Mendel developed the formal rules of genetics ··· 1
  1.2.2　Mendel's laws have extensions and exceptions ··· 6
  1.2.3　Genes are arranged linearly on chromosomes and can be mapped ··· 9
  1.2.4　The nature of genes and how they determine phenotypes was long a mystery ··· 9
1.3　The great breakthrough to molecular genetics ··· 12
  1.3.1　Bacteria and bacteriophage exhibit genetic behavior and serve as model systems ··· 12
  1.3.2　Transformation and transduction allow transfer of genetic information ··· 13
  1.3.3　The Watson-Crick model of DNA structure provided the final key to molecular genetics ··· 13

Chapter 2　Structures and Functions of Proteins and Nucleic Acids ··· 16
2.1　Introduction ··· 16
2.2　Proteins ··· 18
  2.2.1　Amino acids and peptides ··· 18
  2.2.2　Levels of structure in the polypeptide chain ··· 20
  2.2.3　Protein folding ··· 32
  2.2.4　Protein destruction ··· 35
2.3　Nucleic acids ··· 39
  2.3.1　Chemical structure of nucleotides ··· 39
  2.3.2　Physical structure of DNA ··· 41
  2.3.3　Physical structures of RNA ··· 52
  2.3.4　One-way flow of genetic information ··· 53
  2.3.5　Methods used to study nucleic acids ··· 53

Chapter 3　Recombinant DNA: Principles and Applications ··· 58
3.1　Introduction of homologous recombination and cloning ··· 58
  3.1.1　The beginnings of recombinant DNA technology ··· 58
  3.1.2　Homologous recombination and cloning ··· 59
3.2　Construction of recombinant DNA molecules ··· 61
  3.2.1　Major classes of restriction endonucleases ··· 61
  3.2.2　Recognition sequences for type II restriction endonucleases ··· 62
  3.2.3　DNA ligase joins linear pieces of DNA ··· 64
  3.2.4　Sources of DNA for cloning ··· 65
3.3　Vectors for cloning ··· 66
  3.3.1　Genes coding for selectable markers are inserted into vectors during their construction ··· 66
  3.3.2　Plasmid DNA were the first cloning vector ··· 66
  3.3.3　Recombinant bacteriophages can serve as vectors ··· 67
  3.3.4　Cosmids and phagemids expand the repertoire of cloning vector ··· 70

3.4 Expression of recombinant genes ⋯ 71
   3.4.1 Expression vectors allow regulated and efficient expression of cloned genes ⋯ 71
   3.4.2 Expression systems ⋯ 72
3.5 Introducing recombinant DNA into host cells ⋯ 72
   3.5.1 Numerous host-specific techniques are used to introduce recombinant DNA molecules into living cells ⋯ 72
   3.5.2 Transient and stable transfection assays ⋯ 73
3.6 Constructing DNA libraries ⋯ 74
   3.6.1 Type of different libraries ⋯ 74
   3.6.2 Library screening and probes ⋯ 74
3.7 Sequencing of entire genomes ⋯ 76
   3.7.1 Genomic libraries contain the entire genome of an organism as a collection of recombinant DNA molecules ⋯ 76
   3.7.2 There are two approaches for sequencing large genomes ⋯ 77
3.8 Practical application of recombinant DNA technologies ⋯ 77
   3.8.1 Gene therapy ⋯ 77
   3.8.2 Delivering a gene into sufficient cells within a specific tissue and ensuring its subsequent long-term expression is a challenge ⋯ 77

Chapter 4 Tools for Analyzing Gene Expression ⋯ 81
4.1 Introduction ⋯ 81
4.2 Gene isolation and detection ⋯ 81
   4.2.1 Agarose gel electrophoresis of DNA ⋯ 81
   4.2.2 Gradient centrifugation ⋯ 82
   4.2.3 Nucleic acid hybridization ⋯ 83
   4.2.4 Polymerase chain reaction ⋯ 85
   4.2.5 *In vitro* mutagenesis ⋯ 87
   4.2.6 DNA sequencing ⋯ 88
4.3 Analysis at the level of gene transcription: RNA expression and localization ⋯ 89
   4.3.1 Reverse transcription PCR ⋯ 89
   4.3.2 Northern blotting ⋯ 90
   4.3.3 RNase protection assay (RPA) ⋯ 90
   4.3.4 *In situ* hybridization ⋯ 90
   4.3.5 DNA microarrays ⋯ 91
4.4 Analysis of the transcription rates ⋯ 91
   4.4.1 S1 nuclease protection and primer extension ⋯ 91
   4.4.2 Rapid amplification of cDNA ends ⋯ 92
   4.4.3 Reporter genes ⋯ 93
   4.4.4 DNA footprinting ⋯ 94
   4.4.5 Electrophoretic mobility shift assay ⋯ 94
   4.4.6 Chromatin immunoprecipitation ⋯ 96
4.5 Analysis at the level of translation: protein expression and localization ⋯ 98
   4.5.1 Western blotting ⋯ 101
   4.5.2 Enzyme-linked immunosorbent assay (ELISA) ⋯ 101
   4.5.3 Immunohistochemistry (IHC), immunocytochemistry (ICC) and immunofluorescence (IF) ⋯ 102
   4.5.4 Flow cytometry ⋯ 103
4.6 Antisense technology ⋯ 104
   4.6.1 Antisense oligonucleotides ⋯ 104
   4.6.2 RNA interference (RNAi) ⋯ 105

Chapter 5 The genetic Code, Genes, and Genome ... 107
- 5.1 Introduction ... 107
- 5.2 The genetic code, genes, and genomes ... 108
  - 5.2.1 Genes as nucleic acid repositories of genetic information ... 108
  - 5.2.2 Relating protein sequence to DNA sequence in the genetic code ... 109
  - 5.2.3 Discovery from the eukaryotic cell: introns and splicing ... 112
  - 5.2.4 Genes from a new and broader perspective ... 118
  - 5.2.5 Comparing whole genomes and new perspectives on evolution ... 121
- 5.3 Physical structure of the genomic material ... 123
  - 5.3.1 Prokaryotic and viral genome ... 123
  - 5.3.2 Eukaryotic chromatin ... 125
  - 5.3.3 Lateral gene transfer in the eukaryotic genome ... 134

Chapter 6 Protein-Nucleic Acid Interactions and Protein-protein Interactions ... 137
- 6.1 Introduction ... 137
- 6.2 Protein-nucleic acid interactions ... 137
  - 6.2.1 DNA-protein interactions ... 139
  - 6.2.2 RNA-protein interactions ... 146
  - 6.2.3 Studying protein-nucleic acid interactions ... 149
- 6.3 Protein-protein interactions ... 150
  - 6.3.1 Protein-Protein Interactions Essentials ... 150
  - 6.3.2 Analysis of structural analysis of proteins ... 151
  - 6.3.3 Methods to investigate protein-protein interactions ... 154
  - 6.3.4 The proteome and protein interaction networks ... 156

Chapter 7 Mechanism of Transcription ... 163
- 7.1 Introduction ... 163
- 7.2 Overview of transcription ... 163
  - 7.2.1 There are aspects of transcription common to all organisms ... 165
  - 7.2.2 Transcription requires the participation of many proteins ... 166
  - 7.2.3 Transcription is rapid but often interrupted by pauses ... 167
  - 7.2.4 Transcription can be visualized by electron microscopy ... 168
- 7.3 Transcription in bacteria ... 169
  - 7.3.1 Transcription initiation begins with a multi-subunit polymerase holoenzyme assembly ... 169
  - 7.3.2 The initial step of elongation is frequently aborted ... 172
  - 7.3.3 Elongation in bacteria must overcome topological problems ... 173
  - 7.3.4 There are two mechanisms for transcription termination in bacteria ... 175
- 7.4 Transcription in eukaryotes ... 176
  - 7.4.1 Overview of transcriptional regulation of eukaryotes ... 176
  - 7.4.2 Transcription by RNA polymerase II ... 180
  - 7.4.3 Transcription by RNA polymerase I and III ... 195
  - 7.4.4 Transcription in eukaryotes: pervasive and spatially organized ... 199
  - 7.4.5 Methods for Studying Transcription ... 202
- 7.5 Understanding transcription is useful in clinical practice ... 205
  - 7.5.1 Some common antibiotics that act by inhibiting bacterial transcription ... 205
  - 7.5.2 Drugs currently used to treat AIDS act on the viral reverse transcriptase ... 206

Chapter 8 Transcription Regulation ... 209
- 8.1 Introduction ... 209
- 8.2 Regulation of transcription in bacteria ... 209

- 8.2.1 General models for regulation of transcription 210
- 8.2.2 Specific regulation of transcription 211
- 8.2.3 Transcriptional regulation of operons important to bacterial physiology 214
- 8.2.4 Other modes of gene regulation in bacteria 220
- 8.2.5 Coordination of gene expression in bacteria 223
- 8.3 Regulation of transcription in eukaryotes 225
  - 8.3.1 Regulation of transcription initiation: regulatory regions and transcription factors 225
  - 8.3.2 Regulation of transcriptional elongation 233
  - 8.3.3 Transcription regulation chromatin structure 234
  - 8.3.4 Regulation of transcription by histone modification variants 235
  - 8.3.5 DNA methylation 250
- 8.4 Long noncoding RNAs in transcriptional regulation 256
  - 8.4.1 Noncoding RNAs play surprising roles in regulating transcription 256
  - 8.4.2 The sizes and genomic locations of noncoding transcripts are remarkably diverse 256
- 8.5 X chromosome inactivation 258
  - 8.5.1 Random X chromosome inactivation in mammals 258
  - 8.5.2 Molecular mechanisms for stable maintenance of X chromosome inactivation 259
- 8.6 Methods for measuring the activity of transcriptional regulatory elements 259

Chapter 9 Transcription Regulation in the Human Genome 263

- 9.1 Introduction 263
- 9.2 Rapid full-genome sequencing allows deep analysis 263
- 9.3 Basic concepts of ENCODE 264
  - 9.3.1 ENCODE depends on high-throughput, massively processive sequencing and sophisticated computer algorithms for analysis 264
  - 9.3.2 The ENCODE project integrates diverse data relevant to transcription in the human genome 264
- 9.4 Regulatory DNA sequence elements 266
  - Seven classes of regulatory DNA sequence elements make up the transcriptional landscape 266
- 9.5 Specific findings concerning chromatin structure from ENCODE 267
  - 9.5.1 Millions of DNase I hypersensitive sites mark regions of accessible chromatin 267
  - 9.5.2 DNase I signatures at promoters are asymmetric and stereotypic 267
  - 9.5.3 Nucleosome positioning at promoters and around TF-binding sites is highly heterogeneous 269
- 9.6 ENCODE insights into gene regulation 270
  - 9.6.1 Distal control elements are connected to promoters in a complex network 270
  - 9.6.2 Transcription factor binding defines the structure and function of regulatory regions 272
  - 9.6.3 TF-binding sites and TF structure co-evolve 272
- 9.7 ENCODE overview 273
  - 9.7.1 What have we learned from ENCODE, and where is it leading? 273
  - 9.7.2 Certain methods are essential to ENCODE project studies 274

Chapter 10 RNA Processing 276

- 10.1 Introduction 276
- 10.2 Processing of eukaryotic mRNA: end modifications 277
  - 10.2.1 Eukaryotic mRNA capping is co-transcriptional 277
  - 10.2.2 Polyadenylation at the 3′-end serves a number of functions 277
- 10.3 Processing of eukaryotic mRNA: splicing 279
  - 10.3.1 The splicing process is complex and requires great precision 279
  - 10.3.2 Splicing is carried out by spliceosomes 280
  - 10.3.3 Splicing can produce alternative mRNAs 281

10.3.4	Tandem chimerism links exons from separate genes	283
10.3.5	*Trans*-splicing combines exons residing in the two complementary DNA strands	286
10.3.6	Regulation of splicing and alternative splicing	286
10.3.7	Self-splicing: introns and ribozymes	291
10.4	Methylation of mRNA: N^6-methyladenosine	294
10.4.1	The m^6A methylation and demethylation reaction of mRNA	294
10.4.2	The biological functions of mRNA m^6A methylation	295
10.5	Overview: the history of an mRNA molecule	296
10.5.1	Proceeding from the primary transcript to a functioning mRNA requires a number of steps	296
10.5.2	mRNA is exported from the nucleus to the cytoplasm through nuclear pore complexes	297
10.5.3	RNA sequence can be edited by enzymatic modification even after transcription	298
10.6	Processing of constitutive noncoding RNA, tRNA and rRNA	299
10.6.1	tRNA processing is similar in all organisms	299
10.6.2	All three mature ribosomal RNA molecules are cleaved from a single long precursor RNA	299
10.7	Biogenesis and functions of small silencing RNAs	301
10.7.1	All ssRNAs are produced by processing from larger precursors	301
10.7.2	LncRNAs are produced by processing from larger precursors	305
10.7.3	CircRNAs is formed by backsplicing	306
10.8	RNA quality control and degradation	307
10.8.1	Bacteria, archaea, and eukaryotes all have mechanisms for RNA quality control	307
10.8.2	Archaea and eukaryotes utilize specific pathways to deal with different RNA defects	308

Chapter 11 Mechanism of Translation · 312

11.1	Introduction	312
11.2	Messenger RNA	313
11.2.1	Structure of mRNA	313
11.2.2	Overall translation efficiency depends on a number of factors	316
11.3	Ribosome structure and assembly	317
11.3.1	Structure of ribosomes	317
11.3.2	Ribosome biogenesis	317
11.4	Transfer RNA	318
11.4.1	tRNA molecules fold into four-arm cloverleaf structures	318
11.4.2	tRNAs are aminoacylated by a set of specific enzymes, aminoacyl-tRNA synthetases	319
11.4.3	Proofreading activity of aminoacyl-tRNA synthetases	320
11.5	The process	321
11.5.1	Initiation of translation	321
11.5.2	Elongation and events in the ribosome tunnel	325
11.5.3	Termination of translation	328
11.6	Translational and post-translational control	329
11.6.1	Regulation of translation by controlling ribosome number	329
11.6.2	Regulation of translation initiation	330
11.6.3	mRNA stability and decay in eukaryotes	330
11.6.4	Phosphorylation	331
11.7	Protein processing and modification	332
11.7.1	Structure of biological membranes	332
11.7.2	Protein translocation through biological membranes	332
11.7.3	Proteolytic protein processing: cutting, splicing, and degradation	334
11.7.4	Post-translational chemical modification of side chains	334
11.7.5	The genomic origin of proteins	334

Chapter 12 DNA Replication — 337
- 12.1 Introduction — 337
  - Features of DNA replication shared by all organisms — 337
- 12.2 Bacterial DNA replication — 340
  - 12.2.1 DNA replication in bacteria — 340
  - 12.2.2 The processes of Bacterial DNA replication — 350
  - 12.2.3 Bacteriophage and plasmid replication — 355
- 12.3 DNA replication initiation in eukaryotes — 357
  - 12.3.1 Replication initiation in eukaryotes proceeds from multiple origins — 357
  - 12.3.2 Eukaryotic origins of replication have diverse DNA and chromatin structure depending on the biological species — 358
  - 12.3.3 There is a defined scenario for formation of initiation complexes — 358
  - 12.3.4 Re-replication must be prevented — 358
  - 12.3.5 Histone methylation regulates onset of licensing — 359
- 12.4 Histone removal at the origins of replication — 360
- 12.5 Replication of chromatin — 360
- 12.6 The DNA end-replication problem and its resolution — 362
- 12.7 Alternative modes of DNA replication — 365
  - 12.7.1 Replication in viruses that infect eukaryotes — 366
  - 12.7.2 Models for organelle DNA replication — 367

Chapter 13 DNA Recombination and DNA Repair — 371
- 13.1 Introduction — 371
- 13.2 Homologous recombination — 371
  - 13.2.1 Homologous recombination in bacteria — 371
  - 13.2.2 Holliday junctions are the essential intermediary structures in HR — 373
  - 13.2.3 Homologous recombination in eukaryotes — 373
- 13.3 Nonhomologous recombination — 377
  - 13.3.1 Transposable elements or transposons are mobile DNA sequences that change positions in the genome — 377
  - 13.3.2 Many transposons are transcribed but only a few have known functions — 378
  - 13.3.3 There are several types of transposons — 378
  - 13.3.4 DNA class II transposons can use either of two mechanisms to transpose themselves — 381
  - 13.3.5 Retrotransposons, or class I transposons, require an RNA intermediate — 383
- 13.4 Site-specific recombination — 384
  - 13.4.1 Bacteriophage λ integrates into the bacterial genome by site-specific recombination — 384
  - 13.4.2 Immunoglobulin gene rearrangements also occur through site-specific recombination — 386
- 13.5 Types of lesions and repair in DNA — 389
  - 13.5.1 Natural agents, from both within and outside a cell, can change the information content of DNA — 389
  - 13.5.2 Introduction to pathways and mechanisms of DNA repair — 391
  - 13.5.3 Nucleotide excision repair is active on helix-distorting lesions — 393
  - 13.5.4 Base excision repair corrects damaged bases — 394
  - 13.5.5 Mismatch repair corrects errors in base pairing — 395
  - 13.5.6 Mismatch repair pathways in eukaryotes may be directed by strand breaks during DNA replication — 396
  - 13.5.7 Repair of double-strand breaks can be error-free or error-prone — 397
  - 13.5.8 Homologous recombination repairs double-strand breaks faithfully — 398

13.5.9	Nonhomologous end-joining restores the continuity of the DNA double helix in an error-prone process	399
13.5.10	Translesion synthesis	400
13.5.11	Many repair pathways utilize RecQ helicases	401
13.5.12	Histone variants and their post-translational modifications are specifically involved in DNA repair	403

Chapter 14 Genetically Modified Organisms: Use in Basic and Applied Research ··· 407
14.1 Introduction ··· 407
14.2 Model organisms and genetically modified organisms ··· 407
14.3 Transgenics, gene targeting, and genome editing ··· 410
   14.3.1 Transgenic mouse ··· 410
   14.3.2 Gene-targeted mouse models ··· 413
   14.3.3 Genome editing by engineering nucleases ··· 417
14.4 Applications of genetically modified mouse in medical field ··· 420
   14.4.1 Gene function research ··· 420
   14.4.2 Animal models of human diseases ··· 421
   14.4.3 Gene knockout-based immunodeficient mice for PDX model ··· 421
   14.4.4 Fully human antibody discovery and pharmacodynamic evaluation ··· 421
14.5 Applications of other transgenic animals ··· 422
   14.5.1 Transgenic nonhuman primates ··· 422
   14.5.2 Gene pharming ··· 422
   14.5.3 Transgenic livestock ··· 422

Chapter 15 Disease Genes ··· 424
15.1 Introduction ··· 424
15.2 Genome and gene abnormalities ··· 424
   15.2.1 Abnormal gene and genome structures show various types ··· 425
   15.2.2 Many factors can cause gene and genome abnormalities ··· 426
15.3 Molecular outcome of genetic abnormalities ··· 427
   15.3.1 Genetic abnormalities lead to loss of gene function ··· 428
   15.3.2 Genetic abnormalities lead to gain of function of genes ··· 429
15.4 Clinical outcomes of gene and genomic abnormalities ··· 430
   15.4.1 The genes in autosomes can be mutated and lead to the genetic diseases ··· 430
   15.4.2 Diseases by genetic abnormalities in sex chromosomes and mitochondria are related to gender ··· 431
   15.4.3 Common and complex diseases are the results of the interaction between multiple genes and environmental factors ··· 432
15.5 Principles of identification and positional cloning of disease genes ··· 433
   15.5.1 The key to identify and clone disease-related genes is to determine the disease phenotype and the substantial relationship between genes and phenotypes ··· 434
   15.5.2 Identification and cloning of disease-related genes requires a comprehensive strategy of multi-discipline and multi-approach ··· 434
   15.5.3 The determination of candidate genes is the intersection of a variety of methods for cloning disease-related genes ··· 435
15.6 Strategies and methods for identification and cloning of disease-related genes ··· 435
   15.6.1 The identification and cloning of disease-related genes can adopt a strategy that does not depend on chromosome location ··· 435
   15.6.2 Using animal models to identify cloned disease-related genes ··· 439

15.6.3 Positional cloning is a classical method for identifying disease ·················· 439
15.6.4 The process of disease-related gene location and cloning includes three steps ·················· 441
15.6.5 Identifying genes for common diseases require genome-wide association analysis and whole exon sequencing ·················· 442
15.6.6 Bioinformatics database stores abundant disease-related gene information ·················· 442

Chapter 16 Gene Diagnosis and Gene Therapy ·················· 445

16.1 Introduction on gene diagnosis ·················· 445
16.2 Gene diagnose of inheritance disease ·················· 445
 16.2.1 Hemoglobinopathy ·················· 445
 16.2.2 Hemophilia ·················· 446
16.3 Gene diagnose of cancer ·················· 446
 16.3.1 Oncogene in gene diagnose ·················· 446
 16.3.2 Gene mutation in gene diagnose ·················· 447
 16.3.3 Virus gene detection in cancer ·················· 447
16.4 Basic techniques used for gene diagnosis ·················· 448
 16.4.1 Molecular hybridization of nucleic acids ·················· 448
 16.4.2 Polymerase chain reaction ·················· 448
 16.4.3 DNA Sequencing ·················· 448
 16.4.4 Zymogram analysis ·················· 448
 16.4.5 Analysis of single strand conformation polymorphism ·················· 449
 16.4.6 Biochip/DNA microarray ·················· 450
 16.4.7 Restriction fragment length polymorphism ·················· 450
16.5 Introduction on gene therapy ·················· 451
16.6 Main strategies of gene therapy ·················· 451
 16.6.1 Gene replacement ·················· 451
 16.6.2 Gene augmentation ·················· 451
 16.6.3 Gene interference ·················· 451
 16.6.4 Suicide gene ·················· 451
 16.6.5 Gene immunotherapy for cancer ·················· 452
16.7 Main strategies of gene delivery ·················· 452
 16.7.1 Liposome vectors ·················· 452
 16.7.2 Retrovirus vectors ·················· 452
 16.7.3 Adenovirus vectors ·················· 453
 16.7.4 Adeno-associated virus vectors ·················· 454
16.8 Application of gene therapy ·················· 454
 16.8.1 Gene therapy for inherited immunodeficiency syndromes ·················· 454
 16.8.2 Cystic fibrosis gene therapy ·················· 455
 16.8.3 HIV gene therapy ·················· 455
16.9 The future of gene therapy ·················· 456
 16.9.1 IMLYGIC™ in cancer ·················· 456
 16.9.2 Precision medicine initiative ·················· 456

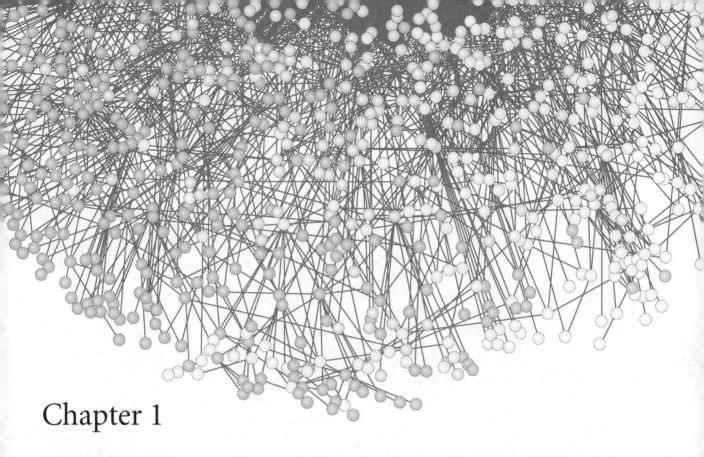

Chapter 1

From Classical Genetics to Molecular Genetics

1.1 Introduction

In this chapter we will visualize the machinery of life at the molecular or atomic level and use that ability to explore the fundamental mechanisms and processes of life. There is, however, more to biology than structure and biochemistry. A true molecular biology must also explain how the information needed to produce biological structures and processes is stored, expressed, and transmitted from one generation to the next. This last task lies in the province of genetics. Much of genetics, or what we now call classical genetics, was developed before anything was known of the molecular processes involved, yet it provided a vital and important impetus and direction for the new science of molecular genetics.

1.2 Classical genetics and the rules of trait inheritance

Genetics is the study of inheritance. From earliest times, philosophers and plant and animal breeders have recognized that there are similarities between parents and their offspring, but any quantitative explanation for how traits are transmitted and which are favored was long in coming. It was not until the middle of the nineteenth century that the first rigorous analysis was carried out. Remarkably, this was not done by any of the eminent biologists of the day but by a little-known monk, Gregor Mendel.

1.2.1 Gregor Mendel developed the formal rules of genetics

As Box 1.1 shows, Mendel was an unusual person to be doing such scientific research. As an Augustinian friar, he tended gardens at the monastery at Brno and studied, over a period of about a decade, the genetics of garden peas. These were a good choice because peas can be raised rapidly and exhibit clearly recognizable traits that can breed true for many generations. Also, they are capable of either self-fertilization

> **Box 1.1 Two unusual scientists who defined classical genetics**
>
> Although many scientists contributed to the formulation of classical genetics, there are two names that stand out: Gregor Mendel and Thomas Hunt Morgan. Each was, in a sense, unusual among the scientists of his day and deserves special recognition. Their backgrounds and modes of doing science were entirely different.
>
> *Gregor Mendel*
> Mendel was born to a poor family in 1822 as Johann Mendel, in an area then part of the Austro-Hungarian Empire. He took the name Gregor when he entered the Augustinian abbey at Brno, presently in the Czech Republic, in 1843, where he spent most of his life. Mendel's scientific training appears to have been limited to two years at the University of Vienna, where he was sent by the abbey to study physics. Most of his genetic research was conducted in the gardens of the abbey, during the period 1856—1863. During that time, he is said to have raised and personally examined about 28 000 pea plants and kept careful quantitative records of their propagation. The results of this work, essentially Mendel's laws of genetics, were published in an obscure journal in 1866. Although Mendel sent reprints to most of the prominent biologists of the time, including Charles Darwin, he seems to have had little response. In fact, in the 35 years following publication, there were exactly three citations of his paper. In the present, frenzied climate of Science Citation Index, Mendel would not have fared well. Essentially, the work was neglected and forgotten and was only rediscovered around 1900.
>
> *Thomas Hunt Morgan*
> Morgan was born in 1866 to a prominent Kentucky family and, in contrast to Mendel, enjoyed the best of education. He obtained his Ph.D. in developmental biology from Johns Hopkins University in 1890 and spent the next decade on the faculty of Bryn Mawr College. In 1904, he accepted a professorship at Columbia University, where he spent his most productive years. Rather than working alone, as did Mendel, Morgan gathered about him an unusually talented group of students and postdoctoral researchers, many of whom also went on to make major contributions to genetics. These included Alfred Sturtevant, who constructed the first genetic map of a chromosome; George Beadle, who with Edward Tatum first related genes to proteins; Theodosius Dobzhansky, who showed how mutations could drive evolution; and Hermann Muller, who discovered the mutational effects of short-wavelength radiation. The list could be extended for further academic generations. Like Mendel, Morgan was distinguished by an ability to see the important general law in a mass of complex data.
>
> According to some, Thomas Morgan can be credited with revolutionizing the way in which research groups operate. Up to Morgan's time, the European model, with a rigid, hierarchical structure dominated by an almost inaccessible professor, was emulated throughout the scientific world. In stark contrast, Morgan's fly lab had an informal, relaxed ambience; visiting scientists were often shocked to find all participants on a first-name basis. Gradually, this became the norm in American universities and to some extent elsewhere. If you find the lab in which you do graduate studies is a socially pleasant, relaxed place, you may owe a debt to Thomas Hunt Morgan.

or cross-fertilization (Figure 1.1 and Table 1.1). The unusual aspect of Mendel's work, for this field at the time, was his careful quantitation of the outcome of every breeding experiment.

In his experiments, Mendel would first choose a pair of stocks exhibiting contrasting traits in a particular character, yellow versus green seeds, for example, each of which he knew to breed true in self-fertilization. These are referred to as the parental phenotypes, the P1 generation. When these were cross-fertilized, it was always observed that only one of the two alternate traits was expressed in the progeny, called the F1 or filial 1 generation. In our example, the trait expressed in F1 is yellow (Figure 1.2; see Table 1.1). This showed Mendel that one trait, yellow seeds, was dominant over the other, green

Table 1.1 Examples of pairs of contrasting traits in peas studied by Mendel.

Part of plant	Traits	Phenotype of F1	Phenotypic ratios observed in F2
seeds	round/wrinkled	round	2.96 round/wrinkled
pods	green/yellow	green	2.86 green/yellow
flowers	violet/white	violet	3.15 violet/white
height of plant	tall/dwarf	tall	2.84 tall/dwarf

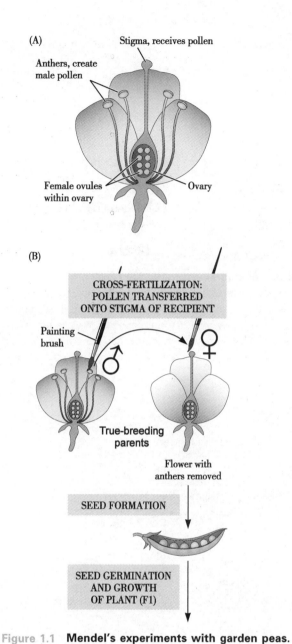

Figure 1.1 **Mendel's experiments with garden peas.** (A) Anatomy of the pea flower. Peas are self-fertilizing but can also be cross-fertilized. Clear-cut alternative, or antagonistic, forms of particular traits exist: seed color and shape, flower color, pod color and shape, stem length, and flower position. Crossing of plants that have different forms of these traits allowed the formulation of Mendel's laws of inheritance. (B) Mendel isolated and perpetuated true-breeding, or pure-breeding, pea lines, in which a trait that he was studying had remained constant from generation to generation for eight generations. He cross-fertilized antagonistic forms to produce hybrids; for each experiment he also did reciprocal crosses. The first filial or F1 generation was hybrid; it was then allowed to self-fertilize to produce second, third, and further filial generations: F2, F3, etc. Mendel then followed the inheritance patterns of the traits over several generations and quantified the data, leading to the most profound scientific understanding of inheritance laws.

seeds. Why this occurred was shown by the next experiments, in which members of the F1 generation were self-hybridized. Now the other trait, the recessive trait, reappeared but in only one-fourth of the F2 progeny. This meant that there must be units of heredity, now called genes, that could be distributed according to fixed rules and dictated the dominant and recessive states. Most eukaryotes are diploid, carrying two copies of each gene, now referred to as alleles, and therefore two alleles for each trait in their somatic cells (Box 1.2). In the case of seed color, the dominant allele is denoted Y and the recessive allele y. We can now understand the whole process by recalling the fact that, at fertilization, each parent donates one gamete, either sperm or ovum, to the fertilized egg. Because the gametes are haploid, this means that each parent donates one or the other of the two alleles, chosen at random. The F1 generation, obtained from cross-fertilization of two opposite homozygotes, each carrying two copies of just one of the alternate alleles, must be heterozygous, and only the dominant trait is expressed. In the F2 generation, however, random combination of gametes will lead to the possibilities YY, Yy, yY, and yy. Because Y is dominant, this yields the 3:1 ratio of yellow/green phenotypes observed. Mendel's experimental results, which are based on thousands of crosses, are summarized for several traits in Table 1.1.

From these experiments, Mendel formulated the two laws, which are really hypotheses, that constitute the basis of classical genetics. Mendel's first law, the law of segregation, states that the two alleles for each trait separate or segregate during gamete formation and then unite at random, one from each parent, at fertilization. The first law can be expressed in a number of ways. Here we choose to break it into several statements, in modern nomenclature.

- Variation in phenotype is explained by the existence of alternate versions of genes. These versions are called alleles.
- The alleles of each gene segregate, independently, one to each gamete.
- Every individual inherits two alleles of each gene, in one gamete from each parent.

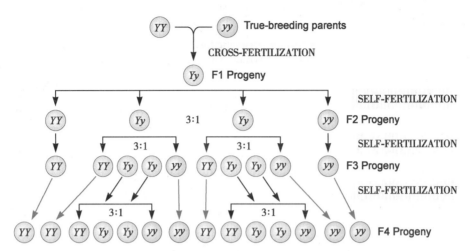

Figure 1.2 **Mendel's results showing the distribution of a pair of contrasting traits.** Seed color after four rounds of fertilization is illustrated. Capital *Y* stands for yellow color, whereas lowercase *y* stands for green color. For each trait, the plant carries two copies of a unit of inheritance, or two alleles of a gene in contemporary understanding. The trait that appears in all F1 hybrids is dominant; the antagonistic trait that remains hidden in F1 but reappears in F2 is recessive. In this specific example, yellow color is dominant and green is recessive. The schematic also illustrates the difference between homozygous and heterozygous individuals. Homozygous individuals of each generation breed true, as shown by red arrows, whereas heterozygous do not, as shown by black arrows.

Box 1.2 **Sexual reproduction, mitosis, and meiosis**

"Birds do it, and bees do it. Indeed, researchers estimate that over 99.99% of eukaryotes do it, meaning that these organisms reproduce sexually, at least on occasion." —Sarah Otto

One of the oldest arguments for why sexual reproduction is so commonplace goes back to biologist August Weismann, who suggested, in the late 1880s, that sex generated variable offspring upon which natural selection can act. We now know that this notion may be an oversimplification, but its general tenets still hold.

A simple schematic of sexual reproduction, as it relates to humans, is presented in Figure 1. As illustrated in the schematic, two major types of cell division take place during the reproduction cycle: mitosis produces new diploid somatic cells, that is, cells of the body, whereas meiosis, which occurs in germ cells, gives rise to haploid gametes, sperm and ova (Figure 2). Both mitosis and meiosis begin by duplicating the genetic material, DNA (see Chapter 4), within the cell and then compacting it into condensed chromosomes, visible structures that are recognizable under the microscope. Chromosome structure is discussed in more detail in Chapter 8. Then mitosis and meiosis take different paths in the way the chromosomes behave, which leads to the production of either diploid cells or haploid gametes, respectively.

Cells that undergo mitosis go through different phases, each characterized by its own biochemical processes and structural reorganizations. It is beyond the scope of this text to describe them in detail, but the most important information that we will need on numerous occasions

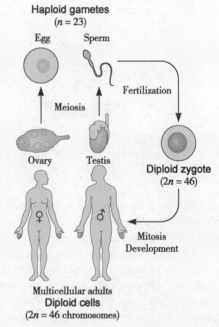

Figure 1 **Conceptualized schematic of the sexual reproduction cycle in higher eukaryotes.** Note the transitions between cells containing the diploid number of chromosomes, 46 in humans, and those with the haploid number of chromosomes, 23 in humans. The first transition occurs during the formation of the female and male gametes, egg and sperm, respectively, and involves a reduction division known as meiosis. The haploid gametes then reunite during fertilization to form the diploid zygote or fertilized egg. All cells of an adult organism are formed from further divisions known as mitosis.

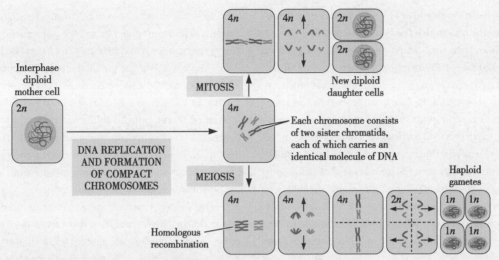

Figure 2 **Conceptualized schematic of mitotic and meiotic cell divisions.** Both mitosis and meiosis start with replication of DNA, formation of compact chromosomes, and disappearance of the nuclear membrane. In mitosis, the chromosomes line up individually at the equatorial plate and the two sister chromatids that make up each chromosome are moved toward the two opposite poles of the mitotic spindle. By contrast, in meiosis, the homologous chromosomes pair and the pairs align at the equatorial plate; two divisions then follow. During the first division in meiosis, each chromosome of a pair moves toward one of the opposite poles of the spindle; during a second division, the two sister chromatids of a chromosome move toward the opposite poles of new spindles formed in a direction perpendicular to the first one. As a result, four cells are formed, each with a haploid chromosome number.

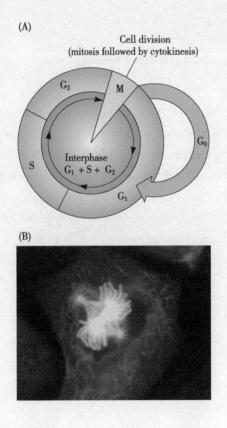

throughout the book is presented in Figure 3. The cell cycle is also described in more detail in Chapter 20, where we discuss eukaryotic DNA replication.

Figure 3 **Mitotic cell cycle.** (A) Schematic showing the traditional phases characteristic of cycling cells: G_1, cell growth and synthesis of components required for DNA synthesis; S, synthetic phase, DNA replication; G_2, preparation for mitosis; M, mitosis followed by cytokinesis or cell division. The G_0 detour is the special phase following mitosis during which differentiated cells perform their special function(s). This phase is not part of the normal cycle of continuously proliferating cells, in which M is immediately followed by G_1, the phase in which the cell prepares for DNA replication. Note that this rule holds true for most differentiated cells, but there are exceptions: some differentiated cells can perform their specialized function while actively proliferating. Note also that nonproliferating cells can re-enter the cell cycle under certain conditions: cancer cells are an *in vivo* example. (B) Fluorescent micrograph of a newt lung epithelial cell in the metaphase stage of mitosis. The cell was fixed and stained for immunofluorescence localization of microtubules, shown in green/yellow, and keratin filaments, shown in red; condensed mitotic chromosomes are in blue. By metaphase, the bipolar mitotic apparatus is fully formed and shaped like a spindle of thread. This structure supports the production of mechanical forces required for segregating the replicated chromosomes into daughter nuclei. In epithelial cells the spindle and its associated chromosomes are surrounded by a cage of keratin filaments, which prevents the motion of other cell organelles into the region containing the chromosomes. (B, courtesy of Conly L. Rieder, New York State Department of Health.)

Mitosis permits development of a wide variety of differentiated somatic cells from a few types of stem cells

Stem cells are cells that can both self-renew and generate progeny capable of following more than a single differentiation pathway. Of all of the stem cells in an organism, embryonic stem cells or ES cells have drawn the most attention. ES cells arise from the inner cell mass of the mammalian blastocyst (Figure 4) and can be maintained in culture in a pluripotent state. Pluripotency is defined as the ability to differentiate into every single cell type that is found in adult organisms. By contrast, multipotency is characteristic of lineage-restricted stem cells that are already committed to a certain pathway. For example, hematopoietic stem cells give rise to all kinds of blood cells but cannot become neurons or liver cells; neural stem cells can generate only neuronal and glial cells.

ES cells can be induced to differentiate *ex vivo* by culture conditions and can, under certain circumstances, develop an entire organism. This latter property has created considerable controversy in the public domain. The potential of ES cells to generate differentiated cells, tissues, and organs for use in clinical practice for the treatment of disease is deemed useful, but the ability to create whole human beings is seen as dangerous human cloning and is prohibited by law in most developed countries. By contrast, cloning of animals and plants has been accepted and provides the potential to improve food production.

Meiosis follows a path somewhat similar to mitosis; however, following DNA replication there occur two successive cell divisions. Therefore, a single cell will yield four haploid gametes, each carrying only one copy of each gene.

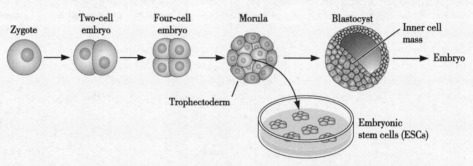

Figure 4 **Schematic depicting how embryonic stem cells are derived.**

- If the alleles differ, one will be dominant and one recessive. If the individual is heterozygous for an allele, only the dominant allele and trait will be expressed in the first generation.

But what happens if one cross-breeds peas that differ in two traits? Does the segregation of alleles for one trait affect the other? Mendel also carried out such experiments and derived what is called Mendel's second law, the law of independent assortment. This law states that, during gamete formation, the segregation of the alleles of one allelic pair is independent of the segregation of the alleles of another allelic pair. In other words, traits segregate independently; there is no linkage between genes for different traits (Figure 1.3). This happens not to be always true, as we shall see.

1.2.2 Mendel's laws have extensions and exceptions

As with many great scientific breakthroughs, the true situation has proved to be more complicated than the initial insight suggested. There are many examples of exceptions to simple Mendelian genetics. In general, the extensions to Mendel's laws can be classified in two groups, depending on whether a trait is encoded by a single gene or by many genes, known as multifactorial inheritance.

In the single-gene inheritance group, there are three major extensions.

(1) Three major extensions in the single-gene inheritance

First, dominance is not always complete. In incomplete dominance, the hybrid resembles neither parent. In co-dominance, neither allele is dominant, with the F1 hybrid showing traits from both true-breeding parents. These relationships are schematically depicted in Figure 1.4.

Second, a gene may have more than two alleles. There are numerous examples of such genes, including those that determine human blood groups and the genes that code for human histocompatibility antigens, which are cell surface proteins that participate in proper immune responses. The latter proteins are encoded by three genes, each with between 20 and 100 alleles; each allele is co-dominant with every other allele at the molecular level. The most extreme example known to date is the olfactory genes, which have ~1300 alleles, only one of which is expressed. This monoallelic gene expression is strictly regulated by mechanisms that remain largely unknown. Usually cells express both alleles of a gene.

Third, one gene may contribute to several visible characteristics; this phenomenon is known as pleiotropy, from the Greek pleion, meaning more, and tropi, meaning to turn or convert. A classic example of pleiotropy is found in sterile males among the aboriginal Maori people of New Zealand. These men are sterile and have respiratory problems. The gene's

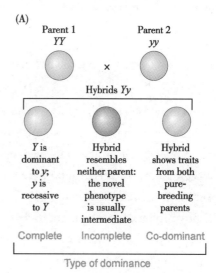

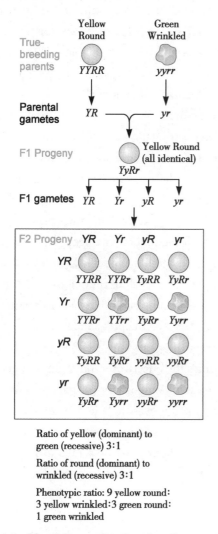

Figure 1.3 **Mendel's results showing the segregation of two independent traits.** Capital Y stands for yellow color, and lowercase y stands for green color; capital R stands for round shape, and lowercase r stands for wrinkled shape.

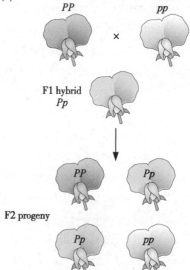

Figure 1.4 **Dominance relationships between alleles.** (A) Different dominance relationships between pairs of alleles are revealed by the phenotype of the heterozygote. Variations in dominance relationships do not detract from Mendel's law of segregation; rather, they reflect differences in the way in which gene products control the production of phenotype. (B) In this example of incomplete dominance, the petal color of the F1 heterozygote *Pp* is unlike that of either homozygote parent. The phenotypic ratios in F2 are an exact reflection of the genotypic ratios. The explanation of this behavior lies in the biochemistry of pigment production. *p* does not encode a functional enzyme for pigment production, hence a white color of the petals results. Conversely, the *PP* homozygote produces a double dose of the enzyme and hence a bright pink color is seen.

normal dominant allele specifies a protein needed in both cilia and flagella; in men who are homozygous for the recessive allele, cilia and flagella do not function properly, affecting their abilities to both clear mucus from their respiratory tract and produce motile sperm.

(2) Three major extensions in multifactorial inheritance

In a multifactorial inheritance group, two or more genes can interact to determine a single trait, and each type of interaction produces its own signature of phenotypic ratios. In this group, there are three major extensions to Mendel's laws.

First, novel phenotypes can emerge from the combined action of the alleles of two genes. The genes either complement each other or are epistatic to each other. In complementation, a wild-type offspring is produced from crosses between strains that carry different homozygous recessive mutations in different genes; the phenotype of the two homozygous parents is the same, that is, they are not distinguishable by appearance. Complementation, or reversal to wild-type phenotype, can occur only if the mutations occur in different genes, so that the genome of the offspring carries one wild-type allele that complements the mutated allele of the same gene. In epistasis, one gene's alleles mask the effects of another gene's alleles. Biochemically, this situation arises when several genes participate in succession in a single biochemical pathway. The inactivity of a gene at the beginning of the pathway will hide the fact that subsequent genes may be expressed.

Second, a given genotype does not always produce the same phenotype: phenotype often depends on penetrance and expressivity. Penetrance describes how many members of a population with a particular genotype show the expected phenotype. Penetrance can be complete or incomplete. A frequently cited example of incomplete penetrance concerns the disease retinoblastoma: only 75% of people carrying a mutant allele for the retinoblastoma protein develop the disease. In addition, in some people who have retinoblastoma, only one eye is diseased: expressivity refers to the degree or intensity with which a particular genotype is expressed in a phenotype. It is important to understand that chance can affect penetrance and expression. For example, in the case of retinoblastoma, every cell carries the inherited mutation in one allele of the retinoblastoma gene, but a second chance event is needed for the disease phenotype to appear. Damaging radiation or errors in DNA replication in retinal cells provide the second hit, creating a mutation in the second copy of the retinoblastoma gene within one or more cells. Such situations gave rise to the two-hit hypothesis for the origin of cancer proposed by Alfred Knudson in 1971.

Third, there are also quantitative traits that vary continuously over a range of values. A good example is height and skin color in humans. These traits are polygenic and show the additive effects of a large number of genes and their alleles.

Here we mention the concept of modifier genes, which have secondary, more subtle effects on a trait. We also introduce the concept of allele frequency within a population. This is the percentage of the total number of gene copies in a population comprised of any one allele. The most prevalent allele, the one with the highest frequency in a population, is defined as the wild-type allele. In evolution, new alleles appear as a result of mutations.

Finally, it is now clear that the environment can affect the phenotypic expression of a genotype. When environmental agents cause a change in phenotype that mimics the effects of a mutation in a gene, this is known as phenocopying. A painful example of this phenomenon was the effect of the sedative drug thalidomide. If taken by pregnant women, this drug produced a phenocopy of a rare dominant trait called phocomelia, which disrupts limb development in the fetus.

Perhaps the most significant exception is the fact that Mendel's second law is not generally correct. There are many cases in which genes are linked, and discovery of this led to the second great advance in classical genetics.

1.2.3 Genes are arranged linearly on chromosomes and can be mapped

The next major step in genetic research was accomplished by Thomas Hunt Morgan, who spent most of his career at Columbia University (see Box 1.1).

In the early part of the twentieth century, Morgan began working with fruit flies, *Drosophila melanogaster*. This was an even better choice than Mendel's peas because the flies breed very rapidly and could be raised in large numbers in very little space. This was important because Morgan was dependent, for much of his work, on the occurrence of rare, spontaneous mutations. These were the source of the changes in the genes that produced modified alleles and hence phenotypes. Thus the discovery, by Hermann Joseph Muller in Morgan's lab, that mutations could be induced by X-rays or other damaging radiation was very helpful. Muller was awarded the 1946 Nobel Prize in Physiology or Medicine "for the discovery of the production of mutations by means of X-ray irradiation."

Contrary to Mendel's observations, Morgan found that a number of traits of the flies appeared to be genetically linked. The difference with Mendel's conclusion may have lain in the fact that many of the pea traits studied by Mendel corresponded to genes on different chromosomes, which would be expected to segregate independently, whereas Morgan was initially concentrating on genes on the same chromosome, the female X chromosome. In any event, Morgan observed linked transmission of a number of genes. He hypothesized that the lack of linkage in some cases must result from recombination of alleles (Figure 1.5). Furthermore, he noted that the probability of such recombination must increase with the distance between the two genes in the chromosome. Thus, the degree of linkage must measure gene separation. Then something wonderful happened. Alfred Sturtevant, a student working in Morgan's laboratory, realized that this fact allowed mapping of genes on chromosomes. He skipped his assigned homework one night to produce the first genetic map. Soon this was extended to many *Drosophila* genes, and a new paradigm emerged: genes are arranged linearly on

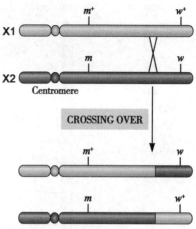

Figure 1.5 **Recombination between the two X chromosomes of the female *Drosophila* fly.** Chromosome X1 carries two wild-type alleles: m+ codes for normal wings and w^+ determines red eyes. Chromosome X2 carries two mutant alleles, m coding for miniature wings and w coding for white eyes. During egg formation, a crossing-over or recombination event occurs somewhere between these two genes on the two chromosomes, resulting in two recombinant chromosomes, each of which carries a mixture of the parental alleles. This process creates a new combination of alleles, hence the name recombination.

chromosomes. A sample genetic map is presented in Figure 1.6. In 1933, Morgan was awarded the Nobel Prize in Physiology or Medicine "for his discoveries concerning the role played by the chromosome in heredity."

1.2.4 The nature of genes and how they determine phenotypes was long a mystery

Despite rapid advances in classical genetics in the early years of the twentieth century, the nature of genes and their mode of function remained obscure. Indeed many geneticists preferred to think of them as abstract entities. Genes gained some substance through the examination of polytene chromosomes. These are parallel aggregates of a number of identical chromosomes, found in the salivary glands of some insects, including *Drosophila* (Figure 1.7 and Figure 1.8). Upon proper staining and under the light microscope, they show a banded pattern, and some bands can be correlated with genes mapped by the Sturtevant-Morgan technique. Such studies reinforced the image of genes as physical objects, but of what were they made?

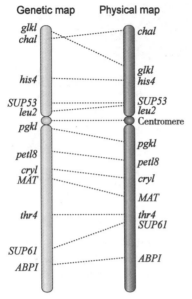

Figure 1.6 Genetic and physical maps of *S. cerevisiae* chromosome III. The genetic map was constructed by determining the frequency of recombination in genetic crosses; the physical map was determined by DNA sequencing. Despite some discrepancies between the two maps, their overall similarity is impressive. Note that the order of the upper two markers or identifiable genes has been incorrectly assigned on the genetic map; the relative positions of some markers are also somewhat different on the two maps. (Adapted from Oliver SG, van der Aart QJM, Agostoni-Carbone ML, et al. [1992] *Nature*,357:38–46. With permission from Macmillan Publishers, Ltd.)

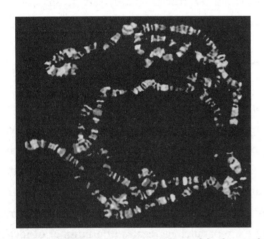

Figure 1.7 Polytene chromosomes with the typical banding patterns. The banding patterns of *Drosophila* polytene chromosomes were depicted by Calvin Bridges as early as 1935 and are still widely used to identify chromosomal rearrangements and deletions. A fluorescent image of the *Drosophila* salivary gland chromosomes stained for two different proteins, BRAMA in green and Pol II in red, is presented. The two proteins show a high degree of overlap. (From Armstrong JA, Papoulas O, Daubresse G, et al. [2002] *EMBO J*,21:5245–5254. With permission from John Wiley & Sons, Inc.)

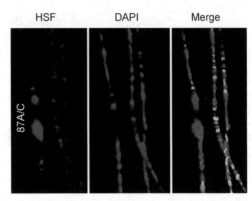

Figure 1.8 Banding patterns of polytene chromosomes reflect the transcriptional activity of chromosome regions. A fluorescent image of heat-shock-induced puffs, loci 87A/C, in *Drosophila melanogaster* chromosomes is shown. Inactive genes that are located in compact chromatin form condensed bands. In puffing, the material of the bands loosens and becomes decompacted, and local swelling of the chromosome region occurs. DNA is stained with 4′,6-diamidino-2-phenylindole or DAPI, shown in blue, while the heat-shock factor (HSF), a transcriptional activator of genes induced by heat shock, is shown in red. The merged image on the right indicates that the transcriptional activator is located in the puffed, decondensed regions. (From Armstrong JA, Papoulas O, Daubresse G, et al. [2002] *EMBO J*,21:5245–5254. With permission from John Wiley & Sons, Inc.)

This same period saw the beginnings of protein studies, and the function of proteins as enzymes was well established. Probably for this reason, and because of misunderstanding of the nature of nucleic acids, most researchers before about 1940 assumed that genes were proteinaceous in composition. It was also becoming evident, however, that genes must dictate protein structure. Certain disease states that depended on the loss of one enzyme function exhibited Mendelian inheritance; this early observation led to the dictum one gene, one enzyme. After the brilliant work in several laboratories on sickle cell anemia (Box 1.3), this dictum had to be modified to one gene, one polypeptide chain. Today we realize that not only protein sequences but also non-protein-coding RNA sequences are dictated by genes, leading to further attempts to properly define a gene. As we see in Chapter 7, the definition of the term gene is still evolving, with a new definition recently proposed in 2012 as a result of studies on the human genome.

Box 1.3 Sickle cell anemia: a key to molecular genetics

A genetically transmitted disease called sickle cell anemia or SCA affects millions of people around the world, especially those of African origin or ancestry. It can be deadly and is often debilitating. Red blood cells, known as erythrocytes, are normally discoid in shape, but those in patients with SCA tend to be elongated or sickle-shaped, especially under conditions of oxygen deprivation (Figure 1A). Such distorted cells have two deleterious effects. First, they tend to block venous capillaries, causing pain and tissue damage. Second, they are more fragile than normal erythrocytes, being easily broken and releasing their hemoglobin; this leads to anemia.

It was first shown in 1949 by James Neel and E. A. Beet that sickle cell anemia is inherited as a Mendelian trait with two alleles. In individuals homozygous for the trait, the disease is fully expressed, with significant impairment of life quality or even early mortality. Heterozygous individuals, carrying one copy of the SCA allele, are often able to function quite normally, except in situations of oxygen deprivation or stress, where they will exhibit some sickling. Such individuals are often called carriers of the trait: they are not very sick, but their progeny will develop the disease if they are born homozygous for the SCA allele.

In the same year that Neel and Beet demonstrated the genetics of SCA, Linus Pauling and co-workers made a remarkable discovery. They found that the hemoglobin of people with SCA differed in electrophoretic mobility from that of healthy individuals. Furthermore, heterozygous individuals showed both electrophoretic bands (Figure 1B). This led Pauling to the conclusion that this was a molecular, genetic disease, the first clearly identified.

This observation was followed, in the 1950s, by protein sequencing studies in the laboratory of Vernon Ingram. It was known by then that hemoglobin contains two types of protein subunits, called α and β. Ingram found that a single amino acid substitution in the β chain constituted the sole difference between normal and sickle cell hemoglobin. Thus, it could be deduced that a single mutation was responsible for the trait. This and Pauling's work strengthened the one gene, one protein hypothesis.

But why does this mutation cause sickling? Numerous studies had shown that the mutant protein induces the formation of long filaments of hemoglobin, which pack side by side within the erythrocyte (Figure 1C). Because hemoglobin is highly concentrated in these cells, this aggregation deforms the cells in the manner observed. The way in which the disease was expressed in homozygous and heterozygous individuals now became clear. In the former, no normal protein can be made; all of the hemoglobin can aggregate.

In a heterozygous individual, both normal and aggregating hemoglobin would be mixed. Sickling could occur, but not so easily. Why did low oxygen have an effect? It was eventually shown that hemoglobin undergoes a conformational change as oxygen levels change. The low-oxygen form is more prone to aggregation.

There is one further twist to the story. Being heterozygous for the sickle cell trait can actually be advantageous in tropical regions where malaria is common. The malarial parasite must spend part of its life cycle within an erythrocyte. Sickled cells are inhospitable to the parasite because they are so fragile.

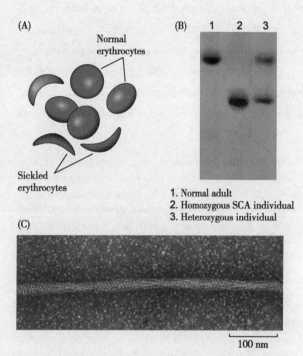

Figure 1 **Sickle cell anemia, a molecular disease.** (A) Shapes of normal and sickled erythrocytes. (B) Electrophoretic behavior of hemoglobin from normal individuals and individuals homozygous and heterozygous for the mutation in the β chain of hemoglobin. (C) Electron microscopic image of an aggregate of hemoglobin molecules in sickle cell anemia. The amino acid that is mutated in the β chain forms a protrusion that accidently fits into a complementary site on the β chain of other hemoglobin molecules in the cell. Instead of remaining in solution, these mutated molecules aggregate and become rigid, precipitating out of solution. (A, adapted, courtesy of Darryl Leja, National Human Genome Research Institute. B, courtesy of Michael W. King, Indiana University. C, from Dykes G, Crepeau RH, Edelstein SJ [1978] *Nature*, 272:506–510. With permission from Macmillan Publishers, Ltd.)

1.3 The great breakthrough to molecular genetics

1.3.1 Bacteria and bacteriophage exhibit genetic behavior and serve as model systems

In Chapter 4, we describe in detail the remarkable experiments that established once and for all that DNA is the genetic material. This recognition, together with the Watson-Crick structure of DNA, meant that by the mid-1950s a molecular theory of genetics had begun to develop. Much of the early work in this direction depended on the use of bacteria and viruses, particularly those viruses that infect bacteria, the bacteriophages or phage in scientific jargon.

The genetics of phage and bacteria had long been neglected, for neither follows the laws of classical Mendelian genetics. Both are haploid under most circumstances. However, in 1943, microbiologist Salvador Luria and physicist Max Delbrück provided convincing evidence of mutations in bacteria. In the same year, a major technical breakthrough was accomplished by Joshua Lederberg and Edward Tatum in the discovery of bacterial conjugation (Figure 1.9). In this process, one bacterium inserts all or part of its DNA into another, followed by recombination of the two DNA molecules. With some strains, called high frequency of recombination strains, practically all of the donors are active, and conjugation can be synchronized. If conjugation is halted at a series of different times, different amounts of DNA

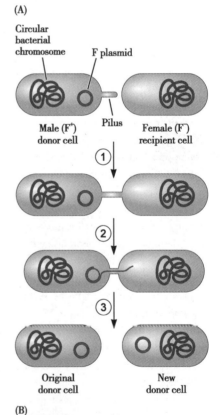

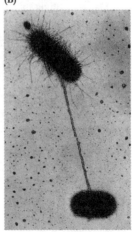

Figure 1.9 **Bacterial conjugation.** Conjugation between two bacterial cells can occur only when one of the partners carries the F or fertility plasmid; these cells are known as F-positive or F$^+$ cells. The F plasmid exists as an episome, that is, independently of the main bacterial chromosome. It carries its own origin of replication, an origin of transfer where nicking occurs to initiate transfer to a recipient F$^-$ cell, and a whole battery of genes responsible for formation of the pilus and attachment to the recipient cell. (A) Steps in the process of conjugation are as follows: (Step 1) Pilus attaches to recipient cell and brings the two cells together. Most probably the pilus is not directly used as a transfer channel. The channel is formed through the action of a specific enzyme at the base of the pilus, which initiates membrane fusion. (Step 2) The F plasmid is nicked and the nicked strand is unwound from the intact strand; transfer to the recipient cell begins. (Step 3) Single-stranded DNA is transferred to the recipient and copied to produce a double-stranded F plasmid; the single stranded F plasmid in the donor cell is simultaneously copied to produce a double-stranded F plasmid. Sometimes the F plasmid is integrated into the genome of the donor; these strains are known as high frequency of recombination or *Hfr* strains. In such cases, the entire bacterial chromosome or a part of it can be transferred into the recipient cell. The amount of chromosomal DNA transferred depends on how long the two conjugating bacteria stay in contact: transfer of the entire chromosome normally requires ~100 minutes. Homologous recombination allows for integration of the transferred chromosome into the genome of the recipient cell. (B) Electron microscopic view of two bacterial cells in the process of conjugation. (B, courtesy of Charles Brinton and Judith Carnahan, University of Pittsburgh.)

will be transferred, permitting recombination of only those genes that have been transferred. This provided a convenient way to map genes on the bacterial chromosome before powerful sequencing methods were available.

1.3.2 Transformation and transduction allow transfer of genetic information

Conjugation is by no means the only way in which bacteria can acquire foreign DNA. As early as 1928, Frederick Griffith had demonstrated the phenomenon of transformation of bacterial strains by the transfer of some factor. However, he did not know the nature of that substance. It was not until 1944 that Oswald Avery, Colin MacLeod, and Maclyn McCarty showed that genetic transformation was caused by the transport of DNA, through solution, into the transformed bacterium.

Finally, DNA can be transferred between bacteria via bacteriophage, by a process called transduction (Figure 1.10). This usually involves temperate phage, which can exhibit two alternative life cycles: lytic and lysogenic. When in the lytic cycle, the virus can enter the bacterial cell, replicate, lyse the host bacterium, and go on to infect other bacteria. In the alternative lysogenic cycle, it can integrate its DNA into the bacterial chromosome. There it may remain for many bacterial generations in a dormant state, until some stimulus, such as radiation or chemical insult, causes it to be released from the host genome. It will then form new viruses and kill the bacterium. Packaging of the viral DNA into viral particles is a low-fidelity process: small pieces of bacterial DNA may become packed, alongside the phage genome, and thus transferred to the newly infected cell. At the same time, phage genes can be left behind in the bacterial chromosome into which they had been integrated. The phage can be modified by addition of foreign DNA, and can act as a vector to insert this into bacteria.

1.3.3 The Watson-Crick model of DNA structure provided the final key to molecular genetics

By 1955, two years after Watson and Crick's publica-

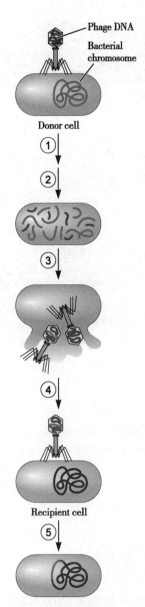

Figure 1.10 **Conceptual schematic of transduction.** Steps involved in transduction are as follows: (Step 1) A phage infects a bacterial cell. (Step 2) Phage DNA enters cell and replicates; phage proteins are made; bacterial chromosome is broken down. (Step 3) Occasionally, pieces of bacterial DNA are packaged into the phage head; some viral particles released during cell lysis contain bacterial DNA. (Step 4) A phage carrying bacterial DNA infects a new bacterial cell. (Step 5) Recombination between the donor bacterial DNA and the recipient bacterial DNA can occur; the recombinant cell is different from both the donor and the recipient cell.

tion of the DNA double-helix model, the molecular basis of genetics was clear (Figure 1.11). Genes are made of DNA, which is carried in chromosomes. Bacteria and phage have haploid chromosomes, constituted of one or a few double-helical DNA molecules. These

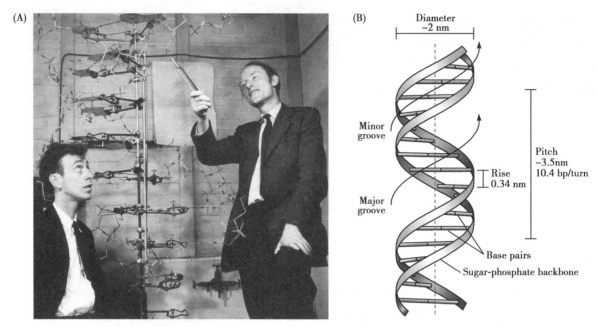

Figure 1.11 **The Watson-Crick model of DNA structure.** (A) Watson and Crick. (B) Watson and Crick's DNA double-helix model.

simple systems provided the entry into molecular genetics and key ideas for molecular biology. Most eukaryotes have two copies of double-helical DNA in each somatic cell but only one in each gamete (see Box 1.2). When cells replicate, DNA is duplicated by copying each of the strands of the double helix. As DNA replication occurs independently in each cell of a cellular population, at any point in time some cells will have DNA content characteristic of the G_1 phase, before replication begins; other cells will be in the process of replicating their DNA, in S phase; and still others will have their DNA replicated, in G_2 phase. The distribution of cells in the different stages of the cell cycle can be monitored by flow cytometry. Mutations occur by modification of DNA sequences, and exchange of alleles occurs by recombination. Almost none of the details or control of these processes were understood in 1955. Much has been learned since then, as we show in following chapters.

Key Concepts

- Gregor Mendel formulated the basic laws of genetics from experiments with garden peas.
- Mendel's first law states that alleles of a gene segregate independently.
- If two alleles are different, one dominates and determines the phenotype in the first hybrid progeny.
- Mendel's second law states that two traits, as dictated by pairs of alleles, will assort independently; genes are not linked.
- Thomas Hunt Morgan, in experiments on fruit flies, showed that genes are in fact sometimes linked.
- Linkage is proportional to the proximity of genes on a chromosome. This observation allows chromosome mapping.
- Experiments with bacteriophage and bacteria, plus the discovery of DNA structure, led to modern molecular genetics.

Key Words

allele（等位基因）
diploid（双倍体）
dominant（显性）
epistasis（上位）
expressivity（表现度）
gene（基因）
genotype（基因型）
linkage（连锁）
Mendel's law（孟德尔定律）
phenotype（表型）
penetrance（外显率）

phenocopying(拟表型)
pleiotropy(多效型)
recessive(隐性)
trait(性状)
transformation(转化)
transduction(转导)
vector(载体)

Questions

1. Could you know why Mendel's work was neglected and forgot and only rediscovered around 1900?
2. Describe briefly Mendel's first law. And how did Mendel formulate the law?
3. Describe briefly Mendel's second law. Give an example to explain the second law.
4. Describe briefly the discovery by Thomas Morgan.
5. Describe briefly the great breakthrough to molecular biology.

References

[1] Bradley A (2002) Mining the mouse genome - We have the draft sequence - but how do we unlock its secrets? *Nature*, 420:512–514.
[2] Brenner S (1974) New directions in molecular biology. *Nature*, 248:785–787.
[3] Goffeau A, Barrell BG, Bussey H, et al. (1996) Life with 6000 genes. *Science*, 274:546, 63–67.
[4] Luria SE (1966) The comparative anatomy of a gene. *Harvey Lect*, 60:155–171.
[5] Sawin KE (2009) Cell cycle: Cell division brought down to size. *Nature*, 459:782–783
[6] Zhaxybayeva O, Doolittle WF (2011) Lateral gene transfer. *Curr Biol*, 21:R242–R246.
[7] Zhimulev IF, Belyaeva ES, Semeshin VF, et al. (2004) Polytene chromosomes: 70 years of genetic research. *Int Rev Cytol*, 241:203–275.

Shemin Lyu

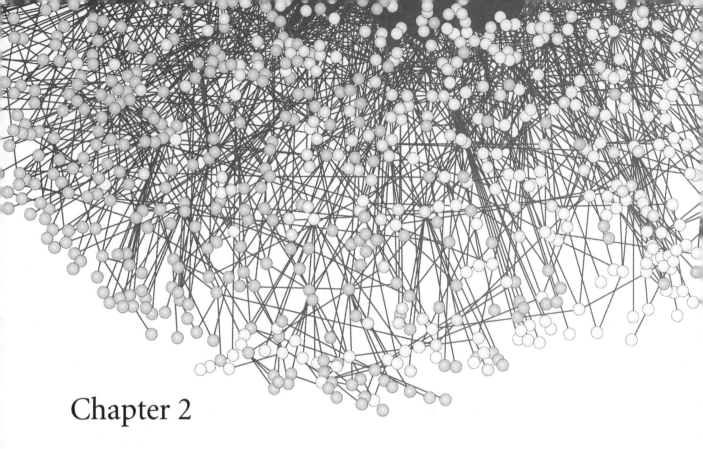

Chapter 2

Structures and Functions of Proteins and Nucleic Acids

2.1 Introduction

One class of macromolecules, protein, is almost everywhere in a cell. Proteins are so widespread in biology because they are exceedingly versatile components, exhibiting enormous variety in both structure and function (Table 2.1). They are the most versatile and important working molecules in the cell.

All proteins are large molecules: some are very large, with masses in the millions of Daltons. Some are fibrous and highly extended and thus have mainly structural roles. Similarly, scaffold proteins connect and hold in place proteins whose functions are interconnected. A vast class of more compact proteins includes molecules that act as signals within and between cells, as transporters of small molecules, as regulators of cellular processes, and as enzymes. Enzymes are the catalysts responsible for facilitating the myriad chemical reactions that a living cell or organism utilizes in its metabolism and growth. The first part of this chapter will present detailed information regarding protein structures and biological functions.

Each protein accomplishes its roles of structural and functional by having a unique amino acid sequence, which determines its secondary, tertiary, and quaternary structures. The information that dictates these sequences must somehow be stored in the cell, expressed in proteins, and transmitted through generations of cells and organisms. These vital functions are provided by biopolymers called nucleic acids, or polynucleotides, of which there are two kinds: ribonucleic acid (RNA) and deoxyribonucleic acid (DNA). The second part of this chapter will be devoted to describing these nucleic acids, their structures and possible conformations, and the multiple ways in which they can store and transmit information in the cell.

Table 2.1 **Major functions of proteins.**

Group of proteins	Function	Characteristic structural features	Examples
enzymes	catalyze more than 4000 biochemical reactions, often with high specificity and enormous efficiency	the active site may have only a few amino acid residues in direct contact with the substrate, and usually only 3–4 residues are involved in actual catalysis	enzymes of the energy producing cycles, enzymes involved in maintenance and flow of genetic information in the cell; digestive enzymes
Structural proteins	organize intracellular structures; extracellularly, provide mechanical support to cells and tissues	often multisubunit proteins, in which individual subunits interact with each other to form fibers	Inside cells, tubulin forms microtubules, actin forms actin filaments that support the plasma membrane; in the nucleus, histones form octameric protein cores around which DNA wraps to form structures known as nucleosomes; collagen and elastin are extracellular proteins that form fibers in tendons and ligaments
Scaffold proteins	hold together proteins that are part of a signaling or catalytic pathway	multidomain; many apparently evolutionarily distinct types	Scaffold proteins are essential to the signaling pathway that allows yeast to sense pheromones; Hsp70 and Hsp90 regulate sequential protein folding
Transport proteins	carry small molecules or ions; those embedded in membranes carry substances through the membranes	Changes in conformation of the binding site usually accompany binding of the transported molecule or ion	Albumin in the bloodstream carries lipids; hemoglobin in red blood cells carries oxygen; transferrin carries iron; protein calcium pumps transport Ca^{2+} into muscle cells to trigger muscle contraction
Motor proteins	move along fibers of proteins or molecules of nucleic acids to transport organelles or substances or to synthesize macromolecules	Molecular motors are usually ATPases, enzymes that use energy from the hydrolysis of ATP or other nucleoside triphosphates to move or carry cargo along molecular tracks	Myosin in muscle cells participates in muscle contraction by sliding on actin filaments; kinesin moves along; microtubules to transport organelles or substances around the cell; dynein makes cilia and flagella beat or rotate; DNA and RNA polymerases move along DNA. Strands during synthesis of DNA or RNA
storage proteins	serve as depot for storage of small molecules and ions in certain cells or as repositories of amino acids for synthesis of other proteins	Often resemble enzymes, having multiple binding sites, but lack catalytic function	Ferritin stores iron in liver cells, ovalbumin in egg white of birds and casein in the milk of mammals are sources of amino acids for embryos or newborns; endosperm proteins in plant seeds feed germinating and developing embryos
signaling proteins	provide communication between and within cells	Often highly specific in their interactions with cell surfaces or intracellular structures; belong to the class of proteins that are intrinsically disordered	Hormones and growth factors circulate in the bloodstream to coordinate functions of individual cell types and tissues: insulin controls glucose levels in blood; epidermal growth factor stimulates cell growth and division in epithelial cells
receptor proteins	usually membrane proteins that detect signals (environmental or developmental) and transmit them to the cell interior to elicit appropriate cellular responses	Usually dimerize in the membrane in response to signals	Rhodopsin in the retina detects light; insulin receptor mediates cellular response to glucose by interacting with insulin

2.2 Proteins

It has been estimated that a typical human cell contains about a billion protein molecules, of over 20 000 different kinds. Some are present as only a few molecules per cell, while others are present in the millions. The directions for producing this vast array of complicated molecules are encoded in every cell of an organism, in its DNA. The pathways through which DNA communicates this information to direct the formation of protein molecules, how it selects which proteins are to be produced in each cell, and how that information is preserved from cell to cell and generation to generation is a major focus of molecular biology. Proteins are one, but not the only, end product of information stored in DNA, but they are vital to the cell. Thus, it is important that we begin by finding out how proteins are structured, and how they use that structure.

2.2.1 Amino acids and peptides

(1) Amino acids are the building blocks of proteins

Proteins are polymers of α-amino acid monomers. The general structure of an α-amino acid is shown in Figure 2.1. These are amino acids in which the amine group is attached to the α-carbon or Cα, the one next to the carboxyl group. All amino acids contain this core, but they differ in the side chain R that is also attached to the Cα. Glycine is an unusual amino acid because its side chain R is a single hydrogen atom and thus its Cα is a center of symmetry. The Cα of all other amino acids is an asymmetric center and therefore they have stereoisomers, designated D and L. Only the L-isomer is found in native proteins, although some D-isomer amino acids do exist in living cells. Why nature chose only the L-isomer to make proteins remains a mystery.

The structures of the 20 classical amino acids that DNA codes for and specifies in the proteins of all living organisms are shown in Figure 2.2. Recently it has been shown that selenocysteine and pyrrolysine are also DNA-coded amino acids that are included in proteins, albeit rarely. The variety of side chains in the different amino acids allows a great variety of interactions with the solvent environment, other protein molecules, or other groups within the same protein molecule. Some side chains are either aliphatic or aromatic; these will tend to be hydrophobic and so are typically packed into the interior of protein molecules. Others, such as those with acidic or basic groups, or their carboxamides such as glutamine, will be hydrophilic, preferring to associate with the water surrounding a protein. Other amino acids contain specialized groups, such as hydroxyl or sulfhydryl. Altogether, the variety in the side chains of their amino acids provides protein molecules with a formidable tool kit for molecular interactions.

(2) Amino acids are covalently connected to form polypeptides of proteins

The bonds that connect amino acids together into a protein polymer are called peptide bonds, and the polymer is a polypeptide. These bonds are formed between two monomers by a process that effectively

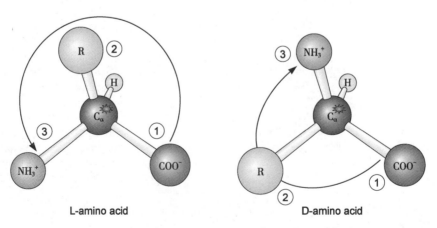

Figure 2.1 **Chemical structure of α-amino acids.**

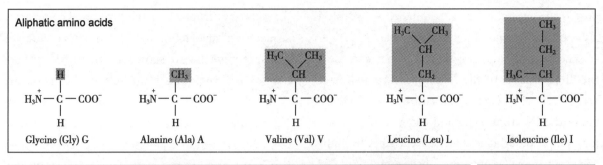

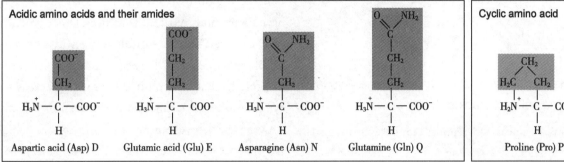

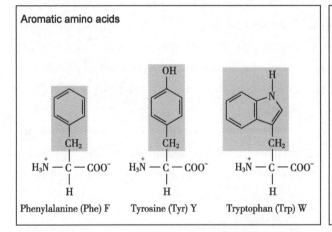

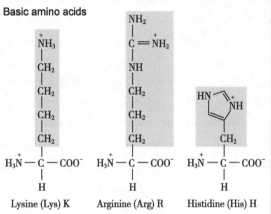

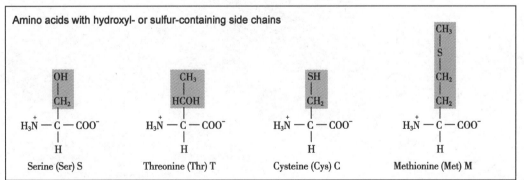

Figure 2.2 Classical amino acids found in proteins. The amino acids are grouped according to the chemical properties of their side chains. Each amino acid is presented by its full name, its three-letter abbreviation, and its one-letter code.

eliminates a water molecule (Figure 2.3A), although in cells the reaction is actually much more complex and indirect. The remainder of the amino acid that remains within the polypeptide chain is called an amino acid residue. The reverse of this process, cleaving the peptide bond by adding a water molecule, is called proteolysis and is a specialized form of hydrolysis, or splitting by adding water. In aqueous solution, the hydrolysis of polypeptides occurs spontaneously, although the process is slow unless it is catalyzed by proteolytic enzymes. Consequently, the cell must use a special energy source to make proteins from amino acids. Spontaneous hydrolysis also explains why no protein lasts forever.

The peptide bond is rigid and planar because of the directionality of the N- and C-carboxyl orbitals (Figure 2.3B). However, rotation is relatively free about the C-carboxyl–Cα and Cα–N bonds, thus, a long polypeptide chain could have a great flexibility and many possible conformations.

2.2.2 Levels of structure in the polypeptide chain

(1) The primary structure of a protein is a unique sequence of amino acids

Proteins are linear polypeptides that have distinguishable ends (Figure 2.4). One end is an unreacted NH2 group, which is called the N-terminus. The other end with an unreacted COOH is the C-terminus. At physiological pH, these groups are usually charged: NH_3^+ and COO^-. The remarkable and extremely important characteristic of proteins is that each has a defined and unique sequence of amino acid residues. This sequence is, by convention, written from the N-terminus to the C-terminus, using either the three-letter or one-letter abbreviated codes given in Figure 2.2. This sequence of amino acid residues constitutes the primary structure of the protein. The primary sequences of myoglobin from the muscles of humans and sperm whales are shown as an example in Figure 2.5. If myoglobins from millions of people are sampled, most will have exactly the same sequence, although an occasional mutant variety, differing in one or more residues, might turn up. On the other hand, myoglobins from other animal species will have significantly different sequences. Other proteins, with completely different functions, will have very different sequences. There are examples, however, of proteins that have quite different sequences carrying out the same function in different organisms. An example is myohemerythrin from certain primitive invertebrates. Like myoglobin, myohemerythrin stores oxygen in tissues, but its sequence is different from that of myoglobin. Evolution has sometimes found different routes to the same function. Conversely, there are also many examples of proteins that have similar sequences but carry out quite different functions.

Myoglobin, with only 153 residues, is a small protein. Most proteins contain about 200–300 residues, and some are as large as 4000 residues. The number of possible proteins is almost infinite; theoretically,

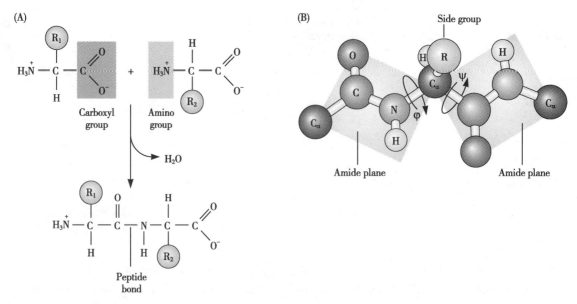

Figure 2.3 **The peptide bond.** (A) A peptide bond is formed between two amino acids by elimination of a molecule of water. (B) The peptide bond is planar and rigid. Rotation is possible only around the N–Cα bond (angle φ), and around the Cα–C bond (angle ψ). The group of atoms about the peptide bond could exist in two possible configurations, cis and trans. The trans configuration, in which the adjacent Cα atoms are placed further apart, is favored in proteins, as it avoids steric clashes between bulky R groups.

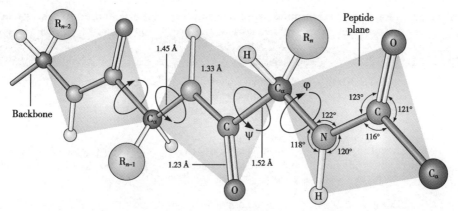

Figure 2.4 **Extended polypeptide chain showing typical backbone bond lengths and angles.** R_{n-2}, R_{n-1}, and R_n represent different side chains.

```
          1                                           30
Human   G L S D G E W Q L V L N V W G K V E A D I P G H G Q E V L I
Whale   V L S E G E W Q L V L H V W A K V E A D V A G H G Q D I L I
          31                                          60
Human   R L F K G H P E T L E K F D K F K H L K S E D E M K A S E D
Whale   R L F K S H P E T L E K F D R F K H L K T E A E M K A S E D
          61                                          90
Human   L K K H G A T V L T A L G G I L K K K G H H E A E I K P L A
Whale   L K K H G V T V L T A L G A I L K K K G H H E A E L K P L A
          91                                          120
Human   Q S A T K H K L I P V K Y L E F I S E C I I Q V L Q S K H P
Whale   Q S A T K H K L I P I K Y L E F I S E A I I H V L H S R H P
          121                                         153
Human   G D F G A D A Q G A M N K A L E L F R K D M A S N Y K E L G F Q G
Whale   G N F G A D A Q G A M N K A L E L F R K D I A A K Y K E L G Y Q G
```

Figure 2.5 **Amino acid sequences of myoglobin from two sources: human and sperm whale.** Myoglobin is highly conserved between these two species: 84% of the residues are identical. Another 16 of the amino acid substitutions are conservative, meaning an amino acid is replaced by an amino acid of the same chemical type; these are shown by blue boxes. Only nine substitutions are nonconservative, designated by red letters.

200 residues can be arranged into 20 200 different polypeptides. In the 1920s, the discovery that proteins were indeed large, covalently linked molecules came as a shock to many scientists. The later discovery that proteins have unique sequences was of equal or even greater significance. A number of methods were originally used to sequence proteins directly, whereas today, almost all protein sequences are deduced from their encoding gene sequences. One technique that remains of importance in analyzing proteins is gel electrophoresis.

(2) A protein's secondary structure involves regions of regular folding stabilized by hydrogen bonds

Although the polypeptide chain is capable of great flexibility (see Figure 2.4), it also exhibits levels of defined folding, which are known as secondary and tertiary structure. Essential to the stability of these structures are noncovalent interactions between portions of the polypeptide chain. These interactions are summarized in Table 2.2. All are weaker than covalent bonds, with the hydrogen bond being the strongest. Electrostatic interactions between positive and negative charges on side chains can also be important. At physiological pH, arginines, lysines and most histidines are positively charged, whereas aspartic and glutamic acids carry negative charges. Van der Waals interactions include a number of ways in which neutral molecules can attract one another via dipoles or charge fluctuations.

The term secondary structure is reserved for regions of regular folding within the protein chain that are stabilized by hydrogen bonds between amide hydrogens and carbonyl oxygens in the polypeptide

Table 2.2 Molecular interactions determining protein secondary, tertiary, and quaternary structure.

Type of interaction	Participants in interaction effect	Typical amino acid side chains	Range of interaction (nm)	Comment
hydrogen bonds	Hydrogen donors (D) and hydrogen acceptors (A): (D)-H-(A)	asparagine, glutamine, serine, threonine, lysine, arginine; also backbone H-N, O=C	About 0.3 nm (D-A)	Strongest of the non-covalent interactions
charge-charge interactions	Positively and negatively charged groups on side chains: ...(+)...(−)...	lysine (+), arginine (+), glutamic acid (−), aspartic acid (−)	0.5–2.0 nm "long" range	May be either stabilizing (+/−) or destabilizing (−/− +/+)
van der Waals interactions	All molecules and groups; arises from electron inhomogeneity in molecules	Most important for small, aliphatic amino acids	about 0.3 nm	become repulsive at very short distances
hydrophobic effect	All amino acids with hydrophobic side chains	phenylalanine, leucine, isoleucine	not defined; close packing of residues	packing of hydrophobic residues within the protein, away from water

backbone. These secondary structures have been confirmed repeatedly in structures of real proteins. Linus Pauling proposed a number of possible secondary structures that both satisfied the planarity of the peptide group (see Figure 2.3B) and allowed a maximum of hydrogen bonding. These have since been confirmed repeatedly in structures of real proteins. The most important secondary structures are the α-helix and β-sheet, as shown in Figure 2.6. In the α-helical conformation, hydrogen bonds are made between residues along the α-helix; in the β-sheet structure, hydrogen bonds are between parallel or antiparallel folds of the chain. The side chains of the residues do not play a direct role in forming these structures, but by steric effects they may favor one or the other.

(3) Each protein has a unique three-dimensional tertiary structure

The most important component of structure in determining a protein's function is the next level of folding, termed the tertiary structure. The whole polypeptide chain, including elements of defined secondary structure, is folded into a unique three-dimensional configuration. There would be an almost limitless variety of tertiary structures that can be formed by polypeptide chains of different sequence. A few examples are shown in Figure 2.7.

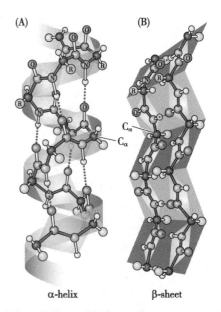

Figure 2.6 **α-Helix and β-sheet: the two most common secondary structures in proteins.** These structures differ mainly in how their stabilizing hydrogen bonds are formed. (A) In the α-helix, the hydrogen bonds are within a single length of coiled polypeptide chain and are oriented almost parallel to the helix axis. The number of residues in this helix is 3.6/turn of the helix, which corresponds to 0.54 nm/turn. (B) In the β-sheet, the hydrogen bonds occur between spatially adjacent chains. These could belong to the same polypeptide chain folded back upon itself or to two individual chains of proteins that have multisubunit structures, as in quaternary structure. In the β-sheet, the hydrogen bonds are almost perpendicular with respect to the chain direction.

Figure 2.7 **Gallery of protein structures.** The convention used here employs ribbons to represent secondary structures that are twisted in α-helices or that lie parallel or antiparallel in β-sheet structures; thin lines represent regions that do not have a simply defined secondary structure.

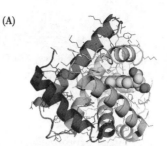

(A) Sperm whale myoglobin

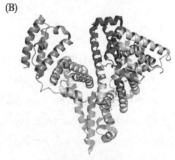

(B) Human serum albumin

(C) Dihydrofolate reductase

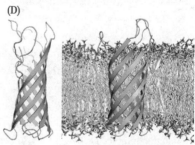

(D) OmpA   OmpA embedded in a membrane

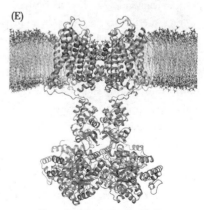

(E) Potassium channel embedded in a membrane

In recent years, it has become possible to determine such structures, often to atomic detail, by the techniques of X-ray diffraction and multidimensional nuclear magnetic resonance (NMR). To date, thousands of structures have been determined and can be found in publicly available protein data banks. The question that must be asked is: if so many structures are possible, why does each protein have its unique one?

The specific tertiary structure that a protein will adopt is dictated by noncovalent interactions, primarily between side chains but also with the surrounding water molecules. Characteristics of these interactions are given in Table 2.2. Of equal importance is the fact that hydrophobic residues tend to pack into the core of the molecule, away from water; this is known as the hydrophobic effect (Figure 2.8).

The hydrophobic effect is not truly a force, like van der Waals interactions, but a consequence of the effect of hydrophobic groups on water structure. If inserted into water, they tend to arrange surrounding water molecules into rigid cages. Taking hydrophobic groups out of water into a protein interior releases these water molecules into their natural freedom. The increase in entropy associated with such freedom drives hydrophobic residues into the interiors of proteins. Hydrophilic residues, on the other hand, tend to be located on the surface of the protein molecule, in contact with surrounding water molecules (Figure 2.8). A very important consequence of these interactions is that the primary structure - that is, the amino acid sequence - dictates both the secondary and tertiary structure of each protein molecule. Thus, the enormous diversity of protein structures is a direct consequence of the enormous number of possible sequences and will be determined by whatever determines those sequences. Today, wo know that the DNA sequence of each protein-coding gene

(A)
```
VLSEGEWQLV LHVWAKVEAD VAGHGQDILI RLFKSHPETL EKFDRFKHLK
TEAEMKASED LKKHGVTVLT ALGAILKKKG HHEAELKPLA QSHATKHKIP
IKYLEFISEA IIHVLHSRHP GDFGADAQGA MNKALELFRK DIAAKYKELG
YQG
```

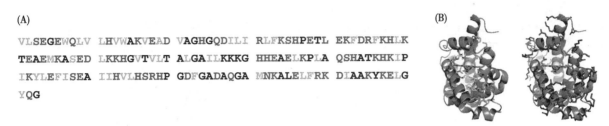

Figure 2.8 **Distribution of hydrophilic and hydrophobic residues in sperm whale myoglobin.** (A) Distribution in sequence: note the two types seem to be randomly scattered. (B) Distribution in tertiary structure. Hydrophilic residues, shown in red, are mainly on the surface, in contact with water, while hydrophobic residues, shown in green, are largely buried.

determines the protein sequence. In some cases, the secondary or tertiary structure of a protein may be modified by interactions with small molecules or other macromolecules. For example, we note that the ionic composition of the medium may promote conformational changes, as can the binding of certain small molecules, known as allosteric modulators (effectors) (Box 2.1). Even changing the protein environment around a particular sequence may cause that sequence to adopt a different secondary structure, as for so-called chameleon sequences in proteins. Finally, chemical modification of the protein side chains can affect conformation. However, because the structure of the protein defines and allows these interactions, the genetic information is still dominant.

Box 2.1 Allostery

Most proteins and other macromolecules are capable of specifically binding one or more kinds of small molecules, called ligands. There are many cases in which the binding of one ligand influences, either positively or negatively, the strength with which another ligand is bound. This phenomenon is called allostery, from the Greek *allos*, other, and *stereos*, space, and is a very common mode of biochemical regulation. Two classes of allostery can be distinguished. In homeotropic allostery, all ligands are the same. An example is hemoglobin, which has four binding sites on each molecule for the ligand oxygen. Binding even one oxygen molecule facilitates the binding of more oxygen molecules to the same hemoglobin molecule. By contrast, in heterotropic allostery, the ligands are different. For example, the affinity of many enzyme molecules for their substrates is modified by the binding of effector molecules. In some cases, the final product of a long metabolic pathway may allosterically inhibit the first enzyme in the pathway. Thus, a cell that already has a surplus of that product is saved from the wasteful process of making still more. Both positive and negative allosteric controls are important regulators of cell processes. Sometimes both occur on the same protein; for example, some transcription factors that regulate the expression of genes are subject to both positive and negative controls.

How does allostery work at the molecular level? For simplicity, imagine an enzyme that has a binding site for its substrate and a second site for an effector. It is tempting to assume that the binding of the effector molecule simply forces the protein into a conformation that favors binding of the substrate. But the most successful explanations have come from a slightly different model (Figure 1). It is postulated that the allosteric protein has two different conformations, called R and T, and can oscillate between these two states.

The R-state has higher affinity for substrate and thus better enzyme function. Suppose effector binding favors formation of the R-state. Then, in the presence of effector, the protein will be more likely to be found in the R-state and thus enzymatic activity will be higher. Of course, in other cases, effector binding might favor the T-state; the effector is then an allosteric inhibitor.

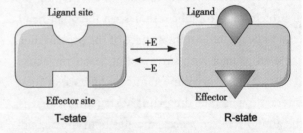

Figure 1 **Simple model for allostery.** Note that in this case both the effector and the ligand site have greater binding affinity when the protein is in the R-state. This results in positive allostery. If the T-state had greater affinity for the ligand, the effector (E) would diminish affinity for ligand by favoring formation of the R-state; this is the case with the Lac repressor, for example.

(4) The tertiary structure of most proteins is divided into distinguishable folded domains

If the tertiary structures of many proteins are examined, it becomes evident that they can often be divided into folded domains. Although the precise definition of a domain is still a matter of contention, the common opinion is that they are units of protein structure that could fold autonomously, that is, independently of other domains in the protein. Domains are often quite compact and are connected by relatively unstructured segments of the polypeptide chain. Their connectivity may be enhanced by metal ion binding or occasionally by covalent disulfide bonds formed by oxidation of sulfhydryl side chains. Domains range from 25 to ~500 residues in length. It seems likely that many domains have evolved independently and have been connected to form specifically functional proteins later in evolution. Domain structure is common; about 90% of all known human proteins contain distinguishable domains.

Protein evolution seems to have proceeded from a relatively small number of domain types, which then further evolved and connected into the multidomain proteins. This may account for the fact that features of protein tertiary structure seem to be even more conserved than sequence and can be used as clues to recognizing distant protein relatives, whose relatedness would not be evident from their sequences alone. A gallery of some common domain types is shown in Figure 2.9. Note that domains can be largely classified by their combinations of α and β secondary structure elements.

(A)

β-Catenin armadillo-repeat domains

(B)

Cadherin transmembrane domains

(C)

Immunoglobulin-like domains

(D)
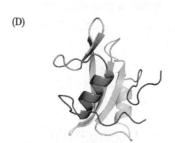
Src homology 2 domain (SH2)

(E)
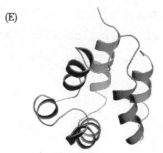
DED (death effector domain)

(F)
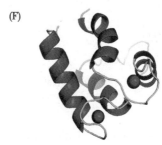
EF-hand motif

Figure 2.9 **Representative examples of individually folded protein domains and their functions.** (A) β-Catenin is made up of a number of repeats of the armadillo domain. (B) Cadherin repeats are named after cadherin proteins that play a fundamental role in calcium-dependent cell-cell adhesion. (C) Immunoglobulin-like domains were originally described in immunoglobulins but have since been found widely distributed among diverse proteins. (D) The SH2 domain contains a large β-sheet flanked by two α-helices, shown in orange and blue. (E) The death effector domain, DED, is a protein interaction domain that allows protein-protein binding by DED-DED interactions. (F) The EF-hand motif consists of two perpendicular 10–12-residue α-helices with a 12-residue loop region between, forming a single calcium-binding site as a helix-loop-helix.

Very often, the combination of several domain types within a protein molecule allows it to carry out multitasking, with each domain having a specific function. An example is pyruvate kinase (Figure 2.10A), a key enzyme in the utilization of glucose as an energy source. This protein has three domains: an all-β regulatory domain and two α/β domains. One of these α/β domains catalyzes the dephosphorylation of phosphoenolpyruvate to pyruvate, and the other accepts the adenosine diphosphate, ADP, that is converted to adenosine triphosphate, ATP, in the reaction. The regulation in this case is allosteric, mediated by a number of metabolites that bind to the regulatory domain and modify the catalysis.

There are also proteins in which one domain type is repeated. Often, this is an economical way to build a very large protein, such as the giant muscle protein titin (Figure 2.10B), which consists of hundreds of repetitions of immunoglobulin or Ig like domains (Figure 2.9). This also demonstrates the versatility of domains; the Ig domains were originally noted in the structure of antibodies, which have no common function with titin.

(5) Algorithms are now used to identify and classify domains in proteins of known sequence

Only a few decades ago, the elucidation of a single protein structure was a huge challenge. Today, the number of known structures is so large that the challenge has shifted to making sense of the great variety of tertiary structures in terms of their function and evolution. Recognition of the importance of domains as structural building blocks and potential elements of evolutionary change has led to the development of automated and semiautomated methods for the identification of domains and their classification. None of these methods is perfect because of enormous difficulties in identifying discontinuous domains (see Figure 2.10) or domains that are tightly associated. Nevertheless, the potential value of such methods for understanding protein function and evolution is undeniable. In the following, we shall describe one such method called CATH: Class, Architecture, Topology or fold, and Homologous superfamily. CATH is a semiautomated method; it has one strong competitor in SCOP, or Structural Classification of Proteins, which relies more heavily on human expertise and manual classification of domains.

CATH classifies protein domains at four levels: classes, which are the predominant secondary structure organization, α, β, or α/β; architecture, which is the overall shape; topology, which is the fold, spatial arrangement, and connectivity of secondary structural elements; and homology, which is the highest level of families and superfamilies (Figure 2.11). Figure 2.12 illustrates the distribution of domain structures among the three major protein classes. How are domains distributed overall among proteins? A comprehensive study in 2005 examined this question on a database consisting of the total genome sequences of 150 organisms. Approximately 1 million protein-coding genes were identified, of which 850 000 could be clas-

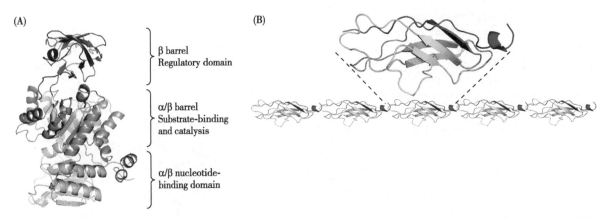

Figure 2.10 **Two examples of multidomain proteins containing either distinct domains or closely related repetitive domains.**

sified into 50 000 families, each of unique domain organization. The remainder were mainly singletons and small proteins. By use of CATH and similar algorithms, 80% of the domains could be grouped into only 5000 families, with the remainder falling into a category of new families. Thus, a relatively small number of domains has been used to construct an enormous variety of proteins.

The next step is to predict protein function from the many sequences that we recognize in the genome but have not yet isolated as proteins. Unfortunately, the protein sequence does not in itself allow us, as yet, to predict the tertiary structure, except by analogy to proteins of very similar sequence. However,

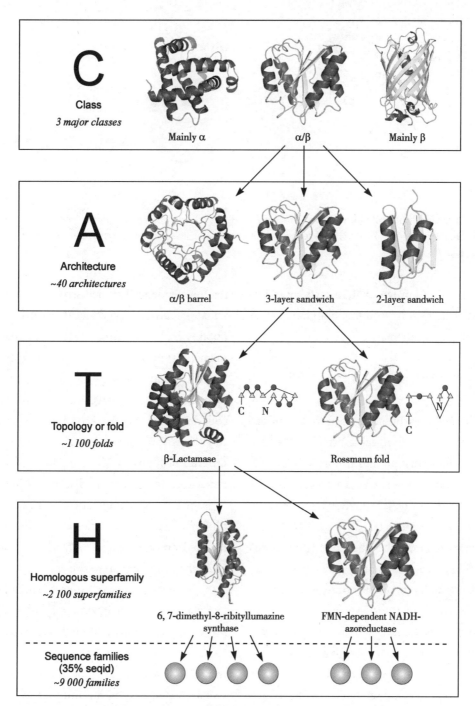

Figure 2.11 **Structural classification of proteins by CATH.** CATH is one of several existing databases that use algorithms to automatically assign domains in proteins with known sequences. The four major hierarchical levels for classifying proteins according to their structural features are Class, Architecture, Topology or fold, and Homologous superfamily. Seqid refers to sequence identity between the domains.

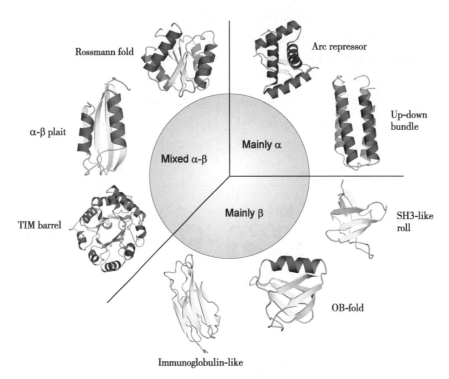

Figure 2.12 **Structural classification of proteins by CATH.** Distribution of domain structures among the three major classes: mainly α, mainly β, or mixed α-β, with representative examples in each class. α-Helices are colored in red; β-sheets are shown in yellow.

similarities between known protein tertiary structures provide better hints to function and evolution than do the sequences themselves.

(6) Some domains or proteins are intrinsically disordered

After the remarkable early successes in X-ray diffraction studies of easily crystallizable proteins, the idea became fixed that every protein probably had well-defined secondary and tertiary structures. As more and more proteins of diverse functions were investigated, however, it became clear that the situation is more complicated. Researchers began to find regions in polypeptide chains that simply did not show up in the crystallographic maps. These regions differed from those portions of the chain that do not have any standard type of secondary structure but are still well defined in the map, such as loops and turns. The sections that are not found on crystallographic maps instead correspond to chain segments that have such mobility that they differ in conformation in different molecules in the crystal or at different times. These are referred to as intrinsically disordered proteins or regions. Evidence for such structural features of proteins in solution also comes from other techniques, notably high-resolution NMR. It became clear that many parts of many proteins have intrinsic flexibility. Indeed, we now know some cases in which an entire protein molecule behaves very much like a random coil, devoid of regular secondary and tertiary structure. These are often referred to as intrinsically disordered proteins or ID proteins. It is important to remember that intrinsic disorder is just as much a consequence of amino acid sequence as is order. ID proteins are often characterized by a combination of low hydrophobicity and high charge, both of which favor their opening up to solvent.

It is believed that the high flexibility of ID proteins facilitates their interaction with multiple other proteins; their conformation can fit to different partners. Figure 2.13 presents one well-studied example, the p53 tumor suppressor protein. Mutations in p53 are associated with a number of human malignancies. p53 participates in a number of biological pathways, interacting with a different partner in each case. A key to this lies in the disordered N- and

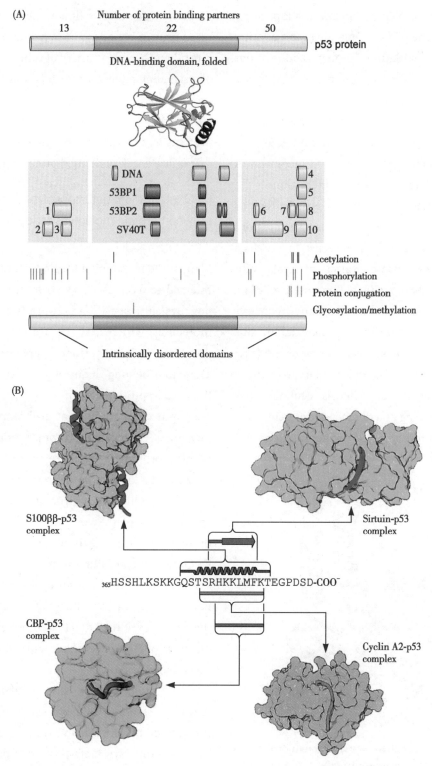

Figure 2.13 **Tumor suppressor protein p53: a protein that contains both a structured region and two intrinsically disordered domains.** (A) p53 domain structure and protein partners that bind to the different domains. More than 70% of the protein partners interact with the nonfolded regions of p53, which comprise only 29% of the length of the polypeptide. Some proteins, as well as DNA, interact with the structured DNA-binding domain. The exact portions of the p53 polypeptide chain that interact with the respective partners are depicted by boxes of different colors. Note that some partners interact with more than one region in p53. The boxed, numbered proteins have been co-crystallized with p53; the crystal structures with cyclin A (4), sirtuin (5), CBP (8), and S100ββ (10) are presented in part B. Post-translational modification sites are shown by colored vertical bars. Note that the majority of these sites are located in disordered regions. (B) Comparison of experimentally determined X-ray structures of the same C-terminal disordered region, in shades of red or in green, which adopts different structures upon interacting with different partners.

C-terminal portions of p53, which can adopt many conformations. The p53 example illustrates another point: most post-translational protein modifications occur in these disordered regions, presumably to tailor the protein to specific partners. Figure 2.14 presents an example of a protein known to recognize and react with methylated lysines and with other groups. Again, flexible peptides can adapt to the geometry of the binding site, promoting strong interactions.

Functional uses for disordered protein regions have also been found in enzymes, where such regions may promote promiscuity, in the sense that a given enzyme may accommodate a variety of similar substrates. This may have facilitated enzyme evolution. The disordered regions correspond more frequently to regulatory sites rather than to the catalytic site. This is believed to be because catalysis requires a fairly rigid molecular structure, complementary to the transition state in the reaction, whereas regulation may occur through binding of a number of different effectors, which a disordered region can accomplish.

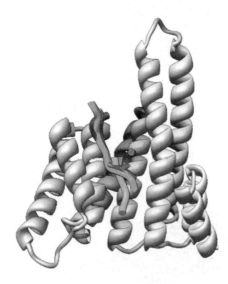

Figure 2.14 **Protein segments with different disordered sequences use their flexibility to adapt to a common binding site.** Crystal structures of five intrinsically disordered peptides bound to the highly structured binding pocket of protein 14-3-3ζ. The protein, shown in light blue, binds to various protein partners by interacting with their disordered segments: peptides from serotonin N-acetyltransferase, shown in green, and histone H3, shown in purple, as well as three other distinct peptides, shown in orange, red, and cyan. Each peptide has a distinct sequence and modifications and follows a different path through the 14-3-3ζ peptide-binding pocket.

Perhaps the most striking utilization of unstructured protein regions is in signaling proteins, which must interact with a variety of other proteins as part of a signaling pathway. To date, no signaling protein-protein interactions that involve structured domains have been identified. Thus, there is a vast class of proteins with functions in signaling and regulation that utilize a lack of defined protein structure.

(7) Quaternary structure involves associations between protein molecules to form aggregated structures

Many proteins exhibit a fourth level of organization, referred to as quaternary structure. This involves noncovalent association between individual protein molecules, which may themselves have multiple domains, to form defined aggregated structures. These may be long chains of globular proteins, as in the muscle protein actin (Figure 2.15). More usual are symmetrical arrays of a small number of monomer units, often called protomer units: arrays of 2, 4, and 6 protomers are common, but cases of up to 48 units are known. Some examples are shown in Figure 2.16. Quaternary structures can be considered special cases of the very general phenomenon of protein-protein interactions. Such interactions may be quite stable, resulting in what we call quaternary assemblies or multisubunit proteins. But they may, in other cases, be transitory but nonetheless vital to the dynamics of the cell. How complex such patterns can be is only now being recognized, in a discipline called bioinformatics.

The protomers comprising a quaternary assembly may be the same or of several kinds; if only one type is involved, they are usually arranged in some variety of point-group symmetry, with one or more axes of symmetry. In Figure 2.16, topoisomerase shows twofold symmetry; rotation of the molecule 180° about an axis in the plane of the page will yield exactly the same image. Proliferating cell nuclear antigen, PCNA, has a threefold symmetry axis, perpendicular to the plane of the page. Hemoglobin is said to have pseudo-twofold symmetry, because the two kinds of subunits are almost, but not exactly, identical.

Figure 2.15 **Proteins can form filaments of identical or nonidentical subunits.** (A) Actin, the monomer globular actin, G-actin, and the helical polymer of actin filaments (F-actin). G-actin is globular in structure; the polymer is a two-stranded helical arrangement of G-actin. Polymerization is induced upon binding of ATP to G-actin; ATP hydrolysis occurs, but ADP stays bound to the filament. The asymmetry of the individual subunits leads to polarity of the filament, defined as + and − ends; the two ends grow at different rates. (B) Tropocollagen, the basic unit of collagen fibers, is a helix of three polypeptide chains, each ~1000 residues in length. Each chain is a left-handed helix with ~3.3 residues/turn. The three helices wrap around each other in a right-handed sense; the structure is stabilized by hydrogen bonds. The structure imposes constraints on the primary sequence, with glycine-proline and hydroxyproline forming a repetitive motif. The image of collagen fibers on the right is taken by atomic force microscopy, AFM. (C) Microtubule, a hollow cylinder. Along the microtubule axis, α- and β-tubulin heterodimers are joined end-to-end to form protofilaments with alternating α- and β-subunits. Staggered assembly of 13 protofilaments yields a helical arrangement of tubulin heterodimers in the cylinder wall.

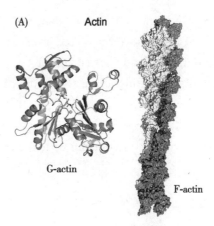

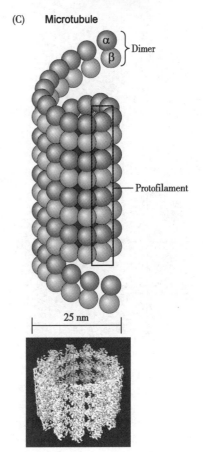

Only rarely is molecular size itself a reason for a protein to have a defined quaternary structure. More often than not, such arrangements facilitate allosteric regulation of protein function. Consider the difference between hemoglobin (see Figure 2.16) and myoglobin, which is very much like a protomer of hemoglobin. Myoglobin will bind oxygen, and does so in tissues. But hemoglobin is much better suited to the delivery of oxygen because it binds and releases oxygen cooperatively: the presence of oxygen on one protomer facilitates its binding to another. Furthermore, the binding of other allosteric effectors to hemoglobin modifies oxygen affinity, allowing adaption to varied circumstances.

In some cases, quaternary structure becomes so elaborate and extensive that a molecule is described as a multiprotein complex. Later chapters feature many examples of quaternary complexes involving as many as a dozen or more different kinds of protomers, grouped together to fulfill some complex function and its regulation. Such complexes are often associated with fundamental cell processes, like replication of DNA, transcription of DNA into RNA, or synthesis of proteins. Many of the protomer types in such complexes bind allosteric effectors, which in turn modulate the function of the whole complex. More and more, it is becoming clear that the cell is a network of interacting molecules. In many respects, quaternary structure and domain structure are complementary ways of achieving the same goals. Quaternary assembly, however, has the advantage that functional units can be exchanged for varied needs.

Quaternary structure is stabilized by the same kinds of noncovalent interactions that account for secondary and tertiary structure in proteins (see Table 2.2). As these involve specific groups on the surfaces of the protomers, we know that the quaternary arrangement is dictated by the protomer tertiary structure, and ultimately by the sequence of the polypeptide chain or chains.

2.2.3 Protein folding

(1) Protein folding can be a problem

As we see in later chapters, proteins are synthesized as unstructured polypeptide chains. The protein-synthesizing machinery of the cell does not impose structure. To become fully functional, the chain must fold into its appropriate secondary and tertiary conformations and, in some cases, associate into quaternary structures. How this happens has long intrigued molecular biologists. Surely it cannot happen by a random search through possible conformations: even if there are only two orientations at each residue, a 100-residue polypeptide would have 2^{100} conformations. Searching only a fraction of these, at nanoseconds per search, would take billions of years. Does each kind of protein have a template protein that shows it how to fold? Then what folds the template?

In reality, it appears that protein folding often can proceed in a stepwise, orderly manner dictated by the primary structure, the information the protein is provided with as it is made. This is referred to as the self-assembly principle. Evidence for this comes from many experiments that show unfolded proteins spontaneously attaining their functional conformation after having been denatured; that is, having had their native structure unraveled. A classic example is shown in Figure 2.17A, which depicts the thermal denaturation of ribonuclease, a small globular protein. If ribonuclease in solution is heated above 40°C, all evidence for defined secondary and

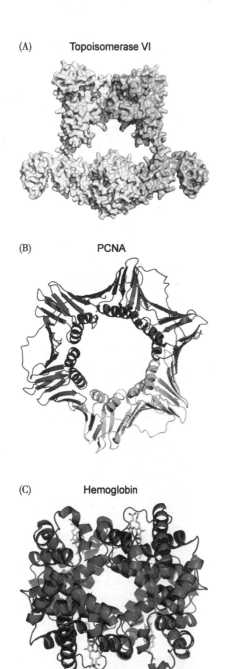

Figure 2.16 **A gallery of proteins whose quaternary structures contain identical or highly similar protein subunits.** (A) Topoisomerase VI is an example of proteins that contain two identical subunits. (B) PCNA, proliferating cell nuclear antigen, is formed by three identical subunits. (C) Hemoglobin contains two α- and two β-subunits.

tertiary structure, as well as enzymatic activity, is lost. Upon recooling, all of these features are quantitatively regained. Furthermore, comparison of data from various techniques indicates that this is an all-or-none transition: at any intermediate temperature, a mix of native and denatured molecules is present. The argument is as follows: different techniques for studying protein structure are sensitive to different aspects of the structure, which relate to various parts of the protein. If proteins are folded piecewise, then tracking the folding by a range of methods should lead to distinguishable curves in different experiments. Instead, the curves obtained by different methods are superimposable (Figure 2.17B). Thus, we know that folding, at least in this case, is a fast all-or-none process, occurring in seconds or minutes.

Such experiments have shown that for many proteins, the amino acid sequence, and thus the DNA gene sequence that dictates it, can define the conformation and functionality without external aid. The native structure represents a free energy minimum, the state of optimum stability in physiological conditions (Figure 2.17C). This is a very important

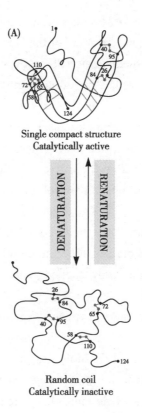

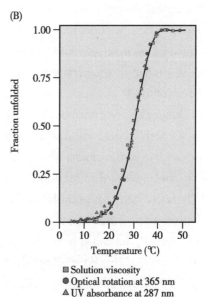

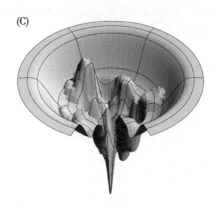

Figure 2.17 **Protein folding.** (A) Thermal denaturation of ribonuclease A. When ribonuclease is heated above a certain temperature, it undergoes a conformational transition from a highly ordered, catalytically active structure into a catalytically inactive random coil that experiences continuous fluctuation among a large number of extended conformations. When the enzyme is inactivated by elevated temperature, its disulfide bonds remain intact, allowing for faster renaturation upon reduction of the temperature; other denaturation treatments may lead to rupture of these bonds and hence to slower renaturation. If disulfide bonds are reduced concomitant with denaturation, reoxidation will lead to scrambled structure with incorrect bonds. However, removal of denaturant previous to reoxidation yields correct structure. (B) A range of physical methods has been used to study the unfolding (denaturation) and the refolding (renaturation) of proteins. (C) Hypothetical free energy map for a protein. The vertical axis represents free energy, and the x- and y-axes represent different conformations; actually there should be about 2^{100} such dimensions, but that is hard to draw. The map illustrates that there is one true free energy minimum, the most stable conformation or native state, and many pseudominima at slightly different conformations in which the molecule could, theoretically, become temporarily trapped.

principle, one of the cornerstones of molecular biology. However, it should be aware of the fact that the folding of many proteins, especially in the crowded environment of the cell, is undoubtedly more complicated and difficult than in the test tube. Interactions with other cellular molecules, including proteins, may block correct folding. Or there may exist other states, distinct from the native state, in which the protein can be trapped. A way of looking at this is shown in Figure 2.17C. The deepest point in the free energy landscape is the native conformation. In the complex kinetic process of becoming folded, the protein may be temporarily trapped in an adjacent, shallow energy well.

It has recently been proposed that multiple conformations close in energy to the native conformation may actually serve biological purposes. A protein that exhibits such behavior might be promiscuous in function, serving more than one purpose or allowing easy evolvability into new functions. But such alternative available states may also lead to misfolding, which can have serious consequences.

Misfolded proteins are generally useless and may even be harmful to the cell, especially through their tendency to aggregate: they must be dealt with. There are two major ways to deal with misfolded proteins. First, they can be given a chance to avoid misfolding or they can be allowed to refold to their native, functionally active state. Macromolecular structures that can help to accomplish these activities are called chaperones. Second, misfolded proteins may be sent to specialized cellular machineries for degradation (Figure 2.18). Both refolding and specific degradation require energy expenditure through ATP hydrolysis, which is believed to lead to conformational transitions in the respective proteins.

(2) Chaperones help or allow proteins to fold

Chaperones are huge protein complexes usually considered as protein folding machines. Chaperones enclose the protein, protect it from the cellular environment, and give it a chance to fold properly. One might say they serve as protein incubators. More than 50 chaperone families have been described to date. We must note that chaperones have a number of additional cellular functions, sometimes seemingly opposite to their role in folding. For example, proteins destined to function in or be transported through membranes should not fold before being inserted into the membrane, and their premature folding is prevented by special chaperones. Chaperones also interact with nascent polypeptide chains emerging from their site of synthesis on ribosomes, so that they do not associate (aggregate) with chains synthesized nearby.

As a specific example, we consider the best-studied bacterial chaperone system, GroEL/GroES, also known as chaperonin 60/chaperonin 10. The GroEL/GroES system is indispensable for the survival and growth of the bacterial cell as numerous proteins, both small and large, depend on it for proper folding. The subunit composition and structure of this large protein complex are shown in Figure 2.19, and the reaction cycle is schematically presented in Figure 2.20.

Another important member of the chaperone family is Hsp90 (heat-shock protein 90) (Figure 2.21). This protein, originally identified as one of the proteins induced by stress conditions including heat, is the most abundant heat-shock protein under normal physiological conditions. It has roles in both protein folding and protein degradation and has been implicated in multiple biochemical pathways including signal transduction. However, its mode of function is still not fully understood, nor do we fully understand why it can exhibit such diverse functions.

ATP hydrolysis by Hsp90 is rather slow: it hydrolyzes one ATP molecule in about 20 min in humans. The slow hydrolysis indicates that the complex conformational rearrangements that occur in the Hsp90 dimer during its action are coupled to ATP hydrolysis. Such coupling is a recurring theme in many reactions that involve ATP hydrolysis. The current view is that ATP binding and hydrolysis are instrumental in shifting the protein from one conformation to another; these conformational transitions, in turn, determine the activity of the protein. The equilibrium between different protein conformations is

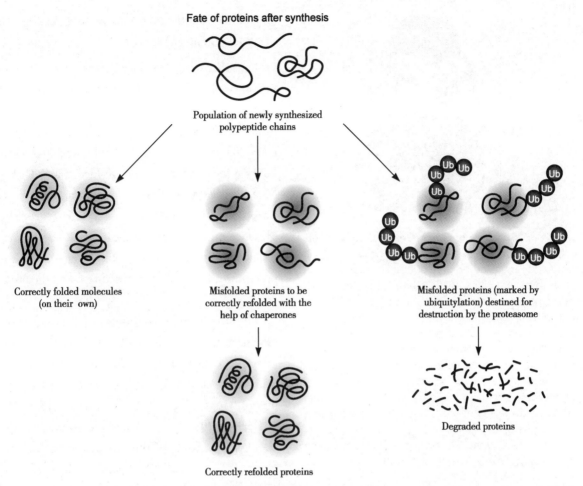

Figure 2.18 **Three possible fates of proteins after synthesis.** Some proteins can correctly fold by themselves. The cell has evolved two alternative pathways to take care of misfolded proteins: they are either refolded with the help of specialized folding machines, known as chaperones, or sent for degradation by a destruction machine, the proteasome.

usually regulated in the cell; in the case of Hsp90, there are a large number of co-chaperones that bind to the C-terminus of Hsp90 and regulate its ATPase activity. If the conformational transitions are inhibited one way or another, the protein is inactivated. Drugs that inactivate Hsp90 by blocking the ATP-binding site are widely used in clinical practice as anti-tumor agents.

2.2.4 Protein destruction

All proteins are eventually destroyed by hydrolysis; this accounts for the fact that proteins in fossils cannot last forever. But hydrolysis is slow unless it is catalyzed by proteases, and in general proteases cannot discriminate between proteins that are damaged or no longer needed in the cell and those that are vital. Therefore, there must be pathways that can mark and then destroy some proteins while not affecting others.

(1) The proteasome is the general protein destruction system

The proteasome is responsible for the degradation of hundreds and thousands of badly misfolded or otherwise aberrant (mutated) proteins. It also destroys regulatory proteins, such as transcription factors or cell-cycle regulators. Transcription factors that regulate the expression of genes in response to temporary stimuli are needed only over a short period of time. Cyclins, regulating the progression of the cell through the cell cycle, need to be present only during specific phases of the cycle and then must be degraded to allow the cell to move to the next phase.

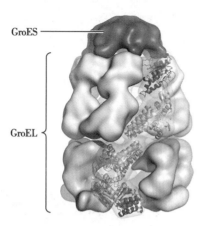

Figure 2.19 **Structure of GroEL/GroES.** The two heptameric rings of GroEL have a characteristic double-doughnut structure, as modeled from electron microscopic maps. Ribbon representations of the atomic structure of two GroEL protomers are superimposed on the model. Each GroEL subunit consists of an equatorial domain shown in light blue in the top molecule, which contains the nucleotide-binding site, an intermediate hinge domain shown in jade, and an apical domain shown in royal blue. The apical domain is located at the opening of the GroEL cylinder and contains the binding site for both GroES and the chaperoned polypeptide. The polypeptide binds at a hydrophobic groove in the apical domain facing the central channel. GroES, also termed the lid, is a dome-shaped heptamer that consists almost exclusively of β sheets.

(2) Ubiquitin-mediated degradation is another way of protein destruction

Proteins that are destined for degradation by the proteasome will be marked by the covalent attachment of the small protein molecule ubiquitin. The structural feature recognized by the ubiquitylation system is the stable patches of hydrophobic amino acid residues on the surface of the misfolded chain. They are usually buried within the protein rather than present on proteins with normal tertiary structure. Their presence on the surface indicates that something is wrong. This ubiquitylation "kiss of death" leads the protein to a multisubunit structure called the proteasome.

The proteasome degradation process is shown schematically in Figure 2.22A. The eukaryotic proteasome consists of a 20S core particle formed by four heptameric rings, two outer α-rings and two inner β-rings (Figure 2.22B and Figure 2.22C), and one or two 19S regulatory particles of very complex composition. Some of the proteins in the core particle possess activities that hydrolytically degrade the client proteins. The main function of the 19S regulatory particle is to first recognize the ubiquitylated

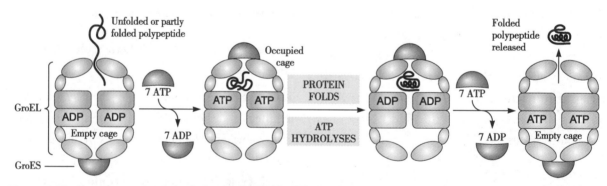

Figure 2.20 **Reaction cycle of the bacterial GroEL/GroES chaperonin system.** The nucleotide-binding abilities of the two GroEL rings are mutually exclusive: thus ATP or ADP can be bound to only one ring at a time. The GroES lid binds to the ring that has the nucleotide bound, creating a cage. The binding of unfolded or partly folded substrate protein to the open GroEL ring is followed by binding of ATP and GroES. The ring cavity now closes and the substrate is released into the cage, where it can fold. Following ATP hydrolysis, the binding of ATP to the opposite ring triggers the dissociation of GroES and the dissociation of folded substrate protein from the complex. The chaperonin complex is now ready to accept another polypeptide.

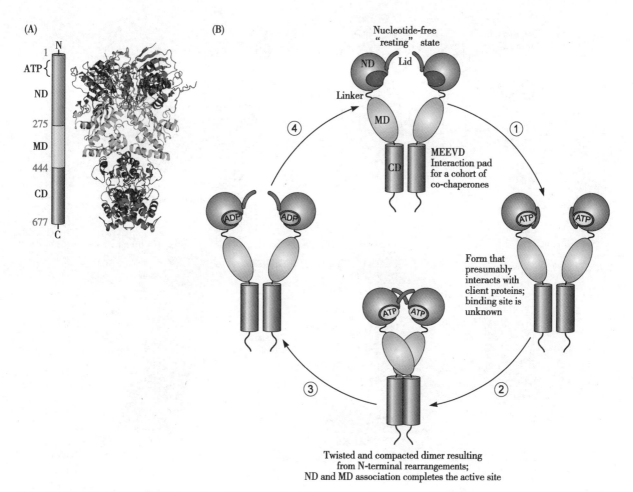

Figure 2.21 Chaperone Hsp90 and its ATPase cycle. (A) Structure of chaperone Hsp90 from yeast. The active protein is a dimer, with each subunit consisting of three distinct domains. The N-terminal domain ND, shown in red, possesses a deep ATP-binding pocket, where ATP binds in an unusual kinked conformation. The N-terminal domain in eukaryotes is connected to the middle domain MD, shown in green, by a long flexible linker sequence. The C-terminal domain CD, shown in royal blue, is the dimerization domain and contains, in eukaryotes, an amino acid motif MEEVD that binds to a variety of co-chaperones whose function is to regulate the ATPase activity of the N-terminus. (B) ATPase cycle of Hsp90. (Step 1) Following ATP binding, the lid flaps over the ATP-binding pocket. (Step 2) Lid binds to N-domain of the other subunit, producing a strand-swapped, transiently dimerized conformation. (Step 3) ATP hydrolysis; NDs dissociate from MDs; monomers separate N-terminally; lid opens. (Step 4) ADP and inorganic phosphate are released. The exact mode of client protein binding remains unknown, as does the actual mechanism of protein folding.

protein and then regulate access of the client protein to the 20S degradation chamber. The opening to this chamber is tiny, only 1.3 nm in diameter, and only unfolded polypeptide chains can pass through it. Thus, access to the degradation chamber is regulated by the 19S particle first unfolding the proteins. In addition, specific proteins in the 19S complex act as ubiquitin receptors, whereas others function to remove the polyubiquitin chains once the client protein is recognized and bound.

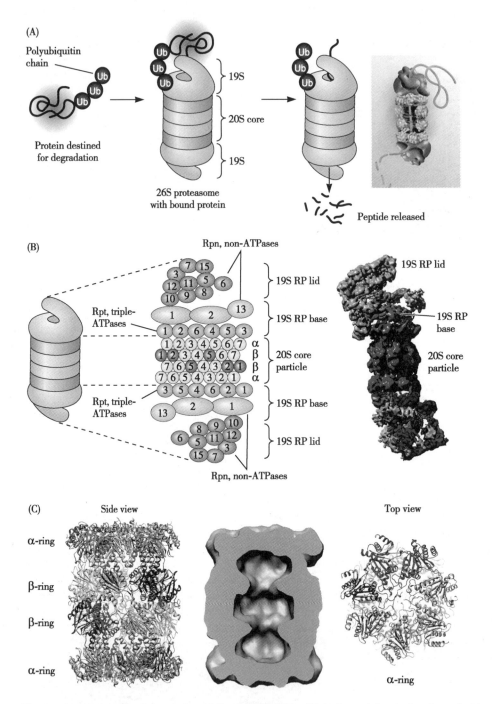

Figure 2.22 **The proteasome: structure and protein degradation.** (A) Left, proteins to be degraded by the proteasome are usually modified by polyubiquitin chains, which are recognized and bound by components of the 19S regulatory particle. The misfolded protein needs to be unfolded to pass through the narrow gate of the 20S core particle. Right, cross section of the proteasome in action. (B) Left, schematic of subunit structure of the 26S proteasome. RP, regulatory particle, consisting of base and lid components; Rpn, regulatory particle non-ATPases; Rpt, regulatory particle ATPases. Right, three-dimensional reconstitution of the proteasome from single particles visualized by electron microscopy. (C) Left, architecture of the 20S core particle from yeast, ribbon diagram. Middle, volume of the particle calculated from atomic coordinates, cut in half to show three chambers within the particle. Right, top surface of the α-ring: the N-termini of the individual subunits form a closed gate that blocks access to the inner chamber.

2.3 Nucleic acids

2.3.1 Chemical structure of nucleotides

(1) The monomeric unit is composed of five-carbon sugar; nitrogenous base and phosphate group

The general structure for the monomeric units for DNA and RNA - the nucleoside monophosphates, or nucleotides - involves a five-carbon sugar attached to a base unit and a phosphate (Figure 2.23). The mononucleotides are members of a whole class of nucleoside phosphates, many of which have other biological roles; for example, adenosine triphosphate (ATP) and guanosine triphosphate (GTP) are important energy currencies in the cell. In each case, the phosphate is attached to the 5′-carbon of the sugar, and in the polynucleotide, there is a phosphodiester link to the 3′-carbon of the next residue (Figure 2.24). The sugars involved are ribose in RNA and 2′-deoxyribose in DNA.

The bases are all derived from either purine or pyrimidine, and the bases found in nucleic acids are shown in Figure 2.24A. The basic groups of nucleic acids are always attached to the 1′-carbon of the sugar. Four different kinds are found in DNA: adenine (A), guanine (G), cytosine (C), and thymine (T). RNA contains the same bases except that uracil (U) substitutes for thymine. A ribose or deoxyribose with an attached base is called a nucleoside (Figure 2.24B). When a phosphate group is attached to the 5′-carbon of the nucleoside, a nucleotide is obtained, which is also called as nucleoside 5′-phosphate.

As depicted in Figure 2.25A, the sugar moiety of nucleic acids is in the ring or β-furanose form. To a first approximation, four of the five atoms of the sugar ring lie in the same plane; the position of the fifth atom, either C-2′ or C-3′, with respect to the plane determines whether the conformation is that of an endonucleoside, with the fifth atom on the same side of the plane as the C-5′ atom, or an exonucleoside, with the fifth atom on the opposite side of the plane. Actually the situation is more complex, with a variety of slightly different sugar conformations found. Nucleosides also have two conformations determined by the non-free rotation around the glycosidic bond (Figure 2.25B). The preferred conformation in nucleic acids is the *anti* conformation, in which the base and the sugar ring are as far away from each other as possible. As an example, Figure 2.25C shows the structure of the entire nucleoside monophosphate unit of deoxyguanosine monophosphate (dGMP).

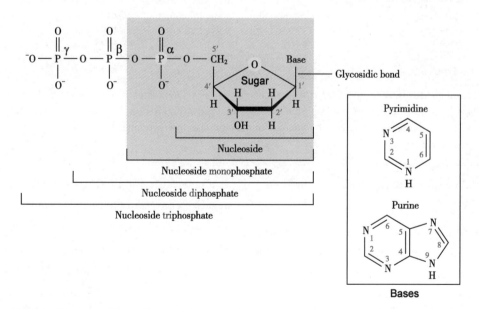

Figure 2.23 **Chemical composition of nucleotides.** Nucleotides contain a five-carbon (pentose) sugar—either deoxyribose in DNA, as shown here, or ribose in RNA (see Figure 4.2)—as well as a nitrogenous base and phosphate group(s). The base and sugar together form a nucleoside; depending on the number of phosphate groups attached to a nucleoside, it may be designated a nucleoside mono-, di-, or triphosphate. Nucleic acids are made of nucleoside monophosphates, shaded in blue, which contain nitrogenous bases that are derivatives of either pyrimidine or purine.

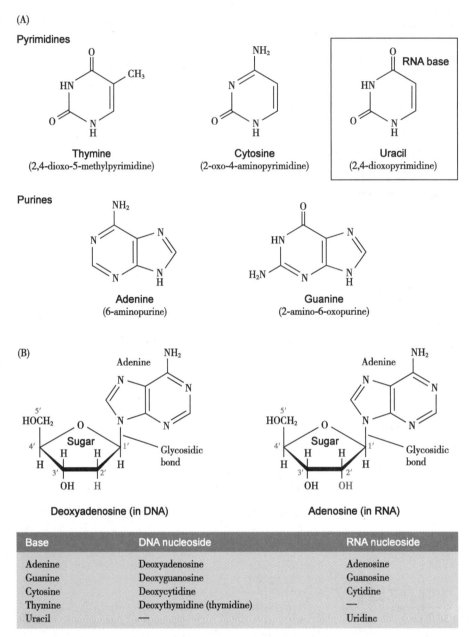

Figure 2.24 **Nucleosides in RNA and DNA.** (A) Chemical formulas of pyrimidines and purines. Uracil, shown in the box, is present only in RNA. (B) Differences between the nucleosides in DNA and RNA are highlighted in red. The table shows the nomenclature of nucleosides formed by the addition of each base to the respective pentose.

(2) Nucleic acids (polynucleotides) are polymers of nucleotides

Polynucleotides, or nucleic acids, can be visualized as being formed by the removal of a water molecule between the 5′-phosphate of one nucleotide and the 3′-hydroxyl of another. Addition of a nucleoside triphosphate with elimination of pyrophosphate more closely describes the process (Figure 2.26). The chains can be very long, involving millions or even billions of nucleotide residues. Nucleic acids are acidic, because the linking phosphate groups carry negative charges. Like proteins, most nucleic acids have unique sequences. Note that there is an unreacted phosphate on one end of the chain, called the 5′-end, and an unreacted hydroxyl at the other end known as the 3′-end, except in cases where the molecule is circular. Thus, each polynucleotide chain has polarity (Figure 2.27).

The sugar-phosphate backbone of any nucleic acid is repetitious and uniform; the sequence unique-

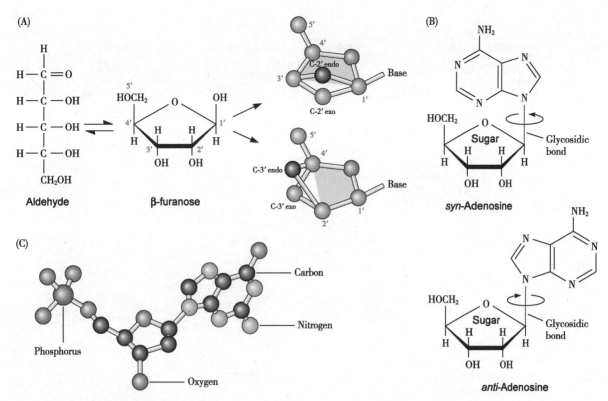

Figure 2.25 Conformations of nucleotide components. (A) Conformation of the ribose ring. In solution, there is equilibrium between the straight-chain or aldehyde form and the ring or β-furanose form of ribose; RNA contains only the ring form. The ribofuranose rings in nucleotides exist in two different puckered conformations: four of the five atoms in the ring lie in a single plane, while the fifth atom, C-2' or C-3', is either on the same side of the plane as the C-5' atom, in endo-nucleosides, or on the opposite side, in exonucleosides. (B) *Syn* and *anti* conformations of nucleosides. Rotation around the glycosidic bonds allows the existence of two nucleoside conformations: *syn* and *anti*. In solution, purines rapidly equilibrate between the two conformers, while in pyrimidines the *anti* predominates. In nucleic acids, the preferred conformation is *anti*. (C) The three-dimensional structure of deoxyguanosine 5'-monophosphate (dGMP); hydrogen atoms have been omitted for clarity. The plane of the purine ring is almost perpendicular to that of the furanose ring. The purine base is in the *anti* conformation. The sugar ring is puckered in the 3'-exo conformation. The phosphoryl group attached to the C-5' atom is positioned well above the sugar and far away from the base.

ness is carried by the order of the attached basic groups. Thus, we can compactly write the unique sequence of a DNA molecule as in this example, AGTCCTAAGCCTT, starting with the base at the 5'-end on the left according to convention. In RNA, uracil substitutes for thymine and so the corresponding RNA polynucleotide chain would read AGUCCUAAGCCUU.

2.3.2 Physical structure of DNA

(1) Discovery of the B-DNA structure was a breakthrough in molecular biology

By the early 1950s, interest in DNA as a candidate for the genetic material had reached a high level of excitement. Critical experiments had strongly indicated this to be the long-sought genetic substance. Scientists still had no idea about DNA's structure, although there were already some provocative clues. For example, it had been shown that a peculiar relationship existed between the stoichiometry of bases in natural DNA: the amount of A was always about the same as the amount of T, with a similar result for G and C. These general rules concerning the base composition of any DNA were deduced by Edwin Chargaff from analysis of the nucleotide composition of DNA samples from different organisms (Table 2.3). The first of Chargaff's rules states the equal representation of A and T and of C and G in any DNA molecule; in other words, the sum of purines, A + G, equals the sum of pyrimidines, C + T.

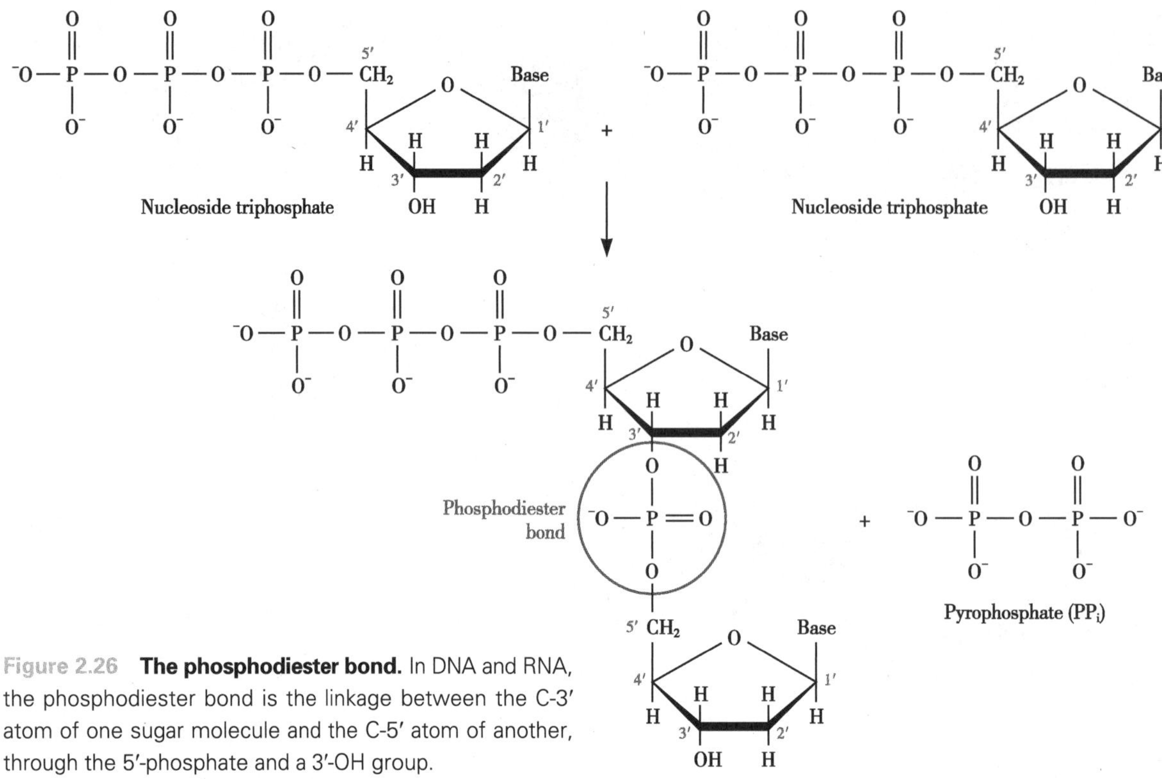

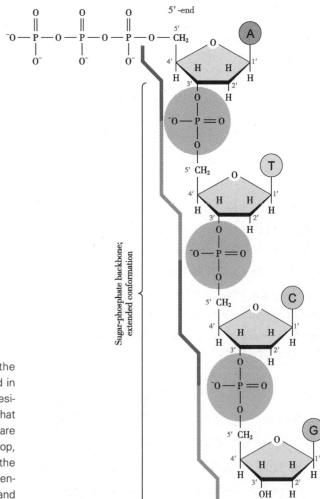

Figure 2.26 **The phosphodiester bond.** In DNA and RNA, the phosphodiester bond is the linkage between the C-3' atom of one sugar molecule and the C-5' atom of another, through the 5'-phosphate and a 3'-OH group.

Figure 2.27 **A tetranucleotide.** The backbone of the chain is formed by the phosphates and sugars, marked in alternating blue and red for the successive nucleotide residues; bases are indicated by the circled letters. Note that the chain possesses polarity: the 5'-end and the 3'-end are chemically different. Reading the sequence from the top, in the 5'→3' direction, will give ATCG; reading from the bottom, in the 3'→5' direction, will give GCTA. By convention, nucleotide sequences in nucleic acids are written and read from 5' to 3'.

Table 2.3 **Base composition of DNA (mole %) and ratios of bases that led to the formulation of Chargaff's rules.** The slight deviations in the A/T, G/C, and purines/pyrimidines ratios from 1.00 reflect uncertainties in experimental measurements. Bacteriophage φX174 genome is single-stranded and thus does not follow Chargaff's rules.

Organism	A	G	C	T	C+G	A/T	G/C	Purines/Pyrimidines
bacteriophage φX174	24.0	23.3	21.5	31.2	44.8	0.77	1.08	0.89
Escherichia coli	23.8	26.8	26.3	23.1	53.2	1.03	1.02	1.02
Mycobacterium tuberculosis	15.1	34.9	35.4	14.6	70.3	1.03	0.99	1.00
yeast	31.7	18.3	17.4	32.6	35.7	0.97	1.05	1.00
Drosophila	30.7	19.6	20.2	29.5	39.8	1.03	0.97	1.01
corn/maize	26.8	22.8	23.2	27.2	46.1	0.99	0.98	0.98
calf	27.3	22.5	22.5	27.7	45.0	0.99	1.00	0.99
pig	29.8	20.7	20.7	29.1	41.4	1.02	1.00	1.01
human	29.3	20.7	20.0	30.0	40.7	0.98	1.04	1.00

The second rule concerns the overall composition of DNA in different organisms: it varies from one species to another. Any proposed structure must explain Chargaff's rules and clarify the even more mysterious problem of how the genetic material could be accurately replicated when cells divide or organisms reproduce.

It is now widely known that the solution came from the brilliant insights of James Watson and Francis Crick, guided by the experimental work of others, including that of Chargaff and the crystallographic results of Rosalind Franklin and Maurice Wilkins (Box 2.2). The right-handed double-helix model proposed by Watson and Crick (Figure 2.28) fitted all the requirements of a genetic material so perfectly that it just had to be correct. First, the A-T and G-C base pairings allowed strong hydrogen-bonding between the two DNA strands (Figure 2.28) and accounted for the peculiar base stoichiometry, A = T and G = C. This structure also put the 1'-carbons in either pair exactly the same distance apart, 1.08 nm, allowing for a uniform, unstrained double helix. Finally, and most importantly, the model provided a strong hint, noted by Watson and Crick, as to how DNA could be replicated. Note that the two strands in the double helix, which are antiparallel, meaning they run in opposite directions, are exact complements of one another. If the strands were separated and each was used as a template to specify a new strand, two new double-stranded molecules would be obtained, each an exact copy of the original duplex (Figure 2.29). This mode of copying, called semiconservative replication, was demonstrated to be correct by Matthew Meselson and Frank Stahl.

Box 2.2 **Franklin, Watson, Crick, and the structure of DNA**

The tale of how, in 1953, James Watson and Francis Crick discovered the double-stranded helical structure of DNA is familiar to many. But the way in which it actually happened is a bit more complicated than the popular version and tells much about how science really works. In the first place, Watson and Crick were an unlikely pair for such a groundbreaking effort. Watson came to the Cavendish Laboratory in Cambridge, England, as a young postdoctoral biologist, with little knowledge of DNA. Crick was still working on his Ph.D.; his studies had been interrupted by the Second World War.

Both shared the strong belief that the structure of DNA was important and could be solved with the aid of X-ray diffraction studies of DNA fibers. But Watson had no experience with the method, and Crick's experience was largely theoretical. Furthermore, they lacked the appropriate equipment to obtain such data and the director of the Cavendish Laboratory, Sir Lawrence Bragg, was unenthusiastic about their embarking on such a project. On the other hand, excellent equipment and long experience in diffraction studies of fibrous polymers existed in the laboratory headed by Maurice Wilkins at King's College, London. There, a young researcher named Rosalind Franklin had, for several years, been attempting to get better and better diffraction patterns from DNA fibers prepared under various circumstances. Franklin was a very careful worker, unwilling to publish or share results until she was convinced of their validity. By 1952, however, she was obtaining good patterns from two forms of DNA, an A-form found at low humidity and the B-form found in wet samples. The patterns strongly indicated that the B-form was helical, with about 10 residues per turn of the spiral and a spacing of 0.34 nm between residues. These data still left the overall structure unclear: there could have been one or more chains in the molecule, and the bases could have been on the outside or the inside of the chain. Franklin apparently wished to wait for more data before publication.

Meanwhile, Watson and Crick were attempting to build reasonable models based on the skimpy older data and stereochemical constraints. Linus Pauling had proceeded in much the same way in deducing the structure of the α-helix of proteins and was by now also trying to do the same with DNA. Both came up with first models that were soon realized to be unlikely, putting the bases on the outside of the helix and crowding the negatively charged phosphates in the center.

January 30, 1953, was a pivotal day in the history of molecular biology. Watson went to London and stopped to see Rosalind Franklin and Maurice Wilkins. The meeting with Franklin did not go well, but the subsequent chat with Wilkins was quite another matter. Wilkins showed Watson the very fine diffraction patterns that Franklin had obtained for B-DNA. Watson was astounded and immediately realized the clear implications: that B-DNA was indeed helical, with 10 units per turn. This provided a framework on which models could be built. Franklin had come to the same conclusion but was opposed to model building.

Watson and Crick were not in the least inhibited and immediately began building three-dimensional models using Franklin's findings. Available data on the density and water content of DNA fibers convinced them that the structure involved two DNA strands. Furthermore, models showed that the phosphate backbones could not lie crowded at the center of the double helix but must be at the periphery with the bases inside. This initially caused problems: the bases were of different dimensions, the purines bigger than the pyrimidines. Fitting them together would yield an irregular bumpy helix. A further difficulty came from the fact that Watson and Crick were initially using the wrong tautomers of G and T; Jerry Donohue, a postdoctoral associate, corrected this. Watson soon realized that there was one way in which a smooth helix could be produced: if a purine was always across from a pyrimidine and vice versa. Furthermore, it became evident that only pairing A with T and G with C allowed the formation of a number of hydrogen bonds between the conjugate pairs (see Figure 2.28). But then A = T and G = C, exactly following Chargaff's rules. Suddenly everything fell into place, even a mechanism for transmitting genetic information between generations: if the chains must be complementary, then separating and copying each of the separated strands would lead to two double helices identical to the original.

In ascribing credit for this remarkable discovery, one must remember that the critical data were obtained by Rosalind Franklin during years of hard, careful work. But the inspiration that put all the pieces of the puzzle together was clearly that of Watson and Crick. When the Nobel Prize in Physiology or Medicine was awarded in 1962, it was divided between Watson, Crick, and Wilkins. Franklin had died a few years earlier without comparable honors: the Nobel Prize is never awarded posthumously.

The original Watson-Crick model was based on fiber X-ray diffraction patterns. Such a pattern does not provide the kind of unambiguous structure that can be obtained from X-ray diffraction patterns from single crystals. Many years passed before true crystals of small DNA molecules were obtained and studied by this technique.

The results turned out to be remarkably close to those predicted by the Watson-Crick model (Figure 2.30). There are in fact nearly, but not exactly, 10 base pairs for each turn of the helix, and the bases are almost, but not exactly, perpendicular to the helix axis (Table 2.4).

(2) A number of alternative DNA structures exist

DNA is, in fact, capable of adopting more than one type of duplex structure, depending on the base sequence in certain regions of the polynucleotide chain and the environmental conditions. The one that Watson and Crick studied is termed B-form DNA and is favored at high humidity; it is therefore the most common form *in vivo* and in solution *in vitro*. There is also A-form DNA, existing especially under conditions of low humidity, and a left-handed Z-form DNA, which requires special base composition. These are shown in Figure 2.31, and the detailed structural parameters characteristic of each form are

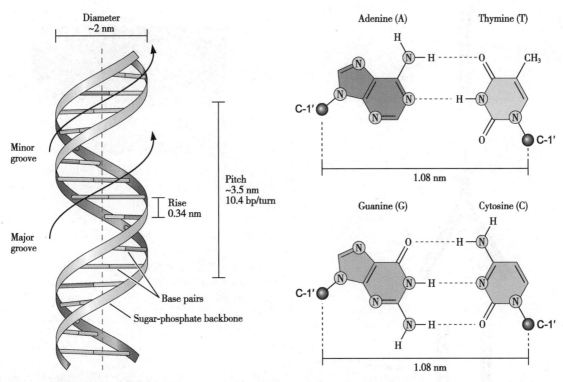

Figure 2.28 **Schematic of the double-helix structure of B-form DNA.** Base pairing occurs between a purine on one strand and the corresponding pyrimidine on the other strand; the pairing between these complementary bases allows the distances between the C-1′on the sugar moieties to be exactly the same for both adenine-thymine and guanine-cytosine base pairings. Hydrogen-bonding between the complementary bases and stacking interactions between successive pairs are both important in stabilizing the helix. The original Watson-Crick model closely resembles this but with 10.0 base pairs/turn.

Table 2.4 Structural parameters of the three forms of double-helical DNA.

	B-form	A-form	Z-form
Handedness (helical sense)	right-handed	right-handed	left-handed
diameter(nm)	~2.0	~2.6	~1.8
Base pairs per helical turn	10.4	11	12
Helix rise per base pair (nm)	0.34	0.26	0.37
Base tilt normal to the helix axis (deg)	6	20	7
Sugar pucker conformation	2′-endo	3′-endo	2′-endo for pyrimidines;3′-endo for purines
Glycosidic bond conformation	anti	anti	Anti for pyrimidines; syn for purines

listed in Table 2.4. Occasionally, single and unpaired DNA strands are found in nature, but for most purposes, the B-form is the biologically most important form, and the Watson-Crick model is a good description of its predominant helical structure.

(3) DNA can undergo reversible strand separation

At elevated temperature, DNA will denature in the sense that the duplex will dissociate into single strands (Figure 2.32). This process often occurs at a sharply defined temperature and is therefore also called melting. The melting point is increased by high G/C content or high salt, facts that are often useful in experimental protocols. Denatured DNA can reconstitute double-stranded base-paired duplexes upon slow cooling or annealing. In addition, the single stranded DNA or RNA from different samples can also be annealed to form heteroduplexes, including hybrids of DNA-DNA, DNA-RNA or RNA-RNA. This process is termed nucleic acid molecule hybridization,

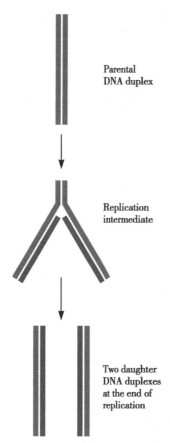

Figure 2.29 **The double-helical structure of DNA allows for a simple semiconservative mechanism of DNA replication.** The unwinding of the two strands of the parental duplex, shown in brown, is accompanied by synthesis of two complementary daughter strands, shown in red, each using one parental strand as a template. This model of replication of the genetic material was envisaged by Watson and Crick in 1953, as soon as they deciphered the structure of DNA.

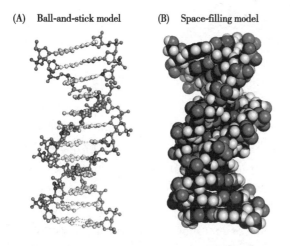

Figure 2.30 **Structure of B-DNA as derived from crystallographic studies.** (A) Ball-and-stick model. (B) Space-filling model. The base stacking is especially clear in this model, in which each atom is represented by a sphere of its van der Waals radius.

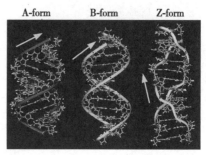

Figure 2.31 **Alternative forms of DNA double helices.** The three major forms of the DNA double helix, A, B, and Z, are shown. Note the handedness of the sugar-phosphate backbone, shown by arrows: it is right-handed in the A and B structures but left-handed in Z. Note that the B-form has a wider major groove and a narrower minor groove. These grooves are important in DNA-protein interactions (see Chapter 6). In the A-form, both grooves are nearly the same width.

which is one of the important techniques of contemporary molecular biology.

(4) DNA can also form folded tertiary structures

Special DNA sequences can engender three-dimensional DNA structures. One example is found in palindromic sequences. In linguistics, a palindrome is a sequence of letters or words that reads the same way in either direction: "Able was I ere I saw Elba," Napoleon might have said. Consider the sequence shown in Figure 2.33. Because of its palindromic nature, it can exist either in the normal linear duplex or in the cruciform structure. Cruciforms are usually extruded when a topologically constrained DNA molecule is subjected to high levels of negative supercoiling stress. A single-stranded palindrome can form a hairpin structure.

A more exotic triple helix structure, H-DNA, results from the formation of triple-strand regions that are stabilized in part by unusual hydrogen-bonding, known as Hoogsteen base pairing (Figure 2.34). This structure requires strand regions that are either all-purine or all-pyrimidine and was dubbed H-DNA because it requires high concentrations of protons to form the triple helix. Triple helices are also important in the interaction of single-stranded RNA with duplex DNA.

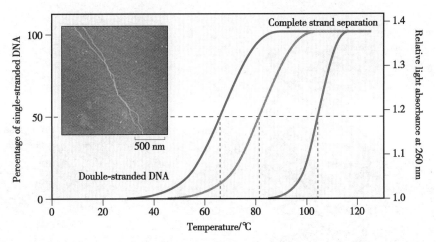

Figure 2.32 **Melting of double-stranded DNA.** Absorbance at 260 nm is usually measured in high-salt buffers, 100 mmol/L Na⁺, as the temperature of the solution is slowly raised at a rate of 1 °C/min. The melting temperature is defined as the temperature at which 50% of the DNA is still in the form of a duplex. The three different curves illustrate the dependence of melting behavior on the base composition of the DNA. Poly[d(AT)] is represented by the red line, naturally occurring DNA by black, and poly[d(GC)] by blue. The melting temperature, T_m, is indicated by the corresponding dotted lines. (Inset) Electron micrograph of partially denatured DNA.

A great variety of complicated structures can, in fact, be generated by synthesizing specific single-stranded oligonucleotide sequences and letting them interact or anneal with each other to form double-helical structures in their complementary regions. The ability to design such artificial structures is now being utilized in the construction of complex DNA-based nanostructures for applications in nanotechnology (Figure 2.35).

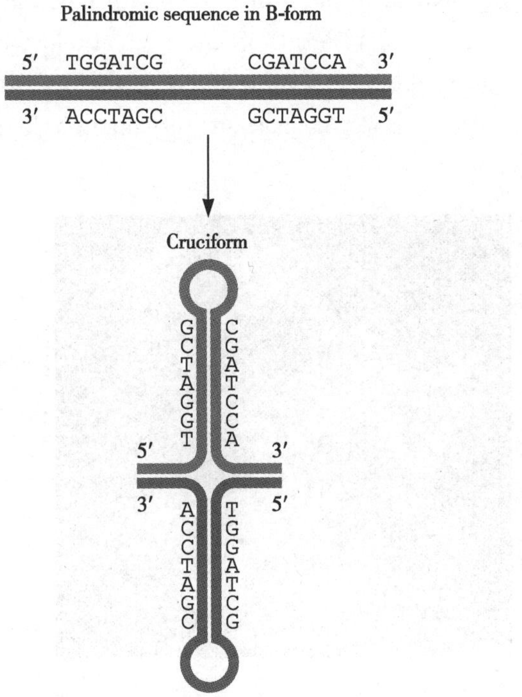

(5) Closed DNA circles can be twisted into supercoils

The range of possible DNA structures becomes even more remarkable when we consider closed circular duplexes. Many DNA molecules, especially those found in bacterial cells, are closed circles. In addition, most of the DNA of higher organisms exists as linear molecules that form loops through attachment to insoluble protein networks in the eukaryotic nucleus. Such constrained loops formally behave as circles; that is, they are topologically constrained. All such molecules have a unique property: they can be supercoiled, which means that the axis of the coiled double helix is coiled about itself. A simple example of supercoiling can be seen in a telephone cord. The cord is coiled, but chances are very good that the coil has become twisted and supercoiled about itself.

To explain more fully what this signifies, a few thought experiments are helpful. First, suppose that we have a linear DNA, containing 105 base pairs, with exactly 10.5 base pairs/turn. Assume it is lying on a flat surface, and with ultramicro-tweezers we push it into a circle, with the ends touching (Figure 2.36A). We

Figure 2.33 **Cruciforms (double hairpins).** These structures are extruded when palindromic sequences of double-stranded DNA are subjected to high negative superhelical stress.

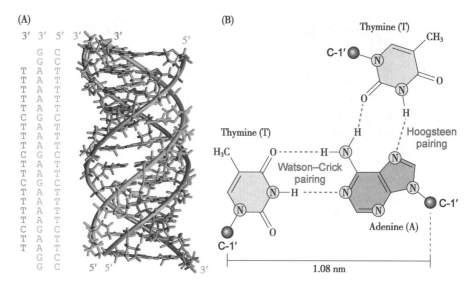

Figure 2.34 **Triplex DNA.** (A) A third pyrimidine strand, shown in green, binds to the major groove of pyrimidine-purine duplex DNA to form a triple-stranded DNA. (B) Hoogsteen-type base pairing occurs in addition to the normal Watson-Crick base pairing in these triplexes.

now, by some clever enzyme, join the ends, forming a closed circular DNA. It will be a relaxed circle that remains flat on the surface because it had an integral number of turns: 105 bp divided by 10.5 bp/turn = 10 turns. In such a molecule the 5′-end of each strand will exactly meet with the 3′-end of the same strand and could be easily joined into a circle by our smart enzyme (Figure 2.36B). But now suppose we had unwound the DNA by one turn, by rotating one end counterclockwise while holding the other end stationary, before joining the ends. Now we have a circular DNA containing 105 bp in 9 turns rather than in 10. It is said to be underwound, and its linking number, the number of times one strand crosses the other, has been decreased by one. This DNA is not relaxed and is under strain. It can accommodate that strain in various ways. As in Figure 2.36C, it could spread the strain over the whole circle, making a DNA molecule with 11.67 bp/turn, that is, 105 bp with 9 turns. This DNA circle can also lie flat on the surface. Another more likely possibility is that the DNA axis would make one turn about itself, produc-

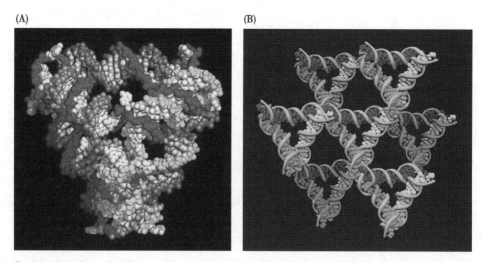

Figure 2.35 **Rational design of self-assembled three-dimensional DNA crystals.** A number of complicated artificial structures have been created by annealing single-stranded oligonucleotides that possess complementary sticky ends. These ends associate with each other preferentially, assume B-DNA structure, and guide the formation of the three-dimensional crystal. (A) DNA tetrahydron, manufactured in the laboratory of Andrew Turberfield. (B) A structure known as the tensegrity triangle has been manufactured, crystallized, and resolved to 4 A.

ing a negatively supercoiled structure; note that the actual crossing of the DNA duplexes is right-handed or positive; this can be seen in Figure 2.36D. This writhing allows the DNA to gain back one right-hand twist, which is compensated by the supercoil. The DNA is now back to 10 turns, at 10.5 bp/turn. The supercoiled DNA, because it crosses itself, cannot lie flat on a planar surface.

DNA supercoiling can be modeled by using a piece of elastic belt or rubber band. A relaxed band joined to form a circle corresponds to a relaxed circular double-stranded DNA molecule, with edges marking the two strands. We can model supercoiling by holding one end of the band, twisting the other end, and then joining the two ends. If the starting model is a closed circle, like a rubber band, it must first be cut before twisting and resealing. As the band represents a B-form helix, the coiling we see is a secondary level of coiling. When the band is twisted even more before joining, the supercoiled structure becomes more complex and more compact at the same time, and we can observe branches forming. This principle of supercoiling is clearly evident in the series of lightly supercoiled DNA molecules shown in the micrographs of Figure 2.37. The relationship between linking number Lk, twist Tw, and writhe Wr can be put on a quantitative basis as described in Box 2.3.

At high levels of supercoiling, the DNA can become quite contorted (Figure 2.37). Two general kinds of supercoils may be formed: plectonemic and toroidal (or solenoidal) structures (Figure 2.38). Note that the level of overall compactness of the

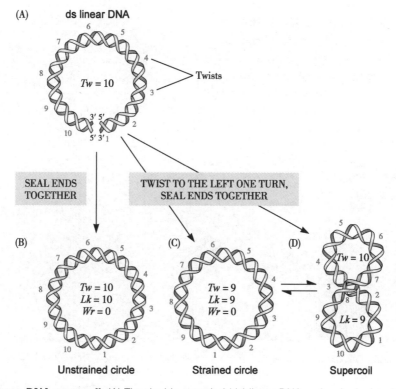

Figure 2.36 **Forming a DNA supercoil.** (A) The double-stranded (ds) linear DNA molecule depicted here has 105 bp and a pitch of 10.5 bp/turn, giving Tw (twist) = 10. (B) Because the number of turns is an integer, that is, the last turn is completed, the 5'- and 3'-ends are oriented such that the molecule can be closed into a circle that is relaxed and lies flat on the surface. In an unstrained circle, Lk is 10, Tw is 10, Wr (writhe) is 0, and the pitch is 10.5 bp/turn. (C) If the number of turns is reduced by one before the ends are joined, the newly formed circle will have only nine turns. For this circle to lie flat, the pitch must change from 10.5 to 11.67 bp/turn. Lk is 9, Tw is 9, and Wr is 0. This molecule is said to be underwound, as more base pairs are needed to form each turn of the helix than is normal for B-DNA. The molecule is strained because its helix is untwisted with respect to the thermodynamically stable B-form. (D) Rather than changing its twist, a strained molecule may writhe or supercoil. Here, the pitch is 10.5 bp/turn, Lk is 9, Tw is 10, and Wr is –1. A molecule that is supercoiled cannot lie flat on the surface. When a DNA molecule is strained, it initially absorbs the stress by changing its twist, but soon after that it undergoes a transition to a supercoiled structure. This transition has been dubbed the buckling transition.

Box 2.3 A Closer Look: Supercoiling, linking number, and superhelical density

It is possible to describe the configurations of closed circular duplex DNA in a quantitative manner. First, note that once a DNA molecule has been closed into a circle, it has an invariant topological quantity that can only be changed by cutting, twisting, and resealing. This is the linking number, Lk, which is defined as the number of times two circles are interlinked. In the case of DNA, the two circles are the two individual DNA strands in a closed circular molecule. It must be an integer, and its sign is positive for right-handed crossings and negative for left-handed ones (Figure 1).

The linking number is distributed between twist, Tw, the number of times each strand twists around the DNA axis, and writhe, Wr, the number of times the DNA axis crosses itself.

$$Lk = Tw + Wr \quad \text{(Equation 2.1)}$$

These relationships are shown in Figure 4.14. Note that the forms shown in Figure 4.14 parts C and D both have a linking number of 9 but it is distributed differently between twist and writhe.

The amount of supercoiling present in a circular DNA molecule is measured in terms of the supercoil density, σ:

$$\sigma = (Lk - Lk_0)/Lk_0 \quad \text{(Equation 2.2)}$$

where Lk_0 is the linking number of the molecule if relaxed, which equals Tw_0 because a relaxed molecule has zero writhe. A typical σ value for DNA molecules *in vivo* is about –0.05, which means that there are about five negative supercoils for every 100 turns, or about 1000 bp, of DNA. Negative supercoiling is most common; most naturally occurring circular DNAs are underwound.

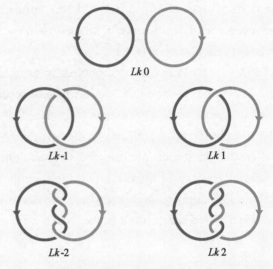

Figure 1 Linking number. Any two closed curves in space can be moved, cut, and resealed to be linked in a number of ways; a few simple examples are shown. The way the two curves interlace determines the linking number and its sign: Lk is positive for right-handed crossings and negative for left-handed ones.

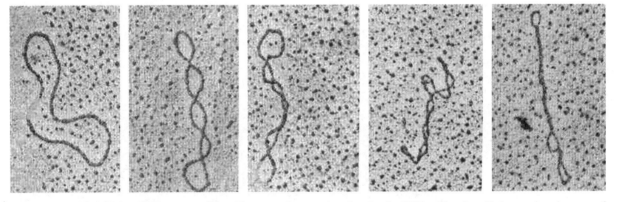

Figure 2.37 Visualizing DNA supercoiling. Electron micrographs of a circular DNA molecule with increasing degrees of supercoiling from left to right.

same molecule differs between the two forms, with toroids forming more compact structures than plectonemes. Negative values of writhe correspond to left-hand coiling in a toroid but right-hand crossing in plectonemic structures. Although difficult to visualize, this can be demonstrated by wrapping a piece of tubing in several toroidal turns about a cylinder and then removing the cylinder while the ends of the tubing are held. Toroidal supercoiling is common in eukaryotic DNA, which is often packed in very large amounts into a small nucleus. In eukaryotic chromosomes in the interphase state, DNA toroids are supported by proteins that are associated with chromatin. However, cells usually use both forms of supercoiling to compact their DNA, as exemplified by the *Escherichia coli* chromosome (Figure 2.39). Although bacteria usually possess only one giant circular DNA, this circle is partitioned into individual

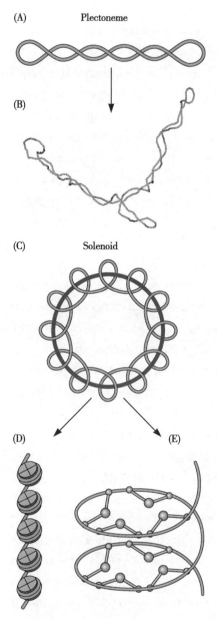

Figure 2.38 **Three-dimensional trajectories of superhelical DNA.** Schematic presentation of the two forms adopted by superhelical DNA: (A) plectoneme and (C) solenoid, also called toroid. (B) Typical conformation of supercoiled plectonemic DNA: computer simulation of a supercoiled molecule of 3500 base pairs in length with a supercoil density σ of –0.06 under physiological ionic conditions. Two forms of solenoidal supercoil are shown: (D) Wrapping of DNA around histone octamers, shown as red cylinders, compacts the DNA, which forms a left-handed solenoid. (E) Condensin proteins, shown as red ball-and-stick structures, that participate in forming mitotic chromosomes affect global DNA writhe by forming large positive solenoids.

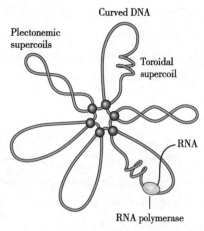

Figure 2.39 **Supercoiled DNA domains in the *E. coli* chromosome.** The drawing shows the presence of both plectonemic loops and loops containing toroidal or solenoidal structures; the proteins involved are not depicted. This is a simplified cartoon of the chromosome; in real life, there are perhaps as many as 400 different domains. The loops exist due to the presence of specific structural proteins bound at the bases of the loops. Intrinsically curved DNA that bends spontaneously because of a specific sequences of bases tends to be localized at the tips of supercoils.

domains that possess different structures and different, regulated degrees of supercoiling.

Cells possess enzyme molecules that can regulate the superhelicity of their DNA. To do this, the linking number must be changed, and this requires cutting and resealing the duplex. These enzymes are called topoisomerases, and there are two general classes: type I and type II. Type I topoisomerases change the linking number in steps of one, whereas type II topoisomerases change it in steps of two. This difference mechanistically stems from the number of DNA strands being broken and resealed in individual enzymatic steps: one DNA strand for type I topoisomerases and both strands for type II enzymes.

(6) Rigid double helix can be bent by bound proteins

Most depictions of B-DNA suggest that it is a rigid rodlike molecule, but this is not an exact description. DNA duplexes are bendable, and in some cases they contain built-in bends. The flexibility of any polymer molecule can be expressed in terms of a quantity called the persistence length. This may be understood from the following thought experiment: imagine that you are holding one end of a molecule rigidly in your hand. How accurately can you predict the position of the other end? If the molecule is extremely rigid, you can make an exact prediction. Such a molecule is said to have a very large persistence length. On the other hand, if the molecule is extremely flexible and floppy, you can make no good prediction as to the position of the other end. This molecule has a very small persistence length. Many studies of B-DNA in solution have resulted in a value of about 45 nm for the persistence length, corresponding to ~130 bp. This limited bendability places constraints on the structures that DNA can form by interacting with proteins. Thus, for example, although it is possible to wrap 147 bp of DNA 1.67 turns about the protein core of a nucleosome, this wrapping requires significant bending energy, as well as some changes in twist and some dislocations in the helix.

In addition to this basic flexibility, DNA can have intrinsic bends, dictated by base sequence. The stacking together of base pairs can be asymmetric; some pairs will stack more tightly on one side of the helix than on the other. To take one example, a series of regularly spaced dA/dT pentamers [d(A/T)$_5$ tracts] will produce significant bending. The existence of intrinsically bent DNA may be important in facilitating the binding of certain proteins to DNA. Intrinsic DNA bending can be detected by its effect on the electrophoretic mobility of DNA in gels: bent DNA will migrate more slowly than linear DNA of the same length. An additional method for estimating the degree of bending involves the circularization of small pieces of DNA by use of a ligase, an enzyme that is capable of joining two pieces of DNA into a single uninterrupted molecule. The experiments are done with pieces of DNA that are so short, and hence rigid, that they would not be able to form circles on their own if they do not contain built-in bends. Circularization will be greatly enhanced if the molecule is bent, with the probability of circularization going up with larger or sharper or more numerous bends. Gel electrophoresis is a convenient method to estimate the probability of circularization, as it readily separates linear molecules from circular ones.

2.3.3 Physical structures of RNA

RNA can adopt a variety of complex structures but not the B-form helix

Ribonucleic acid, RNA, differs chemically from DNA in only two respects. First, RNA contains the sugar moiety ribose instead of the 2′-deoxyribose in DNA. Second, RNA contains the base uracil, U, instead of thymine, which has the chemical structure 5-methyluracil. These differences may seem small, but they serve to qualify the two kinds of nucleic acids for very different functions in the cell. One direct consequence is that RNA cannot form the double-stranded B-form found in cellular DNA: the hydroxyl group in ribose gets in the way, whereas the smaller hydrogen atom in deoxyribose does not. Complementary RNA strands can form double helices, but these are invariably in the A-form.

In fact, most of the RNA found in cells is generated as single-stranded copies of one strand of DNA. In such copying, which is called transcription, the base A in DNA pairs with U in the newly forming RNA strand; other pairings are as described for DNA. However, chemical modification of bases is much more common in RNA than in DNA. We have much more to say about transcription in other chapters; here we present only a brief overview that will define the roles of RNA in biology.

In terms of three-dimensional structure, some of any single-stranded RNA will be in a disordered, random-coil conformation. However, interactions between complementary RNA regions that may be in distant parts of the chain promote specific and often complex folds of many RNA molecules. The simplest folds are hairpin structures or half-cruciforms formed by palindromes, but much more complicated structures are common. In fact, the structural potential of RNA may be almost as diverse as that of proteins. Extended regions of defined secondary structure, such as α-helices and β-sheets in proteins and A-form duplexes in RNA, are found in both types of macromolecule, and in both cases these regions can further fold to form tertiary structure. These structures appear, in many cases, to have evolutionarily adapted to specific functions. Some RNA molecules bind to proteins and modify their function; others, called ribozymes, have catalytic functions like protein enzymes. Thus, the information passed from DNA to RNA can have two kinds of function: to code for proteins or to produce functional RNA molecules. It is becoming evident that the latter expresses the largest fraction of the genome.

2.3.4 One-way flow of genetic information

In the years following the elucidation of the Watson-Crick DNA structure, a series of studies revealed how the transfer of genetic information from DNA actually occurs. Actually, the exploration of this huge and complex problem is still a major effort in molecular biological research and the subject of much of this book. But to provide a perspective and clarify the roles of DNA and RNA, a brief overview is useful at this point. The essentials were termed by some the central dogma of molecular biology:

$$DNA \xrightarrow{\text{Transcription}} RNA \xrightarrow{\text{Translation}} Protein$$

RNA molecules were at first seen to function primarily as intermediates in the transfer of information from DNA sequence to protein sequence. The amino acid sequence information for any protein, coded in the DNA sequence, is transcribed into a messenger RNA (mRNA) during the copying of one of the two DNA strands of a protein-coding gene. This mRNA must be translated, via a code, from the four-letter language of the four bases in polynucleotides to the 20-letter language of the 20 amino acids in proteins. It is clear that a single nucleotide or even a doublet would not work; 4 or even $4 \times 4 = 16$ does not provide enough combinations to code for 20 amino acids. It was found that each amino acid corresponds to one or more nucleotide triplet(s). The genetic code is very nearly universal over all of life.

Translation of the message carried by mRNA occurs on cellular particles called ribosomes, which are complexes of ribosomal RNA and specific proteins. Each nucleotide triplet in the messenger RNA is matched to a specific amino acid that will be added to the growing peptide chain through the function of a small class of transfer RNAs (tRNAs). Each transfer RNA will bind to a specific amino acid, and each carries an anticodon triplet that will base-pair with the appropriate codon on the mRNA when both mRNA and tRNA are on the ribosome. The ribosome then helps to catalyze the joining of the amino acid carried by the tRNA to the growing polypeptide chain. Thus, while DNA can be thought of as the cellular repository of genetic information, RNA plays a role in every step of translating that information into proteins.

In addition, it becomes increasingly clear that many RNA molecules have regulatory roles in many of the processes of information transfer and that RNA can serve a wide variety of cellular functions. This and our present recognition that information can sometimes pass from RNA back to DNA requires a revision of the simple scheme written above:

$$DNA \underset{\text{Reverse transcription}}{\overset{\text{Transcription}}{\rightleftarrows}} \begin{matrix} mRNA \\ ncRNA \end{matrix} \xrightarrow{\text{Translation}} Protein$$

In this new diagram, ncRNA stands for all of those noncoding RNAs that do not code for proteins but do other things. In higher organisms, this utilizes by far the larger part of the genome, and we have only glimpses as to what functions these ncRNAs may be performing. The change reminds us that the dogma was never a dogma anyway, for that term connotes an unchanging truth. Nothing in science is unchanging.

2.3.5 Methods used to study nucleic acids

Over the years, numerous methods have been developed to study nucleic acids. Some of these are rather simple in terms of instrumentation and execution; others require sophisticated and expensive machines. The most common methods—gel electrophoresis (Box 2.4), gradient centrifugation, nucleic acid hybridization, and DNA sequencing (Box 2.5)—are presented here in a series of boxes. Additional methods are detailed in relevant chapters, mainly in Chapter 4.

Box 2.4 Gel electrophoresis of circular and supercoiled nucleic acid molecules

Because the sieving effect in gel electrophoresis depends on molecular dimensions, it should not be surprising that the technique can be used to separate linear from closed circular molecules of the same length. Moreover, topoisomers of a given circular DNA molecule can be resolved. Recall that the more supercoiled the DNA molecule is, the more compact it is (see Figure 2.37). As one might expect, the mobility of the same circular molecule increases with the absolute value of the linking number, until a limit is approached for highly supercoiled molecules (Figure 1).

This very simple analysis has one drawback, however: it cannot distinguish positive from negative supercoils. But there is a clever trick to accomplish this. First, the mixture of molecules is electrophoresed in one dimension. This gives a series of bands, each of which may contain both positive and negative supercoils of a given absolute Lk. The one-dimensional gel is then soaked in a buffer containing a DNA intercalating dye [such as ethidium bromide (EtBr) or chloroquine]. Intercalators are cyclic organic compounds with planar structures that place themselves between adjacent base pairs (Figure 2), thus increasing the distance between them: this leads to unwinding of the double helix. The degree of unwinding depends on the nature of the intercalator. Ethidium bromide unwinds (or untwists) the duplex by approximately −26° per molecule intercalated, whereas the unwinding angle of chloroquine is approximately −8°. Thus, the overall degree of unwinding will depend on the number of molecules intercalated, which in turn is a function of the concentration of intercalator in the buffer. The intercalator-mediated change in twist must be compensated by changing writhe. This makes the negative isomers less supercoiled and the positive ones more so. Electrophoresis again, in a perpendicular direction, produces the kind of result shown in Figure 3.

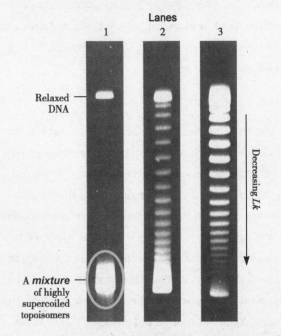

Figure 1 Separation of individual topoisomers of supercoiled DNA by one-dimensional agarose gel electrophoresis. Lane 1, supercoiled DNA isolated from bacterial cells; lanes 2 and 3, same DNA treated with type I topoisomerase, which partially relaxes the supercoils, for 15 and 30 min, respectively. The individual bands between relaxed and supercoiled DNA bands represent topoisomers of decreasing Lk; the Lk of the DNA molecules in neighboring bands differs by one.

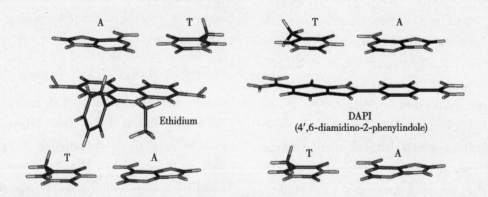

Figure 2 **Theoretical modeling of stacking interactions for two widely used intercalators: EtBr and DAPI.** Although 4′,6-diamidino-2-phenylindole (DAPI) typically binds to the minor groove, it can also, under certain circumstances, intercalate between DNA bases. DAPI is widely used as a counterstain to visualize DNA with respect to another molecule of interest, usually a protein that is stained with a different dye in cellular/nuclear cytological preparations.

Figure 3 **Discrimination between positively and negatively supercoiled DNA by two-dimensional electrophoresis.** The dashed line indicates that a single band in the first dimension, if it happens to contain both positive and negative supercoils, now produces two spots in the second dimension. The spot numbers indicate the writhe of the molecule in the second-dimension gel. Remember, the crossing of the two DNA helices in an overwound molecule with positive writhe is actually negative (see Figure 2.36). The magnitude of change depends on the concentration of the intercalator in the buffer: In the specific example illustrated here, the chloroquine concentration, and hence the number of molecules intercalated, is such that it unwinds two helical turns in the DNA molecule, which is compensated by creation of two positive writhes. Thus, the intercalation of chloroquine into the +1 topoisomer in the original topoisomer mixture has changed its geometry; it now contains three rather than one negative crossings. At this chloroquine concentration, the −1 topoisomer in the original topoisomer mixture has changed its geometry from one positive crossing to one negative crossing. As a result of these changes, the two topoisomers that were indistinguishable in the first dimension can now be separated from each other in the second dimension because of the difference in compaction imposed by chloroquine intercalation. Recall that the *Lk* of a given molecule is invariant and cannot change without breaking and resealing of the DNA backbone, which does not occur here. Thus, what changes following chloroquine intercalation is the geometry of the molecule, not its *Lk*.

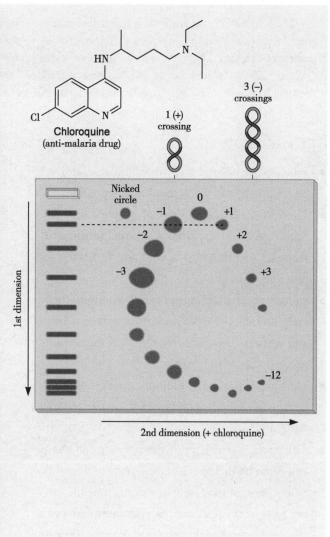

Box 2.5 DNA sequencing

Methods developed in the late 1970s led to the revolutionary capability to determine the nucleotide sequence of any DNA. Now a second revolution has occurred. These methods have subsequently been automated and coupled to sophisticated data analysis to allow unprecedented accuracy and speed in sequencing entire genomes, including the genome of *Homo sapiens*.

Two major methods have been used for DNA sequencing: chemical sequencing, developed by Maxam and Gilbert, and enzymatic sequencing, developed by Sanger. Both methods share the same underlying principle: the nucleotide sequence is reconstructed by determining the size of nested sets of DNA molecules. The DNA molecules in each set share a common 5′-end and terminate at a single type of base at the 3′-end, either A, T, C, or G. The DNA molecules in any given set differ, however, in their lengths, which are determined by the distance of these respective bases from the 5′-common end; that is, by their position in the sequence. The processes that create the A (or T or C or G) sets of molecules are random in terms of which particular A out of all the As present in the DNA fragment is being used for the chemical cleavage reaction of Maxam and Gilbert or as a site of termination of elongation in the enzymatic method of Sanger. We will illustrate how the method works for the enzymatic method, as the chemical method is no longer used.

The enzymatic method is also called the chain-termination method. In this method, the nested sets of DNA fragments are created by *in vitro* reactions that closely mimic chain elongation during the normal DNA replication process. However, elongation of the nascent daughter strand is purposefully terminated, at random sites along the sequence, by supplying the reaction mixture with nucleoside triphosphate derivatives that lack the OH group at the 3′-position of the sugar . Separate runs are made with each of the four dideoxy derivatives or ddNTPs. If, during synthesis, the polymerase incorporates the ddNTP derivative instead of the dNTP, then synthesis is terminated at that point, because the 3′-OH group is needed for further elongation (see Chapter 13).

Walter Gilbert, Frederick Sanger, and Paul Berg received the 1980 Nobel Prize in Chemistry. Gilbert and Sanger were given the award "for their contributions concerning the determination of base sequences in nucleic acids." Paul Berg was recognized for "his fundamental studies of the biochemistry of nucleic acids, with particular regard to recombinant-DNA."

Key Concepts

- Proteins are polymers of l-α-amino acids, joined by peptide bonds. Thus, they are polypeptides.
- There are millions of kinds of proteins in living organisms, and each has a unique amino acid sequence or primary structure, which dictates its function.
- Certain regular foldings of the polypeptide chain are preferred: the most important are the α-helix and β-sheet. These are referred to as the secondary structure of the protein.
- Most proteins have, in addition, a preferred three-dimensional folding, referred to as the tertiary structure.
- The tertiary structure of the polypeptide chain can often be divided into recognizable domains. Some proteins have only one domain, while others have many. Occasionally, domains or even whole proteins lack any regular secondary or tertiary structure. These are called intrinsically disordered domains or proteins.
- Sometimes folded polypeptide chains associate noncovalently with one another to form more complex structures. These associated complexes are referred to as the quaternary level of structure.
- In many cases, protein folding and association processes are spontaneous and dictated by the primary structure.
- In the complex milieu of the cell, proteins can sometimes misfold. Chaperone complexes may engulf a misfolded protein and allow it to refold correctly. Alternatively, protease complexes may digest the aberrant protein back to its constituent amino acids.
- Recent proteomic studies are beginning to unravel the very complex interactions among proteins in the cell.
- Two kinds of biopolymers are utilized for information storage and transmission in the cell: ribonucleic acid, RNA, and deoxyribonucleic acid, DNA. Both polymers are built on a repetitive backbone of sugar moieties linked by phosphodiester bonds. In RNA the sugar is ribose; in DNA the sugar is 2′-deoxyribose. Attached to the 1′-atom of each sugar is a basic unit, either a purine or a pyrimidine. In DNA, the purines are adenine (A) and guanine (G) and the pyrimidines are cytosine (C) and thymine (T). The same bases are found in RNA except that uracil (U) substitutes for T. The sequence of bases along the polynucleotide chain provides unique identity to each DNA or RNA molecule.
- DNA and RNA have significantly different structures and functions *in vivo*. The B-form DNA structure first proposed by Watson and Crick consists of antiparallel, double-stranded helices, with base pairing of A with T and G with C in the complementary strands. The helix is right-handed with ~10 bp/turn. Other structures are possible under special circumstances.
- The B-form DNA structure is quite stable, allowing it to serve as a repository of genetic information, and its complementary double-stranded nature allows for accurate copying and thus information transfer from generation to generation.
- Some DNA molecules are circular, and these allow further conformational variation, especially supercoiling. A supercoiled DNA is characterized by its linking number, Lk, the total number of times one strand is interlinked with the other. The linking number is a topological invariant and can be changed only by cutting and resealing the molecule. The linking number is distributed between the twist of one chain about the other and the writhe of the double-helix axis. Highly supercoiled molecules can be very compact.
- A class of enzymes called topoisomerases can

change the linking number of DNA molecules in the cell. The two classes of topoisomerases, topo I and topo II, use different mechanisms to do so.
- RNA molecules are usually found *in vivo* as single-stranded molecules that have been copied from one strand of genomic DNA. They can, however, exhibit complex tertiary structures, which allows them to play multiple roles in the cell, from participation in the mechanism of translation to regulation of transcription to the enzyme-like functions of ribozymes.

Key Words

active center(活性中心)
allostery(变构)
amino acid(氨基酸)
amino acid sequence(氨基酸序列)
base(碱基)
catalysis(催化)
chaperon(分子伴侣)
domain(结构域)
double-helix(双螺旋)
enzyme(酶)
folding(折叠)
glycosidic bond(糖苷键)
hairpin-structure(发夹结构)
melting(融解)
nucleic acid(核酸)
nucleoside(核苷)
nucleotide(核苷酸)
palindromic sequence(回文序列)
peptide(肽)
peptide bond(肽键)
phosphodiester bond(磷酸二酯键)
proteasome(蛋白酶体)
protein(蛋白质)
primary structure(一级结构)
quaternary structure(四级结构)
ribose(戊糖)
secondary structure(二级结构)
α-helix(螺旋)
β-sheet(片层)
subunit(亚基)
supercoil(超螺旋)
tertiary structure(三级结构)
ubiquitin(泛素)

Questions

1. Describe briefly how the secondary and tertiary structure of a protein is stabilized.
2. What are the basic structural features of the secondary structure of a protein?
3. Describe briefly how a newly synthetic peptide becomes folded.
4. Describe briefly the functions of Chaperons?
5. Describe briefly the mechanisms of protein destruction.
6. Why are so important the enzymes in biological system?
7. Describe briefly the characteristics of enzyme.
8. Describe briefly the physical structure of DNA discovered by Watson and Crick.
9. Is it necessary for DNA duplex to be separated under normal conditions?
10. Describe briefly the supercoil features of a circular genome.

References

[1] Anfinsen CB, Scheraga HA (1975) Experimental and theoretical aspects of protein folding. *Adv Protein Chem*, 29:205–300.
[2] Crick FHC, White JH, Bauer WR (1980) Supercoiled DNA. *Sci Am*, 243:100–113.
[3] Ellis RJ (2006) Protein folding: Inside the cage. *Nature*, 442:360–362.
[4] Frank-Kamenetskii MD, Mirkin SM (1995) Triplex DNA structures. *Annu Rev Biochem*, 64:65–95.
[5] Orengo CA, Thornton JM (2005) Protein families and their evolution: A structural perspective. *Annu Rev Biochem*, 74:867–900.
[6] Rich A, Zhang S (2003) Z-DNA: The long road to biological function. *Nat Rev Genet*, 4:566–572.
[7] Sanger F, Nicklen S, Coulson AR (1977) DNA sequencing with chainterminating inhibitors. *Proc Natl Acad Sci USA*, 74:5463–5467.
[8] Tanaka K (2009) The proteasome: Overview of structure and functions. *Proc Jpn Acad Ser B*, 85:12–36.
[9] Wandinger SK, Richter K, Buchner J (2008) The Hsp90 chaperone machinery. *J Biol Chem*, 283:18473–18477.
[10] Watson JD, Crick FHC (1953) Molecular structure of nucleic acids: A structure for deoxyribose nucleic acid. *Nature*, 171:737–738.
[11] Zheng J, Birktoft JJ, Chen Y, et al. (2009) From molecular to macroscopic via the rational design of a self-assembled 3D DNA crystal. *Nature*, 461:74–77.

Yifu Guan

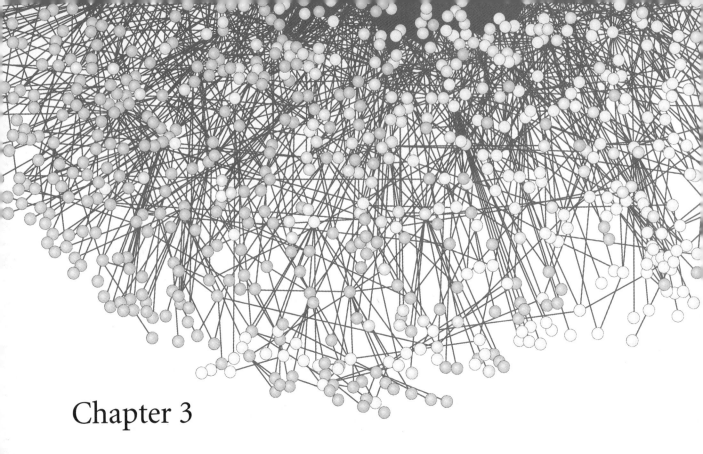

Chapter 3

Recombinant DNA: Principles and Applications

3.1 Introduction of homologous recombination and cloning

3.1.1 The beginnings of recombinant DNA technology

Recombinant DNA technology, introduced in the early 1970s, allow the isolation and copying of any gene, thus generating the amount and purity of DNA necessary to determine the gene's nucleotide sequence (Box 3.1). Moreover, they provide a way to express genes at the high level needed for thorough investigation of their products, most importantly proteins. Finally, recombinant DNA technology makes it possible to introduce any desired mutation at a predetermined position in any gene, thus allowing the production of proteins with desired properties that can be rather different from the properties of the original native proteins. Modified genes can be introduced into living cells as a way of studying their biochemical and physiological roles. For the first time, we are able to modify at will the genetic content of cells and whole organisms.

From a purely practical viewpoint, recombinant DNA technology has given rise to the contemporary biotechnology industry that produces and markets hundreds of new drugs and vaccines. It also underpins the production of genetically modified crops with desirable characteristics, such as improved nutritional value or resistance to insects, herbicides, drought, and salinity. Finally, it has given rise to gene therapy projects that aim to cure genetic diseases. Although these projects are not yet mature enough for general application in clinical practice, they are making rapid and significant progress.

The importance of these technologies for both basic research and other human activities is enormous and will be referred to again and again in the coming chapters. Therefore, in this chapter, we introduce the benchmark techniques that collectively form the basis for the creation and use of recombi-

> **Box 3.1 The first recombinant DNA molecules**
>
> In 1972, Paul Berg and his group at Stanford University carried out the first experiments to modify natural DNA specifically. They used the *Eco*R I restrictase, newly discovered by Herbert Boyer at the same institution, to cleave the DNA chromosome of the mammalian virus SV40 at one specific point, opening the closed circle into a linear form. They then used the enzyme terminal transferase to add 3′ extensions of either poly(dA) or poly(dT) onto such molecules, after resection of the 5′-ends with exonuclease. Such molecules could then be annealed together through A-T base pairing.
>
> Single-stranded regions were filled in by DNA polymerase and ligated by ligase. In this way, Berg and colleagues were able to construct dimers and higher oligomers of SV40 DNA and to insert foreign DNA fragments into the viral DNA. Because SV40 can infect eukaryotic cells, they envisioned their procedure as a gateway for mammalian DNA modification.
>
> This was a powerful and groundbreaking result: it was the first time that a natural DNA molecule had been specifically modified in a premeditated way. The technique was somewhat awkward, however, especially in the requirement for poly(dA) and poly(dT) tails, which necessarily involved inserting some extraneous dA/dT tracts into the product.
>
> Shortly thereafter, it was recognized that the very nature of the cut made by *Eco*R I, as well as many other restriction endonucleases, necessarily left overlapping, complementary ends on the cleaved product. In 1973, Stanley Cohen and Herbert Boyer took advantage of this in a seminal paper that laid the groundwork for recombinant DNA research.
>
> Cohen and Boyer worked with plasmids that could be introduced into bacteria and contained all of the elements needed for their replication in the bacterial cell. They also contained expressible genes for resistance to various antibiotics, which allowed rapid and efficient screening for colonies of bacteria containing each plasmid. Finally, each contained an *Eco*R I site, placed so as not to interfere with plasmid replication or gene expression. Cohen and Boyer were able, by taking advantage of the self-complementarity of the DNA ends produced by *Eco*R I cleavage, to make recombinant combinations of antibiotic-resistant plasmids and then "clone" these in the bacterial host.

nant DNA molecules. We begin with a general introduction to the concept and the individual steps of the fundamental technique in recombinant technology: the making of multiple copies of a DNA sequence by the technique called cloning.

3.1.2 Homologous recombination and cloning

A clone is a group of identical copies of some biological entity; a colony of bacteria grown from a single bacterium is a clone. When we speak of the cloning of DNA, we mean the production, in some organism, of multiple copies of a particular sequence. The development of such techniques in the 1960—1970 period solved the problems noted above.

As illustrated in Figure 3.1, all cloning projects begin with two essential components: the piece of DNA to be cloned and the vector molecule that will be used to introduce the sequence of interest into a living host cell. By definition, vectors are relatively small DNA molecules that can incorporate the foreign nucleotide sequence, be introduced into a cell, and then be maintained stably in that cell.

The next step is insertion of the donor DNA segments into the vector. This is essentially a cut-and-paste process that requires several enzymes, including restriction endonucleases for cutting and DNA ligases for resealing. Once the donor sequences are inserted into a vector, the resulting recombinant vectors are introduced into living cells. There are a variety of ways to do this, each optimized for the specific host, be it a bacterium or a eukaryotic cell. This step results in the production of several different types of host cells: those that did not take up any vector, those that contain the unaltered self-ligated vector, and those that contain recombinant vectors. The majority of this last group of host cells may contain sequences in which the investigator is not interested; a relatively small proportion will have the recombinant of interest. In order to retrieve the clone of interest from this heterogeneous cell population, various selection or screening methods are applied, as illustrated in Figure 3.1 and Figure 3.2. The reader will learn about these methods along the way, when we describe the properties of the cloning vectors by using specific examples.

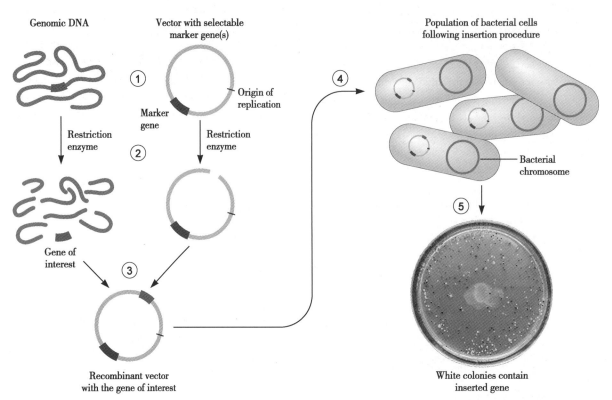

Figure 3.1 **General scheme for DNA cloning in bacteria.** DNA cloning can be divided into several steps: (Step 1) Isolate DNA from an organism and prepare cloning vector with marker gene. (Step 2) Produce recombinant DNA molecules via treatment of both donor and vector DNA with a restriction endonuclease and a ligase. The restriction endonuclease treatment will digest the organismal DNA into a large set of linear DNA fragments of heterogeneous length. Note that the type of fragment ends and the length of the fragments will depend on the restriction endonuclease used. (Step 3) Ligate the DNA fragments produced into the vector linearized by restrictase treatment. This step results in three different kinds of vectors: those that carry the gene of interest, those that have other DNA fragments from the original mixture inserted, and those that self-ligate without incorporating any foreign DNA. For simplicity, only a recombinant vector carrying the gene of interest is depicted. (Step 4) Introduce the population of vector molecules into bacterial cells by transformation or transduction, plate the cell population onto solid agar plates, and allow individual cells to proliferate and form colonies on the plate. It is important to dilute the bacterial cultures before plating so the individual colonies can be observed and further tested. Four different kinds of bacterial cells are obtained following the vector insertion procedure, reflecting the three different kinds of vectors described above as well as cells that have not taken up any vector. (Step 5) Screen for clones that contain the gene of interest. The step is performed in three stages. First, cells that did not take up the plasmid are eliminated by growing the cell population on solid medium, such as agar, that contains an antibiotic. Cells that do not contain a vector will die, because the antibiotic resistance gene is carried only by the vector. Second, one should be able to discriminate between cells that carry recombinant vectors and those that carry original vector, self-ligated with no insert. This is usually done by monitoring a gene on the vector that becomes inactivated when a foreign fragment is inserted into it. If the insertion is into a gene that confers resistance to a second antibiotic, these cells will become sensitive to this drug; these are the desired cells. Other genes, like the one carrying the enzyme β-galactosidase, can be used for insertional inactivation. Third, there should be a way to find the bacterial clone that carries the gene of interest against the background of all other recombinant bacteria. This is done by colony hybridization with a labeled probe that contains the gene of interest or a portion of it. (Petri dish, courtesy of Stefan Walkowski, Wikimedia.)

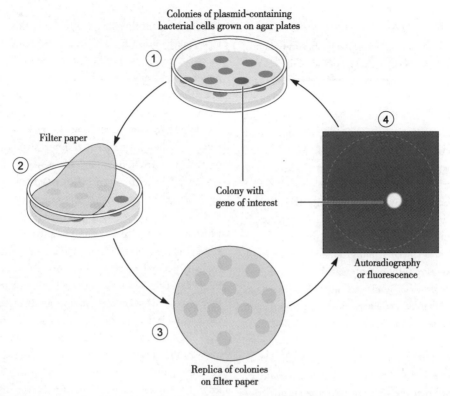

Figure 3.2 **Nucleic acid hybridization identifies the colony that carries the gene of interest.** (Step 1) Grow colonies of plasmid-containing bacterial cells on agar plates. The colony containing the gene of interest is shown in red. (Step 2) Press a filter against colonies to create a replica of the colonies where some cells from each colony adhere to the filter. (Step 3) Wash the filter with an alkali solution to denature the DNA. The resulting ssDNA is then hybridized with the gene of interest labeled in some way; only the colonies that contain the gene of interest will hybridize to the probe and become labeled. (Step 4) Detect the label in an appropriate way: by autoradiography if the probe is radioactively labeled or by fluorescence if the probe carries a fluorophore. Then compare the pattern with probe to the pattern on the original plate to identify the colony containing the gene.

3.2 Construction of recombinant DNA molecules

To insert a given DNA sequence into a vector, one must first cleave the vector, allow recombination, and then ligate the recombinant vector. Two classes of DNA enzymes are essential for creating any recombinant DNA molecule: restrictases and ligases. DNA polymerases constitute another very important class of DNA enzymes used to manipulate nucleic acids, but as they encompass a wide variety of enzymes, each with its own role and mechanism of action, we describe them in relevant chapters elsewhere in the book.

3.2.1 Major classes of restriction endonucleases

The discovery of restriction endonucleases, also called restrictases, is one of the most important developments in molecular biology; for a historic account, see Box 3.2.

The majority of these enzymes catalyze the cleavage of DNA at specific nucleotide sequences, producing defined double-stranded DNA fragments suitable for numerous applications in basic research and biotechnology.

More than 3000 restrictases have been described to date. They are all DNA endonucleases, that is, they hydrolyze internal phosphodiester bonds within the polynucleotide chain (Figure 3.3). Exonucleases, by contrast, hydrolyze the chain from either its 3'- or 5'-end, removing one nucleotide at a time. The name of each restrictase carries information about the bacterial species and strain in which it was originally discovered; a Roman numeral is used to differentiate between several enzymes from the same strain. For example, *Eco*R I designates one of the two known

> **Box 3.2** **The discovery of restriction endonucleases**
>
> How basic research has implications far beyond science. The existence of enzymes that cleave nucleic acids was recognized as early as 1903, but the fact that some could make specific cleavages came only half a century later. The first hint of such activity came from seemingly unrelated genetic studies in the laboratory of the pioneering molecular geneticist Salvatore Luria. In 1952, Luria and Mary Human published studies on the comparative susceptibility of various bacterial strains to several bacteriophages. The results were curious: some strains were almost completely resistant to some phage but vulnerable to others. The word almost here is significant; even in resistant strains, there were a few bacteria in which the phage could grow, and these phages, when harvested, were fully potent against the previously resistant strain of bacteria. Similar results, with other phage, were soon obtained in other laboratories. The phenomenon was first termed host-controlled variation and later host restriction.
>
> The puzzling result waited a decade for explanation. In 1962, Werner Arber and associates at the University of Geneva proposed that the resistant bacterial strains contained an enzyme that degraded the incoming phage DNA. But why did such an enzyme not degrade host DNA? The answer must be that the host DNA had somehow been modified by another enzyme to resist this degradation. DNA methylation by a methylase was suspected as the source of host resistance by Arber as early as 1965 but was not experimentally demonstrated until 1972. The postulate of a protecting enzyme also explained the occasional observation of acquired infectivity: in rare cases, the DNA of invading phage could itself be methylated.
>
> Early studies on the restriction process centered largely on type I restriction endonucleases, which do not cleave at a specific site. It was a major breakthrough when, in 1970, Hamilton Smith and colleagues at Johns Hopkins University discovered type II restriction endonucleases, which both recognize and cleave at a sequence-specific site. A year later, Kathleen Danna and Daniel Nathans, also at Johns Hopkins University, used such an enzyme to digest the DNA of the virus SV40 into discrete fragments, which they separated by gel electrophoresis. This led in turn to the routine isolation and cloning of specific fragments of DNA. Restriction endonucleases play fundamental roles in many of the most important techniques in molecular biology. Recognition of the enormous potential of this work led in 1978 to the award of the Nobel Prize in Physiology or Medicine to Arber, Smith, and Nathans "for the discovery of restriction enzymes and their application to problems of molecular genetics."

restrictases in *Escherichia coli* strain R; *Hae* III (see Figure 3.3) denotes the third enzyme present in *Haemophilus aegyptius*. There are three general classes of restrictases. Type II enzymes are of primary interest to us because their ability both to recognize a particular DNA site and to cut at a defined point within that site makes them ideal tools for recombinant DNA technology.

3.2.2 Recognition sequences for type II restriction endonucleases

Each restriction enzyme is characterized mainly by the specific sequence it recognizes and the site where it cuts the double helix (Table 3.1). Most recognition sequences are 4, 5, or 6 bp in length, although longer sites have also been described. The longer the recognition site, the less frequently it occurs in DNA. For instance, enzymes that target 4-bp sites cut about once every 256 or 4^4 nucleotide pairs, whereas 6-bp sites will be found on average once in 4096 or 4^6 nucleotide pairs. As illustrated in Figure 3.3A, the enzyme can attack either one of the two ester bonds in the phosphodiester linkage, producing different products. In addition, different enzymes can cut the opposite strands of the DNA helix, either in a staggered way to produce overhangs, also known as sticky ends, or exactly opposite each other to produce blunt ends, also known as flush ends (see Figure 3.3B). Both kinds of products can be ligated to form uninterrupted molecules, but the ligation efficiency of blunt-ended fragments is much lower than that of fragments containing overhangs, because the latter can actually base-pair with each other, increasing the probability of the two ends staying in close proximity and in proper orientation for ligation to occur.

Type II restriction endonucleases have proven to be of great value in recombinant technologies, as they possess well-defined specificity for the cleavage sequence, and thus the products of their action are very well defined. These enzymes usually recognize palindromic sequences, which read the same on both complementary DNA strands, 5′→3′ and 3′→5′,

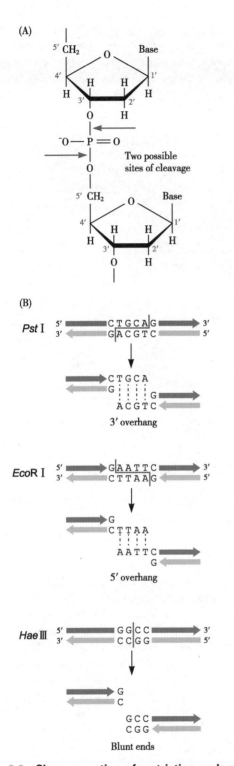

Figure 3.3 Cleavage action of restriction endonucleases. (A) In general, endonucleases that cleave polynucleotide chains internally can be specific with respect to which ester bond in the phosphodiester linkage they cleave, thus producing DNA fragments of different termini, as depicted. Type II restriction enzymes produce 5'-phosphates and 3'-OH groups only. (B) Cleavage sites on the two DNA strands can be offset, by as many as four nucleotides, or occasionally even more, to produce overhangs, also known offset of the cutting positions, blunt-end products are created.

and bind as homodimers. The cleavage patterns of type II restrictases can be rather diverse. Figure 3.4 illustrates some frequent cleavage patterns and introduces two important concepts: isoschizomers and methylation-sensitive isoschizomers. Isoschizomers have the same recognition site but come from different bacterial species; they may cut the sequence at different places, thus adding to the diversity of cleavage patterns and DNA ends produced. Methylation-sensitive isoschizomers may or may not cleave a sequence, depending on whether the sequence carries methyl groups. These pairs of enzymes have proven instrumental in studying the methylation patterns of specific sequences of interest. The initial binding of type II restrictases occurs at random sites, and the subsequent search for the recognition sequence involves one-dimensional translocation along the DNA. Numerous restrictases have been crystallized and their structures have been determined in their free forms, bound to noncognate (nonspecific) DNA fragments, and bound to cognate sequences.

Type I restrictases are multifunctional enzymes that combine both nuclease and methylase activity in a single protein trimeric complex. One subunit, R, carries the restriction function, cleaving the DNA at random locations, sometimes very far, 10 kilobase pairs or more, from the enzyme's recognition sequence; it also has an ATP-dependent motor activity that is involved in translocating the DNA. The second subunit, M, carries out the DNA methylation reaction, which occurs in the recognition site itself. The third subunit, S, recognizes the specific sequence to which the complex binds. For cleavage to occur, the recognition/binding site has to be brought into proximity with the cleavage site; that is, the intervening DNA has to loop out (at the same time it is supercoiled). Therefore, these enzymes are unusual molecular motors that bind specifically to DNA and then move the rest of the DNA through this bound complex.

Type III restriction endonucleases are very similar to type I, but they do not require ATP, they methylate just one of the strands, and their cleavage site is relatively close to the recognition sequence.

Table 3.1 **Classification and properties for restriction/modification systems.** All enzymes require Mg^{2+} for cleavage; only type I enzymes require ATP for functioning of the motor. The letter N stands for any nucleotide.

	Type I	Type II	Type III
example	EcoB	EcoRI	EcoPI
recognition site	$TGAN_8TGCT$	GAATTC	AGACC
cleavage site	random, up to 10 kb away from recognition site	between G and A on both strands	24–26 bp 3' to recognition site
methylation site	$TGA\underline{N}_8TGCT$ $ACTN_8\underline{A}CGA$	GA\underline{A}TTC CTT\underline{A}AG	AG\underline{A}CC TCTGG only one strand is methylated
nuclease activity	yes	yes	yes
methylase activity	yes	no; methylation of recognition site by a separate enzyme	yes

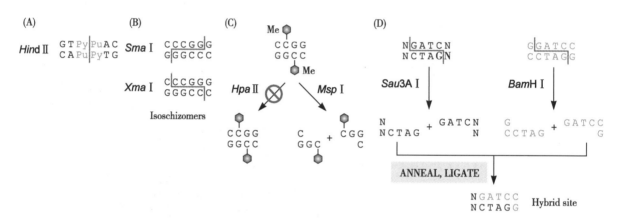

Figure 3.4 **Diversity of restriction endonuclease cleavage patterns.** (A) Ambiguity of recognition sequences, as illustrated for the example of HindII, the first type II enzyme to be described. Py stands for either pyrimidine, C or T; Pu stands for either purine, A or G. All possible sequences of this kind are cleaved by HindII. (B) Isoschizomers are enzymes from different sources that recognize the same target. Some pairs of isoschizomers cut their targets at different places, as in the example shown. (C) Methylation-sensitive isoschizomers: if the recognition sequence is methylated, in this case on the C, one enzyme of the isoschizomer pair will cut it while the other one will not. In the specific example shown, HpaII is methylation-sensitive and will not cut when C is methylated, while MspI will cut whether C is methylated or not, that is, MspI is methylation-indifferent. Such pairs of enzymes are extensively used in research to probe the methylation status of a particular sequence. (D) Production of a hybrid site by cohesion of complementary sticky ends generated by two different enzymes. One of these enzymes (Sau3AI) recognizes a 4bp-sequence that is embedded in the hexanucleotide sequence recognized by a different enzyme, BamHI in this case. The sticky ends produced by these enzymes can base pair to produce a hybrid site. The new site is sensitive to Sau3AI but may not constitute a target for BamHI, depending on the nucleotides adjacent to the original Sau3AI site.

3.2.3 DNA ligase joins linear pieces of DNA

DNA ligases are also among the most important enzymes in the cell. They are indispensable for numerous cellular processes, including DNA replication, recombination, and repair, whenever there is a need to seal nicks in the DNA double helix or join broken helices. Of interest to us here is their essential role in joining fragments of double-helical DNA to form intact molecules in recombinant DNA technologies. Ligases form phosphodiester bonds between the 3'-OH and 5'-phosphate termini of DNA fragments (Figure 3.5A). The reaction involves several steps. Adenylation of the active-site lysine uses ATP in phages and eukaryotes but uses NAD+, nicotinamide adenine dinucleotide, in bacteria. Figure 3.5B and Figure 3.5C depict the domain structure of human ligase I and the crystal structure of this enzyme complexed with DNA.

For the purposes of recombinant DNA technology, a distinction should be made between ligases that can only join DNA ends on overhangs and those that can ligate blunt-end DNA fragments. Most use is made of the first type: it is mechanistically easier to join the phosphodiester backbones of fragments that are kept in close proximity by base pairing of complementary overhangs. The *E. coli* DNA ligase is a representative of this type. Joining together blunt ends is much more challenging, and the enzymes that are capable of doing this are much less efficient. An example of such an enzyme is the DNA ligase from bacteriophage T4. The efficiency of blunt end ligation can be increased by using small oligonucleotides that carry restriction sites for sticky-end production, known as DNA linkers. The strategy for using linkers is depicted in Figure 3.6.

3.2.4 Sources of DNA for cloning

DNA for cloning can be chemically synthesized, as well as obtained from genomic DNA and cDNA libraries, or from PCR products. For chemical synthesis method, the length limitation of synthesized DNA sequence is to 100 nt. Two complementary DNA strand will be synthesized and annealed to achieve the double DNA strands for cloning. PCR or RT-PCR (reverse transcription-polymerase chain reaction) is the commonly used method for obtaining objective DNA sequences for cloning. RT-PCR templates are cDNAs which are reversely transcribed from mRNAs extracted from mammalian cells. Genomic DNA library and cDNA library also can act as templates to achieve PCR products as the DNA sources for cloning, especially with long length.

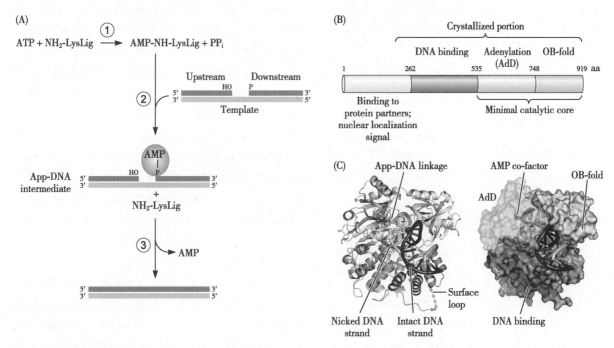

Figure 3.5 **DNA ligation.** (A) Steps in the reaction catalyzed by DNA ligase. (Step 1) Enzyme-AMP is formed by attack of lysine on the α-phosphate of ATP, releasing inorganic pyrophosphate. (Step 2) The 5′-phosphate of the nicked DNA strand, downstream side, attacks the Lys-AMP intermediate to form an App-DNA intermediate, where pp indicates a pyrophosphate linkage, 5′-P to the 5′-phosphate of AMP. (Step 3) The 3′-OH end of the nicked strand, upstream side, attacks the 5′-P of App-DNA, covalently joining the DNA strands and liberating AMP. Thus, ligation pays an energy price of one ATP. (B) Domain organization of human ligase I. OB-fold is the oligonucleotide/oligosaccharide-binding domain. (C) Left, the three domains, colored as in part B, fully encompass the App-DNA reaction intermediate. The intact DNA strand is shown in black, the nicked strand is shown in gray, and the App-DNA linkage is shown in blue. Gray spheres indicate a poorly ordered surface loop at residues 385–392. Right, molecular surface: the adenylation domain, AdD, is semitransparent to highlight the AMP cofactor held within its active site. (Adapted from Pascal JM, O'Brien PJ, Tomkinson AE, et al. [2004] *Nature*, 432:473–478. With permission from Macmillan Publishers, Ltd.)

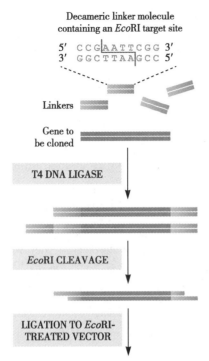

Figure 3.6 **DNA linkers for joining of blunt-end DNA fragments.** Linkers can be chemically synthesized to contain any restriction target site. Blunt-end ligation is performed by the DNA ligase from phage T4, as the *E. coli* DNA ligase will not catalyze blunt-end ligation except under special reaction conditions of macromolecular crowding. Linkers can be ligated to both ends of the blunt-end DNA fragment to be cloned; upon treatment with the restriction endonuclease that recognizes the linker sequence, sticky ends are produced. This treatment is then followed by standard cloning techniques that use sticky ends for efficient ligation of the foreign gene into a vector treated with the same enzyme.

3.3 Vectors for cloning

3.3.1 *Genes coding for selectable markers are inserted into vectors during their construction*

As indicated in Figure 3.1, the construction of vectors involves an important step: the introduction of genes that can be used for the selection of cells that carry the vector. The most commonly used selectable markers are antibiotic resistance genes, conferring resistance to antibiotics such as tetracycline, ampicillin, chloramphenicol, and kanamycin. The protein products of these genes act by a variety of pathways, some preventing the antibiotic from entering the cell and others destroying or enzymatically modifying the antibiotic to inactive forms. A very large repertoire of cloning vectors based on plasmids, viruses, or combinations thereof are used in recombinant DNA technology. Which vector one would choose depends on the target cell, the amount of DNA to be incorporated, and the purpose of incorporation.

3.3.2 *Plasmid DNA were the first cloning vector*

Plasmids are extrachromosomal nucleic acid molecules carried by bacteria that replicate independently of the main bacterial chromosome. All plasmids of a given type within a bacterial cell will have exactly the same size. These properties distinguish them from the heterogeneous, circular extrachromosomal elements that exist in certain bacterial species, such as Bacillus megaterium. Most, but not all, plasmids are double-stranded circular DNA molecules. If both strands are continuous, the molecules are described as covalently closed circles or CCC; if only one strand is intact, the molecules are known as open circles or OC. CCC extracted from cells are usually negatively supercoiled. Plasmids possess genetic elements that can maintain them in just a few copies per cell (these plasmids are known as stringent plasmids), or as multicopy populations, known as relaxed plasmids. Plasmids are generally dispensable for the host cell but may carry genes that are beneficial under certain circumstances. There is a long list of genes carried by known plasmids, most of which confer antibiotic resistance or are responsible for antibiotic production. Some plasmids are capable of inducing malignant tumors in plants.

A prerequisite for using plasmids as cloning vectors is the ability to isolate them in large quantities and sufficient purity. The first step toward achieving this goal is to gently lyse the bacterial host, so that the chromosomal DNA stays of high molecular weight and can be readily removed by high-speed centrifugation, producing a cleared lysate that contains the plasmids. Further purification of plasmid DNA may involve centrifugation in CsCl gradients containing ethidium bromide, EtBr. Incorporation of EtBr into the double helix unwinds the DNA. CCC plasmid forms can unwind only to a limited extent because there are no free ends to rotate, the result being that the amount of intercalated EtBr is relatively low. On the other hand, fragmented linear pieces of the chromosome can incorporate many more EtBr molecules, creating DNA-EtBr complexes of lesser density. The density difference makes it possible to separate the two kinds of molecules in density gradients.

The alternative method depends on the different denaturation properties of linear and CCC DNA forms. There exists a narrow range of high pH, 12.0–12.5, in which linear molecules denature but circular molecules remain intact. Upon neutralization of the solution, chromosomal DNA partially renatures, forming a highly insoluble network that is easily removed by centrifugation.

Plasmids that are chosen to be cloning vectors have several desirable characteristics: low molecular weight, ability to confer selectable traits on the host cell, and availability of single sites for restrictase cleavage. The first requirement stems from the ease of purifying small plasmids that remain intact during the isolation procedure. In addition, smaller plasmids are usually present in multiple copies, allowing for stable high-level expression of the genes they carry; this becomes especially important for genes that confer selectable phenotypes. The number of derivative plasmids that have been used for cloning is enormous. These are commercially available and allow the researcher to choose the appropriate vector for a specific project. Figure 3.7 presents a few classical examples of plasmid cloning vectors and introduces some of the strategies used for each vector to easily select the transformed recombinant bacteria. The multiple cloning site splits the *lacZ* gene, allowing for white-blue screening (see Figure 3.7). pBluescript is an expression vector, as the cloning site is flanked by promoter sequences derived from phages T3 and T7.

3.3.3 Recombinant bacteriophages can serve as vectors

Plasmid vectors are often used for cloning small DNA segments, up to 10 kbp. Cloning larger fragments, such as individual large genes or fragments to be used for the construction of genomic libraries, requires the use of alternative vectors, such as derivatives of bacteriophages. The most widely used phage for cloning is phage λ and its derivatives. There are two types of λ vectors: insertion and replacement. In insertion vectors, the linear phage DNA molecule is cut with a restriction enzyme that cleaves the sequence only once. The two fragments generated are joined to the foreign DNA fragment, and the resulting recombinant is packaged into infectious mature viral particles (Figure 3.8). This approach, as easy as it sounds, imposes a strict limitation on the size of foreign DNA that can be cloned. This is because the capacity of the phage head is limited; it does not allow packaging of molecules whose size deviates from the wild-type λ genome by more than ±5%. In order to overcome this problem, replacement vectors were introduced. These vectors make use of the fact that about a third of the λ genome is not necessary for replication and can thus be replaced by any piece of foreign DNA without affecting the infectivity of the recombinant phage. The construction of recombinant replacement phages is illustrated in Figure 3.9. These were named Charon vectors by their creators, Fred Blattner and colleagues, after Charon, the ferryman in Greek mythology, who ferried the dead to the underworld across the rivers Styx and Acheron. The analogy may seem far-fetched but it does confer the idea of transfer: of dead souls in Greek mythology and of foreign DNA in cloning.

Another bacteriophage often used as a cloning vector for subsequent DNA sequencing through Sanger's method is the filamentous phage of *E. coli*, M13 (Figure 3.10A). The life cycle of M13 is depicted in Figure 3.10B. The M13 genome is a 6.4 kbp single stranded circular DNA encapsulated in a coat consisting of ~2700 copies of a small coat protein. Additional minor proteins participate in the attachment of the phage to host *E. coli* cells. The coat is amazingly flexible, changing its size to accommodate DNA molecules of very different sizes, ranging from a few hundred nucleotides to around twice the size of the native genome. The genome of M13 has been modified to serve as a cloning vector as depicted in Figure 3.10C.

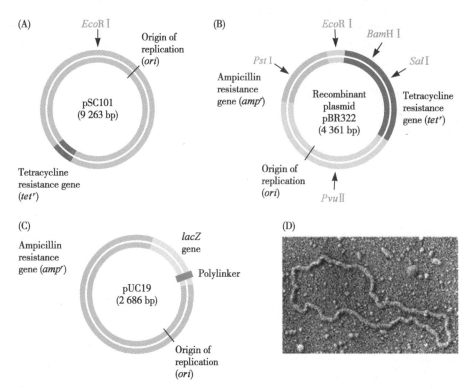

Figure 3.7 **Some bacterial cloning vectors.** (A) pSC101, the first bacterial plasmid used for cloning. p stands for plasmid, and SC stands for Stanley Cohen. (B) pBR322, in which BR stands for Bolivar Rodriguez, the postdoctoral researcher who constructed the plasmid, is an early cloning vector that is itself a recombinant DNA molecule. It contains an origin of replication sequence ori from a naturally occurring plasmid, ColE1, and two selectable marker genes from two other plasmids: an ampicillin resistance gene, shown in green, and a tetracycline resistance gene, shown in red. If the restriction site to be used for cloning lies within one of these genes, that specific gene is inactivated upon insertion of the foreign sequence. This is a useful property since it can be used for selection of the bacteria that carry the foreign gene. All bacteria that carry the vector, whether recombinant or not, will be resistant to the antibiotic specified by the intact resistance gene. Bacteria carrying the foreign gene will become sensitive to the antibiotic whose gene was inactivated due to the insertion. (C) pUC19, in which UC stands for University of California, where the vector was constructed by Joachim Messing and colleagues. It contains, in addition to the standard antibiotic selectable marker and origin of replication, the *lacZ* gene, shown in yellow, which encodes the enzyme β-galactosidase. The *lacZ* gene is split by a poly linker, an array of multiple restriction sites for cloning, also known as a multiple cloning site or MCS. Thus, transformed cells containing the plasmid with inserted gene can be distinguished from cells containing the original, nonrecombinant plasmid by the color of the colonies they produce in an appropriate medium. Nonrecombinant cells will turn blue because the presence of active β-galactosidase, from the unsplit *lacZ* gene, will transform a colorless substrate present in the medium to a blue compound. Recombinant cells will be white because the insert has inactivated the *lacZ* gene. (D) Electron microscopic image shows the typical appearance of an isolated plasmid. (D, from Cohen SN [2013] *Proc Natl Acad Sci USA*,110:15521–15529. With permission from National Academy of Sciences.)

M13-derived vectors can be used in a technique called phage display. This method allows large peptide and protein libraries to be expressed on the surface of the filamentous phage and thus presented to potential interaction partners. It permits the selection of expressed peptides and proteins, including antibodies, that have high affinity and specificity for almost any target, including specific proteins or DNA sequences. The direct link between the experimental phenotype, that is, the properties of the peptides or proteins on the surface, and the genotype encapsulated within the vector allows the selection of molecules that are optimized for binding. In an important application, phage display facilitates the engineering of antibodies to amend their size, affinity, and effector functions.

How does phage display work? DNA fragments are inserted into the middle of gene 3, the protein product of which is crucial for attachment of the phage to the bacterial pilus through which it enters

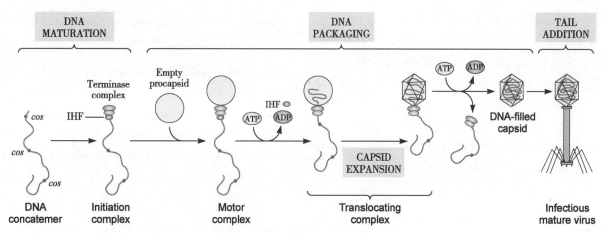

Figure 3.8 Packaging of λ phage DNA into infectious mature viral particles. The heterodimeric terminase enzyme complex and the *E. coli* integration host factor (IHF) bind cooperatively to a cos site in a DNA concatemer. This is the form of the viral genome that is produced during replication; it contains multimers of genome-size fragments connected by cos sequences. The terminase possesses cos-site cleavage endonuclease, helicase, and ATPase activities that work in concert to assemble the packaging machinery on the concatemer and prepare the end for packaging. The same enzyme also possesses a DNA- and ATP-dependent translocase activity that packages DNA into an empty procapsid, thus serving as a molecular motor. The procapsid is an empty protein shell composed of four different proteins; the tail is composed of 11 different viral glycoproteins. The terminase-concatemer complex, named the initiation complex, binds to the portal, a ring like structure in the procapsid, to form the motor complex which pumps the viral DNA into the capsid. DNA packaging triggers a process of expanding the procapsid, with additional viral proteins binding to it. Once the packaging motor encounters the next cos site in the concatemer, terminase again cuts the duplex, and its helicase activity helps in packaging the genome to near-liquid-crystalline density. Following the addition of some more proteins to prevent DNA release through the portal, the preassembled tail is added to complete virion assembly. It is now possible to assemble infectious viral particles *in vitro* starting with seven individual proteins, purified procapsids and tails, and mature λ DNA. Importantly, the same system can be used to produce infectious viruses containing recombinant DNA. (Adapted from Gaussier H, Yang Q, Catalano CE [2006] *J Mol Biol*,357:1154–1166. With permission from Elsevier.)

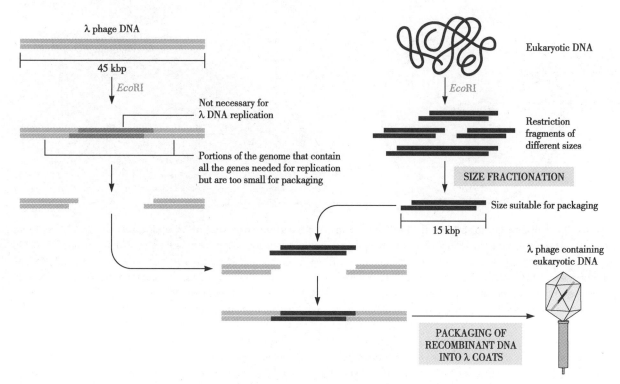

Figure 3.9 Use of λ phage as a vector. The λ phage genome contains a large middle region that does not participate in DNA replication and therefore can be replaced by a piece of foreign DNA. The size of the insert should be ~15 kbp to allow efficient packaging of the recombinant DNA into infectious viral particles.

the cell (see Figure 3.10B). When the p3 gene is expressed, it produces p3 fusion proteins that contain the amino acid sequence encoded by the insert. The fusion proteins become incorporated into the phage coat in the same way as wild-type p3. In this way, the fusion proteins are accessible and can be readily probed for interactions with desired targets. The method is outlined in Figure 3.11.

3.3.4 Cosmids and phagemids expand the repertoire of cloning vector

Cosmids are modified plasmids that carry cos sites from phage λ in addition to the standard origin of replication and selectable marker genes. They are capable of cloning large, >40 kbp fragments and are introduced into bacterial cells by standard protocols.

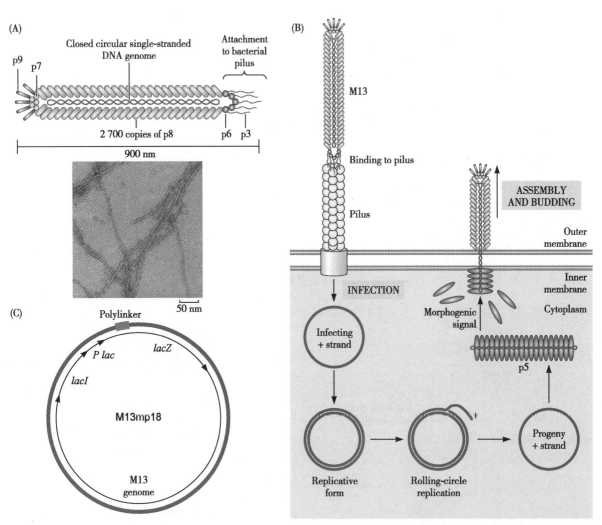

Figure 3.10 **Bacteriophage M13, its life cycle, and its derivative constructed as a cloning vector.** (A) Structure of M13. The single-stranded DNA (or ssDNA) circular genome is encapsulated by a flexible capsid consisting of ~2700 molecules of one of the phage-encoded proteins, p8. (B) M13 life cycle. M13 enters bacterial cells through pili; following uncoating, the ssDNA circle is replicated to the so-called replicative form, which is double-stranded. This form is then replicated by the rolling-circle mechanism (see Chapter 19), producing single-stranded DNA, the form that is incorporated into the viral particle. The ssDNA is temporarily packaged, with the help of p5, into a form that is finally transformed into a mature particle. The viral proteins taking part in this process are all membrane-attached. Assembly is followed by budding of the phage from the infected cell, without lysis of the host cell. (C) M13 as a cloning vector. The following elements have been inserted into the genome: the *lacZ* gene, to allow white-blue selection for plasmids carrying foreign DNA fragments; a *lac* promoter upstream of the *lacZ* gene; and the gene for the Lac repressor, *lacI*, to allow regulation of the lac promoter. The derivatives of M13 used as cloning vectors are named after the specific polylinker region they contain. (A, electron microscope image, from Murugesan M, Abbineni G, Nimmo SL, et al. [2013] *Sci Rep* 3 [10.1038/srep01820]. With permission from Macmillan Publishers, Ltd.)

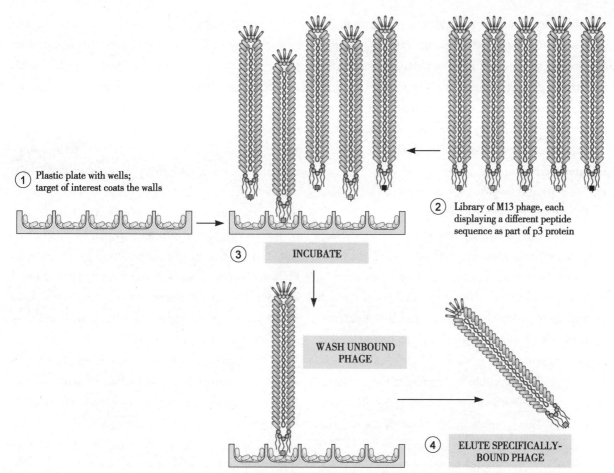

Figure 3.11 Phage display identifies DNA sequences that encode peptides or proteins of desired characteristics. As an example, we present the sequence of steps in phage display screening to identify polypeptides that bind with high affinity to desired target protein or DNA sequence. (Step 1) Target proteins or DNA sequences are immobilized to the wells of a plastic microtiter plate. (Step 2) A library of DNA sequences that code for peptides/proteins is prepared by inserting these sequences into the gene encoding coat protein 3. Other coat proteins can also be used, including the major protein p8 (see Figure 3.10). Thus, the products expressed by the recombinant genes are fusion proteins between p3 and the peptide/protein encoded by the inserted DNA sequence. The fusion proteins end up on the phage surface. (Step 3) This phage-display library is added to the plate and allowed time to bind. Then a buffer wash removes the unbound phages, while those carrying the fusion proteins that interact with the immobilized target remain bound. (Step 4) The specifically attached phages are eluted with excess known ligand for the target or by lowering pH. The eluted phages are used to generate more phages by infection of suitable bacterial hosts. The new phage mixture is enriched, containing considerably fewer irrelevant (nonbinding) phages than were present in the initial mixture. (Step 5) Steps 1–4 are repeated to achieve further enrichment in phages carrying binding proteins. Following further bacterial-based amplification, the DNA in the interacting phage is sequenced to identify the interacting proteins or protein fragments. This sequence can be further mutagenized *in vitro* by site-specific mutagenesis to optimize binding characteristics.

Cos sites enable the DNA to be packed into infectious viral particles *in vitro* (see Figure 3.8), thus facilitating their purification. Like plasmids, cosmids perpetuate in bacteria because they do not carry the genes for lytic infection and thus do not kill their bacterial hosts.

Phagemids are constructs that combine the features of phages and plasmids. One commonly used phagemid is pBluescript II, which is derived from the plasmid pUC19 and contains two origins of replication, an ampicillin resistance gene, and a *lacZ* gene.

3.4 Expression of recombinant genes

3.4.1 Expression vectors allow regulated and efficient expression of cloned genes

If the aim of a project is to produce large amounts of protein, whether native or mutated, specialized vectors called expression vectors must be used. As their name implies, these vectors are capable not only of carrying an inserted gene but also of expressing the protein product. They must ensure that the gene will be transcribed, and in a form that can be translated into protein, by use of the host cell's machinery.

Expression vectors ideally have the following general characteristics:

- They should be present and maintained in the host cells at high copy number.
- They should be able to integrate into the host genome to ensure stable propagation of the gene of interest from cell generation to cell generation.
- They should carry a promoter that is functional within the given host. Promoters are elements where RNA polymerases bind to initiate gene transcription.
- It should be possible to regulate the promoter to ensure that the gene is expressed only when desired. This is especially important if the protein happens to be toxic to the host cell.

There are also desirable characteristics for the cloned gene itself. As redundancy exists within the genetic code, the same amino acid can be encoded by more than one codon. Some organisms make more frequent use of some of these codons than the others specifying the same amino acid. In order to optimize the expression of a foreign protein, it is desirable for its gene to be optimized for codon usage in the host cell. Current technologies make this possible by using codon exchange rather routinely.

It is also mandatory to use intron-free genes. Introns are noncoding regions found within most eukaryotic genes. In the eukaryotic nucleus, introns are spliced out, but bacteria cannot accomplish this. Even if a eukaryotic gene is expressed in a eukaryotic host such as yeast, the use of intron-free constructs is preferable, as it avoids complications that could arise from alternative splicing.

3.4.2 Expression systems

One of the most useful applications of recombinant DNA technology is the ability to artificially synthesize large quantities of natural or modified proteins in a host cell such as bacteria or yeast. The recombinant expression system includes prokaryotic and eukaryotic expression vectors. For a foreign to be expressed in a bacterial cell, it must have particular promoter sequences upstream from the coding region, to which the RNA polymerase will bind prior to transcription of the gene. The choice of promoter is vital for correct and efficient transcription, since the sequence and position of promoters are specific to a particular host such as bacteria. It must also contain a ribosome binding site, placed just before the coding region. Unless a cloned gene contains both of these sequences, it will not be expressed in a bacterial host cell. If the gene has been produced via cDNA from a eukaryotic cell, then it will certainly not have any such sequences. In some cases, the protein is expressed as a fusion with a general protein such as β-galatosidase or glutathione S-transferase (GST) to facilitate its recovery. It is not only possible but essential to use cDNA instead of a eukaryotic genomic DNA to direct the production of a functional protein by bacteria. One useful eukaryotic expression system is based on the monkey COS cell line. These cells each contain a region derived from a mammalian monkey virus termed simian virus 40 (SV40). A defective region of the SV40 genome has been stably integrated into the COS cell genome. This allows the expression of a protein, termed the large T antigen that is required for viral replication. When a recombinant vector having the SV40 origin of replication and carrying foreign DNA is inserted into the COS cells, viral replication takes place. This results in a high level of expression of foreign proteins.

3.5 Introducing recombinant DNA into host cells

3.5.1 Numerous host-specific techniques are used to introduce recombinant DNA molecules into living cells

The techniques for introducing recombinant DNA molecules into living cells for further propagation are highly specific for both the type of host cell, bacteria or eukaryote, and the type of cloning vector used. Plasmid vectors are introduced into bacterial cells by transformation techniques that involve prior treatment of bacteria with calcium chloride at low temperature followed by a brief heat shock. It is believed that calcium chloride affects the structure of the cell wall and may help in the binding of DNA to the cell surface. The heat shock stimulates the actual DNA uptake.

The efficiency of introducing recombinant DNA molecules into bacterial cells is greater with phage-based cloning vectors. The development of *in vitro* packaging techniques to pack the modified DNA in the phage head represented a major improvement in this stage of the process.

Introduction of DNA into eukaryotic cells is more complicated, especially for yeast and plant cells, which possess, in addition to the plasma membrane, outer protective polysaccharide layers. These layers must be removed, usually by enzymatic treatments, to produce protoplasts, which survive and are competent for transformation. Transformed plant protoplasts are grown on selective medium to produce large masses of undifferentiated cells known as calli, plural for callus. These can then be transferred to nutritional medium containing specific growth hormones that allow regeneration of whole fertile transformed plants.

The search for efficient and more universal methods of introducing cloned genes into a wide variety of microbial, plant, and animal cells led to the development of purely mechanical methods. One such technique is electroporation. When subjected to an electric shock-for example, brief exposure to a voltage gradient of 4000–8000 V/cm-cells take up exogenous DNA through holes created in the plasma membrane. Drugs such as colcemid that arrest cells at metaphase increase transformation efficiency. This increase is attributed to the lack of a nuclear membrane during this stage of the cell cycle, as the nuclear membrane also needs to be crossed for stable integration of the gene into the genome.

Another method for transformation of animal cells or the protoplasts of yeast, plant, and bacterial cells is liposome-mediated gene transfer. Liposomes are unilamellar (single-bilayer) vesicles prepared from cationic lipids that readily and spontaneously form complexes with DNA in solution. The positively charged liposomes also bind to cultured cells and presumably fuse with the plasma membranes. The use of liposomes for transformation or transfection is known as lipofection.

Researchers have developed gene guns to facilitate gene transfer into cells that are notoriously difficult to transform. The method can be applied not only to unicellular organisms but also to plant leaves or entire animals, such as *Drosophila* and mice.

It has been particularly useful for chloroplast transformation, as there are no other known methods to introduce foreign DNA into these organelles. The efficiency of the gun system varies, with skin cells being the most susceptible to transformation. It has already been used for delivering hepatitis B DNA-based vaccine to mice and humans. The gene gun method relies on the ability of nucleic acids to adhere to biologically inert metal particles, such as gold or tungsten. The DNA-particle complex is accelerated under partial vacuum and hits the target tissue in its acceleration path. There are numerous ways to create acceleration, such as the use of pneumatic devices, magnetic or electrostatic forces, and sprayers. In a variant of the method, uncoated metal particles are shot through a solution containing the cells and the DNA. Presumably, they pick up DNA in transit and carry it into the cell.

3.5.2 Transient and stable transfection assays

Mammalian cell expression systems are important for the production of heterologous proteins with a full complement of posttranslational modifications. A number of established cell lines have been developed for this purpose. Cells derived from African green monkey kidney (COS), baby hamster kidney (BHK), and human embryonic (HEK-293) are used for short-term (transient) gene expression for either rapid production of small various stages of heterologous protein or testing the integrity of constructs during various stages of vector development. Chinese hamster ovary (CHO) cells are commonly used for long-term (stable) gene expression and when high yields of heterologous proteins are required. Although hundreds of mammalian expression vectors have been developed, they all tend to have the same shared features and are not very different in design from other eukaryotic expression vectors.

For the most part, the systems are used to select transfected mammalian cells are the same as those for other host cells. In fact, a number of bacterial marker genes have been adapted for eukaryotic cells. For example, the bacterial *Neo* gene that encodes neomycin phosphotransferase is often used to select transfected mammalian cells. However, in eukaryotic cells, G-418 (Geneticin), which is phosphorylated by neomycin phosphotransferase, replaces neomycin as the selective agent because neomycin is not an effective inhibitor of eukaryotic protein synthesis.

3.6 Constructing DNA libraries

3.6.1 Type of different libraries

(1) Genomic libraries

The cloning and sequencing of individual genes of interest was the first application of recombinant DNA technology in research. It was soon realized that it would be possible, by use of recombinant technology, to prepare large sets of recombinant DNA molecules-vectors plus inserts-that can contain entire genomes in the form of overlapping sequences. These genomic libraries can be multiplied in bacteria and stored indefinitely. The availability of such libraries was the most important prerequisite for the sequencing of whole genomes, first for viruses, then bacteria and yeast, then higher eukaryotes, and culminating in 2001 with the complete human genome.

Creating complete libraries is often technically challenging. The size of the library needed to guarantee representation of all the sequences of a genome is a function of the size of DNA fragments that can be accommodated in the cloning vector and the size of the genome itself. Simple calculations indicate that, in order to include any desired sequence with a probability of 99%, the number of recombinant clones in a human library should be close to 700 000. These calculations are based on the assumption that the genome has been fragmented randomly, for example, by mechanical shearing, as in sonication. The cloning of sheared fragments is relatively inefficient, so usually an acceptable, albeit not perfectly random, population of DNA fragments is obtained by limited digestion with restrictases.

Suitable vectors for creating genomic DNA libraries should allow the cloning of large DNA fragments, to reduce the requisite number of clones for full representation, and should possess high cloning efficiency. λ phage vectors, cosmids, and bacterial artificial chromosomes (BACs) fulfill these requirements. A major advantage of such vectors is that they can be efficiently packaged *in vitro* to produce highly infectious viral particles.

(2) cDNA Library

The best way to study which protein genes are transcribed, and later translated, in a given cell type is to isolate the population of mature mRNA from the cytoplasm and then, by use of in vitro methods such as reverse transcription, to produce double stranded DNA copies of this population. These are known as cDNAs or copy DNAs.

Cloning this entire population of cDNAs into bacterial vectors produces cDNA libraries. These can be maintained stably in bacteria and stored indefinitely, exactly as genomic libraries are. The existence of such libraries has practical applications: because they lack introns, cDNA clones are a convenient source of sequences to be directly expressed in bacterial cells.

3.6.2 Library screening and probes

(1) Types of DNA and RNA probes

The increasing accumulation of nucleic acid database entries and availability of custom synthesis of oligonucleotides has provided a relatively straightforward means for designing and producing gene probes and primers for PCR. Gene probes and primers are usually designed using bioinformatics software and nucleic acid databases or gene family-related sequences. However, there are many gene probes that have traditionally been derived from cDNA or from genomic sequences and which have been cloned into plasmid and phage vectors. These require manipulation before they may be labeled and used in hybridisation experiments. Gene probes may vary in length from 100 bp to a number of kilobases, although this is dependent on their origin. Many are short enough to be cloned into plasmid vectors and are useful in that they may be manipulated easily and are relatively stable both in transit and in the laboratory. The DNA sequences representing the gene probe are usually excised from the cloning vector by digestion with restriction enzymes and purified. In this way vector sequences that may hybridise non-specifically and cause high background signals in hybridisation experiments are removed. There are various ways of labeling DNA probes and these are described in the next section.

It is also possible to prepare cDNA probes or riboprobes by *in vitro* transcription of gene probes into a suitable vector. A good example of such a vector is the phagemid pBluescript SK, since at each end of the multiple cloning site where the cloned DNA fragment resides are promoter for T3 or T7 RNA polymerase. The vector is then made linear with a restriction enzyme and T3 or T7 RNA polymerase is used to transcribe the cloned DNA fragment. Provided a labelled NTP is added in the reaction a riboprobe labeled to a high specific activity will be produced. Provided a labeled NTP is added in the reaction a riboprobe labeled to a high specific activity will be produced. One advantage of riboprobes is that they are single stranded and their sensitivity is generally regarded as superior to the cloned double-stranded probes. They are used extensively *in situ* hybridization and for identifying and analyzing mRNA.

(2) Labeling of probes

An essential feature of a gene probe is that it can be visualized or labeled by some means. This allow any complementary sequence that the probe binds to be flagged up or identified.

There are two main types of label used for gene probes. Traditionally labeling has been carried out using radioactive labels, but non-radioactive labels are gaining in popularity.

Perhaps the most common radioactive label is 32-phosphorus (^{32}P), although for certain techniques 35-sulphur (^{35}S) and tritium (^{3}H) are used. These may be detected by the process of autoradiography, where the labeled probe molecule, bound to sample DNA, located for example on a nylon membrane, is placed in contact with an X-ray-sensitive film. Following exposure the film is developed and fixed just as a black-and-white negative. The exposed film reveals that precise location of the labeled probe and therefore the DNA to which it has hydridised. Non-radioactive labels are increasing being used to label DNA gene probes. Until recently, radioactive labels were more sensitive than their non-radioactive counterparts. However, recent developments have led to similar sensitivities, which, when combined with the improved safety of non-radioactive labels, have led to their greater acceptance.

The labeling systems are termed either direct or indirect. Direct labeling allows an enzyme reporter such alkaline phosphatase to be coupled directly to the DNA. Although this may alter the characteristics of the DNA gene probe, it offers the advantage of rapid analysis, since no intermediate steps are needed. However indirect labeling is at present more popular. This relies on the incorporation of a nucleotide that has a label attached. At present three of the main labels in use are biotin, fluorescein and digoxygenin. These molecules are linked covalently to necleoties using a carbon spacer arm of 7, 14 or 21 atoms. Specific binding proteins may then be used as a bridge between the nucleotide and a reporter protein such as an enzyme.

(3) Phage display

As a result of the production of phagemid vectors and as a means of overcoming the problems of screening large numbers of clones generated from genomic libraries of antibody genes, a method for linking the phenotype or expressed protein with the genotype has been devised. This is termed phage display, since a functional protein is linked to a major coat protein of a coliphage whilst the single-stranded gene encoding the protein is packaged within the virion. The initial steps of the method rely on the PCR to amplify gene fragments that represent functional domains or subunits of a protein such as an antibody. These are then cloned into a phage display vector that is an adapted phagemid vector and used to transform *E. coli*. A helper phage is then added to provide accessory proteins for new phage molecules to be constructed. The DNA fragments representing the protein or polypeptide of interest are also transcribed and translated, but linked to the major coat protein gIII. Thus, when the phage is assembled, the protein or polypeptide of interest is incorporated into the coat of the phage and displayed, whilst the corresponding DNA is encapsulated.

(4) Screening of expression libraries

Once a cDNA library has been prepared, the next task is the identification of the specific fragment of interest. In many cases this may be more problematic than the library construction itself, since many hundreds of thousands of clones may be in the library. One clone containing the desired fragment needs to be isolated from the library and therefore a number of techniques based mainly on hybridisation have been developed.

Colony hybridization is one method used to identify a particular DNA fragment from a plasmid gene library. In many cases it is now possible to use the PCR to screen cDNA libraries constructed in plasmids or bacteriophage vectors. This is usually undertaken with primers that anneal to the vector rather than the foreign DNA insert. The size of an amplified product may be used to characterize the cloned DNA and subsequent restriction mapping is then carried out. The main advantage of the PCR over traditional hybridisation-based screening is the rapidity of the technique. In some cases the protein for which the gene sequence is required is partially characterized and in these cases it may be possible to produce antibodies to that protein. This allows immunological screening to be undertaken rather than gene hybridisation.

3.7 Sequencing of entire genomes

3.7.1 *Genomic libraries contain the entire genome of an organism as a collection of recombinant DNA molecules*

The cloning and sequencing of individual genes of interest was the first application of recombinant DNA technology in research. It was soon realized that it would be possible, by use of recombinant technology, to prepare large sets of recombinant DNA molecules-vectors plus inserts-that can contain entire genomes in the form of overlapping sequences. These genomic libraries can be multiplied in bacteria and stored indefinitely. The availability of such libraries was the most important prerequisite for the sequencing of whole genomes, first for viruses, then bacteria and yeast, then higher eukaryotes, and culminating in 2001 with the complete human genome.

Creating complete libraries is often technically challenging. The size of the library needed to guarantee representation of all the sequences of a genome is a function of the size of DNA fragments that can be accommodated in the cloning vector and the size of the genome itself. Simple calculations indicate that, in order to include any desired sequence with a probability of 99%, the number of recombinant clones in a human library should be close to 700 000. These calculations are based on the assumption that the genome has been fragmented randomly, for example, by mechanical shearing, as in sonication. The cloning of sheared fragments is relatively inefficient, so usually an acceptable, albeit not perfectly random, population of DNA fragments is obtained by limited digestion with restrictases.

Suitable vectors for creating genomic DNA libraries should allow the cloning of large DNA fragments, to reduce the requisite number of clones for full representation, and should possess high cloning efficiency. λ phage vectors, cosmids, and bacterial artificial chromosomes (BACs) fulfill these requirements.

A major advantage of such vectors is that they can be efficiently packaged *in vitro* to produce highly infectious viral particles.

3.7.2 There are two approaches for sequencing large genomes

In the whole-genome shotgun approach, the entire genome is cloned as a series of recombinants, and each clone is sequenced. Computational methods are then used to reconstruct the entire genome, through alignment of clones that contain overlapping sequences. Such overlapping clones always exist, as the initial random fragmentation of the genome is performed on a population of cells, each of which will produce its own random set of fragments. The second strategy for genome sequencing is the hierarchical shotgun sequencing approach, also called BAC-based or clone-by-clone sequencing. This approach involves generating and organizing, through restriction nuclease mapping, a set of large-insert clones, typically 100–200 kbp each, and then separately performing shotgun sequencing on the appropriately chosen clones. The latter approach has been selected, following "lively scientific debate," for sequencing the human genome.

3.8 Practical application of recombinant DNA technologies

3.8.1 Gene therapy

Gene therapy combines recombinant techniques with the aim of correcting defective genes or gene functions responsible for disease development. Administration of the functional gene, instead of the protein itself, is performed because proteins are quickly degraded, whereas a properly integrated gene will continue to be expressed.

There are two major approaches:
- A normal gene may be inserted into a random, nonspecific location within the genome to supplement a nonfunctional gene. This approach is classified as gene addition therapy, since the nonfunctional gene stays in the genome. Examples might be the use of tumor suppressor genes in the treatment of a malignancy or immunostimulatory genes to treat an infectious disease. Gene addition therapy could also be used to provide the cell with transgenes whose expression can control the expression of a mutated gene.
- A mutated gene can be swapped for a normal gene through homologous recombination. This approach is representative of gene replacement therapy.

Gene therapy is a complex multistep process involving several steps in the production of the vector carrying the gene of interest. Once the vector delivers the transgene into the cell, the gene must travel through the cytoplasm and enter the nucleus, where it should be stably integrated into the nuclear genome: only integrated gene copies can be consistently replicated as part of the genome. Finally, properly regulated expression must be achieved. This is not a trivial task, as the majority of vectors will insert the gene they carry at random locations. Such random insertion can create two kinds of problems. In most cases, the gene will end up in a heterochromatic environment that does not permit its transcription. In other cases, the gene may land within other genes or their regulatory sequences, which will lead to the inactivation of these host genes. This may be detrimental to the host cell, to an extent that depends on the importance of the inactivated gene. Researchers are constantly improving existing methods or coming up with new approaches to overcome these problems.

One of the biggest problems in the development of gene therapy is the immunogenicity of the vector. On one hand, the immune response can very effectively eliminate transduced cells. On the other hand, the development of adaptive immunity against viral vectors can prevent their readministration.

3.8.2 Delivering a gene into sufficient cells within a specific tissue and ensuring its subsequent long-term expression is a challenge

Because the anionic nature of DNA hinders transfer across the membrane, the focus of efficient gene therapy is on vector-mediated delivery. There are two types of vectors: viral and nonviral vectors.

Viruses possess the natural ability to enter the cell and nucleus efficiently. Viral vectors differ in several properties, all of which affect the success and safety of the entire process (Table 3.2).

Vector capacity is the size of the transgene(s) that could be packed into the vector. It is determined by the size of the viral genome itself and by the amount of viral nucleic acid, DNA or RNA, that could be excised and replaced by a transgene without affecting viral trafficking ability.

Efficiency of vector production in cell lines used as incubators becomes an issue as more and more viral genes are deleted and replaced with transgenes. The viral vector must be replication-incompetent: that is, it must carry only the essential elements necessary for transgene packaging and expression. The viral genes necessary for viral particle production are removed from the vector and provided in *trans* by another virus in the vector-producing cell lines.

Tropism, or host range, is the ability of the virus to infect different cell types. Viruses enter the cell through endocytosis, which is triggered by binding of the virus to specific receptor proteins on the cell surface. These may or may not be present on all cells. Modifications of the viral genes that encode the envelope and capsid proteins can overcome this problem. Once in the cell, viral DNA is released from the capsid/envelope structures and needs to enter the nucleus. Crossing the nuclear membrane may be an insurmountable obstacle for some viruses, especially retroviruses. This may be the reason retroviruses can infect only rapidly dividing cells; the elimination of the nuclear envelope during mitosis provides any exogenous DNA or RNA with a window of easy entry into the nucleus.

The overall transfection efficiency depends on the efficiency of the steps involved in viral uptake, entry into the nucleus, and escape from degradation. Viruses containing a dsDNA genome have, in general, higher transfection efficiency than ssDNA and RNA viruses; the latter need to first convert their genomes into dsDNA.

Nonviral vectors have been developed in order to overcome the immunogenicity and safety issues linked to the use of viral vectors. Although nonviral

Table 3.2 **Viral vectors for gene therapy.** (Adapted from Waehler R, Russell SJ, Curiel DT [2007] *Nat Rev Genet*, 8:573–587. With permission from Macmillan Publishers, Ltd.)

Feature	Adenoviral vector	AAV vector	Retroviral vector	Lentiviral vector
particle size, nm	70–100	20–25	100	100
cloning capacity, kbp	8–10	4.9	8	9
chromosome integration	no	no; yes if *rep* gene is included	yes	yes
cell entry mechanism	receptor-mediated endocytosis; microtubule transport to nucleus		receptor binding, conformation change of Env, membrane fusion, internalization, uncoating, nuclear entry of reverse-transcribed DNA	
transgene expression	weeks to months; highly efficient short-term expression, appropriate for treatment of cancer or acute cardiovascular diseases	>1 year; medium- to long-term expression for nonacute diseases, onset of transgene expression after ~3 weeks	long-term correction of genetic defects	
emergence of replication-competent vector *in vivo*	possible but not a major concern		risk is a concern	
infects quiescent cells	yes	yes	no	yes
risk of oncogene activation by vector	no	no	yes	yes

Abbreviations: AAV, adenovirus-associated virus; Env, viral envelope protein.

vectors are much safer than viral vectors, they have an important limitation-an impaired ability to enter the nucleus-which has so far limited their use to proliferating cells only. The simplest option is the direct introduction of therapeutic DNA into target cells. The use of liposomes, artificial lipid spheres with an aqueous core containing the DNA, is another option. In still another approach, the DNA is chemically linked to molecules that are capable of binding to special cell-surface receptors or of facilitating nuclear transfer. The list of nonviral vectors and the ways to enhance their transformation efficiency is steadily growing. Prominent among these are human artificial chromosomes, with practically unlimited capacity and stability of gene expression and no immunogenicity.

The administration of gene therapy to humans has been and remains controversial. Following the first excitement around the promise of curing genetic diseases through modern biotechnology approaches, there has been growing skepticism about the efficiency and safety of the procedures. The community in general and clinicians in particular became wary after an 18-year-old boy died from multiple organ failure four days after treatment, and several patients in a Paris-based clinical trial developed leukemia. It became painfully clear that treatment is risky, although in many cases successful and life-saving. More recent efforts aimed at reducing these risks have been promising and there is little doubt that "gene therapy deserves a fresh chance."

Key Concepts

- Recombinant DNA technology allows us to clone, and subsequently sequence, genes or whole genomes. It allows us to manipulate genes at will and to insert or delete them from cells or organisms.
- The essential enzymes for this technology are restriction endonucleases, which specifically cut DNA sequences; ligases, which reconnect them; and polymerases, which extend or copy sequences.
- Cloning vectors are used to insert desired sequences into the genome of a cell or organism. Commonly used vectors include bacterial plasmids, bacteriophage, and artificial chromosomes.
- An expression vector is required for an inserted gene to produce a protein product in the host. If the gene is eukaryotic, introns should be first eliminated from the gene construct.
- Once the desired gene has been incorporated into an appropriate vector, the vector must be delivered to the host cell and often to the nucleus. A wide variety of techniques, chosen according to the vector and host cell, can achieve this.
- In order to assure that the desired gene has in fact been taken up by the host, genes coding for detectable markers are usually incorporated into the vector.
- Cloning and sequencing allow the construction of genomic libraries. Ideally, each library will contain the entire genomic sequence. This has now been accomplished for a wide variety of genomes, from viral to human.
- Construction of transgene animals, such as knockout mice, is a major tool in biological research.
- Recombinant DNA techniques allow the industrial production of a number of proteins that are of direct therapeutic value to humans. They also allow the production of genetically modified plants, incorporating resistance to chemicals or insects or conferring higher nutritional value.
- Gene therapy promises to provide resistance to malignant or disease states in humans.

Key Words

bacteriophage lambda λ(λ噬菌体)
cDNA library(互补DNA文库)
cloning(克隆)
cloning vector(克隆载体)
cosmid(黏粒)
DNA ligase(DNA连接酶)
expression vector(表达载体)
gene therapy(基因治疗)
genomic library(基因组文库)
plasmid(质粒)

recombinant DNA（重组 DNA）
restriction endonuclease（限制性内切酶）
transfection（转染）
transformation（转化）
transgene animal（转基因动物）
vector（载体）

Questions

1. What are type Ⅱ restriction endonucleases? Why are they important for recombinant DNA technology?
2. Described how plasmid pBR322 is used as a cloning vector. What are its special features?
3. Why would you use a plasmid, bacteriophage λ, cosmid, or BAC as a cloning vector?
4. What is a genomic library, or a cDNA library?
5. Suggest several ways that the expression of a cloned gene can be manipulated.
6. Describe the features of a eukaryotic expression vector.
7. Describe at least two selectable marker systems that are used with mammalian expression vectors.
8. Describe the practical applications of recombinant DNA technologies.

References

[1] Beyer P, Al-Babili S, Ye X, et al. (2002) Golden rice: Introducing the beta-carotene biosynthesis pathway into rice endosperm by genetic engineering to defeat vitamin A deficiency. *J Nutr*,132:506S–510S.

[2] Capecchi MR (1980) High efficiency transformation by direct microinjection of DNA into cultured mammalian cells. *Cell*,22:479–488.

[3] Capecchi MR (2005) Gene targeting in mice: Functional analysis of the mammalian genome for the twenty-first century. *Nat Rev Genet*,6:507–512.

[4] Cohen SN, Chang AC, Boyer HW, et al. (1973) Construction of biologically functional bacterial plasmids *in vitro*. *Proc Natl Acad Sci USA*,70:3240–3244.

[5] Cohen SN (1975) The manipulation of genes. *Sci Am*, 233:25–33.

[6] Mullis KB, Faloona FA (1987) Specific synthesis of DNA *in vitro* via polymerase-catalyzed chain reaction. *Methods Enzymol*,155:335–350.

[7] Nathans D, Smith HO (1975) Restriction endonucleases in the analysis and restructuring of DNA molecules. *Annu Rev Biochem*,44:273–293.

[8] Goeddel DV, Kleid DG, Bolivar F, et al. (1979) Expression in *Escherichia coli* of chemically synthesized genes for human insulin. *Proc Natl Acad Sci USA*,76:106–110.

[9] Pingoud A, Fuxreiter M, Pingoud V, et al. (2005) Type Ⅱ restriction endonucleases: Structure and mechanism. *Cell Mol Life Sci*,62:685–707.

Rong Cai

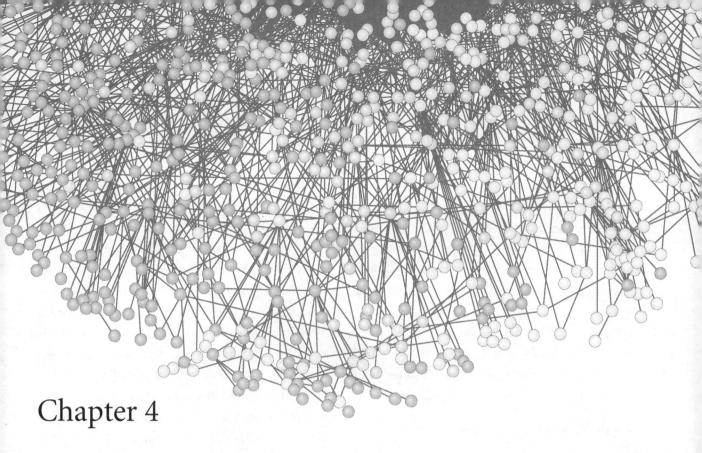

Chapter 4
Tools for Analyzing Gene Expression

4.1 Introduction

The field of molecular biology focuses on the mechanisms by which cellular processes are carried out by the various biological macromolecules in the cell, with a particular emphasis on the structure and function of genes and genomes. The gene is the functional unit of heredity. Each gene is a sequence within the genome that functions by giving rise to a discrete product, which may be a polypeptide or an RNA. And the process by which information from a gene is used to synthesize an RNA or polypeptide product is called gene expression.

As research has advanced, more and more sensitive methods for detecting and amplifying DNA have been developed. Now the methods to assess the content, function, and expression of genes have become commonplace. This chapter discusses some of the most common methods used in molecular biology, ranging from the very first tools developed by molecular biologists, to some of the most recently developed methods now in use.

4.2 Gene isolation and detection

Over the years, numerous methods have been developed to isolate and identify the genes. The most common methods—gel electrophoresis, gradient centrifugation, PCR, nucleic acid hybridization and DNA sequencing are presented here.

4.2.1 Agarose gel electrophoresis of DNA

Gel electrophoretic analysis is central to almost every experimental study of the molecular biology of nucleic acids. It provides a simple, but highly discriminating way to separate mixtures of DNA or RNA according to molecule size. The gel is composed of polyacrylamide or agarose. The agarose gel is a complex network of polymeric molecules whose average pore size depends on the buffer composition and the type and concentration of agarose used. A series of DNA molecules of different lengths will have charge proportional to length, and the friction factor is also roughly proportional to length. Thus, a wide range of DNA and RNA molecules will have about the same free mobility, but their retardation by the gel will increase with increasing length, so that at any finite gel concentration the biggest molecules will move most slowly (Figure 4.1). A set of DNA fragments of exactly known sequence, and hence length, can be run on the same gel as calibrating markers. Using different gel concentrations makes it possible to study nucleic acids in different molecular weight ranges: concentrated gels can give resolution down to the base-pair range, whereas very dilute gels allow migration of molecules up to 100 000 bp or more. Agarose is convenient for separating DNA fragments ranging in size from a few hundred base pairs to about 20kb. Polyacrylamide is preferred for smaller DNA fragments. Special techniques, such as pulsed-field electrophoresis, permit even greater ranges.

4.2.2 Gradient centrifugation

Two techniques that are based on sedimentation in a density gradient are commonly used for the separation of macromolecules, especially nucleic acids. The principles and applications of the two techniques are somewhat different.

Density gradient velocity sedimentation is often referred to as sucrose gradient sedimentation because sucrose is most commonly used to prepare the gradient of solution density, although other substances such as glycerol can be used. The principle is very simple (Figure 4.2). Centrifuge tubes are filled with sucrose solutions from a mixing device, which places the most dense solution at the bottom of each tube with a smoothly decreasing gradient of sucrose concentration, and hence density, toward the top. A thin layer of the macromolecule mixture in a less dense solution is then placed on top, and the tubes are spun in a swinging-bucket centrifuge rotor. The purpose of the gradient is to allow sedimentation to proceed without convective mixing. The mixture will then separate into bands of different sedimentation velocity, which roughly reflects molecular weight, especially for a series of homologous polymers like nucleic acids. The bands can be individually collected by dripping the contents from the tubes, as shown, or by careful pumping. The technique can be used for either analytical or preparative purposes.

In the method of density gradient equilibrium centrifugation, a gradient is developed in a solution of a dense salt—often cesium chloride, but many other dense salts can be used—just by spinning a centrifuge rotor at a very high speed. This is an equilibrium gradient that will remain stable indefinitely as long as the centrifugal field is maintained. If the mixture of macromolecules to be separated is dissolved in the original salt solution, each component will band at a point in the gradient corresponding to its own density. Thus the method separates on the basis of molecular density, not size. It can be sensitive to extremely small differences in density. An example of density gradient equilibrium centrifugation is seen in the Meselson-Stahl experiment.

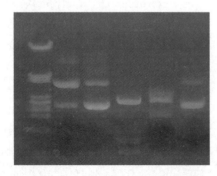

Ethidium bromide (EtBr)

Figure 4.1 Agarose gel electrophoresis of DNA. Example of an agarose gel stained with ethidium bromide (EtBr). EtBr intercalates between the DNA bases and, when exposed to UV light, fluoresces with an orange color. Although EtBr has been used for years, the mutagenic and carcinogenic properties of the intercalator made researchers look for alternatives for visualization of DNA bands on gels, for example, SYBR Green. Some of the new dyes have higher sensitivity.

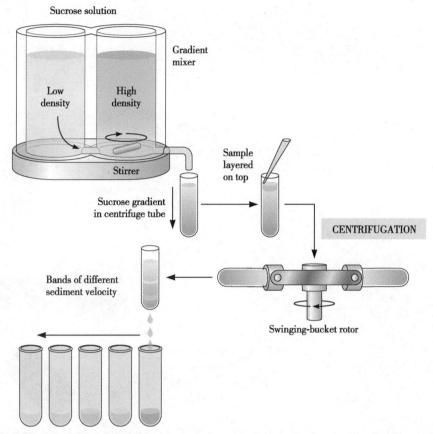

Figure 4.2 **Sucrose gradient centrifugation.** Steps in fractionating mixtures of DNA molecules by sucrose gradient centrifugation are shown. A sucrose gradient is pre-formed in a centrifuge tube by use of a special mixer device. The mixture of molecules is layered on the top of the gradient in a low-density solution. The tube is placed in a swinging-bucket rotor and centrifuged; the different molecules move through the gradient at different velocities that are dependent on the molecular mass and shape of the molecules. Following centrifugation, the content of a tube is collected into a series of tubes, through a hole punched in the bottom of the tube or by a pumping device.

4.2.3 Nucleic acid hybridization

Hybridization is a very useful technique with a myriad of applications in molecular and cellular biology. Hybridization detects sequence complementarity between labeled short probes and DNA target molecules present in a mixture (Figure 4.3). The probe can be either a purified restriction fragment of DNA or a synthetic oligonucleotide that encompasses the sequence of interest and is labeled with radioactivity (Figure 4.4); with digoxigenin or biotin, which are detected by chemiluminescence (Figure 4.5); or with fluorescent dyes that are visualized by fluorimetry.

Southern blotting is invaluable for detection and determination of the number of copies of particular sequences in complex genomic DNA for demonstration of polymorphic DNA in the human genome that corresponds to pathological states (Figure 4.3B). This method was developed in the 1970s by Edwin Southern and named after him. The Southern blotting hybridization protocol involves size separation of the initial DNA mixture by agarose gel electrophoresis, in-gel denaturation, transfer (blotting) of the material onto filters, on-filter hybridization, removal of excess labeled probe, exposure to X-ray film if the probe is radioactively labeled, and film development. Only the DNA fragment(s) that contain sequences complementary to those in the probe will appear as bands on the film. If the probe is labeled by digoxigenin or biotin, the detection method is chemiluminescence, whereas fluorimetry is used to visualize probes labeled with fluorescent dyes.

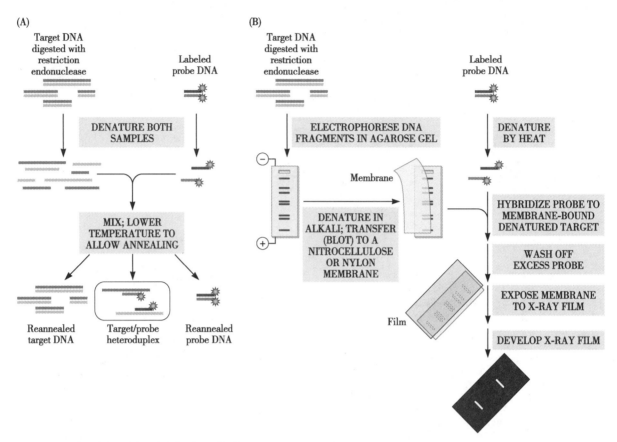

Figure 4.3 **Nucleic acid hybridization.** (A) If the reaction is performed in solution, three possible DNA duplexes will be present after the annealing step. (B) The Southern blotting hybridization. (Adapted from Strachan T, Read A [1999] Human Molecular Genetics, 2nd ed. With permission from Garland Science.)

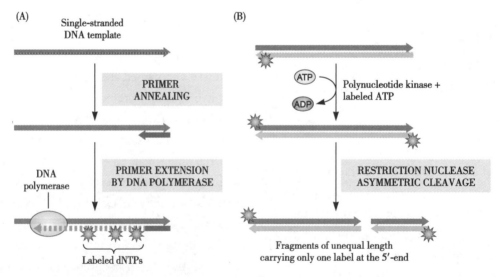

Figure 4.4 **Two common methods for labeling DNA fragments.** (A) Enzymatic copying of DNA single-stranded template in the presence of labeled deoxynucleoside triphosphates (dNTPs); alternatively, normal dNTPs can be utilized in the presence of labeled primer. (B) End-labeling uses the bacteriophage enzyme polynucleotide kinase to transfer a single labeled phosphate from ATP to the 5′-end of each DNA chain. The first method produces DNA fragments that are uniformly labeled along the entire chain; these chains are appropriate for use in hybridization reactions, as they provide easily detectable signals of high intensity.

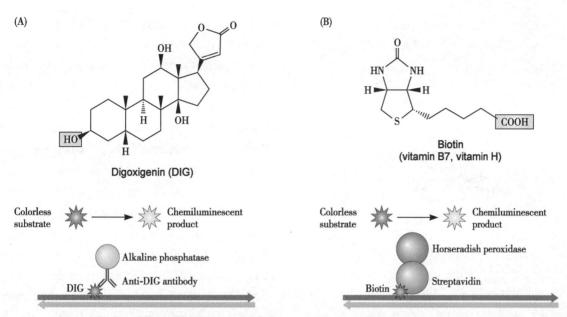

Figure 4.5 **Chemiluminescent detection of digoxigenin- or biotin-labeled DNA samples.** (A) Detection of digoxigenin (DIG)-labeled nucleotides is done by use of anti-DIG antibodies conjugated to alkaline phosphatase. The DIG OH group used for attachment to bases in dNTPs, through linkers, is boxed in gray; the modified dNTPs still have their bases available for base-pairing. When an appropriate substrate is added, alkaline phosphatase breaks it down to a compound that emits a photon of light. (B) Detection of biotinylated DNA is typically done by use of streptavidin or its relatives, such as avidin, proteins with high affinity for biotin. Streptavidin is preconjugated to horseradish peroxidase. The COOH group of biotin used for attachment to bases in dNTPs, through linkers, is boxed in gray. Breakdown of substrate, such as luminol peroxide, results in the release of a photon of light.

4.2.4 Polymerase chain reaction

The polymerase chain reaction (PCR), is a method for obtaining large amounts of highly purified DNA molecules suitable for analysis or manipulation. It was introduced in 1987 by Kary Mullis. Mullis shared the 1993 Nobel Prize in Chemistry with Michael Smith "for contributions to the developments of methods within DNA-based chemistry."

The PCR reaction usually consists of a series of 20–40 repeated cycles, where each cycle consists of three discrete temperature steps: denaturation, annealing, and elongation (Figure 4.6). During the denaturation step, the DNA sample is heated to 94–98°C for 20–30s: the high temperature disrupts the hydrogen bonds between complementary bases and separates the two DNA strands. During the annealing step, the temperature is lowered to 50–65°C for 20–40s to allow annealing of the primers, 15–30-nucleotide synthetic DNA fragments, to the single-stranded DNA template. The primers are complementary in sequence to the ends of the region of interest and possess a free 3′-OH group, which is needed for the action of DNA polymerase at the third step. Typically, the annealing temperature is set to be slightly below the melting temperature of the primer-DNA hybrid duplex, by 3–5°C. Stable DNA-DNA hydrogen bonds are formed only when the primer sequence very closely matches the template sequence. A couple of bases that are not complementary to each other are usually tolerated by the duplex. The ability to form such nonperfect, but still stable, hybrid duplexes is helpful when one tries to amplify a sequence from an organism using knowledge about similar, but probably not identical, sequences from other organisms to design the primers. The exact conditions—temperature, ion concentration, and so on—under which the annealing step is performed determines what is known as the stringency of hybridization. The more stringent the conditions, for example, temperature closer to the melting temperature of the hybrid, the lower the probability that an imperfect, mismatched duplex will be stable enough

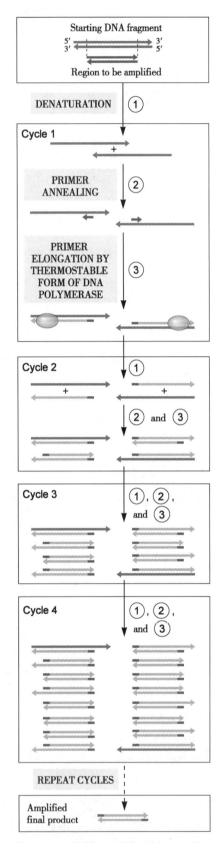

Figure 4.6 **PCR amplification reaction.**

to allow the next step of the reaction cycle to occur, and vice versa.

The Primer extension/elongation step involves the extension of the primer by a thermostable form of DNA polymerase. The temperature at this step depends on the DNA polymerase used. The commonly used polymerase is Taq polymerase, a naturally thermostable enzyme from the thermophilic bacterium *Thermus aquaticus* that works optimally at 75–80°C, but other thermostable DNA polymerases are used for specialized PCR assays. The chemistry of elongation does not differ from that of the normal *in vivo* process. The extension time depends both on the DNA polymerase used and on the length of the DNA fragment to be amplified. The entire cycle is then repeated. Under optimum conditions, the amount of DNA target is doubled at each extension step, leading to exponential amplification of the specific DNA fragment.

The products of PCR can be detected in one of several ways. Following the reaction, the product can be analyzed by electrophoretic techniques. Hybridization to microarrays can be used to detect and analyze more complex products.

An important recent development is quantitative real-time PCR, in which the products of the reaction are monitored during the course of amplification, usually through fluorescence methods. Real-time PCR shows good correlation between that cycle in the entire process during which unambiguous product signals become measurable, usually referred to as the threshold cycle or C_T, and the concentration of template in the starting preparation. Thus real-time PCR provides precious quantitative information on the amount of the sequence of interest in the sample. This unique capability of the method has earned its alternative name, quantitative PCR or qPCR.

Although the PCR process would be quite tedious if performed manually, the process has been automated and programmable PCR machines or thermocyclers are now commonplace in most molecular biology laboratories. The complete PCR reaction can be readily performed in 2–4 hours. Automated PCR is nowadays widely used in forensic cases and to study DNA from fossils. A very recent example is the deciphering of the genome of a human who lived some 4000 years ago: the DNA was recovered and PCR-amplified from a single hair found in a permafrost area in North America. The initial sequencing of four short gene sequences from fossils of Neanderthals reported in 2007 and 2008 was also PCR-based; high-throughput sequencing techniques were later used to compile draft sequences of the entire Neanderthal genome.

4.2.5 In vitro mutagenesis

Introducing defined mutations is critical to our understanding of the biochemistry and physiology of genes, their control regions, and the proteins they encode. By mutating selected regions or single nucleotides within the gene, it is possible to define the role of DNA sequences in gene regulation and amino acid sequences in protein function.

Site-directed mutagenesis is a universal method to mutate any DNA sequence in a desired way. The procedure underwent many modifications, but the principle stays the same (Figure 4.7): Synthesize a short, ~20-nt-long oligonucleotide sequence that complements the DNA sequence of the intended mutation site in the gene of interest but that carries deliberate mistakes, such as single-base substitutions, short deletions, or insertions at its center. Anneal the oligonucleotide to a single-stranded cloned sequence of interest; the product will contain a base or several bases that cannot base-pair with the gene and, depending on the change introduced in the oligonucleotide, may form a loop. The correctly matched bases on both sides of the mutation will keep the annealed complex stable. DNA polymerase will use the annealed oligonucleotide as a primer to synthesize an entire strand complementary to the template, but retaining the modified oligonucleotide; DNA ligase will ligate the nick. The new double-stranded circle is then introduced into bacterial cells and the bacteria are allowed to multiply. Various techniques can be used to get rid of the original molecule.

Site-directed mutagenesis makes it possible, in principle, to insert a protein with a new or modified sequence into an organism and to study its effect on the whole organism. However, it must be remembered that this is the addition of a function; the original variant of the gene is still present and, in most cases, functional. If we want to take away a gene or replace it with an altered version, the challenge is somewhat more difficult.

Linker-scanning mutagenesis can be used to generate clusters of point mutations throughout a promoter region or throughout any other sequence element involved in replication, splicing, or protein structure. For example, it was introduced to study the activity of the promoter of the gene. The linker sequence (10 bp) is systematically substituted for

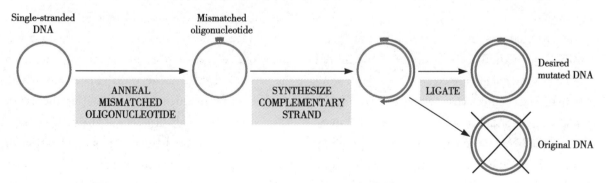

Figure 4.7 **Principle of site-directed mutagenesis.** The single stranded DNA could be that of the single-strand DNA phage M13 containing the cloned gene; alternatively, if the gene had been cloned in a double-stranded vector, the DNA would first need to be denatured, to allow hybridization with the synthetic oligonucleotide. Several rounds of replication of the ligated product introduced into a bacterial cell will result in the formation of a mutated sequence of the original DNA. Convenient methods exist to get rid of the latter. (Adapted from Kunkel TA, Roberts JD, Zakour RA [1989] In Recombinant DNA Methodology Wu R, Grossman L, Moldave K eds, 587–601. With permission from Academic Press.)

10bp-sequences throughout the promoter, and the mutations destroyed promoter activity. The linker sequence actually provides the desired clusters of point mutations as it is moved or "scanned" across the region by its position at the varied endpoints of the deletion mutation series.

4.2.6 DNA sequencing

DNA sequencing methods developed in the late 1970s led to the revolutionary capability to determine the nucleotide sequence of any DNA. Two major methods have been used for DNA sequencing: chemical sequencing, developed by Maxam and Gilbert, and enzymatic sequencing, developed by Sanger. Both methods share the same underlying principle: the nucleotide sequence is reconstructed by determining the size of nested sets of DNA molecules. The DNA molecules in each set share a common 5'-end and terminate at a single type of base at the 3'-end, either A, T, C, or G. The DNA molecules in any given set differ, however, in their lengths, which are determined by the distance of these respective bases from the 5'-common end; that is, by their position in the sequence. The processes that create the A (or T or C or G) sets of molecules are random in terms of which particular A out of all the As present in the DNA fragment is being used for the chemical cleavage reaction of Maxam and Gilbert or as a site of termination of elongation in the enzymatic method of Sanger. We will illustrate how the method works for the enzymatic method, as the chemical method is no longer used.

The enzymatic method is also called the chain-termination method. In this method, the nested sets of DNA fragments are created by *in vitro* reactions that closely mimic chain elongation during the normal DNA replication process. However, elongation of the nascent daughter strand is purposefully terminated, at random sites along the sequence, by supplying the reaction mixture with nucleoside triphosphate derivatives that lack the OH group at the 3'-position of the sugar (Figure 4.8). Separate runs are made with each of the four dideoxy derivatives or ddNTPs. If, during synthesis, the polymerase incorporates the ddNTP derivative instead of the dNTP, then synthesis is terminated at that point, because the 3'-OH group is needed for further elongation.

Walter Gilbert, Frederick Sanger, and Paul Berg received the 1980 Nobel Prize in Chemistry. Gilbert and Sanger were given the award "for their contributions concerning the determination of base sequences in nucleic acids." Paul Berg was recognized for "his fundamental studies of the biochemistry of nucleic acids, with particular regard to recombinant-DNA."

These methods have subsequently been automated and coupled to sophisticated data analysis to allow unprecedented accuracy and speed in sequencing entire genomes, including the genome of Homo sapiens. Second generation sequencing, aided in the automation and scaling up of the procedure. This still required PCR amplification of the starting material, which is first randomly fragmented and then amplified. This technology relies on the detection and identification of each nucleotide as it is added to a growing strand. In one such application, the primer is tethered to a glass surface and the complementary DNA to be sequenced anneals to the primer. Sequencing proceeds by adding polymerase and fluorescently labeled nucleotides individually, washing away any unused dNTPs. After illuminating with a laser, the nucleotide that has been incorporated into the DNA strand can be detected. Other versions use nucleotides with reversible termination, so that only one nucleotide can be incorporated at a time even if there is a stretch of homopolymeric DNA (such as a run of adenines). Still another version, called pyrosequencing, detects the release of pyrophosphate from the newly added base. The availability of automated systems for pyrosequencing is facilitating the use of the technique for high-throughput analyses.

Third generation sequencing is a collection of methods that avoids the problems of amplification by direct sequencing of the material, DNA or RNA, still giving multiple short (but longer than second generation sequencing) reads by using single-molecule sequencing (SMS) templates fixed to a surface for sequencing.

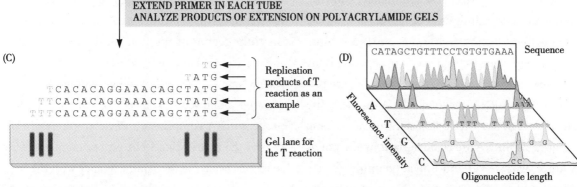

Figure 4.8 Sanger sequencing reaction. (A) Single-stranded DNA is amplified in the presence of fluorescently labeled dideoxy-NTPs (ddNTPs) in a reaction that mimics the normal replication reaction (see Chapter 13). The presence of dideoxythymidine triphosphate (ddTTP) will cause DNA synthesis to abort when the polymerase encounters an A in the DNA template strand. (B) Four reaction mixtures containing DNA template, primer, DNA polymerase, all four dNTPs, and one of the four ddNTPs at a ratio of one ddNTP to 100 dNTP. Each ddNTP is labeled with a different fluorophore. When the ratio of ddNTP to dNTP in the reaction mixture is 1:100, the frequency of termination will be, on average, once per 400 nucleotides. (C) The fragments of DNA are then separated via polyacrylamide gel electrophoresis and visualized with four different laser beams to excite the fluorescence of each individual fluorophore. Each reaction mixture is in a separate lane. (D) Finally, the data are compiled into a sequence with the aid of a computer. An electrophoretogram of a finished sequencing reaction is shown. (D, adapted from *The Science Creative Quarterly* [2009] 4. With permission from Fan Sozzi.)

4.3 Analysis at the level of gene transcription: RNA expression and localization

The process of gene expression may terminate in a product that is either RNA or polypeptide. Usually, gene expression analysis studies can be broadly divided into two levels: transcription and translation. In this section, the classic methods of RNA expression and localization analysis will be discussed.

4.3.1 Reverse transcription PCR

Reverse transcription PCR (RT-PCR) can amplify RNA sequences into DNA. In some cases, researchers are interested in amplifying only genes that are actually transcribed in the cell; that is, those that participate in the flow of information from genes to proteins or that perform different kinds of structural, enzymatic, or regulatory functions. To that end, a reverse transcription (RT) step is performed before the PCR reaction. During this step, pools of RNA molecules extracted from cells and a special kind of polymerase, RNA-dependent DNA polymerase or reverse transcriptase, is used to produce a double-stranded DNA copy of the RNA molecules. This DNA is known as complementary DNA or cDNA which is derived from mRNA. Once the cDNA is at hand, the standard PCR protocol is used. And real-time PCR makes the process quantitative.

4.3.2 Northern blotting

Northern blotting was developed by James Alwine, David Kemp and George Stank at Stanford University in 1977. The technique got its name due to the similarity of the process with Southern blotting. The primary difference between them is that Northern blotting involves the transfer of RNA from a gel to a membrane, instead of DNA.

Northern blotting analysis reveals information about RNA identity, size, and abundance, allowing a deeper understanding of gene expression levels. As all normal blotting technique, Northern blotting starts with the electrophoresis to separate RNA samples by molecular size. In some cases that the target sequence is mRNA, the total mRNA can be firstly isolated through oligo-dT cellulose chromatographic techniques, as mRNA are characterized by the polyA tail. Since agarose gel and polyacrylamide gel are fragile in nature, the separated RNA molecules are transferred to the nitrocellulose membranes or nylon membrane. And then the RNA molecules are immobilized and incubated with specific complementary probes labeled with radioactivity, with digoxigenin or biotin, or with fluorescent dyes. After the removal of excess labeled probe, the signal of bound probes was developed and detected. In this technique, formamide is generally used as a blotting buffer as it can reduce the annealing temperature.

4.3.3 RNase protection assay (RPA)

The RNase protection assay (RPA) is a highly sensitive specific technique for the detection, mapping, and quantitation of specific mRNAs. It is not only can be used to identify and measure the abundance of specific mRNAs in samples, but also can be employed to map the transcription start-sites of specific mRNAs. The assay begins with specific RNA probe that is uniformly labeled by incorporation of [α-32P] NTP, usually [α-32P]UTP. The RNA probe is biosynthesized by SP6, T7, or T3 bacteriophage RNA polymerase *in vitro*, and its sequence can determine the purpose of this assay. If the RNA probe can hybridize to a specific cellular mRNA in solution, it can be used to quantify the abundance of this specific mRNA. And if the RNA probe contains the complementary sequence of the transcription start site for the gene of interest, it is designed to find the start site of the specific gene.

After the radiolabeled probe is annealed to cytoplasmic or total cellular mRNA purified from the cells of interest for several hours or even overnight, RNase A and/or RNase T1 is added to the hybridization reactions to digest the nonhybridizing single-stranded overhang regions of RNA molecules, but RNA-RNA hybrids are resistant to its cleavage. This resistance forms the conceptual basis for the procedure; the region of the RNA probe that anneals to the specific mRNA will be resistant to digestion. And then RNases can be removed by treatment with proteinase K, followed by phenol extraction of the RNA probe: mRNA complexes, and electrophoretic isolation of the hybridizing RNA probe fragments. Since sense RNAs having the same sequence with the target cellular mRNA can be obtained by chemical synthesis, appropriate standard curves can be generated and used to quantitate the changes in tissue mRNA levels. Because the assay requires perfect sequence complementarity for full protection, it also provides conclusive evidence for the existence of a specific mRNA in a given tissue and can be used to identify the sequence of the transcription start site.

4.3.4 In situ hybridization

In situ hybridization (ISH) is a powerful technique for precise localization of specific nucleic acid targets (DNA or RNA) within fixed tissues and cells. It can be used to obtain temporal and spatial information about gene expression and the distribution of genetic loci. The underlying basis of ISH is that DNA and RNA, if preserved adequately within a histologic specimen, can be visualized by the application of a labeled probe. While the basic workflow of ISH is similar to that of common nucleic acid hybridization, the major difference is the greater amount of information can be collected by visualizing the results within the tissue (*in situ*).

4.3.5 DNA microarrays

DNA microarrays, known as chips have been introduced for the measurement of transcription of thousands of genes at once to create a global picture of cellular function. The procedure is based on the use of reverse transcriptase, a specialized RNA-dependent DNA polymerase that is involved in the life cycle of viruses whose genome is RNA-based. RNA from a whole cell, such as a bacterium, or a given cellular compartment, such as the cytoplasm or nucleus of a eukaryotic cell, is first converted into double-stranded (ds) cDNA with the help of reverse transcriptase. cDNA copies are further analyzed by DNA tiling arrays (Figure 4.9).

Microarrays are small, solid supports onto which the sequences from thousands of different genes are immobilized, or attached, at fixed locations. The supports themselves are usually glass microscope slides but can also be silicon chips or nylon membranes. The DNA, whether in the form of DNA, cDNA, or oligonucleotides, may be printed, spotted, or synthesized directly onto the support by use of robots. The RNA samples to be analyzed are fluorescently labeled and allowed to hybridize with the array, and the extent of hybridization is estimated by measuring the intensities of fluorescence in each individual spot. Tiling arrays contain millions of spotted oligonucleotide sequences that cover large portions of a genome or entire genomes. For example, the Affymetrix tiling array provides 6.4 million spots, each containing oligonucleotides of length 25 nucleotides, spaced 10 base pairs apart, across the entire human genome. Such arrays allow unbiased analysis of the entire human transcriptome without any prior knowledge of genic and intergenic regions, introns and exons, etc. Once the microarray data are available, computer algorithms can be used to map these tags back to the reference genome. Microarrays provide a huge amount of noisy data. Interpretation of these data requires a combination of biological knowledge, statistics, machine learning, and the development of efficient algorithms that are able to select features above background noise.

4.4 Analysis of the transcription rates

The amount of the transcripts reflects the efficiency of transcription. And the transcript concentration depends on its rate of synthesis and degradation. Usually, the ability of a gene to be expressed is controlled by a protein that binds to the DNA. RNA can be synthesized when the proteins are bound to control site on the DNA. To measure transcription rates, we can use some methods to analyze the transcriptional regulatory elements and the interaction of the DNA element and DNA binding protein.

4.4.1 S1 nuclease protection and primer extension

Mapping the exact position in a gene where transcription is initiated, the transcription start site or TSS, is important: it helps in the identification of

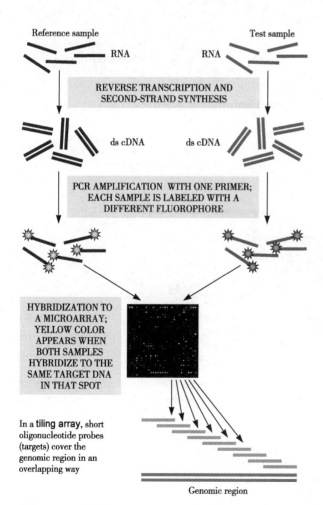

Figure 4.9 **DNA microarrays.** (microarray, courtesy of John Coller, Stanford University.)

promoter sequences and transcribed regions in the genome. There are two classical methods for mapping TSSs in known genes: S1 nuclease protection and primer extension (Figure 4.10).

4.4.2 Rapid amplification of cDNA ends

In addition to mapping TSSs, identifying 5'- and 3'-regions of genes can significantly contribute to our overall understanding of functional elements of genomes. The need to identify the ends of genes led to the development of a variety of rather sophisticated methods. Rapid amplification of cDNA ends, RACE is diagrammed in Figure 4.11. This method is a variation of reverse transcription polymerase chain reaction, RT-PCR, which amplifies unknown cDNA sequences corresponding to the 5'- or 3'-ends of the RNA.

In 5'-RACE, the first-strand cDNA synthesis reaction is primed by use of an oligonucleotide complementary to a known sequence in the gene. After removal of the RNA template, terminal deoxynucleotidyl transferase adds a nucleotide tail to create an anchor site at the 3'-end of the single-stranded cDNA. An anchor primer complementary to the newly added tail is used to synthesize the second cDNA strand. An alternative strategy uses an RNA ligase to add the necessary anchor.

In 3'-RACE, a modified oligo (dT) sequence

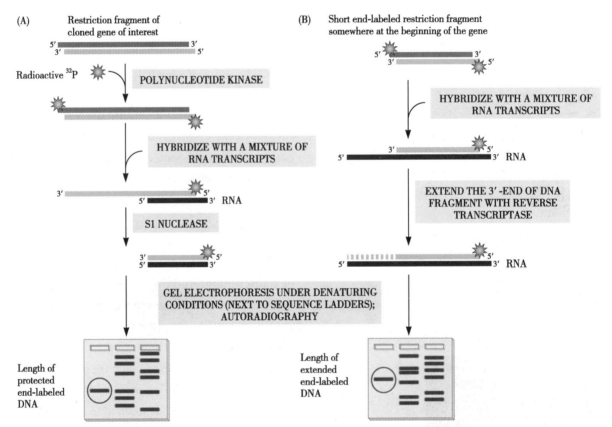

Figure 4.10 **Mapping the position of transcription start sites.** (A) S1 Nuclease protection assay. A restriction fragment from the 5'-end of a cloned gene is end-labeled at the 5'-ends, denatured, and hybridized to total RNA from cells expressing the gene. The DNA-RNA heteroduplexes are treated with the fungal enzyme S1 nuclease, which specifically cleaves single-stranded DNA or RNA, until it reaches the boundaries of the duplex; note that other enzymes with similar action can also be used. High-resolution electrophoresis is run to identify the size of the protected DNA fragment precisely; usually, the gel also contains sequencing lanes to determine the exact base at which transcription is initiated. (B) Primer extension assay. The restriction fragment expected to contain the TSS is deliberately chosen to be small. Hybridization with a cognate RNA will create an overhanging 5'-end, which can be used as a template for reverse transcription. The 3'-OH group of the DNA strand will serve as a primer in the reaction. The DNA will be extended until the 5'-end of the RNA is reached. The size of the extended DNA primer can be precisely determined as it was in the last step in the S1 protection assay.

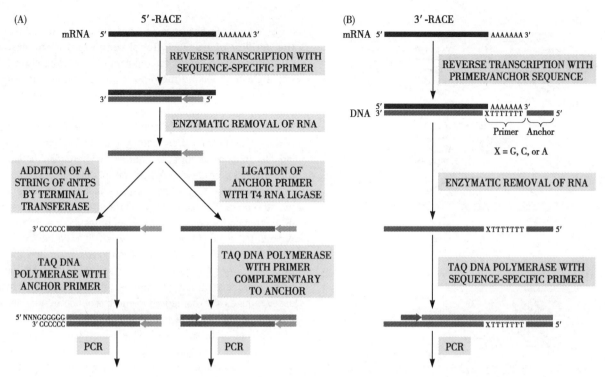

Figure 4.11 **RACE: rapid amplification of cDNA ends.** (A) 5'-RACE. (B) 3'-RACE.

serves as the reverse transcription primer. This oligo (dT) sequence is composed of a primer oligo (dT) sequence, which anneals to the poly(A)+ tail of the mRNA, and an adaptor sequence at the 5'-end. A single G, C, or A residue at the 3'-end of the primer ensures that cDNA synthesis is initiated only when the primer/adaptor anneals immediately adjacent to the junction between the poly(A)+ tail and 3'-end of the mRNA.

4.4.3 Reporter genes

Assays that measure the activity of transcriptional regulatory elements are based on the use of reporter gene constructs, whose transcriptional activity can be easily monitored (Figure 4.12). The most frequently used reporter genes are chloramphenicol acetyltransferase or CAT, β-galactosidase, or luciferase genes. CAT detoxifies the antibiotic chloramphenicol by acetylation, preventing its binding to the ribosome. Cells that express CAT can grow on medium containing the antibiotic. β-Galactosidase catalyzes the hydrolysis of β-galactosides into monosaccharides. X-gal, a colorless modified galactose sugar, is metabolized by the enzyme to an insoluble product, 5-bromo-4-chloroindole, which is bright blue and thus functions as an indicator of enzymatic activity. Finally, luciferase, an enzyme derived from the firefly, is responsible for oxidation of luciferin pigment, a reaction that is accompanied by the production of light or bioluminescence.

One of the most popular reporters that can be used to visualize patterns of gene expression is green fluorescent protein (GFP), which is obtained from jellyfish. GFP is a naturally fluorescent protein that, when excited with one wavelength of light, emits fluorescence in another wavelength. In addition to the original GFP, numerous variants that fluoresce in different colors, such as yellow (YFP), cyan (CFP), and blue (BFP), have been developed. GFP and its variants can be used as reporter genes on their own, or they can be used to generate fusion proteins in which a protein of interest is fused to GFP and can thus be visualized in living tissues.

We can also use reporter genes to detect the translational efficiency after we alter regions of a gene that effect translation.

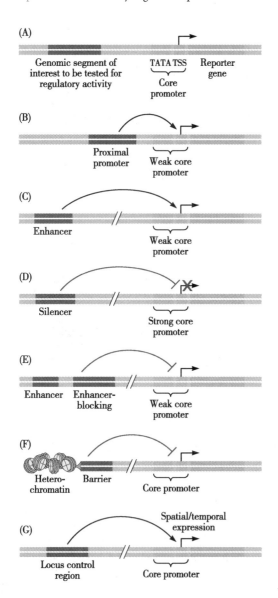

Figure 4.12 **Functional *in vivo* assays to measure the activity of transcriptional regulatory elements.** (A) Regions to be tested for regulatory activity are cloned into a plasmid bearing the reporter gene; the constructs are transfected, either transiently or stably, into cultured cells; and the activity of the reporter gene is monitored. If the segment is tested for core promoter activity, then it is placed immediately upstream of the reporter gene lacking its endogenous promoter, not shown here. (B) Testing for proximal promoter elements: an increase in transcription is expected. (C, D) Testing for enhancers or silencers would require the use of appropriate strength promoters. (E, F) Testing for two different types of insulator activities. A potential enhancer blocking element should be active when it is cloned between an existing enhancer and the gene. A barrier element should block the spreading of heterochromatin structures from neighboring regions. This assay requires stable integration into the genome. A barrier element should insulate the gene construct from position effects, that is, the gene should always be active wherever in the genome the integration occurred. (G) Definitive identification of a locus control region, LCR, also requires stable integration. The LCR should confer regulated expression of the linked gene, independently of where the construct integrates. (Adapted from Maston GA, Evans SK, Green MR [2006] *Annu Rev Genomics Hum Genet*, 7:29–59. With permission from Annual Reviews.)

4.4.4 DNA footprinting

DNA footprinting depends on the fact that proteins bound to DNA will protect that DNA from cleavage in the region of binding (Figure 4.13). High-resolution gels can sometimes locate binding sites with base-pair precision. This method can be used to look for regions in the DNA that are protected by the bound protein from being methylated by DNA methyltransferases.

4.4.5 Electrophoretic mobility shift assay

Electrophoretic mobility shift assay (EMSA/gel shift assay) relies on the fact that a small DNA has a much higher mobility in gel electrophoresis than the same DNA does when it is bound to a protein. When a protein binds to DNA or RNA, the complex exhibits a lower electrophoretic mobility than the free polynucleotide, mainly because of charge neutralization. As Figure 4.14 shows, this can be used in a simple and fast technique that requires very little material. The DNA or RNA fragments studied, termed probes, are labeled either with radioactive isotopes or with fluorescent dyes, both of which can be easily detected. The method suffers from the disadvantage that unstable complexes may fall apart; methods such as chemical cross-linking may be used for added sta-

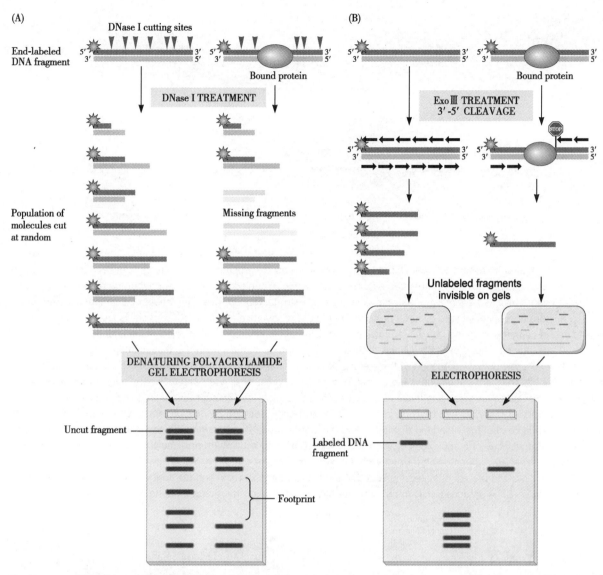

Figure 4.13 DNA footprinting methods. (A) DNase I footprinting. A protein bound to a specific site on the DNA should protect the DNA from enzymatic attack by pancreatic deoxyribonuclease, DNase I. The naked DNA and the protein-DNA complex are treated in parallel with DNase I under conditions in which each naked DNA fragment is cut once on average, at random. The two samples are analyzed side by side on high-resolution sequencing gels: the bands on such gels differ from their neighbors by one nucleotide. The protection by the bound protein gives rise to a footprint: absence of bands, or presence of bands of very low intensity, in the protected region. Since DNase I has some sequence specificity of cutting, alternative cleavage agents, such as the organic compound methidiumpropyl-EDTA-Fe^{2+} (MPE-Fe^{2+}), have been introduced. Another widely used agent produces hydroxyl radicals that also cleave DNA nonspecifically; the technique is known as hydroxyl-radical foot printing. (B) Exo III footprinting. Exonuclease III is an *E. coli* enzyme that cleaves the DNA exonucleolytically, from the ends, until it reaches the bound protein. The direction of cleavage is from 3′ to 5′. The small labeled DNA fragments resulting from the cleavage reaction are analyzed by high-resolution sequencing gels and are detected by autoradiography.

bility. The basic assay can be augmented by adding specific or nonspecific competitors. These competition reactions help to assess the degree of specificity of complexes between the labeled DNA fragment of interest and the protein. In another version of the assay, an antibody specific to the protein of interest is added to the incubation mixture: when the antibody binds to the protein-DNA complex, a supershift is produced. Caution should be taken not to overinterpret the relative intensity of bands, as there is always some possibility of dissociation of the complex during migration.

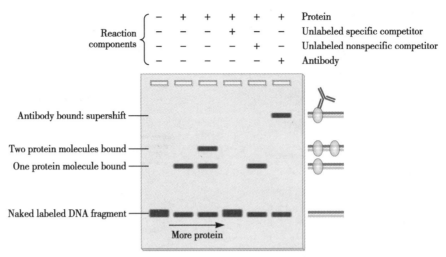

Figure 4.14 **Gel shift assay.** The gel shift assay, also known as electrophoretic mobility shift assay (EMSA) or as gel retardation assay, is performed in three steps: (a) binding reactions, (b) electrophoresis, and (c) probe detection. The DNA probe fragment is labeled either with radioactive isotope or with fluorescent dye. The labeled probe is incubated with a protein mixture to allow binding to occur. If proteins bind to the probe, its electrophoretic mobility is changed: the complex is retarded in comparison to the free DNA fragment. Following electrophoresis, the labeled probe is detected by the usual methods of autoradiography or fluorescence detection. The method suffers from the disadvantage that unstable complexes may fall apart during application of the electric field. The relatively low ionic strength of the electrophoresis buffer helps to stabilize transient interactions, and sometimes chemical cross-linking is used to create covalent bonds between the protein and DNA, thus stabilizing these complexes. The basic assay can be augmented by doing competition reactions in which an unlabeled probe—the DNA fragment itself, as a specific competitor, or a mutant or unrelated DNA fragment, as a nonspecific competitor—is added to the binding reaction mixture. These supplemented reactions help in assessing the degree of specificity of complexes between the labeled DNA fragment of interest and the protein. In another version of the assay, especially when it is performed with a mixture of proteins that presumably contains a protein of interest, an antibody specific to the protein of interest is added to the incubation mixture. When the antibody binds to the protein-DNA complex, a supershift is produced: that is, the protein-DNA complex is further retarded due to antibody binding.

4.4.6 *Chromatin immunoprecipitation*

In recent years, a number of new methods have been developed that give an enormous amount of information about the genomic localization of protein binding sites and that can identify the proteins bound at these sites. Chromatin immunoprecipitation (ChIP) has allowed the rapid identification of hosts of nucleic acid-binding proteins and other proteins that bind to them. The original idea was to fragment some large genomic region that carries bound cross-linked proteins and then use antibodies to precipitate or otherwise select those fragments that have a particular protein attached; the proteins can be bound to nucleosomal or nonnucleosomal portions of eukaryotic genomes. Actually, the same term is used for the method when it is applied to bacterial cells, and we know that chromatin is a characteristic feature of eukaryotic genomes only. After the immunofractionation of the material into protein-bound and unbound fractions, one can quickly assay the DNA sequences bound (Figure 4.15), frequently by very high-throughput methods. Furthermore, it is now becoming clear that many sites on the genome carry not only the protein with affinity to the site but also other proteins attached to it. So one can now hunt for clusters of proteins attached at particular sites. Unfortunately, such a vast array of data can now be quickly massed that the problem becomes one of data handling and interpretation.

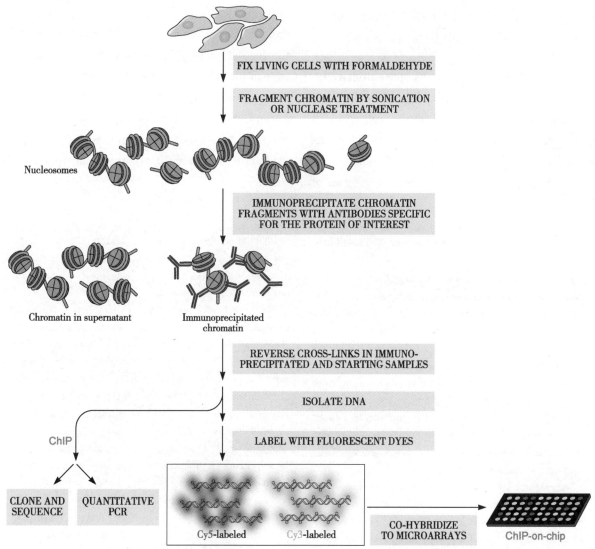

Figure 4.15 **Chromatin immunoprecipitation and ChIP-on-chip.** Genomewide localization analysis (GWLA) is a hybrid of two highly potent individual techniques that allows interrogation of protein-DNA interactions *in vivo*: chromatin immunoprecipitation (ChIP), and high-resolution, high-density DNA microarrays, known as chips. The combination of these two techniques has given GWLA its popular name: ChIP-on-chip or ChIP-chip. Living cells are subjected to a protein-DNA crosslinking agent, usually formaldehyde, to fix the protein of interest to its chromatin location in the genome. Chromatin is then fragmented by sonication or treatment with nucleases, either restriction endonucleases or micrococcal nuclease that cleaves predominantly in the linker regions between nucleosomes to produce core particles. The population of chromatin fragments is incubated with an antibody specific to the protein of interest, and the reacted fragments are immunoprecipitated. Typically, the antibodies are conjugated to beads to facilitate the separation of reacted from unreacted fragments. In classical ChIP, DNA isolated from the immunoprecipitated fraction is cloned and sequenced, or subjected to quantitative PCR (qPCR) amplification to identify the DNA fragments that had been cross-linked by the formaldehyde treatment. In ChIP-on-chip, the starting and immunoprecipitated DNA populations are amplified, labeled by two different fluorophores, mixed, and co-hybridized to microarrays. The microarray platforms are either commercially available or custom made, depending on the specific project. The slides may contain a variety of probes arranged in a variety of ways; they may cover groups of genomic regions, such as open reading frames (ORFs) and intergenic sequences, entire chromosomes, or for organisms of relatively small genome sizes, entire genomes. In addition, the genome coverage can vary in the probe density, from one probe every ~300 bp to probes that are tiled in a way to ensure that each genome region or nucleosome is represented by several overlapping probes.

4.5 Analysis at the level of translation: protein expression and localization

As shown in Chapter 2, proteins are the most versatile and important biomolecules in the cell. It has been estimated that a typical human cell contains about a billion protein molecules, of over 20 000 different kinds. And cells must continually change the profile of the proteins that they contain, in terms of both the types of proteins and their relative abundances. These changes are mainly achieved through the regulation of gene expression. In this section, we will introduce several widely used methods. The ultracentrifuge (Box 4.1) and gel electrophoresis (Box 4.2 and Box 4.3) remains of importance in analyzing proteins. And the other techniques rely on specific interactions between proteins and antibody molecules.

Box 4.1 Svedberg, the ultracentrifuge, and giant protein molecules

In the 1920s, biochemists had begun to realize that proteins were polypeptides, but there was no clear agreement as to their size. In order to investigate this problem, biophysicist Theodor (The) Svedberg developed an entirely new kind of instrument, the analytical ultracentrifuge. This is simply a high-speed centrifuge equipped with an optical system that allows observation of the sedimenting substance. To investigate some smaller proteins, very high rotational speeds are needed, up to 100 000 revolutions per minute, and observation requires very clever design. Even early studies with this instrument revealed that some proteins were very large, much larger than anyone had expected.

Sedimentation velocity is expressed in terms of the sedimentation coefficient, where s = velocity of sedimentation/centrifugal field strength. The s value has the dimension of seconds, and 10^{-13}s is called one Svedberg, abbreviated as S. This gives an approximate measure of molecular size, and we still use the unit today; for example, 20S particles.

Approximately, for spherical molecules:

$$s = M(1 - \bar{v}\rho)/6\pi\eta rN \quad \text{(Equation 4.1)}$$

where M is molecular mass, ρ is solution density, $\bar{v}$ is the partial specific volume ($\approx$1/protein density), η is solvent viscosity, r is radius of the particle, and N is Avogadro's number (6.02×10^{23} molecules/mole).

Svedberg found that some proteins had sedimentation coefficients as large as 100S, which corresponded to molecular masses in the millions of Daltons. More exact values were later obtained by a variation of the technique called sedimentation equilibrium, in which sedimentation at lower speeds was allowed to continue until the protein came to equilibrium against back-diffusion. These studies proved beyond a doubt the immense size of many proteins. They also showed unequivocally that many proteins were homogeneous in molecular mass, all molecules of a given protein having the same size. This was a first hint that the sequences of different proteins might be unique.

Svedberg received the 1926 Nobel Prize in Chemistry for "his work on disperse systems." Svedberg's pioneering method has now been largely replaced by mass spectrometry for the measurement of molecular mass, but it still has many uses in studying macromolecular interactions and quaternary structure.

Box 4.2 Electrophoresis: General principles

The use of electrophoresis is a powerful and versatile method for the analytical separation of biopolymers. It is fair to say that modern molecular biology could not have developed without this technique, and it remains probably the most widely used method. It is extensively employed, in a wide variety of variants, to separate both proteins and nucleic acids.

The basic idea is very simple. If a protein carrying a net electrical charge is placed in an electric field, there will be a net force on it that causes it to move toward the electrode of opposite charge. The molecule will accelerate until, very soon, its velocity is large enough that the drag force from the surrounding medium just balances the electrical driving force: at this point, the molecule is at steady-state velocity. The driving force is given by ZeE, where Z is the number of units of electron charge e, + or −, on the molecule and E is the electrical field strength. The drag force is fv, where v is velocity and f is the frictional coefficient, which depends on the size and shape of the protein molecule.

Thus, at steady-state velocity:

$$fv = ZeE \text{ or } v/E = Ze/f \text{ or } U = Ze/f \quad \text{(Equation 4.2)}$$

The velocity per unit field is termed the electrophoretic mobility U. This definition is much like that of the sedimentation coefficient: each involves the ratio of driving force to frictional resisting force. Nowadays, we do not generally measure absolute mobilities but carry out electrophoresis in some kind of gel matrix and measure the mobility (U_i) relative to that of some small, rapidly moving dye molecule (U_d). The relative mobility of component i is defined as

$$U_{ri} = U_i/U_d \qquad \text{(Equation 4.3)}$$

If protein molecules are migrating in a gel matrix, in an apparatus like that shown in Figure 1, they will be separated into discrete bands and their relative mobilities can be calculated from the distance d_i each has traveled after some time of electrophoresis relative to the distance traveled by the tracking dye band:

$$U_{ri} = d_i/d_d \qquad \text{(Equation 4.4)}$$

The fact that electrophoresis is usually carried out in some kind of gel matrix introduces another complicating but useful factor into the analysis. A gel will have a sieving effect, depending on its concentration and the size of the protein molecules. Small molecules can slip through the gel almost as easily as they move through pure solvent, while large or asymmetric molecules have more impediments. If we experimentally graph the log of relative mobility versus the concentration of the gel, we usually observe a straight line. Such graphs, known as Ferguson plots, are shown in Box 4.2 for several different kinds of molecules. The limit at zero gel concentration corresponds to what would have been observed in pure buffer, where mobility is determined only by Z and f. The slope of the line depends on the protein's molecular size and shape, being steeper for large and asymmetric molecules.

These general principles can be used to devise different kinds of separations for different biopolymers or different problems.

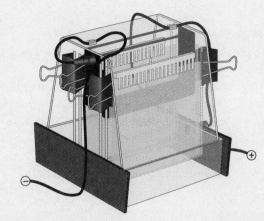

Figure 1 **Typical apparatus for vertical polyacrylamide gel electrophoresis.** The gel is cast between the two glass plates, which are separated by thin plastic spacers and held together by the paper clips. The comb at the top will be removed, leaving wells into which the samples will be loaded.

Box 4.3 **Electrophoresis: techniques for protein electrophoresis**

Native gels

The simplest technique is to simply load a solution containing a mixture of proteins onto a gel like that shown in Box 4.2. The gel is saturated with a buffer that maintains the proteins in their native, undenatured state, providing the possibility of detecting associations between molecules that might be of significance *in vivo*. By staining the gels with a quantitative stain, relative protein concentrations can be measured by the intensity of the stained bands. The disadvantage of native gels is that they tell virtually nothing about the molecule. Ferguson plots can be used to obtain a protein's free mobility (Figure 1), but that parameter depends on both charge and size. Even the order in which proteins appear after migration may depend upon gel concentration.

SDS gels

Detergents such as sodium dodecyl sulfate, SDS, destroy the secondary, tertiary, and quaternary structures of proteins to produce elongated micelles, each incorporating an extended protein molecule. The size of the micelle depends on the molecular weight of the protein. Because each attached detergent molecule carries one negative charge, and the number of attached detergent molecules is proportional to the protein chain length, the charge will also be proportional to protein molecular weight, as the residual charge on the protein is usually negligible compared to the micellar charge. Thus, SDS separates particles whose charge and size, and hence frictional coefficient, are both proportional to the molecular weight of the protein in the micelle. So according to Equation 4.3, all such particles should have about the same free mobility at a gel concentration of zero. Nevertheless, their Ferguson plots will be very different, with slope increasing with molecular weight (see Figure 1C). Thus, at any given gel concentration, we can sort proteins by molecular weight (see Figure 1D). Furthermore, if we include on the gel a lane containing a mixture of known proteins, we can calibrate the scale and thus measure the approximate molecular weights of the unknown proteins.

Note, however, that some information has been lost by denaturing the proteins. In particular, the multichain quaternary structure of proteins—other than multisubunit proteins that are held together by covalent bonds, such as disulfide bonds—will have been destroyed. This can be checked by repeating the experiment in the presence of a reducing agent, which will cleave these bonds.

SDS gel electrophoresis has been an important tool in molecular biology but is now being supplanted rapidly for many purposes by mass spectrometry, MS, which is much more accurate and versatile. Nevertheless, gel electrophoresis is a simple, inexpensive method that can be employed in any lab, whereas MS requires expensive complex equipment.

Isoelectric focusing and two-dimensional gels

Each protein contains a unique constellation of acidic and basic side chains. The charged state of each of these, +, –, or 0, will depend on the solution pH and will change as the pH is adjusted. There will be some pH at which the + and – charges exactly balance, so the net charge is zero. This is known as the isoelectric point of the protein. It is possible to fractionate proteins according to their isoelectric points by creating a gel across which there is a gradient in pH. In such a gel, each protein will concentrate at its own isoelectric pH where no electrical force is acting on that molecule.

Today, such fractionation is generally used as a part of a more powerful scheme, termed two-dimensional gel electrophoresis. As shown in Figure 2, a mixture of proteins is first separated by isoelectric focusing in a tube or narrow strip. This is then attached to a second-dimension gel slab that contains SDS, and electrophoresis is conducted in the perpendicular direction. Here, each of the proteins will be separated on the basis of molecular weight. Thus, each protein ends up on the slab at a point determined by both size and relative number of negatively and positively charged groups. It is possible, in some cases, to display the whole proteome of a bacterium in this way. Furthermore, points can be picked from such a display for high-resolution mass spectrometry (MS).

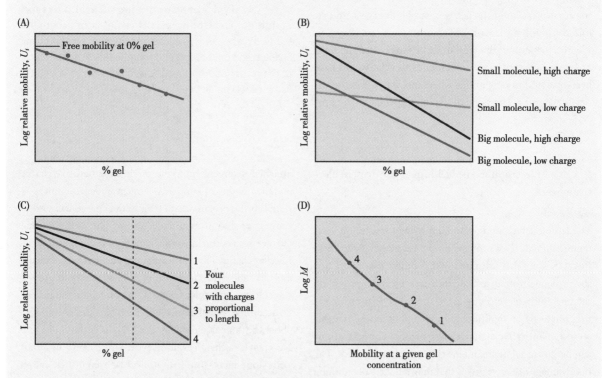

Figure 1 **Mobility of molecules in gel electrophoresis.** A Ferguson plot shows the log of relative mobility U_i as a function of concentration of the gel matrix. (A) Ferguson plot for a single protein: extending the plot to 0% gel gives the theoretical free mobility of that molecule. (B) Ferguson plots for four proteins that differ in size and charge: the free mobility depends mainly on charge, but the slope of the lines depends mainly on size. (C) When the molecular charge is proportional to the length of the molecule, the free mobilities are almost the same but the sieving effect of the gel is more pronounced for longer molecules (the molecules are numbered in order of increasing length and charge). (D) Relationship between molecular weight M and mobility, at a given gel concentration. If log M is plotted against mobility at a given gel concentration for molecules of known M, the standard curve obtained can be used to determine the molecular weight of a molecule of interest. The graph shown is constructed from gel concentration values from part C, black broken line. Points are numbered to correspond to the lines in part C.

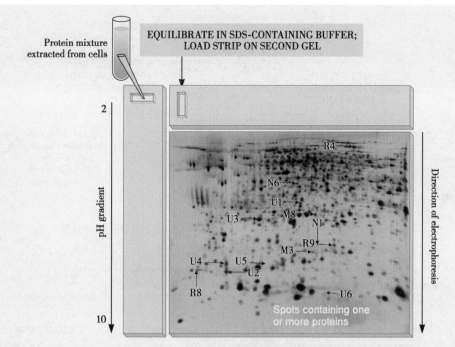

Figure 2 **Two-dimensional gel electrophoresis for separation of complex protein mixtures.** The isoelectric focusing first-dimension gel is not stained for proteins because the stain interferes with the second-dimension gel electrophoresis. (Adapted from Jagadish SVK, Muthurajan R, Oane R, et al. [2010] *J Exp Bot*,61:143–156. With permission from Oxford University Press.)

4.5.1 Western blotting

Western blotting, also known as immunoblotting, was introduced by Towbin, et al. in 1979 and is now an important routine technique for protein analysis in cell and molecular biology. The name Western blotting mimics the name Southern blotting (for DNA analysis), which has been described above. They and Northern blotting (for RNA analysis) have similar procedures and all of them need to transfer nucleic acids or proteins from the electrophoretic gel to membranes. Western blot employs the specificity of the antibody-antigen interaction to separate and identify a protein of interest in the midst of a complex protein mixture. And this method can produce qualitative and semi-quantitative data about the target protein. Western blot uses three elements to accomplish this task: (a) use SDS-PAGE to separate protein samples by molecular weight, (b) transfer proteins to a second matrix, generally a nitrocellulose or polyvinylidene difluoride (PVDF) membrane, (c) visualize the target protein (Figure 4.16). Usually, the transferred protein is probed firstly with an antibody specific to the protein of interest (primary antibody) and then another antibody specific to the host species of the primary antibody (secondary antibody). Most commonly, the secondary antibody is complexed with an enzyme, such as horseradish peroxidase (HRP) or alkaline phosphatase (AP), which when combined with an appropriate substrate, will produce a detectable signal.

4.5.2 Enzyme-linked immunosorbent assay (ELISA)

Enzyme-linked immunosorbent assay (ELISA), also called enzyme immunoassay (EIA), was first developed in the early 1970s as a replacement for radioimmunoassay. It is a polystyrene plate-based assay technique designed for quantifying substances such as peptides, proteins or antibodies.

The ELISA procedure begins with a coating step, where the first layer, either an antigen or an antibody, must be immobilized on a solid surface of plate wells. After that, the microplate surface is blocked

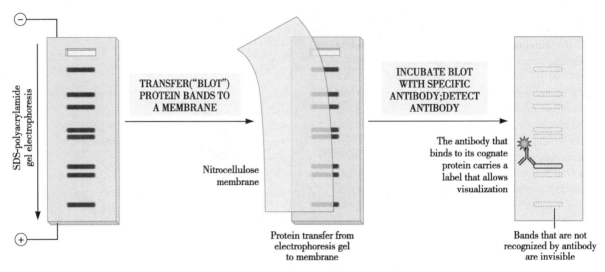

Figure 4.16 **The schematic representation of western blotting.**

with irrelevant protein such as BSA. And then an enzyme-conjugated antibody binds directly or indirectly to the immobilized material. Finally, detection is accomplished by assessing the conjugated enzyme activity via incubation with a substrate to produce a measureable product (Figure 4.17). The most crucial foundation of the detection strategy is a highly specific antibody-antigen interaction.

ELISAs can be performed with a number of modifications to the basic procedure. The key step, immobilization of the protein of interest, can be accomplished by direct attached to the assay plate or indirectly via a capture antibody that has been absorbed to the plate (sandwich ELISA Figure 4.17B). A detection enzyme or other tag can be linked directly to the primary antibody or conjugated to a secondary antibody that recognizes the primary antibody. It can also be linked to a protein such as streptavidin if the primary antibody is biotin labeled. Based on this, ELISAs can be divided into direct ELISA (labeled primary antibody) and indirect ELISA (labeled secondary antibody or avidin) (Figure 4.17A). The most commonly used enzyme labels are horseradish peroxidase (HRP) and alkaline phosphatase (AP).

4.5.3 Immunohistochemistry (IHC), immunocytochemistry (ICC) and immunofluorescence (IF)

Immunohistochemistry (IHC) combines anatomical, immunological and biochemical techniques to image discrete components in tissues by using appropriately-labeled antibodies to bind specifically to their target antigens *in situ*. IHC makes it possible to visualize and document the high-resolution distribution and localization of specific cellular components within cells and within their proper histological context. While Immunocytochemistry (ICC) is performed on samples consisting of cells grown in a monolayer or cells in suspension which are deposited on a slide. It is a technique used to assess the presence of a specific protein in cells by use of a specific antibody that binds to it and make it visible under a microscope. Immunocytochemistry is a valuable tool to study the presence and sub-cellular localization of proteins. Both IHC and ICC can use the same enzyme reactions or the same fluorophores.

Immunofluorescence (IF) is a widely used method of immunostaining and is a form of IHC/ICC methods. This technique uses fluorescent dyes labeled antibody (primary antibody or secondary antibody) to bind directly or indirectly to the proteins of interest within a cell, and therefore make them visible through a fluorescence microscope. It is a particularly robust and broadly applicable method generally used to analyze both the endogenous expression levels of proteins of interest and its distribution within the cell/tissue.

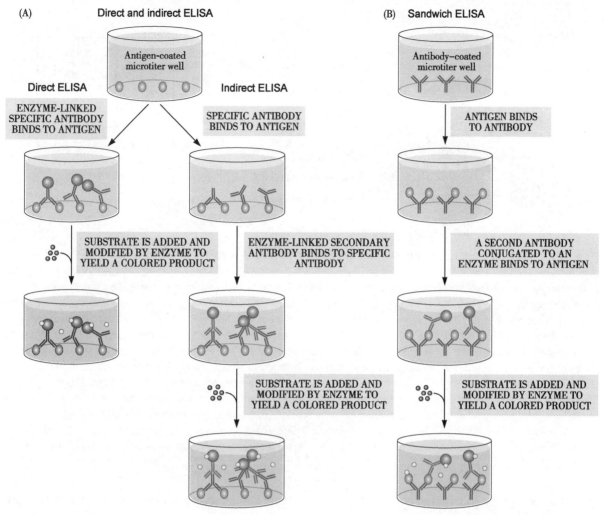

Figure 4.17 **Enzyme-linked immunosorbent assay (ELISA).** The ELISA technique uses microtiter plastic plates that contain 96 even 384 small wells, so that 96 reactions can be carried out simultaneously. The final readout of the assay is visible color, which is automatically quantified by special ELISA reader instruments. Depending on whether the first molecule immobilized on the walls of the microtiter wells is an antigen or an antibody, there are two major types of assays: (A) direct and indirect ELISA or (B) sandwich.

4.5.4 Flow cytometry

Flow cytometry is a popular cell analysis technique that was first used in the 1950s to measure the volume of cells in a rapidly flowing fluid stream as they passed in front of a viewing aperture. Since that time, innovations from many engineers and researchers have culminated in the modern flow cytometer, which is able to make measurements of cells in solution as they pass by the instrument's laser at rates of 10 000 cells per second (or more). Flow cytometry is used to profile a large number of different cell types in a population. The cells are separated on the basis of differences in size and morphology (Figure 4.18A). Additionally, fluorescently-tagged antibodies that target specific antigens on the cell surface can be used to identify and segregate various sub-populations. Today's instruments offer an increased number of detectable fluorescent parameters (from 1 or 2 up to ~30 or more), all measured at the same time on the same cell.

Flow cytometry have numerous applications, including detection and measurement of the protein expression level on the cell surface and even intracellular expression (permeabilization needed), identification of post-translational modifications of proteins, analysis of cell health status and cell cycle status (Figure 4.18B), and so on. An additional capability of specialized flow cytometers is the ability to sort

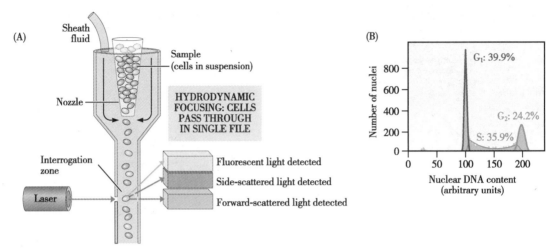

Figure 4.18 **Flow cytometry, multiparameter analysis of individual particles in a population.** (A) Principle of flow cytometry. A sample of cells randomly dispersed throughout the suspending liquid is injected into the cytometer. A fluidic system aligns the individual particles in single file, in a process known as hydrodynamic focusing. As a cell passes through the interrogation zone, it is hit by the light beam, which results in light scatter and possibly fluorescence, depending on whether molecules in or on the cell have been labeled with a fluorescent dye and on the spectral properties of the laser and the dye. A set of light detectors are situated at 90° with respect to each other to catch and quantify the strengths of the respective signals. The detection and analysis of light scatter and fluorescence gives information about each cell. When a cell passes through the interrogation zone, light is scattered in all directions. The forward-scatter detector registers light scattered up to a 20° angle from the excitation beam. The intensity of this signal is proportional to the cell size. The side scatter is detected at 90° from the beam and provides information on granularity and the internal complexity of the cells: the more granular a cell, the greater the side scatter. (B) Nuclear DNA content reflects the position of a cell within a cell cycle: thus, flow cytometric analysis of nuclear DNA content can be used for cell cycle analysis. Distribution of nuclear DNA content of nuclei isolated from broad bean meristem root tip cells prestained with the fluorescent stain DAPI is shown. (A, adapted, courtesy of Abcam, Inc. B, adapted, courtesy of Jaroslav Dolezel, Czech Academy of Sciences.)

cells and recover the subsets for post experimental use, which is called fluorescence activated cell sorter (FACS). Cell sorters can divert a specific population from within a heterogeneous sample into a separate tube, typically based on specified fluorescence characteristics. If the cells are collected under sterile conditions, they can be further cultured, manipulated, and studied.

4.6 Antisense technology

Gene overexpression has been used to analyze the function of genes and their role in disease. However, the phenotypes resulting from overexpression may not reflect the actual gene function. So antisense technology presents an opportunity to knock down gene expression within the cells to study gene function and even treat various diseases.

Highly specific and effective gene silencing of any disease can be achieved by an accurate knowledge of the target mRNA sequence and rational design of its complementary antisense agents for the downregulation of its protein message. This technology has been successfully used for the treatment of cancer, HIV, and other mutating viral diseases. The technology uses agents like antisense oligonucleotides, short interfering RNA (siRNA), (miRNA), and others.

4.6.1 Antisense oligonucleotides

Conceptual simplicity, rational design, relatively inexpensive cost, and completion of the Human Genome Project have led to the use of short fragments of nucleic acid, commonly called oligonucleotides, either as therapeutic agents or as tools to study gene function. And the concept underlying antisense oligonucleotides (AS ODNs) is relatively straightforward: the use of a sequence, complementary by virtue of Watson-Crick bp hybridization, to a specific mRNA can inhibit its expression and then induce a blockade of gene expression. However, although antisense oligonucleotides are commonly in use now both in the laboratory and clinic, and one

AS ODN, Fomivirsen, has been approved by FDA and marketed under the trade name Vitravene by ISIS pharmaceuticals as a local injection to treat retinitis, the critical molecular mechanisms of action of these compounds is still not clear.

4.6.2 RNA interference (RNAi)

RNA Interference (RNAi) is one of the most important technological breakthroughs in modern biology, allowing us to directly observe the effects of the loss of function of specific genes.

In the early 1990s, a number of scientists observed that RNA inhibited protein expression in plants and fungi. And in 1998 Andrew Fire and Craig Mello observed in *Caenorthabditis elegans* (*C. elegans*) that double-stranded RNA (dsRNA) was the source of sequence-specific inhibition of protein expression, which they called "RNA interference". For their discovery of RNA interference, they won the 2006 Nobel Prize in Physiology or Medicine.

Today we have a greater understanding of RNAi. RNAi is a naturally occurring mechanism for gene silencing induced by the presence of short interfering RNA (siRNA). RNAi is an endogenous catalytic pathway that is triggered by dsRNA. The trigger can occur naturally, as in the case of a cellular infection by a dsRNA virus, or by the intentional introduction of dsRNA to induce user-directed degradation of the cognate transcript(s). RNAi is fundamentally a two-step process. The first step involves the master enzyme in the RNAi process, a type III endoribonuclease aptly named Dicer. Dicer is involved in the ATP-dependent digestion of longer dsRNA into 21 to 23 bp siRNA molecules with characteristic 3′ dinucleotide overhangs on both strands. In the second step, siRNA, regardless of the source, becomes part of a multicomponent RNA-induced silencing complex (RISC), and leads to degradation of the target mRNA and down-regulation of the target protein. The use of RNAi as a means of studying the effects of gene expression in a cell or in an organism is occasionally called reverse genetics, the goal of which is to determine the consequences for a cell when a protein is not produced.

The technique has rapidly become an important tool in the ongoing efforts to study gene function, because it can disrupt gene function without creating a mutant organism. RNAi is not restricted to the worm, and it has been used for mid-scale screens in a number of other organisms, including the fruit fly, plants, the mouse, and human cells. And several large-scale RNAi screens have been carried out in mammalian cells. The rapid progress makes RNA interference a field to watch for future medical advances.

Summary

There are more and more methods to assess the content, function, and expression of genes have been developed. DNA or RNA can be separated from the mixtures according to molecule size or density. Specific nucleic acid sequences can be detected using base complementarity. Specific primers can be used to detect and amplify particular DNA targets via PCR. RNA can be reverse transcribed into DNA to be used in PCR. In real time PCR, the quantity of product is monitored throughout multiple reaction cycles. Labeled probes can be used to detect the amount and location of DNA or RNA on Southern, northern blots or *in situ*. Genome-wide transcription analysis is performed using labeled cDNA from experimental samples hybridized to a microarray containing sequences from all ORFs of the organism being used. Classical chain termination sequencing uses dideoxynucleotides (ddNTPs) to terminate DNA synthesis at particular nucleotides. The next generation of sequencing techniques aims to increase automation and decrease time and cost of sequencing. Proteins are detected via the immunological methods, such as western blotting, ELISA, IHC, ICC, IF and flow cytometry, which rely on specific interactions between proteins and antibody molecules. Western blot, ELISA, IHC, ICC, IF and flow cytometry are widely used methods to analyze gene expression at the level of protein. All of them are immunological methods, which rely on specific interactions between proteins and antibody molecules. S1 nuclease protection, primer extension or RACE can be used to identify the regulatory DNA element, while DNA footprinting, EMSA, Chromatin immunoprecipitation (ChIP) allow detection of specific protein-DNA interactions *in vitro* and *in vivo*. Fusion to reporter genes such as luciferase and GFP can be very useful for the study of transcriptional activity. Introducing defined mutations or knock down gene expression via antisense technology is critical to define the role of DNA sequences in gene regulation and amino acid sequences in protein function.

Key Concepts

- Polymerase chain reaction makes multiple copies of a sequence *in vitro*.
- RT-PCR uses reverse transcriptase to convert RNA to DNA for use in a PCR reaction.
- Real-time, or quantitative PCR (qPCR) detects the products of PCR amplification during their synthesis, and is more sensitive and quantitative than conventional PCR.
- The Southern and Northern blotting are used to separate and characterize DNA and RNA, respectively.
- Western blotting, IP and co-IP, ELISA, IHC, ICC, IF and flow cytometry are widely used methods to analyze gene expression at the level of protein. All of them are immunological methods, which rely on specific interactions between proteins and antibody molecules.
- Site-directed mutagenesis permits gene modification at specific sites.
- Reporter genes have been carefully chosen to have products that are very convenient to assay.
- Chromatin immunoprecipitation (ChIP) isolates chromatin-protein complexes by immunoprecipitation with antibodies to the protein for the detection of specific protein-DNA interactions *in vivo*.
- Flow cytometry is used for either counting or separating cell types in a mixture.
- Rapid amplification of cDNA ends (RACE) is used for obtaining the full-length sequence of an RNA transcript. Based on reverse transcription polymerase chain reaction, RT-PCR, which allows the production of amplified cDNA copies of the RNA transcript, which are then sequenced.
- Small interfering RNAs (siRNAs) are Short, 21–26 base pair RNAs derived from double-stranded RNAs by the activity of a specialized cytoplasmic RNase called Dicer. They inhibit gene expression by directing destruction of complementary mRNAs.

Key Words

antisense oligonucleotides（反义寡核苷酸）
chromatin immunoprecipitation（染色质免疫共沉淀）
electrophoretic mobility shift assay（EMSA/gel shift assay）（凝胶阻滞）
enzyme-linked immunosorbent assay（酶联免疫吸附分析）
flow cytometry（流式细胞术）
gel electrophoresis（凝胶电泳）
nucleic acid hybridization（核酸杂交）
PCR（聚合酶链式反应）
quantitative PCR（qPCR），or real-time PCR（实时定量PCR）
reverse transcription PCR（反转录PCR）
reporter gene（报告基因）
RNA interference，RNAi（RNA干扰）
Western blotting（蛋白质印迹）

Questions

1. Describe *in situ* hybridization. When would you use this method, rather than Northern blot?
2. Explain the principle of site-directed mutagenesis, then describe a method to carry out this process.
3. Describe the use of a reporter gene to measure the strength of a promoter.
4. Which methods could be used to detect the interaction of DNA and protein.
5. Explain the principle of real-time PCR.

References

[1] Maxam AM, Gilbert W (1977) A new method for sequencing DNA. *Proc Natl Acad Sci USA*, 74:560–564.
[2] Sanger F, Nicklen S, Coulson AR (1977) DNA sequencing with chainterminating inhibitors. *Proc Natl Acad Sci USA*, 74:5463–5467.
[3] Stellwagen NC (2009) Electrophoresis of DNA in agarose gels, polyacrylamide gels and in free solution. *Electrophoresis*, 30 (Suppl. 1):S188–S195.

Li Yuan

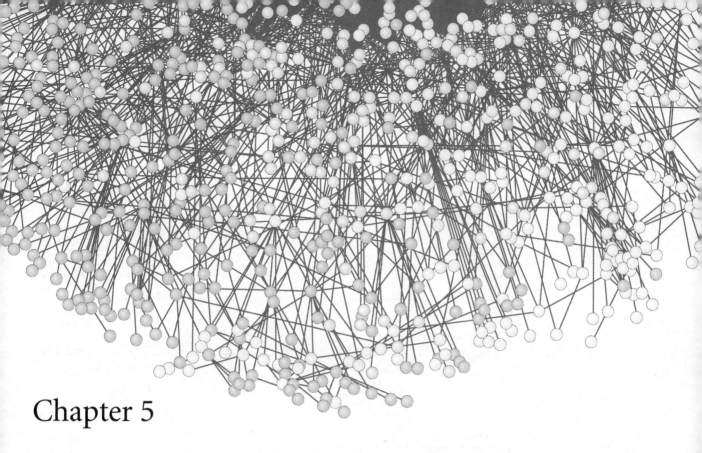

Chapter 5

The genetic Code, Genes, and Genome

5.1 Introduction

This chapter outlines the basic ideas of molecular biology, which are expanded in later chapters, and therefore takes a somewhat historical perspective. These ideas were largely developed in the remarkably short period of about 10 years between 1953 and 1963. Therefore, it seems most interesting to present these fundamental concepts by describing this remarkable burst of discovery. These developments are placed in a broader perspective in Figure 5.1, which also traces the emergence and elaboration of the concept of the gene.

The genome of any organism consists of one or more long polynucleotide chains. In viruses, the polynucleotide chain may be single-stranded or double-stranded, linear or circular, DNA or RNA. Bacteria usually have a single circular double-stranded DNA chromosome. In eukaryotes, each of several chromosomes carries one linear duplex DNA molecule. In each case, the size and informational complexity of all these genomes give rise to two problems that every organism must overcome. These two problems, compaction and specific gene regulation, are dealt with in all organisms by proteins that are bound to the genome. In this chapter, we present an overview of the protein-nucleic acid complexes that are involved, focusing on those that determine the functioning genetic material. Here, we describe the packaging of the genome in organisms of increasing complexity-viruses, bacteria, and eukaryotes-thereby providing background for later discussions of function (Table 5.1).

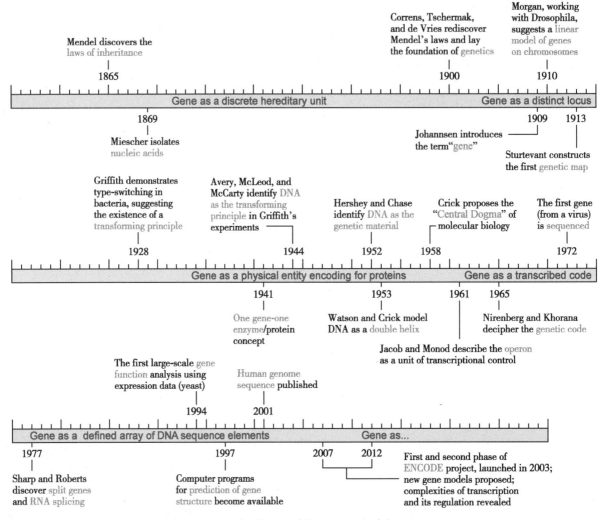

Figure 5.1 **Evolution of the concept of the gene.**

Table 5.1 **Structural aspects of genomes.**

Organismal level	Genome type	Typical genome size (bp)	Localization	Structural proteins
virus	dsDNA, dsRNA, ssDNA, ssRNA; linear or circular	10^3–10^5	viral envelope/capsid	capsid/envelope proteins
bacteria	dsDNA; usually circular	10^6–10^7	cytoplasm	HU, IHF, minor proteins
archaea	dsDNA; usually circular	10^5–10^6	cytoplasm	histone-like proteins (contain histone fold and interact through handshake motif; see main text)
eukaryotes	dsDNA; linear	10^7–10^{11}	nucleus	histones; non-histone proteins

5.2 The genetic code, genes, and genomes

5.2.1 Genes as nucleic acid repositories of genetic information

With the Watson-Crick model of DNA structure of 1953, together with the experiments of Hershey and Chase, it became clear that the genetic substance must be DNA. At this moment in history, the gene became not just an abstract location on a chromosome but a part of a macromolecule that codes for a protein. We still find this to be a correct but far too limited definition. As we shall see, the information in

DNA is indeed used to code for proteins but also has other very important functions.

DNA and proteins contain the same information in two very different languages. In 1958, Crick proposed what has come to be called the central dogma of molecular biology, stating that information flows only from polynucleotide to protein, never in the reverse direction. This was a remarkable, audacious hypothesis, made without a shred of experimental evidence. It is not really a dogma, either, as this is a theological term for something held to be an irrefutable truth. There is nothing irrefutable in science, but in fact the dogma has turned out to be correct, so far as it goes. Furthermore, it is an important idea because if information could leak back from cellular proteins into the genetic material, protein changes induced by environmental effects could be inherited.

5.2.2 Relating protein sequence to DNA sequence in the genetic code

The major breakthroughs in understanding the roles of mRNA, tRNA, and ribosomes made the elucidation of the genetic code an even more important goal (see Box 5.1). It was clear from the start that codons must be groups of at least three bases, known as triplets. One or two bases for each amino acid would not do, as they would code for only 4 or $4^2 = 16$ possible amino acids, respectively. A triplet code allows $4^3 = 64$ permutations, more than enough to code for all 20 amino acids. A number of models of possible codes had been suggested; examples are shown in Figure 5.2A and Figure 5.2B. It soon became clear that neither an overlapping code nor a punctuated code could survive the experimental tests. Crick and Brenner showed by inserting single or multiple base pairs into phage DNA that only an unpunctuated, non-overlapping triplet code could work. A consequence of a non-overlapping, unpunctuated triplet code is that there are three possible reading frames depending on where one begins to read the message (Figure 5.2C). Therefore, a specific start site must be specified to give a particular message. An insertion or deletion of any number of residues that is not a multiple of three will result in a frameshift, specifying a different protein sequence (Figure 5.2D).

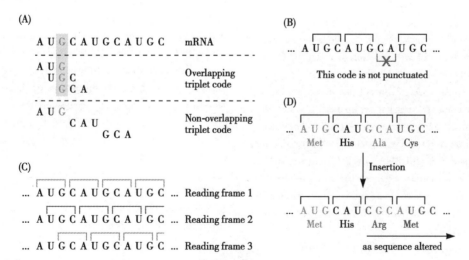

Figure 5.2 Basic features of the genetic code. (A) Schematic of the two possible ways a triplet code can be read: in the overlapping code, the same base can be part of three codons, whereas in the non-overlapping code, each base is part of only one codon. The genetic code is actually non-overlapping. (B) The code is unpunctuated: all nucleotides need to be read in succession and no gaps are allowed between successive codons. (C, D) The concepts of reading frame and frameshift mutations. (C) Each nucleotide sequence has to be read starting from a particular nucleotide. Specification of this first nucleotide defines the succession of codons and thus the primary structure of the polypeptide chain. If a different first nucleotide is specified, the same sequence can be read in a different frame, producing a different polypeptide. (D) Insertion of one nucleotide into the message would change the amino acid sequence in the downstream portion of the polypeptide chain. Such frameshift mutations usually lead to total loss of function of the encoded protein.

But which codons corresponded to which amino acids? A first clue was given when Marshall Nirenberg showed that the homopolynucleotide poly(U) could direct the synthesis of polyphenylalanine. This was quickly confirmed and extended by the experiments of Gobind Khorana using repetitive-sequence synthetic RNAs as messages (see Box 5.1). Elucidation of the specific codons corresponding to each amino acid was a rather tedious undertaking but finally permitted clarification of the entire translational process. The code, as it is now known for most organisms, is summarized in Table 5.2.

Box 5.1 **Cracking the genetic code**

The existence of some kind of code that could translate the language of nucleic acids into the language of proteins had been suspected by some even well before the revolution sparked by Watson and Crick's discovery of the structure of DNA. Indeed, there are hints of it in the 1946 book *What is Life?* by Erwin Schrödinger, an influential volume for early molecular biologists. Nevertheless, even as late as 1961, eight years after Watson and Crick presented their structure, there was not a shred of experimental evidence to suggest how DNA is translated. There were theories and models aplenty for various possible coding schemes, but these only generated debate among researchers.

Then, at the 1961 International Congress of Biochemistry held in Moscow, a paper was presented by Marshall Nirenberg. Because he was young and relatively unknown at the time, he was given about 15 minutes at the end of a session, and few of the better-known scientists were there. Among the exceptions was Matthew Meselson, who recognized the importance of Nirenberg's work and prevailed upon the organizers of the conference to grant Nirenberg another hearing. This time, the repercussions were intense.

Nirenberg, together with postdoctoral fellow Johann Matthaei, had been interested in the mechanism of translation of messenger RNAs, a concept just becoming formulated, into proteins and they had set up a cell-free system to study this process. Such systems, which had been devised by a number of other researchers, included ribosomes; a mix of low-molecular-weight RNAs, probably mostly tRNAs; and an ATP-driven energy-generating system. The low-molecular-weight RNAs could be shown, upon stimulation with viral or crude mRNA, to drive the incorporation of amino acids into protein.

Nirenberg and Matthaei simplified this experiment: they used synthetic polynucleotides containing only one base, added different radiolabeled amino acids to the reaction mixture, and asked which amino acids were incorporated into polypeptides for each polynucleotide message. Homopolyribonucleotides were available as a consequence of Severo Ochoa's development of their enzymatic synthesis in the late 1950s. Their first clear success was the demonstration that poly(U) led, overwhelmingly, to the formation of polyphenylalanine. Thus, codons composed only of U coded for phenylalanine. It is important to note that this experiment gave limited information: it did not say that the codon was UUU, because it had not been shown at this date that the code was triplet. It was silent on the question of punctuation. Rather, the Nirenberg-Matthaei experiment opened the first clear path as to how the code could be studied experimentally. From all accounts, the insiders in the field were intensely irritated at being beaten to this important discovery by relatively unknown researchers.

The Nirenberg-Matthaei approach became much more powerful when synthetic polynucleotides of more complex but regular sequence became available, mainly through the work of Har Gobind Khorana at the University of Wisconsin. As is shown in Figure 1, a homogeneous polynucleotide $(A)_n$ can code only for a polypeptide that is composed of a single amino acid, a repetition of a dinucleotide sequence $(AB)_n$ produces a product of alternating amino acids, and a repetition of a trimer $(ABC)_n$ yields three different homogeneous polypeptides simultaneously. Results of this kind firmly established that the code was triplet and unpunctuated. At the same time, Francis Crick and Sydney Brenner were answering the same question by purely genetic experiments with bacteriophage. In brief, they showed that making one- or two-base-pair insertions or deletions in the genome produced mutant phage because the reading frame for a gene was shifted, while making three closely spaced insertions or deletions led to something very close to the wild-type phenotype.

The genetic code was completely deciphered only in 1967, facilitated by another technical breakthrough by Nirenberg. He discovered that a single trinucleotide could attach to a ribosome and trap the appropriate tRNA. This tRNA would not pass an ultrafine filter and could thus be identified.

It was fitting that the 1968 Nobel Prize in Physiology or Medicine was awarded to Nirenberg, Khorana, and Robert Holley for breaking the genetic code. Holley had determined the sequence of the first tRNA. It is interesting to note that none of the three could be considered members of the inner circle that had dominated the earliest years of molecular biology.

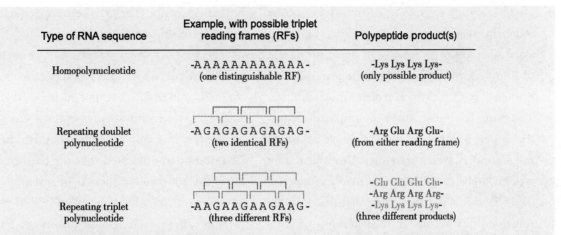

Figure 1 **Properties of the genetic code.** Experiments that used synthetic polynucleotides as templates for protein synthesis in cell-free systems demonstrated that the code was triplet, non-overlapping, and unpunctuated. They also allowed assignments of many codons to specific amino acids.

Table 5.2 **The standard genetic code.** Some characteristic features are as follows. (a) The code is unique. Each codon specifies only one amino acid in a particular organism. (b) The code is degenerate. One amino acid may be specified by multiple codons; for example, serine has six codons, glycine has four, etc. (c) The first two nucleotides are often enough to specify a given amino acid; for example, serine is specified by UC. (d) Codons with similar sequences specify amino acids of similar chemical properties. Codons for serine and threonine differ in the first letter; codons for aspartic acid and glutamic acid differ in the third letter. This ensures that many mutations will result in the incorporation of a similar amino acid that would not significantly affect the structure/function of the protein. (e) There are three stop codons that define the end of the polypeptide chain, and there is an initiation codon, AUG, that determines the point in the mRNA which codes for the first amino acid.

First position (5')	Second position				Third position (3')
	U	C	A	G	
U	Phe	Ser	Tyr	Cys	U
	Phe	Ser	Tyr	Cys	C
	Leu	Ser	Stop	Stop	A
	Leu	Ser	Stop	Trp	G
C	Leu	Pro	His	Arg	U
	Leu	Pro	His	Arg	C
	Leu	Pro	Gln	Arg	A
	Leu	Pro	Gln	Arg	G
A	Ile	Thr	Asn	Ser	U
	Ile	Thr	Asn	Ser	C
	Ile	Thr	Lys	Arg	A
	Met	Thr	Lys	Arg	G
G	Val	Ala	Asp	Gly	U
	Val	Ala	Asp	Gly	C
	Val	Ala	Glu	Gly	A
	Val	Ala	Glu	Gly	G

Several points become obvious:
- All 64 codons are used but three codons-UGA, UAA, and UAG-are commonly used as stop codons to tell the ribosome to stop reading the RNA message and release the polypeptide. In some organisms, these three codons can also code for amino acids (Table 5.3).
- There is a start signal, which always codes for methionine, but not all methionine codons indicate the start of a message that is to be translated into a polypeptide.
- The code is degenerate: most amino acids have more than one codon, with tryptophan and methionine being the only exceptions.

- Degeneracy lies mainly in the third letter of the codon; in other words, the first two letters of codons alone often specify the amino acid. Again, the insight of Francis Crick is important; he proposed the wobble hypothesis, which explains much of this.
- The code is almost, but not quite, universal. Exceptions to the general codon usage appear in mitochondria, which have their own translation systems, and also in a few primitive organisms (see Table 5.3). It is not surprising that the code has remained nearly constant over billions of years of evolution, as any major change would have destroyed most cellular proteins. In fact, it seems remarkable that some changes have been tolerated.

5.2.3 Discovery from the eukaryotic cell: introns and splicing

By 1967, the fundamentals of molecular biology seemed to be firmly established, at least insofar as bacteria and viruses were concerned. Many researchers then turned their attention to the molecular biology of eukaryotes, confidently expecting more of the same. But the great shock was the discovery, first by Phillip Sharp and Richard Roberts in 1977, that the mRNA found in the cytoplasm of eukaryotic cells need not necessarily be complete copies of the genes in the nuclear DNA. Specifically, it was found that within the genes there were stretches of DNA sequence that had no counterparts in the cytoplasmic mRNAs. These intervening sequences, later called introns, did not correspond to any part of the protein sequence. They lay between sequences that did code for protein, now called exons.

The first discovery of introns was in the genes of adenovirus, a virus that infects human respiratory cells, and so it was suspected at first that introns might be some idiosyncrasy of such viruses. This expectation was soon completely dashed when Pierre Chambon demonstrated that the chicken ovalbumin gene contained no less than seven intervening sequences, along with eight exons (Figure 5.3). Furthermore, the total length of introns was far greater than the total length of the exons that contained the cytoplasmic message. Soon, other eukaryotic genes were found to contain introns. We now know that eukaryotic genes without introns are in the minority.

Clearly, something special must take place in the eukaryotic nucleus to prepare mRNA from which the introns have been removed. In subsequent years, an elegant process of splicing was discovered. The gene, it turns out, is first completely transcribed, introns and all, to make a pre-mRNA. This is precisely cut and resealed to make a transcript containing only exon regions, in the proper order. After some further adjustments to the 5′- and 3′-ends, the mature message is delivered to the cytoplasm for translation.

(1) Interspersed elements are primarily transposable elements

Interspersed elements are mostly transposons. The existence of discrete, independent, mobile sequences, known as transposable elements or transposons, was

Table 5.3 **Alternative codon usage.**

First Codon	Codon Common use	Alternative use
CUU/C/G/A	Leu	Thr in yeast mitochondria
AGA, AGG	Arg	Stop in yeast and vertebrate mitochondria
AUA	Ile	Met in mitochondria of yeast, *Drosophila*, and vertebrates
CGG	Arg	Trp in some plant mitochondria
UGA	stop	Trp in mycoplasma and some plant mitochondria; alternatively selenocysteine, Sel, in a few taxa
UAA	stop	Gln in some protozoa
UAG	stop	Gln in some protozoa; pyrrolysine (Pyr) in some bacteria and archaea
CUG	Leu	Ser in *Candida albicans*

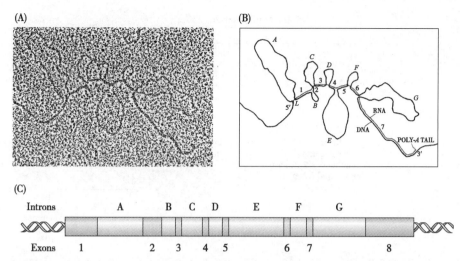

Figure 5.3 **The ovalbumin gene and its mRNA illustrate the concept of split genes.** The protein is 386 amino acids long and could have been encoded by a gene of 1158 base pairs, but the length of the gene is actually 7700 base pairs. (A) This anomaly was explained when ovalbumin mRNA was purified from the cytoplasm and allowed to hybridize with the ovalbumin gene, and the resulting hybrid was examined under the electron microscope or EM. (B) In this schematic representation of the EM image the black line shows the DNA; this is the complete gene including the introns (A to G), regions that are not present in the cytoplasmic mRNA. The mRNA is shown in red; it includes a 5′ leader region (L) and seven exons (labeled 1 to 7). (C) A linear representation of the ovalbumin gene, with the leader plus 7 exons shown in red, and the introns labeled A to G shown in gray. (Adapted from Chambon P [1981] *Sci Am*,244:60–71. With permission from Scientific American.)

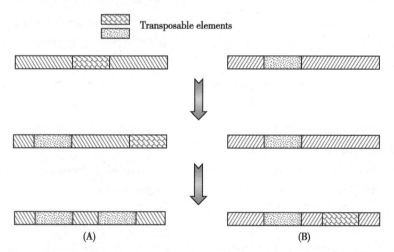

Figure 5.4 **Transposable elements.** Transposable elements are DNA sequences that have the ability to move to a new location in the genome within their cell of origin, sometimes leaving a copy behind. (A) Replicative transposition; (B) conservative transposition.

discovered in maize, Zea mays, in 1948 by Barbara McClintock. It permits a DNA sequence to be lifted or copied from one site and placed into another that exhibits no homology to the original location (Figure 5.4).

Transposons are abundant, scattered throughout the genomes of many plants and animals, and can constitute a considerable portion of the DNA. In human cells, transposons occupy ~32% of the genome (Figure 5.5). But in bacteria, transposons comprise a much smaller portion of the genome and constitute ~0.3% of the genome. The frequency of transposition varies among the various transposable elements, usually between 10^{-3} and 10^{-4} per element per generation. This is higher than the spontaneous mutation rate of 10^{-7}–10^{-5}.

Interspersed repetitive DNA elements are further subdivided into two categories based on their length.

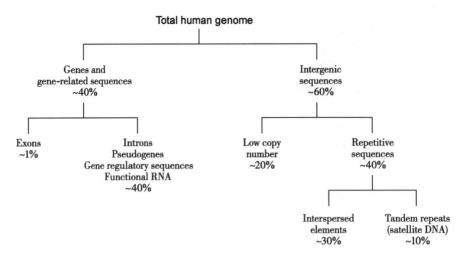

Figure 5.5 **Approximate distribution of DNA functions in the human genome.** Note that the DNA that actually codes for proteins is a very small fraction of the genome. The function of low-copy-number intergenic sequences is not well understood. Interspersed elements are mostly transposons. Satellite DNA can be further subdivided into mini- and microsatellites.

Sequences of fewer than 500 bp are called SINEs (short interspersed nuclear elements) and sequences of 500 bp or more are called LINEs (long interspersed nuclear elements).

By inserting themselves near genes, transposable elements have many different influences on gene regulation, including the potential to rapidly give rise to new regulatory sequences. In general, transposons create rearrangements of the genome. In doing this, they may cause deletions or inversions during the transposition event itself. In addition, they serve as a substrate for HR systems since transposition creates multiple copies on the same or different chromosomes.

(2) Tandem repetitive sequences are arranged in arrays with variable numbers of repeats

About 10% of the total DNA is present as short tandem repeats, also referred to as satellite DNAs (Figure 5.5). These can be further subdivided into minisatellites or short tandem repeats (STRs), which are sequences of 10–15 base pairs repeated up to ~1000-fold, and microsatellites, which are segments of 2–6 bp present in up to 100 copies. These sequences typically are organized as large clusters in the heterochromatic regions of chromosomes, near centromeres and telomeres. They are also found abundantly on the Y chromosome.

Satellite DNA is so named for the following reason: it can be identified as satellite bands, separated from the bulk DNA, in density-gradient equilibrium sedimentation (Figure 5.6). If whole DNA is cleaved with a restriction nuclease that does not cut in the repetitive sequence, the stretches of repeats produced will often have an overall base composition very different from the bulk: consider an $(AT)_n$ repeat, for example. DNA density depends on base composition, so such DNA will band separately from the bulk (Figure 5.6).

Analysis of microsatellites is being widely used in forensics (see Box 5.2).

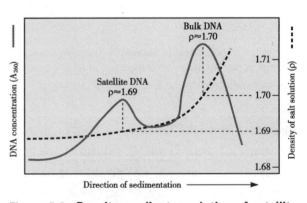

Figure 5.6 **Density-gradient resolution of satellite DNA.** The density gradient is created within the ultracentrifuge cell by the sedimentation-diffusion equilibrium of a dense salt such as CsCl. DNA will stably band at the position where its density matches that of the solution. In this case, an AT-rich satellite bands at lower density than bulk DNA.

Box 5.2 Microsatellites and DNA identification

Certain features of microsatellite DNA have made this genome component especially useful in the identification of individuals from samples of DNA, known as DNA fingerprinting. A particular microsatellite will often differ from one individual to another, especially in the number of repeats. This is believed to result from slippage during DNA replication, because the DNA replication machinery has a hard time moving continuously on repetitive templates. Such satellites are often heterozygous, meaning that the two copies in a diploid individual differ in copy number. These differences are easily detected by gel electrophoresis as shown in Figure 1, which depicts a hypothetical forensic application of the method.

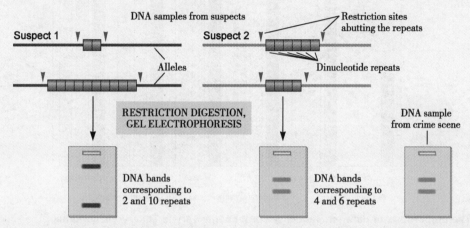

Figure 1 **An oversimplified explanation of how measurements of dinucleotide-repeat lengths in microsatellites can be used in criminal investigations.** Suspect 1 is very likely innocent; suspect 2 may be guilty. Similar tests at other loci are needed to provide definitive evidence.

(3) Genes are generally coded by nonrepetitve DNA

Nonrepetitive DNA consists of sequences that are unique: there is only one copy in a haploid genome.

The proportion of the genome occupied by nonrepetitive DNA varies widely among taxonomic groups. Figure 5.7 summarizes the genome organization of some representative organisms. Prokaryotes contain nonrepetitive DNA almost exclusively. For unicellular eukaryotes, most of the DNA is nonrepetitive: less than 20% fall into one or more moderately repetitive components. In animal cells, up to half of the DNA is represented by moderately and highly repetitive components. In plants and amphibians, the moderately and highly repetitive components can account for up to 80% of the genome, so that the nonrepetitive DNA is reduced to a small component.

What type of DNA corresponds to polypeptide-coding genes? Reassociation kinetics typically shows that mRNA is transcribed from nonrepetitive DNA. Therefore, the amount of nonrepetitive DNA is a better indication of the coding potential than is the size of the genome.

The proportions of these types of sequences are characteristic for each genome, although larger genomes tend to have a smaller proportion of nonrepetitive DNA. Almost 50% of the human genome consists of repetitive sequences, the majority corresponding to transposon sequences. Most structural genes are located in nonrepetitive DNA. The amount of nonrepetitive DNA is a better reflection of the complexity of the organism than the total genome size; the greatest amount of nonrepetitive DNA in genomes is about 2×10^9 bp.

(4) The conservation of exons can isolate genes

Some major approaches to identifying eukaryotic protein-coding genes are based on the contrast between the conservation of exons and the variation of introns. In a region containing a gene whose function has been conserved among a range of species, the sequence representing the polypeptide should have two distinctive properties: (a) It must have an open reading frame. (b) It is likely to have a related (orthologous) sequence in other species.

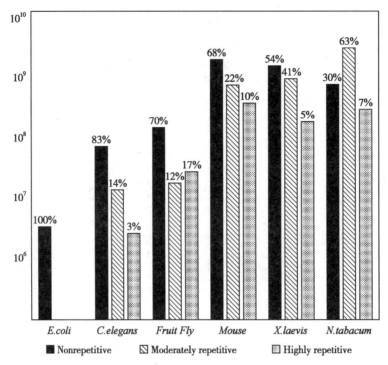

Figure 5.7 **The proportions of different sequence components vary in eukaryotic genomes.** The absolute content of nonrepetitive DNA increases with genome size but reaches a plateau at about 2×10^9 bp.

Researchers can use these features to identify functional genes. After we have determined the sequence of a genome, we still need to identify the genes within it. Coding sequences represent a very small fraction of the total genome. Potential exons can be identified as uninterrupted ORFs flanked by appropriate sequences. What criteria need to be satisfied to identify a functional (intact) gene from a series of exons? Figure 5.8 shows that a functional gene should consist of a series of exons in which the first exon (containing an initiation codon) immediately follows a promoter, the internal exons are flanked by appropriate splicing junctions, and the last exon has the termination codon and is followed by 3′ processing signals; therefore, a single ORF starting with an initiation codon and ending with

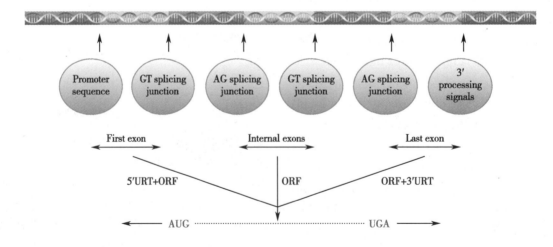

Figure 5.8 **The composition of a functional gene.** Exons of protein-coding genes are identified as coding sequences flanked by appropriate signals (with untranslated regions at both ends). The series of exons must generate an ORF with appropriate initiation and termination codons.

a termination codon can be deduced by joining the exons together. Internal exons can be identified as ORFs flanked by splicing junctions. In the simplest cases, the first and last exons contain the beginning and end of the coding region, respectively (as well as the 5′and 3′untranslated regions). In more complex cases, the first or last exons might have only untranslated regions and can therefore be more difficult to identify.

(5) Human genome

1) The number of human genome. The human genome was the first vertebrate genome to be sequenced. This massive task has revealed a wealth of information about the genetic makeup of our species and about the evolution of genomes in general. Our understanding is deepened further by the ability to compare the human genome sequence with other sequenced vertebrate genomes. Mammal genomes generally fall into a narrow size range, averaging about 3×10^9 bp.

The haploid human genome contains 22 autosomes plus the X and Y chromosomes. The chromosomes range in size from 45 to 279 Mb, making a total genome size of 3235 Mb (about 3.2×10^9 bp). On the basis of chromosome structure, the genome can be divided into regions of euchromatin (containing many functional genes) and heterochromatin, with a much lower density of functional genes. The euchromatin comprises the majority of the genome, about 2.9×10^9 bp. The identified genome sequence represents more than 90% of the euchromatin. In addition to providing information on the genetic content of the genome, the sequence also identifies features that may be of structural importance.

2) 1% of the human genome consists of coding regions. Figure 5.5 shows that a very small proportion (about 1%) of the human genome is accounted for by the exons that actually encode polypeptides. The introns that constitute the remaining sequences of protein-coding genes bring the total of DNA involved with producing proteins to about 25%. As shown in Figure 5.9, the average human gene is 27 kb long with nine exons that include a total coding sequence of 1340 bp. Therefore, the average coding sequence is only 5% of the length of an average protein-coding gene.

3) The number of gene is expressed. Two independent sequencing efforts for the human genome produced estimates of 30 000 and 40 000 genes, respectively. One measure of the accuracy of the analyses is whether they identify the same genes. The surprising answer is that the overlap between the two sets of genes is only about 50%, as summarized in Figure 5.10. An earlier analysis of the human gene set based on RNA transcripts had identified about 11 000 genes, almost all of which are present in both the large human gene sets, and which account for the major part of the overlap between them. So there is no question about the authenticity of half of each human gene set, but we have yet to establish the relationship between the other half of each set. The discrepancies illustrate the pitfalls of large-scale sequence analysis! As the sequence is analyzed further (and as other genomes are sequenced with which it can be compared), the number of actual genes has declined, and is now estimated to be about 20 000.

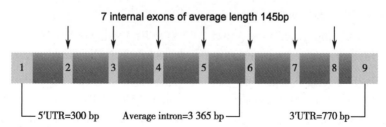

Figure 5.9 **The length and composition of the average human gene.** The average human gene is 27 kb long and has 9 exons usually comprising 2 longer exons at each end and 7 internal exons. The UTRs in the terminal exons are the untranslated (noncoding) regions at each end of the gene. (This is based on the average. Some genes are extremely long, which makes the median length 14 kb with 7 exons.)

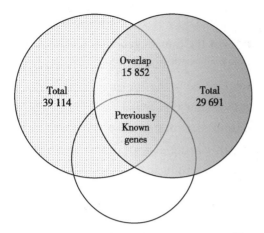

Figure 5.10 **The overlap in human genome.** The two sets of genes identified in the human genome overlap only partially, as shown in the two large upper circles. However, they include almost all previously known genes, as shown by the overlap with the smaller, lower circle.

The number of protein-coding genes is less than the number of potential polypeptides because of mechanisms such as alternative splicing, alternate promoter selection, and alternate poly(A) site selection that can result in several polypeptides from the same gene. The extent of alternative splicing is greater in humans than in flies or worms; it affects more than 60% of the genes (perhaps more than 90%), so the increase in size of the human proteome relative to that of the other eukaryotes might be larger than the increase in the number of genes. A sample of genes from two chromosomes suggests that the proportion of the alternative splices that actually result in changes in the polypeptide sequence is about 80%. If this occurs genome-wide, the size of the proteome could be 50 000 to 60 000 members.

4) **"chip" technology allows to measure the expressed gene number.** Recent technology allows more systematic and accurate estimates of the number of expressed protein-coding genes. One powerful technology uses chips that contain microarrays, which are arrays of many tiny DNA oligonucleotide samples. Their construction is made possible by knowledge of the sequence of the entire genome. In the case of *S. cerevisiae*, each of 6181 ORFs is represented on the micro-array by twenty 25-mer oligonucleotides that perfectly match the sequence of the mRNA and 20 mismatched oligonucleotides that differ at one base position. The expression level of any gene is calculated by subtracting the average signal of a mismatch from its perfect match partner. The entire yeast genome can be represented on four chips. This technology is sensitive enough to detect transcripts of 5460 genes (about 90% of the genome) and shows that many genes are expressed at low levels, with abundances of 0.1 to 0.2 transcript/cell. (An abundance of less than 1 transcript/cell means that not all cells have a copy of the transcript at any given moment.)

The extension of this and newer technologies (e.g., deep RNA sequencing) to animal cells will allow the general descriptions based on RNA hybridization analysis to be replaced by exact descriptions of the genes that are expressed, and the abundances of their products, in any particular cell type. A gene expression map of D. melanogaster detects transcriptional activity in some stage of the life cycle in almost all (93%) of predicted genes and shows that 40% have alternatively spliced forms.

5.2.4 *Genes from a new and broader perspective*

(1) Protein-coding genes are complex

A generalized structure of a eukaryotic protein-coding gene is presented in Figure 5.3. In addition to the introns found in the pre-messenger RNA, there are regions remaining in the mature mRNA that do not code for amino acids. These are appropriately called untranslated regions (UTR), 5' or 3' depending on their location with respect to the coding sequence, and they play a role in the regulation of translation of the message (Figure 5.11). Each gene also contains a defined site where the actual process of transcription is initiated and has an associated promoter sequence, which is the binding site for RNA polymerases. Recognition of the promoter regions by the polymerases could be either direct, as in bacteria, or indirect, with a mechanism that involves general transcription factors, as in eukaryotes. Other regions that are associated with protein-coding genes are

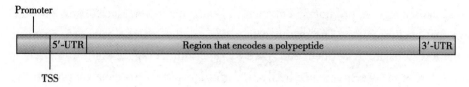

Figure 5.11 Generalized structure of a protein-coding gene. In addition to the portion of the gene that directs synthesis of the respective polypeptide, known as the coding region, each gene has a transcription start site, TSS, and usually two regions, one on each side of the coding region, that are present in the mRNA but are not translated into proteins; these are called untranslated regions or UTRs and perform regulatory functions during protein synthesis. Note that in eukaryotes the majority of protein-coding genes contain introns, as illustrated in Figure 5.3. Finally, each gene must have a promoter, a sequence upstream of the TSS that serves as a landing platform for the RNA polymerase.

called enhancers and silencers; these are involved in the regulation of gene expression even though they may reside at considerable distance from the coding sequence. Thus, a gene cannot be considered as just the string of bases that codes for a protein sequence.

(2) Genome sequencing has revolutionized the gene concept

The entire genetic content of an organism is referred to as its genome, just as the total list of proteins is its proteome. Modern, ultrafast techniques have allowed the sequencing of many whole genomes, in particular the human genome. This, in turn, has completely upset our established notions of what to call a gene. New discoveries of the past decade have made the old molecular definition of a gene obsolete.

The major contribution to analysis of the human genome and its functional elements came from the Encyclopedia of DNA Elements or ENCODE project consortium. The consortium operationally defines a functional element of DNA as a "discrete genome segment that encodes a defined product (for example, protein or non-coding RNA) or displays a reproducible biochemical signature (for example, protein binding or a specific chromatin structure)." The elements mapped across the human genome include all regions transcribed into RNA, protein-coding regions, transcription-factor binding sites, features of chromatin such as DNase I hypersensitivity and histone modifications, and DNA methylation sites. Data production efforts focused on two sets of human cell lines, designated tier 1 and tier 2. Tier 1 cell types were the highest priority set and comprised three widely studied cell lines, two malignant and one embryonic stem cell line; tier 2 included two further malignant cell lines and primary nontransformed cells from umbilical vein endothelium.

ENCODE identified 20 687 protein-coding genes, with an indication that some additional protein-coding genes remain to be found. These genes constitute less than 3% of the entire genome. This proportion is much smaller than had been expected and made it impossible to continue to think that the major role of DNA is to code for proteins, especially since ENCODE data indicate that around 80% of the entire human genome is transcribed. A very large fraction of the human genome is, instead, being transcribed into a number of classes of RNA molecules, most of presently unknown function. The number of small, <200 nucleotide RNA molecules has been estimated at 8801 and the number of long noncoding RNAs (lncRNAs) at 9640. LncRNAs are generated through mechanisms and pathways similar to those that transcribe protein-coding genes, including use of the RNA polymerase specific to protein-coding genes (RNAP II). The project also annotated 11 224 pseudogenes. These are exact or nearly exact copies of genes that are not expressed as proteins. Unexpectedly, and contrary to our present understanding of pseudogenes, a substantial number, 863 or about 7% of these, are transcribed and associated with active chromatin features.

Given this new information, we no longer have a widely accepted definition for a gene. It is clear that the definition must include the ability of a gene to direct the production of either functional RNA or protein molecules. One of the main unresolved issues is whether or not to include in the definition of a gene the regulatory regions-promoters, enhancers, and so on-that accompany it. The problem is that most of these regions are ill-defined, and there is no clear-cut one-to-one correspondence between a single gene and a particular regulatory region. In other words, the same regulatory region can control the transcription of numerous genes; and vice versa, the same gene may be regulated by numerous regions, often at huge distances from each other and from the gene.

(3) Mutations, pseudogenes, and alternative splicing all contribute to gene diversity

The continued evolution of eukaryotes, with ever more diversified cell types and capabilities, has required the development of new protein-coding genes or modifications of old ones. As shown later (Table 5.2), progression from the level of yeast to humans has involved an approximate four-fold increase in the number of protein coding genes. A number of mechanisms can account for this. Obviously mutation is one, but in itself mutation is limited in potential. Frameshift mutations will almost invariably lead to a nonfunctional product. Point mutations in themselves may change the function or properties of a gene product, sometimes in ways that will be evolutionarily favored. However, point mutations will not increase the number of genes; they simply substitute one gene for another of similar or identical function.

Much effective evolution comes about through gene duplication. Mechanisms exist that can result in the duplication of a gene or even a contiguous set of genes. Initially, gene duplication may mean only that extra copies of the gene are present to produce more mRNA. This seems to be the case for the repeated copies of the sets of histone genes (Figure 5.12A), which may be an adaptation to the need for a great burst in histone synthesis to occur concomitantly with DNA replication.

Alternatively, the new copy might not be expressed, perhaps because it lacks a functional promoter or translational controls; it then becomes a pseudogene (Figure 5.13). Pseudogenes can also arise from reverse transcription of processed RNA by the enzyme reverse transcriptase. In either event, mutations can accumulate in the new DNA sequence. If this sequence is, or becomes, expressible, it may now serve a varied or even completely different function. An example of this is seen in Figure 5.12B, which shows the β-globin genes. Copies of an original globin gene have diverged, producing variants of hemoglobin that can function best at different developmental stages.

In recent years, it has been discovered that considerable diversity in proteins is a consequence of alternative splicing. This process, whereby exons are spliced together in the nucleus to produce the mature mRNA, can be regulated so as to allow alternative ways to put the exons together or to omit some entirely (Figure 5.14). An enormous variety of similar but functionally different proteins can be produced in this manner.

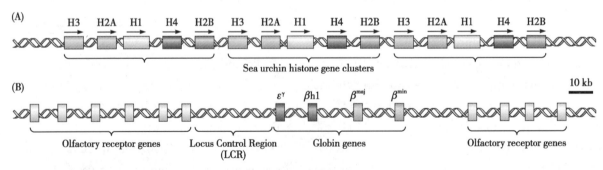

Figure 5.12 **Multigene families consist of either identical or similar member genes.** (A) Histone gene family in sea urchin. The genes for all five histones that participate in organizing chromatin structure are clustered, and the clusters are tandemly repeated numerous times; the number of repeats differs from species to species. (B) Organization of mouse globin genes. These genes differ slightly from each other; they encode slightly different globin proteins, which give rise to slightly different oxygen-bearing hemoglobin molecules. Expression of each of these genes is developmentally regulated during embryogenesis and in the adult organism, with the locus control region (LCR) playing a major role in this regulation.

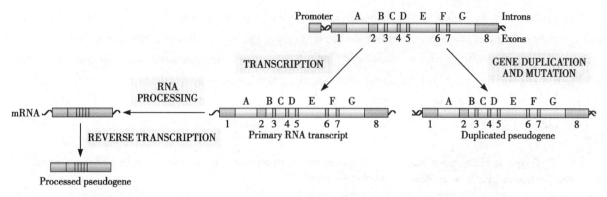

Figure 5.13 Pseudogenes. Pseudogenes are DNA sequences that are very similar to real, functional genes but lack the sequences necessary for gene expression. Duplicated pseudogenes arise by gene duplication events followed by mutations. Processed pseudogenes are formed by reverse transcription and mutation: the primary transcript is first processed to mature mRNA, which is then reverse-transcribed into double-stranded DNA.

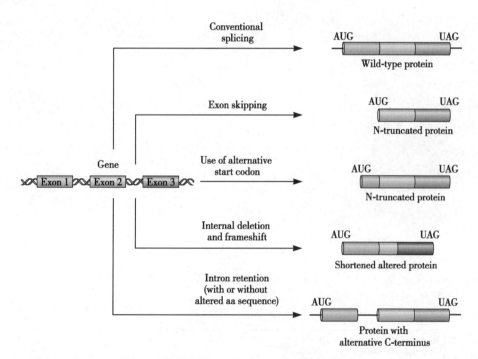

Figure 5.14 Possible pathways of alternative splicing. A hypothetical gene comprising three exons and two introns is shown. Note the different protein products of each pathway.

5.2.5 Comparing whole genomes and new perspectives on evolution

With the development of ultrafast DNA sequencing, it has become possible to determine the entire genome sequences of many organisms. Comparison of even the gross features of these sequences has important implications with respect to evolution. The first genomes to be studied were those of viruses and bacteria, with the landmark publication of the *Escherichia coli* genome, sequenced by Ying Shao and collaborators, in 1997. Many other bacterial genomes have since been sequenced. Extension to the simpler eukaryotes, such as yeast, soon followed, and as of 2013, complete genome sequences are known for a few hundred eukaryotic species, among them at least 130 fungi, 30 insects, 7 nematodes, 21 plants, and 27 mammals, including humans (Figure 5.15 and Table 5.4). The range in genome size is great, even among eukaryotes, with the human genome being 200-fold larger than that of yeast. On the other hand, most metazoans seem to have roughly the same number of protein-coding genes, about 20 000–30 000.

Examination of Table 5.4 raises some curious questions. For example, humans have about the same number of protein-coding genes as the worm *Caenorhabditis elegans*. But humans have about 30 times more total DNA. What are we doing with all that extra DNA that the worm does not seem to need? One postulate is that extra information is needed to program the much more complex development and body plan of humans, and this information is carried in DNA that does not code for proteins. From this perspective, it seems naive to have thought that the major function of genes was to code for proteins. It

Table 5.4 **Genome sizes, gene numbers, and chromosome numbers for several selected organisms whose genome sequences have been completely determined.**

Species	Genome size (× million base pairs)	Number of protein-coding genes	Number of chromosomes
phage λ	0.048	73	1[a]
human immunodeficiency virus	0.009 75	9	2[b]
Haemophilus influenzae	1.83	1743	1
Bacillus subtilis	4.22	4422	1
Escherichia coli	4.6	4288	1
Saccharomyces cerevisiae	13.5	5882	16[c]
Saccharomyces pombe	12.5	4929	3[c]
Caenorhabditis elegans	100	18 424	12
Drosophila melanogaster	139	15 016	8
Arabidopsis thaliana	119	25 498	10
Oryza sativa, rice	390	37 544	24
Mus musculus, mouse	2500	23 786	40
Homo sapiens	3200	20 687	46

[a]Linear double-stranded DNA molecule.
[b]Single-stranded RNA molecules.
[c]Haploid.

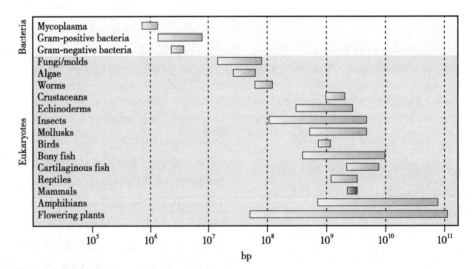

Figure 5.15 **Variation in genome sizes among different taxa.** Note the tremendous variation of genome sizes among eukaryotes and the relatively small range of variation among mammals, shown as a red box. On the basis of completely sequenced genome data as of mid-2009, gene number correlates linearly with genome size in viruses, bacteria, archaea, and eukaryotic organelles, whereas there is no apparent correlation in eukaryotes. Moreover, the number of genes does not correlate with the complexity of the organism.

is similar to thinking that the complete prescription of a factory is given by descriptions of the machines. How they are placed and interconnected, and how they are coordinately controlled, is essential to describing the whole and its functioning.

5.3 Physical structure of the genomic material

5.3.1 Prokaryotic and viral genome

(1) Bacterial genome organization

Most but not all bacteria possess a single, larger circular chromosome, often accompanied by one or more small circular plasmids. The large chromosome is free in the cell cytoplasm, as bacteria have no morphologically defined nucleus. The cytoplasmic complex of DNA and proteins is called the nucleoid (Figure 5.16). It must be tightly folded to fit within the bacterium. For *E. coli*, the genome is 4700 kb in size and exists as one double-stranded circular DNA molecule, with no free 5′ or 3′ ends. The unfolded circumference of the duplex DNA circle would be ~1.6 mm, whereas the dimensions of the bacterium are around 2 μm. This corresponds to a packing ratio of about 1000.

A major factor contributing to compaction is supercoiling of the bacterial DNA. The bacterial DNA is highly negatively supercoiled. The equilibrium superhelical density is about −0.06; given a genome size of 4.6×10^6 bp, or 4.6×10^5 turns of DNA, this indicates about 3×10^4 superhelical twists. This can be seen *in vitro* as a highly branched plectonemic writhing of the DNA. Whether this is the exact mode of supercoiling *in vivo* is unknown. The

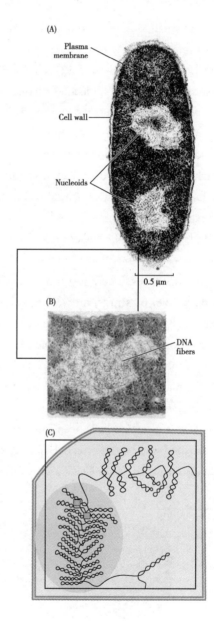

Figure 5.16 **The bacterial nucleoid.** (A) *E. coli* under the electron microscope. The cell has just undergone DNA replication and consequently has two nucleoids. (B) Transmission electron membrane Cell wall microscopy, TEM: osmium fixation of *E. coli* shows a confined nucleoid containing a network of randomly oriented DNA fibers. (C) Schematic representing one of the four DNA topological macrodomains in the nucleoid of *E. coli* and an adjacent less-structured region. Large portions of the circular chromosome occupy the same cellular space, forming macrodomains that are separated by less-structured regions; these regions prevent the domains from colliding. The DNA in the macrodomains is highly packed through interactions with specific protein factors, shown as green squares. (A, from Menge B, Wurtz M. With permission from Photo Researchers, Inc. B, from Eltsov M, Zuber B [2006] *J Struct Biol*,156:246–254. With permission from Elsevier. C, adapted from Boccard F, Esnault E, Valens M [2005] *Mol Microbiol*, 57:9–16. With permission from John Wiley & Sons, Inc.)

supercoiling is created and maintained by several factors. These factors include the binding of structural proteins that are a part of the bacterial nucleoid, processes such as transcription and replication, and the actions of gyrases, which are topoisomerases that introduce negative supercoiling into topologically constrained DNA molecules. Other topoisomerases regulate the level of superhelical stress.

(2) Plasmid DNA

Plasmids are small, double-stranded circular or linear DNA molecules carried by bacteria, some fungi, and some higher plants (Figure 5.17). They are extrachromosomal nucleic acid molecules carried by bacteria that replicate independently of the main bacterial chromosome. They typically range in size from 2 to 100 kb. At least one copy of a plasmid is passed on to each daughter cell during cell division. Their relationship with their host cell could be considered as either parasitic or symbiotic.

Plasmids are generally dispensable for the host cell but may carry genes that are beneficial under certain circumstances. There is a long list of genes carried by known plasmids, most of which confer antibiotic resistance or are responsible for antibiotic production. Some plasmids are capable of inducing malignant tumors in plants.

Bacterial plasmids were the first cloning vectors. The majority of plasmids are circular; however, a variety of linear plasmids have been isolated. If both strands are continuous, the molecules are described as covalently closed circles or CCC; if only one strand is intact, the molecules are known as open circles or OC. CCC extracted from cells are usually negatively supercoiled. Plasmids possess genetic elements that can maintain them in just a few copies per cell (these plasmids are known as stringent plasmids), or as multicopy populations, known as relaxed plasmids.

Plasmids that are chosen to be cloning vectors have several desirable characteristics: low molecular weight, ability to confer selectable traits on the host cell, and availability of single sites for restrictase cleavage. The first requirement stems from the ease of purifying small plasmids that remain intact during the isolation procedure. In addition, smaller plasmids are usually present in multiple copies, allowing for stable high-level expression of the genes they carry; this becomes especially important for genes that confer selectable phenotypes.

(3) Viral genome organization

Viruses need to carry in their genomes only a minimum of information: to construct particles that can enter a cell, replicate their nucleic acid and essential proteins, and then exit the cell. In many cases some information is taken from the host genome; for example, many viruses use host polymerases to replicate their genomes. Viral genomes come in a remarkable variety of forms. The polynucleotide may be DNA or RNA, and it may be single-stranded or double-stranded. The protein shell that surrounds the viral DNA or RNA, known as the capsid, is just sufficient to contain the genome, to permit the virus to get into a host cell, and to allow the newly replicated

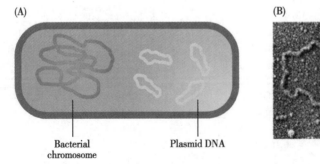

Figure 5.17 Plasmid DNA. (A) Schematic representation of a bacterium containing plasmid DNA. Plasmids are small, circular molecules of DNA that are extrachromosomal and self-replicating within the host bacterium. (B) Electron microscopic image shows the typical appearance of an isolated plasmid. (from Cohen SN [2013] *Proc Natl Acad Sci USA*, 110:15521–15529. With permission from National Academy of Sciences.)

viruses within to escape to infect other cells. These requirements are met in a host of ways in different viruses, each using minimal machinery. As shown in the example of bacteriophage φ29 (Figure 5.18), some viruses are constructed in a way that allows them to inject their DNA into a host cell; in these viruses, the DNA is tightly coiled within the capsid, and energy for injection is stored in the bending of the DNA and in electrostatic repulsion between DNA phosphates. This means that energy must be supplied to force the DNA into the capsid, for example, by ATP hydrolysis.

Other viruses, especially those that attack eukaryotic hosts, use a completely different method: they have an external membrane (envelope) that can fuse with the host's cell membrane to ensure internalization of the viral genome. The proteins or other molecules needed to accomplish this are synthesized by the host cell, using genetic instructions from the viral genome. The compaction of the nucleic acid to virus-compatible dimensions is in this case accomplished by the viral coat proteins, together with the presence of positive ions or small molecules bound to the DNA that help to neutralize the polynucleotide charge. The structural organization of the RNA genome of the human immunodeficiency virus, HIV, as a representative example of such enveloped viruses is presented in Figure 5.19. Many RNA viruses, including those that cause influenza and the common cold, are constructed in this fashion.

5.3.2 Eukaryotic chromatin

(1) Higher condensed DNA-protein complexes

In the great evolutionary leap from bacteria and/or archaea to eukaryotes, a major revision of genome structure occurred. The more complex lifestyles of higher organisms required controls that led to segregation of the genetic material in the cell nucleus and demanded more complex control of processes like transcription and replication.

The chromosomes of eukaryotes are highly condensed complexes of individual DNA molecules and proteins. During the cell cycle, eukaryotic chromosomes go through a series of biochemical processes and structural transitions. In this chapter, we are primarily interested in the interphase state of the

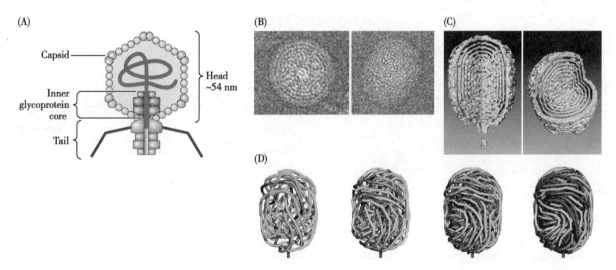

Figure 5.18 φ29 bacteriophage-like viruses. (A) Non-enveloped, head-tail structure. The tail is noncontractile and has a collar plus 12 tail fibers. The genome is double-stranded linear DNA of ~16–20 kb, encoding between 20 and 30 genes. (B) Typical cryoEM images of mature φ29 particles at ~50 000-fold magnification. (C) Three-dimensional reconstruction of the full-length φ29 packaged genome. The degree of order increases from the center toward the capsid, with six concentric DNA shells visible in longitudinal and transverse sections of the volume. (D) Monte Carlo simulation of DNA packaging in φ29: consecutive images show stages at which 30%, 50%, 70%, and 100% of the genome has been packaged. The chain is colored blue at the end entering the confinement and red at the free end within the capsid at full packaging. The capsid is not shown. (A, adapted from ViralZone, Swiss Institute of Bioinformatics. B-D, from Comolli LR, Spakowitz AJ, Siegerist CE, et al. [2008] *Virology*,371:267–277. With permission from Elsevier.)

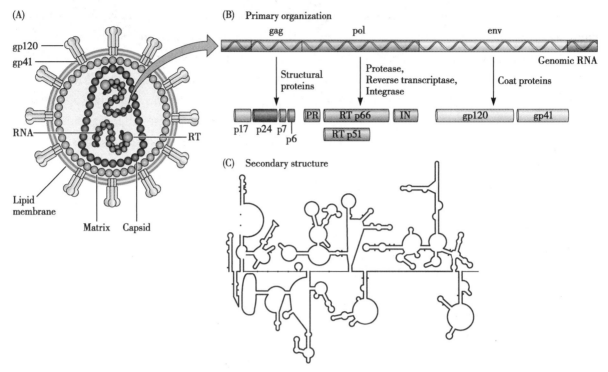

Figure 5.19 HIV and organization of its genome. (A) Anatomy of HIV. The outer coat of HIV is a layer of cell membrane stolen from the infected cell upon viral particle release. The coat is studded with protein spikes composed of two glycoproteins, gp120 and gp41, which serve to dock the virus onto receptor proteins on the surface of the host cell. This membrane layer envelops the matrix and capsid structures that encase the genetic material. The matrix proteins form a coat on the inner surface of the viral membrane and play a central role when new viruses bud from the surface of infected cells. The capsid proteins form a cone-shaped coat around the viral RNA, delivering it into the cell during infection. The proteins that form these structures are colored according to the color scheme representing the primary structure of the genome in part B. Three kinds of proteins that are essential for viral replication—reverse transcriptase (RT), integrase, and protease—are encoded by viral RNA and are present in the virions or mature viral particles. (B) Primary organization of the genetic material along the RNA molecule. (C) The secondary structure of the viral RNA is highly complex and is further coiled in three dimensions. The coiling is so dense that it was practically impossible to see any detail by electron microscopy. (C, adapted from Watts JM, Dang KK, Gorelick RJ, et al. [2009] Nature, 460:711–716. With permission from Macmillan Publishers, Ltd.)

chromosomes, because this is when gene expression and DNA replication occur.

(2) Nucleosomes

The nucleosome is the basic repeating unit of eukaryotic chromatin. It has long been known that the nucleic acid in the interphase eukaryotic nucleus forms a highly compacted complex with basic proteins, now called chromatin. The chromatin was primarily a complex of DNA and histones was the general level of understanding that existed by about 1970. As to how the histones and DNA were co-organized, a prevalent view about held that the DNA was wrapped into some kind of superhelix, which the histones coated like insulation on a wire. But in this period, research in a number of laboratories, using very different approaches, converged on a radically different model (see Box 5.3). The essence of this model is that nuclear DNA is periodically wrapped about histone cores to form particles we now refer to as nucleosomes or, more correctly, nucleosomal core particles. A major key to elucidation of this structure was the observation that cleavage of chromatin with nucleases occurred most strongly at sites about 200 bp apart. Further digestion gave the core particles as a final product. The core of each particle consists of an octamer of histones: one (H3-H4)$_2$ tetramer plus two H2A-H2B dimers (Table 5.5). The DNA wrapping is now known to be a left-handed toroid of 1.67 turns, comprising a stretch of 147 bp. A high-resolution X-ray diffraction structure of the core particle is presented in Figure 5.20.

Table 5.5 **Histones: five major classes.** The linker histones, also known as H1 histones, are a family of closely related members; they bind to the linker DNA between nucleosomal particles. The core histones derive their names from their structural role as organizers of the core particle. The sizes, amino acid compositions, and sequences of these histones vary somewhat from species to species; numbers in the table refer to bovine histones. Still, histones are among the most highly conserved proteins in evolution.

Histone	NCBI Gene ID	Number of amino acids	Number of Lys, Arg	Function
H1[a]	3005	194	56, 6	binds to linker DNA
H2A	92 815	130	13, 13	two molecules take part in formation of histone octamer, the protein core of the nucleosome
H2B	8349	126	20, 8	same as H2A
H3	8350	136	13, 18	same as H2A
H4	121 504	103	11, 14	same as H2A

[a]Member of histone H1 family.

Box 5.3 **Discovery of the nucleosomal structure of chromatin**

Some major scientific breakthroughs come not through a burst of individual genius but from the accumulation of partial leads from a number of scientists. Such was the case for the evolution, over a few years, of the nucleosome model for chromatin. It all began, about 1970, when many researchers became discontented with the kind of uniform supercoil models that had been proposed on the basis of the available electron microscopy and X-ray scattering studies. The latter could be interpreted in various ways, and electron micrographs of condensed chromatin fibers looked, as one scientist quipped, "like an explosion in a spaghetti factory."

A partial breakthrough came when two groups, one led by Chris Woodcock at The University of Massachusetts and the other by Ada and Don Olins at Oak Ridge National Laboratory, began using a new spreading technique for the preparation of electron micrograph samples. Independently, these groups observed a distinctive beads-on-a-string structure, which fit no existing model. This work was presented by both groups at a cell biology meeting in November 1973. The Olins' work was published a few months later; Woodcock's was rejected by a reviewer. Meanwhile, others, working largely without knowledge of the above results or one another's advances, were following different trails. In Australia, Dean Hewish and Leigh Burgoyne were carrying out gentle nuclease digestions of chromatin, *in situ*, which yielded a peculiar pattern of ~200 bp repeats. The laboratory of one of this book's authors, Kensal van Holde, was using nuclease digestion and analytical ultracentrifugation to reveal compact particles of chromatin, containing something over 100 bp of DNA and a roughly equal mass of histone.

Meanwhile, Roger Kornberg and Jean Thomas, at the Medical Research Council Laboratories in Cambridge, had been investigating the interactions between histones using cross-linking methods. They demonstrated the existence of the $(H3/H4)_2$ tetramer. Building upon the Hewish-Burgoyne result and the known histone stoichiometry in chromatin, Kornberg was able to postulate in 1974 a chromatin structure in which stretches of ~200 bp of DNA were each associated with histone octamers to form discrete particles. These were later termed nucleosomes. With some adjustments, the model fitted all of the other data mentioned above and provided a new paradigm for chromatin research.

Further elucidation of the structure of the particle came quickly. Electron micrographs showed individual core particles to be roughly spherical, with a diameter of ~100 Å. Since each particle must contain DNA about 400 Å long, it became obvious that DNA must be coiled. That it was coiled on the outside of the particles was strongly indicated by the DNase I digestion experiments of Marcus Noll at Cambridge University. Definitive evidence that the DNA was outside the protein core was provided by low-angle neutron scattering. Altogether, many people in many laboratories, usually working independently, produced over a period of a few years a new, coherent picture of chromatin structure.

The final definition of the core particle came from X-ray diffraction studies of nucleosome crystals. The pioneer in this work was Aaron Klug at Cambridge University, who was awarded the Nobel Prize in Chemistry in 1982, partly for the first low-resolution 25 Å study, published in 1977. Since then, a series of crystallographic studies, by several laboratories, have steadily refined the structure until we now have the elegant picture shown in Figure 5.20.

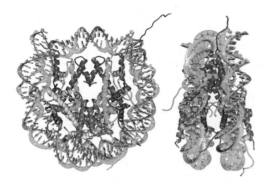

Figure 5.20 **Crystal structure of the core nucleosome at 2.8 Å resolution.** Two views perpendicular to the two-fold axis of symmetry, known as the dyad, are shown. Color scheme: H3, blue; H4, green; H2A, yellow; H2B, red. The main technical difficulty in obtaining this high-resolution structure was getting good crystals. To reduce the inherent heterogeneity in particles isolated from nuclei, it was crucial to use recombinant histones that lack post-translational modifications and are devoid of the N-terminal tails, which seem highly disorganized. The tails in the structure show the presumptive locations of some of the tails. In addition, the DNA sequence used for reconstitution was constructed of two identical halves, connected head-to-tail. (From Luger K, Mäder AW, Richmond RK, et al. [1997] *Nature*,389:251–260. With permission from Macmillan Publishers, Ltd.)

(3) Histones

Histone postsynthetic modifications create a heterogeneous population of nucleosomes. There is a very wide range of postsynthetic or post-translational modifications that can occur on each histone. These include acetylation of lysine residues, methylation of lysine or arginine, phosphorylation of serine, threonine, or tyrosine, and ubiquitylation of lysine as shown in Figure 5.21 and Figure 5.22. The majority of post-translational modifications occur in the histone tails that protrude from the core particle structure, although modifications of residues within the histone fold, which is entirely inside the particle, have been recently recognized (see Figure 5.22).

The modifications of histones in chromatin can potentially give rise to an enormous number of possible distinguishable nucleosome particles that differ in their stability and dynamic properties, and thus in their transcribability. These modifications play vital roles in the specific regulation of the transcription of different genes.

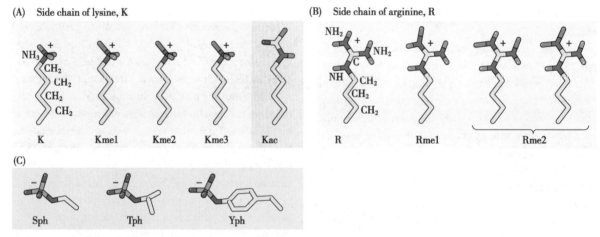

Figure 5.21 **Known histone post-translational modifications and their effect on overall properties of the modified side chain.** The different classes of modified amino acid residues are represented by stick models. Color code: yellow, carbon; blue, nitrogen; pink, polar hydrogen; red, oxygen; orange, phosphorus; green, methyl groups. Background shading denotes the charge of modified side chains at physiological pH: light blue, positive; pink, negative; light green, uncharged. (A) Lysine methylation and acetylation. Charge is ablated upon lysine acetylation, denoted as Kac, whereas all methylated forms of lysine, denoted as Kme, are positively charged or cationic at physiological pH. The incremental addition of methyl groups, shown in green, from K to Kme3 increases the hydrophobicity and the cation radius of the methyl ammonium group; the ability to donate hydrogen bonds concomitantly decreases. (B) Arginine methylation. (C) Phosphorylation of serine, threonine, and tyrosine introduces negative charge. (Adapted from Taverna SD, Li H, Ruthenburg AJ, et al. [2007] *Nat Struct Mol Biol*,14:1025–1040. With permission from Macmillan Publishers, Ltd.)

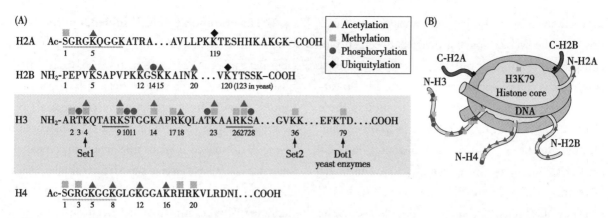

Figure 5.22 Distribution of known post-translational modifications along polypeptide chains of the four core histones. (A) Underlined sequences are motifs that are repeated either in two different molecules, underlined in blue, or within the same molecule, underlined in red. The significance of having such repeats is unclear. The best-studied modifications are those of histone H3. The enzymes that place the same modification on different residues are environment-specific: thus in yeast, different enzymes, Set1, Set2, and Dot1, modify lysines 4, 36, and 79, respectively. (B) The majority of post-translational modifications occur in the histone tails that protrude from the core particle structure, but recent studies have demonstrated the existence of modified residues within the histone fold, which is entirely inside the particle. A good example is the methylation of residue K79 in H3. Note that for clarity only half of the core histones in the particle are shown.

(4) The nucleosome family is dynamic

The level of detail shown in Figure 5.20 might be taken to imply that the nucleosome is a fixed, immutable structure. Nothing could be further from the truth. We now know that the nucleosome can undergo dynamic changes in conformation and even exists stably as a structural family that includes aberrant forms. We shall consider the structure shown in Figure 5.20 as the canonical nucleosome, because most nucleosomes, most of the time, probably look much like this. But there is now evidence that nucleosomes can undergo periods of breathing transitions and opening transitions, in which the DNA is periodically peeled back from the histone surface, to lesser or greater degrees. These changes have been detected in two ways. The treatment of nucleosomes with site-specific nucleases and single-molecule fluorescence resonance energy transfer (FRET) experiments. There also exist, under some conditions, nucleosomal particles that have unusual histone cores, such as only a $(H3/H4)_2$ tetramer. But all of the structures we do not yet know which may have physiological significance.

(5) Nucleosome assembly in vivo uses histone chaperones

Although nucleosomes can assemble spontaneously *in vitro* from DNA and histones, *in vivo* assembly and modification often utilize a number of carriers of histones, called histone chaperones. Histone chaperones have the ability to recognize specifically and bind to specific histone dimers and then to transfer them to other molecules. Importantly, chaperones are not part of the final product. Histone transfer processes include transfer from one chaperone to another; histone transfer onto DNA, known as deposition; histone removal from DNA, known as eviction; histone exchange reactions in already existing nucleosomal particles; and histone transfer to modification enzymes. Chaperones can also serve in chromatin remodeling, in histone storage, for example, in oocytes, and probably in other functions.

(6) Higher-order chromatin structure

1) Nucleosomes along the DNA form a chromatin fiber. Individual nucleosomes are organized into higher-level structures called chromatin fibers. The spacing of nucleosomes on the DNA is generally irregular. As shown by nuclease digestion experiments (see Box 5.4), the average spacing varies between species and cell types, from about 160 to 240 bp. In most cases, this is considerably larger than the 147 bp carried on the nucleosome core particle, and so there must be linker DNA between particles. This is what produces the beads-on-a-string appearance noted by early electron microscopists (see Box 5.3). This linker DNA turns out to be the habitat of the fifth class of histones, the linker histones H1. Together with the core particle, they form what is called the chromatosome. Linker histones bind to the linker DNA entering and exiting the nucleosome in a manner that stabilizes the particle (Figure 5.23). In addition, these histones seem to play important roles in the formation and stabilization of higher-order folding of the chromatin fiber. Such folding is essential to give sufficient compaction of the chromatin in the nucleus. Note that formation of a nucleosome only compacts about 40 nm of DNA into a 10 nm particle. This compaction ratio of 4:1 is far from the values in the thousands necessary for proper compaction in the nucleus.

2) The chromatin fiber is folded, but its structure remains controversial. Does the chromatin fiber have a regular structure, and if so, what is it? A host of studies using a wide variety of physical techniques, ranging from extraction of chromatin fiber from gently digested lysed nuclei to reconstruction of regularly spaced nucleosome arrays on repeated-sequence DNA, have given a superficially similar result: a more or less regular helical fiber of about 30 nm in diameter. This 30-nm fiber has been observed in a number of electron microscopic studies.

We can say a few things about the structure of chromatin fibers with some confidence:

- It probably has physiological significance, since packed structures like this can be occasionally glimpsed in sectioned nuclei.
- It is stabilized by moderately high salt and especially by divalent ions.
- It is stabilized by the linker histones.

Beyond these areas of agreement, the main arguments are on a very fundamental point: what kind of a helix is this? The original proposal, by Mellema and Klug in 1976, was for a left-handed solenoidal

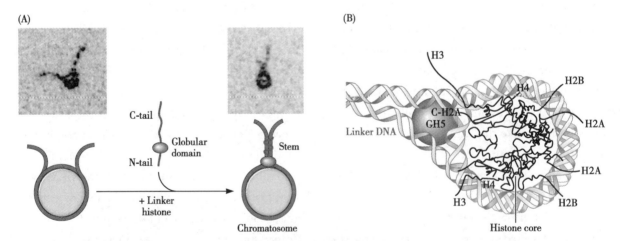

Figure 5.23 **Chromatosome: a particle containing linker histones and more than 147 bp of DNA.** (A) Linker histones, LH, bind to the nucleosome at the entry-exit point of nucleosomal DNA. Each member of the linker histone family possesses a short unstructured N-terminus, a globular domain (winged-helix motif), and a long unstructured C-tail. The binding of an isolated globular domain seals off two full turns of DNA around the histone octamer, compared to the 1.67 turn in the particle lacking LH. Interaction of the C-tail with the incoming and outgoing linker DNA creates the stem structure. The EM images above the two schematics are of nucleosomes reconstituted from naked DNA and histone octamers in the absence or presence of LHs. The stem structure has also been seen *in situ*. (B) To-scale schematic of the chromatosome based on the crystal structure of the core particle, the known dimensions of the globular domain GH5 of linker histones, and the expected trajectory of the two linker DNA segments. (A, EM images from Hamiche A, Schultz P, Ramakrishnan V, et al. [1996] *J Mol Biol*, 257:30–42. With permission from Elsevier. B, adapted from Leuba SH, Bustamante C, van Holde K, et al. [1998] *Biophys J*, 74:2830–2839. With permission from Elsevier.)

Box 5.4 Studying the physical structure of eukaryotic chromatin uses a variety of methods

Many of the methods we described in earlier chapters can be, and have been, applied to chromosome and chromatin studies. There are, however, two techniques that have been especially useful in this field, so we discuss them here.

Nuclease digestion of chromatin

The initial discovery of the nucleosome structure of chromatin involved, among other things, the use of endogenous endonucleases to digest chromatin. Later, the use of two exogenous nucleases, micrococcal nuclease (MNase) and deoxyribonuclease I (DNase I), provided a wealth of information on the organization of nucleosomes in fibers and on the internal structure of the nucleosome particle (Figure 1).

MNase cleaves both strands of DNA. It possesses both endonuclease and exonuclease activities. It first cuts chromatin in the exposed linker DNA, and then it trims the particles from the DNA ends. Unfortunately, MNase possesses some DNA sequence preference, which necessitates including chromatinfree DNA samples as controls. Partial digestion of chromatin, followed by electrophoretic analysis of purified DNA, gives rise to the famous DNA ladders, DNA fragments that are multiples of a constant length (see Figure 1A); this length, which is specific to the cell and tissue type, has been named repeat length.

The use of DNase I provided a tentative answer to the question of where the DNA is located in the particle, inside or outside. DNase I makes single-strand nicks in double-stranded DNA. When DNA from DNase I-digested core particles is analyzed by electrophoresis under denaturing conditions, another type of ladder pattern is observed: the single-strand fragments on the gels are multiples of ~10 nucleotides (see Figure 1B). The simplest explanation of this pattern is that the nuclease cuts preferentially where the DNA is maximally exposed on the nucleosome surface. This early interpretation was soon supported by other studies. Later, high-resolution electrophoretic techniques provided information on subtle differences in the core particle structure that result from the incorporation of histone variants, post-translational histone modifications, or the activity of chromatin remodelers.

Mapping nucleosome positions in chromatin

The precise way in which nucleosomes are arranged on the chromatin fiber is important both for the higher-order structure of that fiber and for the regulation of transcription of chromatin. A number of techniques have been developed for mapping nucleosome positions. As techniques have become more and more precise, it has become evident that many nucleosomes occupy well-defined locations.

The earliest studies of this kind utilized a technique called indirect end-labeling (Figure 2A). Suppose one wishes to examine a genomic region, of known DNA sequence. Whole chromatin is lightly digested—*in situ*, if desired—with MNase to produce a nucleosome ladder pattern. The purified DNA from the MNase digest is cleaved by a pair of restriction endonucleases that cut at defined sites bordering the region of interest. The DNA is then electrophoresed and blot-hybridized to a radioactive probe abutting one

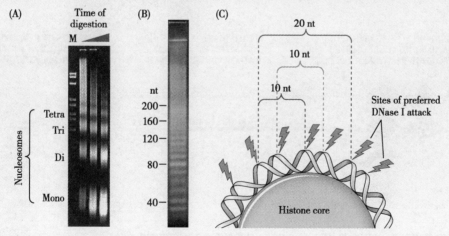

Figure 1 Nuclease digestion of chromatin. (A) MNase digestion pattern of chromatin from chicken erythrocytes: time course. The gel was stained with ethidium bromide and observed under UV light illumination. The lane marked M contains DNA restriction fragments used as size markers. (B) DNase I digestion pattern of core particles isolated from rat liver. (C) Schematic illustrating how the pattern is obtained: the nuclease cuts the DNA at sites of maximal exposure at the surface of the nucleosome. (A, from Mathews CK, van Holde KE, Appling DR, et al. [2012] Biochemistry, 4th ed. With permission from Pearson Prentice Hall. B, from Noll M [1974] *Nucleic Acids Res*, 1:1573–1578. With permission from Oxford University Press.)

end of the DNA. This will reveal only bands corresponding in length to the several oligonucleosomes in the digest that hybridize to this probe sequence. The original method can give, at best, only approximate positions, because the nuclease may cut anywhere within the linker. The precision of locating nucleosomes can be improved by using cleavage reagents that cut naked DNA at almost every base pair, and then performing the electrophoresis on long, high-resolution sequencing gels. Appropriate reagents include DNase I, the reactive intercalator methidiumpropyl-EDTA-Fe^{2+} ($MPE-Fe^{2+}$), and hydroxyl radicals generated by an iron-catalyzed redox reaction. Under optimal conditions, these can give almost singlebase resolution with short oligonucleosomes. However, there is always the possibility that a sample will contain a spectrum of closely related alternate positions, which will blur the results.

A quite different class of techniques involves primer extension, a procedure in which a synthetic oligonucleotide primer is extended by DNA polymerase, using a single-stranded DNA as a template. One such method is shown in Figure 2B.

Recent techniques allow the mapping of nucleosome positions over a whole genome. These rely on the use of microarrays covering wide genomic regions and ultrafast parallel sequencing techniques. In one application, a nuclease digest of whole chromatin is hybridized against such an array, and the captured nucleosomes are then sequenced. With current techniques, millions of nucleosomes can be sequenced and their locations can be mapped against the whole genome. Our knowledge of chromatin structure at the sequence level is suddenly becoming infinitely more comprehensive.

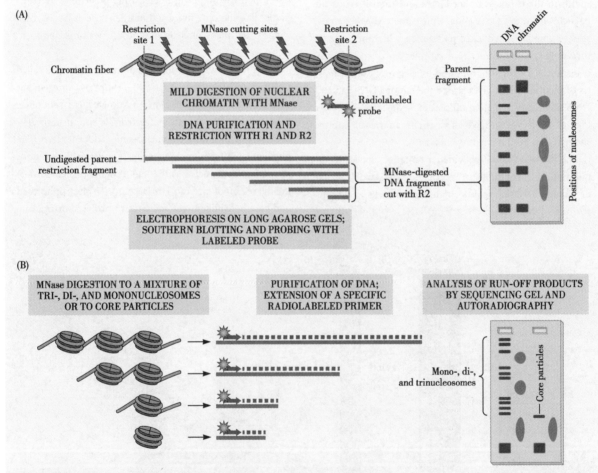

Figure 2 **Two common methods for mapping nucleosome positions *in vivo*.** (A) Indirect end-labeling; (B) primer extension. The autoradiograms of gels show idealized results that assume a single nucleosomal array on the gene of interest. Often, nucleosomes occupy slightly different positions that are spaced ~10 bp apart. (Adapted from Clark DJ [2010] *J Biomol Struct Dyn*, 27:781–793. With permission from Taylor and Francis Group.)

helix with linker DNA passing directly from one nucleosome to the next along the spiral of the helix. But others have proposed two-start helices, or helices with complex internal connections. A few of the many models that have been proposed are shown in Figure 5.24. As one can see from these models, the proposed structures will be difficult to distinguish by simple electron microscopic observation: they look pretty much alike unless observed in fine detail.

There is also the fundamental question as to whether any of the models proposed, or indeed any uniform model, has any relevance to chromatin *in vivo*. Certainly, pieces of 30-nm fiber extracted from nuclei rarely show a smooth, regular helical structure; they are lumpy and bent. This may be what we should expect, as we know that linker lengths are locally heterogeneous, and this fact should argue against any wholly regular structure.

3) The organization of chromosomes in the interphase nucleus is still obscure. Early cytological studies of interphase nuclei revealed the existence of two readily distinguishable forms of chromatin structure: highly condensed heterochromatin and more dispersed euchromatin. These distinctions are still recognized and are thought to correlate with transcriptional activity, with transcription of the condensed heterochromatin being repressed. There also seem to be differences in the overall protein composition of these regions and in the presence of specific histone post-translational modifications (see Chapter 9). Newer techniques are able to show that the linear double-stranded DNA molecules that con-

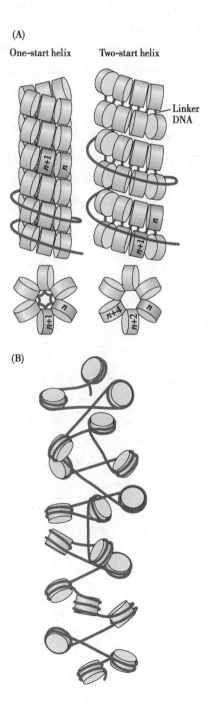

Figure 5.24 **Models depicting the possible arrangement of nucleosomes in the 30-nm fiber. Note that none of the models has been universally accepted.** (A) Side and top views of two major types of models: a one-start helix, or solenoid, and a two-start helix. A two-start helix can be visualized as follows: imagine a zigzag ribbon of nucleosomes, where linkers pass back and forth between two rows of nucleosome particles. Now twist this ribbon into a helix. Nucleosomes are represented in both structures by cylinders. The linker DNA is not shown in the side view of the solenoid, since in most models it is located in the fiber interior; the linkers can be seen from the outside only in the two-start helical ribbon, where they help to create the zigzag ribbon of two rows of nucleosomes. An example of nucleosomes was designated using n and n+1, that are next to each other in the extended chromatin fiber; note that these nucleosomes have different spatial locations, and different neighbors, in the different models for the 30-nm fiber. The red lines are simply to emphasize the helical nature of both structures. (B) Straight-linker model: successive nucleosomes in the fiber are connected by straight linkers that criss-cross the fiber interior. The nucleosomes are situated at the periphery, as in the helical models. (A, adapted from van Holde K, Zlatanova J [2007] *Semin Cell Dev Biol*,18:651–658. With permission from Elsevier. B, adapted, courtesy of Mikhail Karymov, California Institute of Technology.)

stitute individual interphase chromosomes occupy distinct portions of the nuclear volume, forming chromosome territories. The chromatin fiber within each territory can form individual loop domains, and domains that make temporal excursions out of their respective territories can be bridged together, presumably to allow their coordinated regulation through interactions with transcriptional factories. A significant portion of the chromatin fiber is associated with the lamina structure in the so-called lamina-associated domains, which are transcriptionally repressed. Intra and inter chromosomal interactions also form distinct nucleoli that contain both active and inactive copies of tandemly repeated ribosomal genes. A final recognizable structure that may have a role in the spatial and topological organization of the genome is the nuclear matrix, an insoluble meshwork of various skeletal proteins.

A curious, and still poorly understood, phenomenon is the distribution of gene-rich and gene-poor chromosomes within the nuclear volume. The gene poor chromosomes tend to locate at the nuclear periphery, whereas gene-rich chromosomes prefer the nuclear interior.

5.3.3 Lateral gene transfer in the eukaryotic genome

(1) Mitochondrial DNA (mtDNA)

Mitochondria are found in plants, animals, fungi, and protists. The mtDNA encodes essential enzymes involved in ATP production (Mitochondrial DNA and human orientation). mtDNA is usually a circular, double-stranded DNA molecule that is not packaged with histones. There are a few exceptions where mtDNA is linear, generally in lower eukaryotes such as yeast and some other fungi.

(2) Mitochondrial DNA organization varies in different organisms

MtDNA differs greatly in size among organisms. In animals, mtDNA is typically 16–18 kb, while in plants it ranges in size from 100 kb to 2.5 Mb. There are several hundred mitochondria per cell and each mitochondrion has multiple copies of the DNA. The total amount of mitochondrial DNA relative to nuclear DNA is small; it is estimated to be less than 1%. See Table 5.6 and Figure 5.25 for information about the content of the mitochondrial genome and a map of the human mitochondrial genome

Mitochondrial genomes sequenced from many organisms show some general patterns in the representation of functions in mitochondrial DNA. Table 5.6 summarizes the distribution of genes in mitochondrial genomes. The total number of protein-coding genes is rather small and does not correlate with the size of the genome. The 16.6-kb mammalian

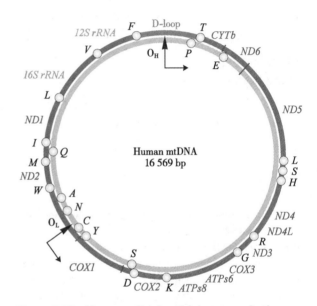

Figure 5.25 **Human mitochondrial genome in the conventional circular representation.** Mitochondrial DNA encodes 13 essential protein components of the respiratory chain, shown in red lettering. Two rRNA genes, 12S and 16S rRNA, and 22 tRNA genes, denoted by blue dots and the single-letter amino acid code, are interspaced among the protein-coding genes. These provide the necessary RNA components for protein synthesis inside the mitochondrion. The D-loop, a 1.1 kb noncoding region, is involved in the regulation of transcription and replication of the molecule and is the only region not directly involved in the synthesis of respiratory-chain polypeptides. The D-loop contains a third polynucleotide strand believed to be the initial segment generated by replication of the heavy strand. Apparently, it arrests shortly after initiation and is often maintained for some period in that state. O_H and O_L are the origins of heavy- and light-strand mtDNA replication. The origin of L-strand replication is displaced by approximately two-thirds of the genome and is located within a cluster of five tRNA genes. (Adapted, courtesy of Center for the Study of Mitochondrial Pediatric Diseases.)

mitochondrial genomes encode 13 proteins, whereas the 60- to 80-kb yeast mitochondrial genomes encode as few as 8 proteins. The much larger plant mitochondrial genomes encode more proteins. Introns are found in most mitochondrial genes, although not in the very small mammalian genomes.

The two major rRNAs are always encoded by the mitochondrial genome. The number of tRNAs encoded by the mitochondrial genome varies from none to the full complement (25 to 26 in mitochondria). This accounts for the variation in Table 5.6.

Table 5.6 **Mitochondrial genomes have genes encoding(mostly complex I-IV) proteins, rRNAs, and tRNAs.**

Species	Size(kb)	Protein-Coding Genes	RNA-Coding Genes
Fungi	19–100	8–14	10–28
Protists	6–100	3–62	2–29
Plants	186–366	27–34	21–30
Animals	16–17	13	4–24

The major part of the protein-coding activity is devoted to the components of the multisubunit assemblies of respiration complexes I-IV. In mammalian mitochondria, the genome is particularly compact. There are no introns, some genes actually overlap, and almost every base pair can be assigned to a gene. With the exception of the D-loop, a region involved with the initiation of DNA replication, no more than 87 of the 16 569 bp of the human mitochondrial genome lie in intergenic regions.

(3) Mitochondrial DNA and human orientation

In 1988, researchers made a link between certain human diseases and mtDNA mutations. Most mtDNA defects lead to degenerative disorders, especially of the brain and muscles, but because of the essential function of mitochondria in cellular ATP production, the effects can be widespread.

One of the first diseases to be linked to a small inherited mutation in a mitochondrial gene was a form of young adult blindness called Leber's hereditary optic neuropathy (LHON). mtDNA mutations such as deletions or duplications that affect many genes at once have also been identified. One example is Kearns-Sayre syndrome, which involves paralysis of eye muscles, progressive muscle degeneration, heart disease, hearing loss, diabetes, and kidney failure. Oxidative phosphorylation provides for most of a cell's ATP. This process is achieved by the electron transport chain-a series of five protein complexes located on the inner mitochondrial membrane. The most common defects associated with LHON occur in genes coding for protein components of complex I of the electron transport chain.

Key Concepts

- The central dogma proposes that there is a one-way flow of information from the nucleotide sequence of DNA to the polypeptide sequences of proteins, and there must exist a code to allow this translation.
- The code is three-letter or triplet, non-overlapping, and unpunctuated.
- One codon, AUG, is used as a start signal for translation, and the three codons UGA, UAA, and UAG are commonly employed as stops.
- The code is almost universal from microbes to humans, with a few exceptions, mostly in mitochondria.
- Bacterial mRNAs are faithful copies of the corresponding genes, but eukaryotic genes contain intervening sequences, known as introns, which must be removed so the functional or coding parts of the message, known as exons, can be spliced together.
- The entire genetic information of an organism is referred to as its genome; its complete protein library is known as its proteome.
- The repeating elements of eukaryotic chromatin are called nucleosomes. Nucleosomes consist of core particles and linker DNA. In each core particle, about 147 bp of DNA is wrapped in 1.67 left-handed toroidal turns about an octamer of histones, two each of four types: H2A, H2B, H3, and H4. H2A and H2B form heterodimers, while H3 and H4 form a tetramer.
- Chromatin fibers tend to fold into a compact 30-nm fiber at physiological salt concentrations.

Key Words

anticodon(反密码)
alternative splicing(可变剪接)
central dogma(中心法则)
degenerate(简并)
exons(外显子)
histones(组蛋白)
introns(内含子)
messenger RNA(信使 RNA)
non-protein-coding RNA(非蛋白编码 RNA)
nucleosomes(核小体)
ribosome(核糖体)
transfer RNAs(转运 RNAs)
translation(翻译)

Questions

1. Brief digestion of eukaryotic chromatin with micrococcal nuclease gives DNA fragments 200 bp long. You repeat the experiment but incubate the samples for a longer period of time while you are in class. This longer digestion yields 146-bp fragments. Why?
2. Do the 10- and 30-nm eukaryotic chromatin fibers exist *in vivo*? Discuss electron microscopic and biochemical evidence in support of your answer.
3. You are asked to characterize the genome of a newly isolated virus and to determine whether it is composed of DNA or RNA. After using nucleases to completely degrade the sample to its constituent nucleotides, you determine the approximate relative proportions of nucleotides. The results of your assay are as follows.

   0% dGTP　　15% GTP　　0% dCTP　　33% CTP
   0% dATP　　22% ATP　　0% dTTP　　30% UTP

   What can you conclude about the composition of the viral genome?
4. You have sequenced the mtDNA from a eukaryotic organism. You find that a mitochondrial gene that codes for an essential subunit of an enzyme involved in cellular respiration is missing; however, the gene encoding the other subunit of the enzyme is present. Provide a possible explanation for your finding and suggest where you might look for the "missing" sequence.

References

[1] Arya G, Maitra A, Grigoryev SA (2010) A structural perspective on the where, how, why, and what of nucleosome positioning. *J Biomol Struct Dyn*,27:803–820.
[2] Dunham I, Kundaje A, Aldred SF, et al. (2012) An integrated encyclopedia of DNA elements in the human genome. *Nature*,489:57–74.
[3] Frazer KA (2012) Decoding the human genome. *Genome Res*,22:1599–1601.
[4] The ENCODE Project Consortium (2012). An integrated encyclopedia of DNA elements in the human genome. *Nature*, 489(7414):57–74.
[5] Yanofsky C (2007) Establishing the triplet nature of the genetic code. *Cell*,128:815–818.

Lianghua Wang

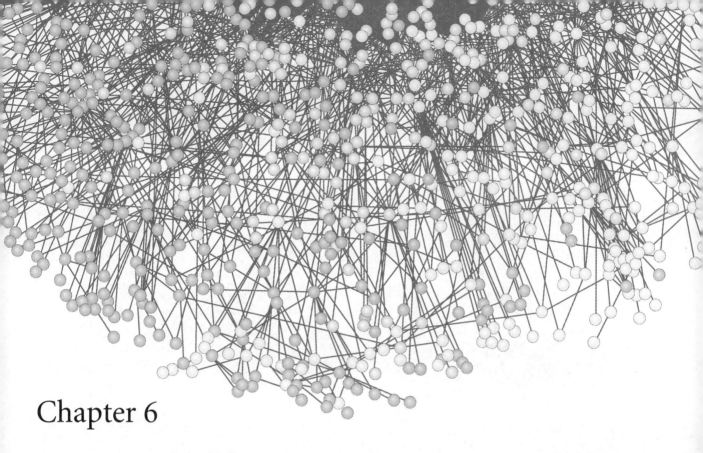

Chapter 6

Protein-Nucleic Acid Interactions and Protein-protein Interactions

6.1 Introduction

In previous chapters, we have provided a structural, and to a small extent functional, overview of two major classes of biopolymers: proteins and nucleic acids. We have seen that proteins have a multitude of roles in cells and tissues; structural proteins and many enzymes usually act alone or with other protein partners. In this chapter, we look at modes and patterns of binding. We look at protein binding to nucleic acid including DNA and RNA and protein-protein interactions. We also briefly describe experimental methods for studying protein-nucleic acid interactions and protein-protein interactions.

6.2 Protein-nucleic acid interactions

But virtually all processes involving nucleic acids also require the participation of proteins. These proteins are almost always bound to nucleic acids by noncovalent bonds, which might include hydrogen bonds, van der Waals interactions with bases, and electrostatic interactions between basic groups on proteins and nucleic acid phosphates. Protein-nucleic acid interactions can be specific, with the protein binding only to a particular nucleotide sequence, or nonspecific, in which case the protein can bind virtually anywhere on a particular kind of nucleic acid. Of course, these definitions apply only to extremes: many specific proteins can also bind nonspecifically and then hunt along the DNA to find their specific sites. Many nonspecific binders also show weak sequence preferences. We find a general, though not exact, distinction between the functions of these two classes of proteins. Nonspecific binders tend to play more structural roles. Maintaining single-stranded DNA is an example where specific binders often act as regulators of nucleic acid function. The oldest and most straightforward of these, which only allow detection of binding, are presented in Box 6.1 and Box 6.2.

Box 6.1 Filter binding

This is the oldest method for detecting an interaction between a given protein and DNA (Figure 1). Although very simple, the method provides information on the strength of binding that cannot be accessed by many other methods. As with every other method, there are drawbacks too. A fundamental problem, especially when monitoring weak interactions, comes from the necessity to wash the filter after the sample is applied. If washing is insufficient, some DNA can be retained adventitiously, but too thorough washing may lose complexes because of dissociation. Thus the K_d values obtained must be regarded with some caution. The method is little employed today but played a significant role in the early days of molecular biology.

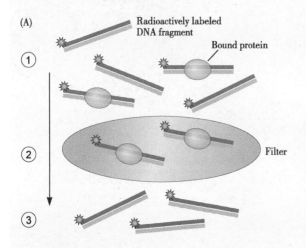
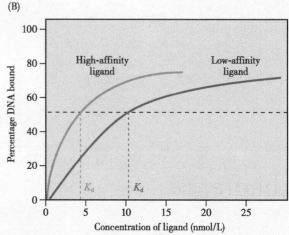

Figure 1 **Filter binding assay.** (A) The assay is performed in three steps: (Step 1) binding reactions, (Step 2) passing the incubation mixture through nitrocellulose paper filters, and (Step 3) detection of radioactively labeled DNA that is retained on the filter. The nitrocellulose filter is negatively charged and will not retain the negatively charged DNA unless it has been bound by protein; most proteins are positively charged. The exact amount of DNA retained on the filter is quantified by measuring the amount of radioactivity. (B) Titration curves, showing DNA fraction bound versus increasing amounts of protein present in the incubation mixture, are used to determine the affinity of protein binding.

Box 6.2 DNA-affinity chromatography

Another simple technique used to search for proteins that bind strongly to a particular nucleic acid sequence is to attach the polynucleotide fragments to silica beads in a chromatographic column and pour a mixture of suspected proteins through (Figure 1). Strongly binding proteins will

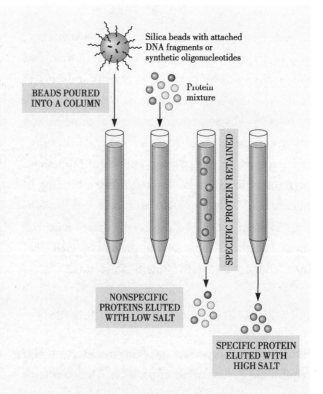

Figure 1 **DNA-affinity chromatography allows purification of DNA-binding proteins.** DNA fragments or synthetic oligonucleotides containing the protein-binding site are immobilized onto silica beads, which are then packed into a chromatography column. A mixture of proteins, usually a nuclear extract, is passed through the column: the binding protein is retained on the column, whereas proteins that do not bind or bind only weakly either pass through directly, without being retained, or can be eluted with low-salt concentration buffers. The specific binding protein can subsequently be recovered by eluting with a high-salt concentration buffer. More recent techniques use magnetic beads, which do not need to be packed into columns but can be easily manipulated by using of magnets external to the tube.

be retained on the column even when dilute salt solutions are run through the column, which will elute weak binders. The investigator can then take advantage of the fact that even strongly binding proteins will dissociate from the nucleic acids under high salt conditions, allowing their isolation.

Neither filter binding nor DNA-affinity chromatography can tell us the exact sequence of the binding site. However, some information can be obtained by using a variety of mutants in the presumed binding region. This is laborious, and there are now easier ways to determine binding sequences.

6.2.1 DNA-protein interactions

(1) DNA-protein binding occurs by many modes and mechanisms

Nonspecific binding of proteins to DNA can proceed very simply through electrostatic interaction between basic groups on the protein and the smooth, regular track of negatively charged phosphates that form the phosphodiester backbone. Virtually any positively charged protein will tend to stick to DNA in solution, at least at low ionic strength. This is often also true for proteins that possess a recognition site for a particular nucleotide sequence: they can cling to the DNA elsewhere but more weakly. This is often a factor in facilitating the access to specific protein-binding sites. Sometimes nonspecific binding is used to coat the DNA, thereby protecting it against unwanted interactions, protecting it from degradation, or compacting it. Some proteins can do all of these things. An important example is found in chromatin. For example, the protein-DNA complex in which DNA is compacted and sequestered in the eukaryotic nucleus. Much more is said about chromatin; for now, suffice it to note that in chromatin, complexes of eight basic protein molecules form spools upon which nuclear DNA can be wound. To a large extent, this binding is nonspecific, as must be expected for an interaction that involves most of the genome. Nevertheless, some elements of nucleotide sequence preference or avoidance for the formation of such structure can be found.

A class of nonspecific DNA-binding proteins preferentially interact with and stabilize single-stranded DNA (ssDNA). These single-strand binding proteins (SSB, or helix-destabilizing proteins) are involved in processes such as genetic recombination, DNA replication, and DNA repair, where it is essential to maintain a region of denatured DNA for some time. SSB proteins are ubiquitous and can function as monomers, as seen in viruses; as homotetramers, as seen in bacteria; or as heterotrimers, such as replication protein A in eukaryotes. It was long believed that such proteins might bind to a transient opening in the helix and then bind cooperatively in a side-by-side fashion to expand the opening. However, recent work in the von Hippel laboratory indicates that this cannot be a universal mode; binding sites for the SSB are often so large that spontaneous openings of sufficient size would be rare. This kinetic block means that a helicase molecule is needed first to unwind a suitable length of DNA. The SSBs can then bind and stabilize the single strand. A well-studied example, the SSB from *Escherichia coli*, is illustrated in Figure 6.1. *In vitro*, the protein binds to ssDNA either as a dimer or as a tetramer, depending on the conditions. SSB binds as a tetramer in the cell, wrapping ~70 nucleotides of ssDNA around the protein subunits. This wrapping leads to an overall reduction in the contour length, as seen in the electron microscope images presented in Figure 6.1C.

(2) Site-specific binding is the most widely used mode

The binding of proteins to specific DNA sites is fundamental to a vast range of functions and structures. It has been estimated that the human proteome contains several thousand site-specific DNA-binding proteins, each with a specific base sequence that it recognizes and binds to. Some binding proteins recognize only a single site in the whole genome, whereas others bind at multiple locations. The most intriguing question is how do protein molecules distinguish particular DNA sequences? Even before we had specific knowledge, it was clear that the B-form

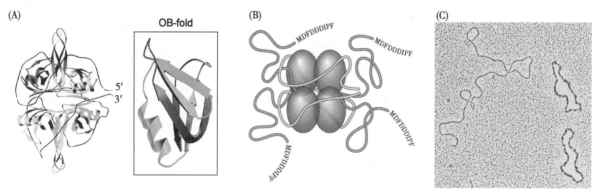

Figure 6.1 *E. coli* **single-strand binding protein.** SSB binds single-stranded regions of DNA to prevent premature reannealing of the two strands during DNA replication and related processes. It also protects ssDNA from nucleolytic cleavage. *E. coli* SSB is composed of four identical 19 kDa subunits that can interact with ssDNA in different modes depending on the environment. In the $(SSB)_{65}$ binding mode depicted, ~65 nucleotides of DNA wrap around the SSB tetramer and contact all four subunits; this mode is favored at high salt concentrations. At lower salt concentrations, the $(SSB)_{35}$ binding mode prevails, with ~35 nucleotides binding to only two of the SSB subunits. (A) Crystal structure of $(SSB)_{65}$ depicting 70 nucleotides of ssDNA, shown as a black line, wrapped around the four SSB subunits. The protein contains a characteristic tertiary structure motif, termed oligonucleotide/oligosaccharide-binding or OB-fold, present in many DNA- or RNA-binding proteins; its general topology is shown in the box. (B) Cartoon representing ssDNA as a ribbon wrapped around the SSB core, corresponding to the structural model in part A, with the addition of the unstructured C-terminal tails, shown as gray lines, that are not observed in the crystal structure. The nine-amino-acid sequence, shown in single-letter amino acid code, at each C-terminus is responsible for the interaction of SSB with other metabolic proteins. (C) Electron microscopic image of naked DNA (left) and SSB-bound DNA (right). The reduction in contour length is due to the wrapping of the DNA around SSB. (A-B, adapted from Kozlov AG, Jezewska MJ, Bujalowski W, et al[2010] *Biochemistry*,49:3555–3566. With permission from American Chemical Society. A, inset, from Agrawal V, Kishan RKV [2001] *BMC Struct Biol*1:5. With permission from Springer Science and Business Media. C, courtesy of Maria Schnos, University of Wisconsin-Madison.)

double helix of DNA contained information that could be read by other molecules. The DNA duplex surface is defined by two deep grooves, known as the major and minor grooves. The edges of the bases are exposed in each of these grooves in a way that presents a unique combination of chemical groups, which can interact with a protein that inserts into the groove. The combination includes, for example, both hydrogen-bond donors and acceptors, and these are presented in a different pattern for a GC base pair than for an AT base pair (Figure 6.2). Thus, with its own hydrogen-bond acceptors and donors, a protein can detect the difference between these two kinds of pairs. In addition to this, there are other kinds of interactions, such as van der Waals interactions between methyl groups on the DNA and nonpolar groups in the protein, that can convey a base pair's identity.

It should be emphasized that many site-specific binders also exhibit a much weaker nonspecific binding mode, involving electrostatic interactions between positive amino acids on proteins and the featureless track of negative charges on the DNA surface. This can help a protein find its specific site by sliding along the DNA helix or hopping from one region to another across DNA loops. Such one-dimensional hunting can be more efficient than wandering through three-dimensional space.

(3) Most recognition sites fall into a limited number of classes

There are many possible specific amino acid-nucleotide interactions that could promote protein recognition of a binding site (Figure 6.3). Analysis of many X-ray diffraction studies of protein-DNA complexes reveals no simple set of rules for such pairing. There is some preference for arginine to pair with G in GC pairs and for glutamine or asparagine to pair with AT pairs (Figure 6.3). Evolution appears to have adopted whatever works, often complicated by distortion of the DNA and/or protein to provide the best fit.

The number of base pairs involved in different DNA recognition sites varies widely, depending on the protein functions. Some restriction endonucleases, for example, have relatively short recognition

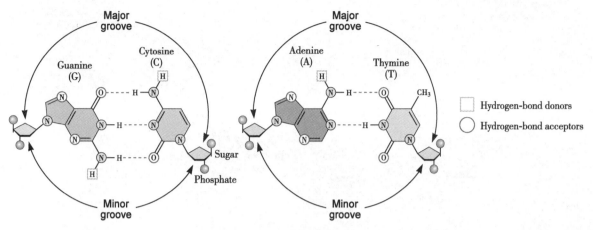

Figure 6.2 Proteins recognize specific sequences on the DNA by predominantly interacting with the major groove. Looking down the DNA helix, one can see the edges of the bases protruding into the major groove. Accessibility of these bases depends primarily on the width of the groove. Clearly, the minor groove is less accessible. A stack of several base pairs along the chain will create a unique constellation of chemical groups into the groove, recognized by DNA sequence-specific binding proteins.

sites, 4–6 base pairs or bp, so that they can cleave genomic DNA in many places. Some transcription activator or repressor proteins targeted against only one or a few sites in the genome, must have larger recognition sites. They need not be excessively large to make unwanted accidental false recognition unlikely. Since the number of possible combinations of bases in a site of size n is 4n, sites 10 bp long will occur only about every million base pairs, giving an expectancy of several thousand such sites in the whole human genome of 3.2 billion bp. Some of these sites may be obscured by other proteins bound to or around them. When n = 20, an astronomical 1012 different sequences are possible; thus, a particular, specific 20 bp site would be unlikely to be found anywhere within the genome.

Some recognition sites, such as those of the Lac repressor binding protein (Figure 6.4), are palindromic and can therefore accept a dimer of the protein, increasing both the specificity and binding affinity. As Figure 6.4 shows, Lac repressor takes this one step further by recognizing two adjacent palindromic pairs and binding as a tetramer. Indeed, as we shall see in further examples, site-specific binding proteins frequently exhibit tandemly repeated binding domains, which then interact with tandemly repeated DNA sites. This mechanism can make it possible to attain high specificity and binding strength, even if there are only a few specific interactions between each domain and the DNA.

Finally, it should be emphasized that distortion of the binding partners, especially bending of the DNA, is a common feature of site-specific binding. Because DNA bending exacts a free energy price, this must be returned by a favorable free energy gain from the interaction process itself. In other words, the binding must be strong to compensate for the bending.

(4) Most specific binding requires the insertion of protein into a DNA groove

DNA binding can gain stability from protein interaction with the uniform phosphate backbone of DNA. However, the recognition of specific sites along the DNA chain demands the insertion of a portion of the protein into one of the grooves in the DNA helix, where the specific base pairs can be sensed. Most DNA-binding proteins use the major groove. This is understandable because the major groove provides more recognition sites and in B-DNA it is significantly wider than the minor groove. The major groove is, in fact, wide enough to accommodate a segment of α-helix snugly, and this is a very common mode of binding. Although less common, there are examples in which two-stranded β-sheets, or the edges of even larger β-sheets, are inserted into the major groove. The good fit of α-helix or β-sheet into

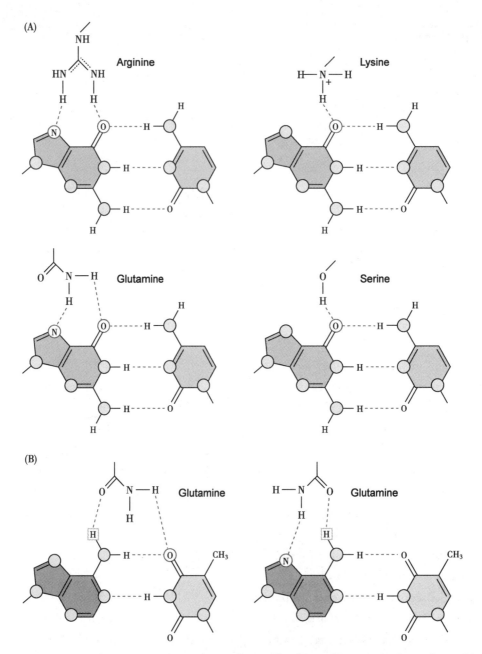

Figure 6.3 **Possible patterns of hydrogen-bonding during recognition of base pairs by amino acid side chains.** (A) Recognition of GC base pairs by arginine, lysine, glutamine, and serine side chains. (B) Recognition of AT base pairs by glutamine side chains. Note the two possible patterns of hydrogen-bonding.

the major groove means that these protein domains can bind to recognition sites with little distortion of the DNA.

Binding into the minor groove is a different matter. In the first place, the minor groove of B-DNA is occupied by a spine of precisely positioned water molecules. In consequence, binding to the minor groove involves releasing this structured water, with a large entropy increase. This appears to be a major free energy source for minor-groove binding. Second, the minor groove is just too narrow to accommodate an element such as an α-helix, so if such binding is to occur, the groove must be stretched open. There are a number of ways in which DNA can accommodate such stretching, but the most common is by compression of the adjacent major groove.

This will have the consequence of bending the DNA helix at the site of protein binding (Figure 6.5). Thus, it is not surprising that minor-groove binders are almost synonymous with DNA benders. The energetic cost of such bending appears to be largely paid by the above-mentioned entropy increase resulting from the release of water. However, there may be an interesting synergy here, in that the water spine contributes to the rigidity of the double helix; so displacing it should make DNA more flexible.

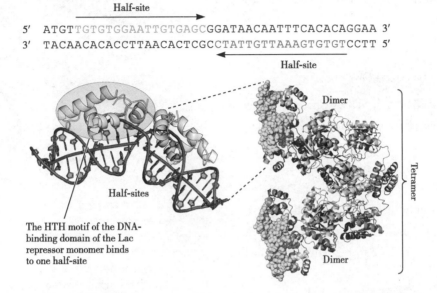

Figure 6.4 **Binding of Lac repressor to DNA.** The repressor binds to DNA via helix-turn-helix (HTH) motifs in the DNA binding domains of two monomers that form a dimer. Note that the two half sites in the DNA recognition sequence form a palindrome, which matches the symmetry of the dimer. The repressor binds to DNA as a tetramer, or dimer of dimers, formed by interactions of the C-termini; each dimer binds two half sites. (Right, from Wilson CJ, Zhan H, Swint-Kruse L, et al. [2007] *Cell Mol Life Sci*,64:3–16. With permission from Springer Science and Business Media. Left, from Rohs R, Jin X, West SM, et al. [2010] *Annu Rev Biochem*,79:233–269. With permission from Annual Reviews.)

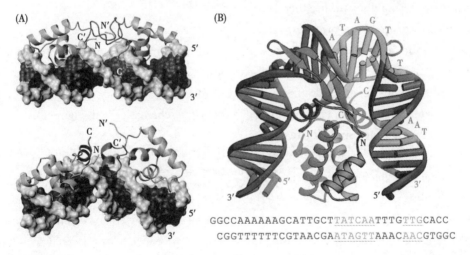

Figure 6.5 **Minor-groove binders cause DNA bending.** (A) DNA-binding domain of Lac repressor bound to nonspecific (top) and specific (bottom) nucleotide sequences. The hinge region is unstructured in both the free protein and the protein bound to nonspecific DNA; it folds into an α-helix that interacts with the minor groove in the specific complex. In the nonspecific complex, the DNA adopts the canonical B-form conformation, whereas in the specific complex it is bent by ~36°. (B) Crystal structure of the integration host factor (IHF) complex with DNA: consensus sequence DNA were indicted. IHF is a small, 20 kDa heterodimeric protein that binds DNA in a sequence-specific manner and induces a large bend of >160°. This bending aids forming of higher-order structures in such processes as recombination, transposition, replication, and transcription. The protein contacts only the phosphodiester backbone and some bases in the minor groove; thus, it represents an example of indirect readout, where the protein relies on sequence-dependent structural features of the DNA, such as backbone conformation and flexibility. This sequence recognition is contrasts with direct readout, where sequences are distinguished through the unique functional groups of DNA bases in the major groove. (A, from Kalodimos CG, Biris N, Bonvin AMJJ, et al. [2004] *Science*,305:386–389. With permission from American Association for the Advancement of Science. B, from Lynch TW, Read EK, Mattis AN, et al. [2003] *J Mol Biol*,330:493–502. With permission from Elsevier.)

It should be emphasized that major-groove binding can also give rise to bending. This can happen when a dimeric or oligomeric protein binds to adjacent sites in the DNA. There exist a number of ingenious ways to study DNA bending and relate it to protein binding.

(5) Some proteins cause DNA looping

We have noted that many DNA-binding proteins exist as non-covalently-bound dimers or even larger oligomers. If such proteins have recognition sites on DNA that are some distance apart, they can to produce DNA loops (Figure 6.6). DNA looping is observed in both prokaryotic and eukaryotic genomes. In many cases, it appears to isolate particular regions of the genome to regulate transcription therein.

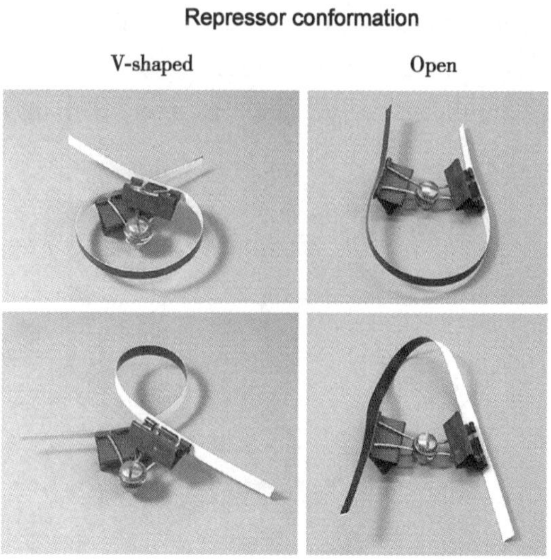

(6) There are a few major protein motifs of DNA-binding domains

Every protein that binds specifically to DNA for a specific function needs at least two domains. One is called the transactivating domain. It either senses an external signal to prompt the protein to bind, as in a transcription factor, for example, or it will carry out some process on the DNA, as a restriction endonuclease, for example. The other domain is the one we have been concerned with so far: the binding domain attaches to the DNA at a specific site or sites. There may be multiple transactivating domains, or complexes of several proteins that are noncovalently attached to the binding domain. In many cases, the transactivating domains can cause conformational changes in the binding domains. This may suggest that there should be caution about the many experiments, especially X-ray structure determinations. Only the purified or cloned DNA-binding domain, bound to DNA, is studied. This structure might not reflect the structure of the entire bound protein.

Surprisingly, a large fraction of binding domains appears to use only a small vocabulary of protein motifs, known as recognition motifs, for the actual binding, even if the binding serves very different functions. This may suggest that the DNA-binding domains of proteins evolved from only a few examples, ages ago. Here, we describe just three of the most frequently encountered motifs; they are depicted schematically in Figure 6.7.

Figure 6.6 DNA looping. Physical models corresponding to possible loop geometries. Photographs show loop models with the V-shaped or crystallographic repressor conformation (left column) and the corresponding configurations that result when the repressor is opened (right column). In each pair, the structure on the right was produced from that on the left by rotating the two half-tetramers, represented by a blue or a red clip, away from each other about the axis of the four-helix bundle, represented by a silver bolt. The paper strip representing the DNA is colored black on one side and white on the other to make any twist in the helix visible. (From Wong OK, Guthold M, Erie DA, et al. [2008] *PLoS Biol*, 6:e232 [10.1371].)

(7) Helix-turn-helix motif interacts with the major groove

Helix-turn-helix (HTH) motifs are common in both prokaryotic and eukaryotic transcription factors. The motif consists of a stretch of ~20 amino acid residues, divided into two α-helices, each about 7–8 residues long, separated by a turn or loop. The second of these two helices, from the N-terminus, is the recognition helix, which lies in the major groove of the DNA. A specific example, CRP activator, is shown in Figure 6.8. Here there are three α-helices. As the α-helix has 3.6 residues per turn, only two turns are presented to the DNA, permitting only a few specific contacts. In many cases, HTH proteins bind as dimers or even tetramers, increasing the number of contacts and hence selectivity (see Figure 6.8). As mentioned above, the binding of more than one monomer can have another consequence: the bending of the DNA. As Figure 6.8 illustrates, the geometry of protein-protein interactions can make it necessary for the DNA to bend to interact with all of the HTH motifs.

(8) Zinc fingers also probe the major groove

Another motif frequently observed in DNA-binding domains is called the zinc finger because of the essential role of zinc in its structure. As shown in Figure 6.9, the most common variant has a short length of α-helix bound to a short β-sheet via the zinc atom. Usually, the zinc is coordinated to two histidine side chains on the α-helix and two sulfhydryl groups of cysteine residues on the β-sheet, but sometimes four sulfhydryls are used. In any event, the compact finger that is produced can be inserted into the major groove without distortion. The number of possible interactions is small, so specificity is often gained by having multiple fingers. These may be all contained in the sequence of one polypeptide chain, as in the TFIIIA transcription factor shown in Figure 6.10, or they may be contained in interacting proteins, as in the steroid receptor proteins in Figure 6.11.

(9) Leucine zippers are especially suited for dimeric sites

Leucine zippers are protein-protein interaction motifs that help the stable formation of protein dimers to increase the specificity of interaction with DNA. Leucine zipper proteins always interact with DNA as homo- or heterodimers, held together by hydrophobic interactions of long α-helices. These

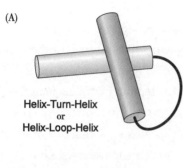
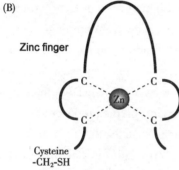
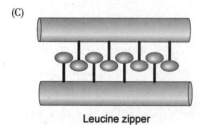

Figure 6.7 **DNA-binding motifs.** (A) Helix-turn-helix, HTH; (B) zinc finger; (C) leucine zipper.

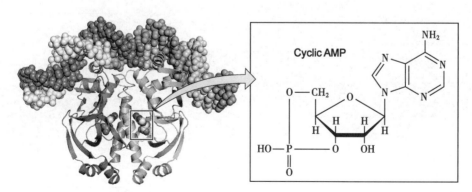

Figure 6.8 **Helix-turn-helix protein bound to DNA.** CRP activator is a cyclic AMP regulatory protein, also known as CAP or catabolite activator protein, that positively controls numerous operons in *E. coli*. Its activity is regulated by cyclic AMP binding. The protein binds as a dimer to a bipartite binding site, causing the DNA to bend ~90°. Each monomer has two structural domains: the N-terminal domain, amino acids 1–140, contains the cyclic AMP nucleotide-binding site, whereas the C-terminal 50–60 amino acid domain contains a helix-turn-helix motif that interacts with the DNA.

interacting protein tails each have leucine or isoleucine residues spaced about 3–4 residues apart. This means that all of these residues will lie on one side of the α-helix, creating a hydrophobic face (Figure 6.12A). This face can be buried away from the solvent by having the helices make a gently coiled coil about one another. The term zipper refers to the fact that isoleucine residues on the two α-helices often interdigitate like teeth of a zipper. At the end of this coil lie the recognition elements themselves, usually α-helix segments that insert into the major groove (Figure 6.12B). Zippers can be homodimers, which interact with a repeated DNA sequence, or heterodimers, which carry different DNA-recognition elements and thus interact with a pair of different DNA sites. This allows a sophisticated kind of control, in which two different protein factors must be present and interact to allow DNA binding.

The DNA-binding elements of proteins described in this section do not represent anything like the totality of modes of protein-DNA interactions. There are many variants on the motifs described above, and there are many proteins that seem to bind in unique ways. Important examples of protein-DNA complexes feature throughout this book.

6.2.2 RNA-protein interactions

Cellular RNA molecules interact with specific proteins, but in ways that are somewhat different than DNA-protein interactions. In the first place, there is a general difference in function between RNA-protein and DNA-protein complexes. As we have seen, the latter often have regulatory functions,

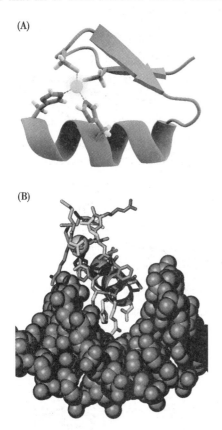

Figure 6.9 **Some proteins bind DNA with multiple zinc fingers.** (A) One of the three zinc fingers of transcription factor Zif268. In this case, the Zn atom is coordinated between two cysteine and two histidine residues. (B) The middle zinc finger of Zif268 bound to DNA, in the major groove, with the Zn(II) atom shown as a sphere. (A, courtesy of Thomas Splettstoesser, Wikimedia. B, from Magliery TJ, Regan L [2005] BMC Bioinf 6 [10.1186/1471-2105-6-240]. With permission from BioMed Central.)

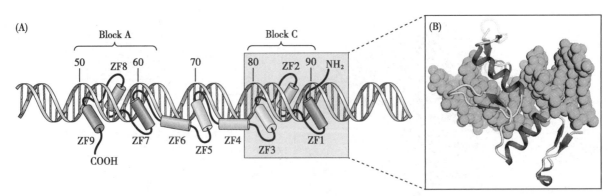

Figure 6.10 **Arrangement of proteins containing multiple zinc fingers along the DNA.** (A) Transcription factor TF III A binds to 5S rDNA, the gene that encodes the 5S RNA component of the ribosome, via multiple zinc fingers (ZF) that insert themselves into the major groove. The two major recognition regions, blocks A and C, are contacted by fingers 7–9 and 1–3, respectively. Fingers 4–6 are used when the protein interacts with RNA. (B) Crystal structure of fingers 1–3 bound to DNA. (A, adapted from Dyson HJ [2012] Mol BioSyst, 8:97–104. With permission from the Royal Society of Chemistry.)

whereas RNA-protein complexes more often carry out catalytic processes. In some cases, the protein moiety is the catalyst, as in RNA-processing complexes. In contrast, in other examples, the proteins appear to primarily function in determining the three-dimensional structure of a complex in which

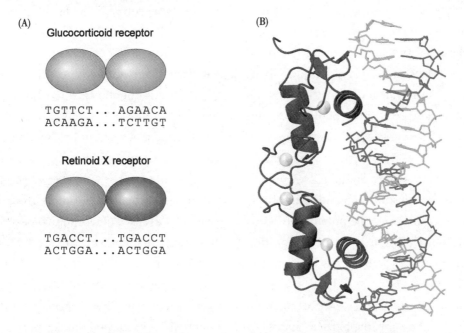

Figure 6.11 **Zinc fingers in steroid receptors.** (A) The DNA sequences that bind steroid hormone receptors, called steroid response elements, consist of two short half-sites that may be palindromic or direct repeats. The receptor proteins bind palindromic sequences as head-to-head homodimers; receptor protein heterodimers bind direct repeats. The schematic shows examples of each binding mode. (B) Crystal structure of the estrogen receptor bound to DNA. The monomer receptors have two Zinc fingers each. The binding of the first finger determines the sequence-specific binding, while the second finger is responsible for dimerization; that is, Zinc fingers can also be protein-protein interaction domains.

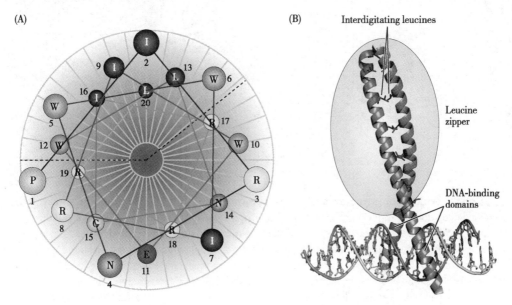

Figure 6.12 **Leucine zippers. Leucine zippers are protein-protein interaction motifs that help the stable formation of protein dimers to increase the specificity of interaction with DNA.** (A) Amphipathic helices can be predicted by helical wheel analysis, in which a stretch of amino acid sequence is imaginarily arranged in a wheel that mimics the arrangement of amino acids in an α-helix. The example given is that of 20 N-terminal amino acid residues of chicken cathelicidin, chCATH-B1. Note that, in addition to leucines, the hydrophobic top side of the helix also contains isoleucine, another hydrophobic amino acid residue. (B) Structure of a leucine zipper protein heterodimer bound to DNA. The DNA-binding domains are usually basic. (A, courtesy of David E. Volk, University of Texas Health Science Center. B, adapted from Goitsuka R, Chen CH, Benyon L, et al. [2007] *Proc Natl Acad Sci USA*, 104:15063–15068. With permission from the National Academy of Sciences.)

the RNA is the catalytic agent, known as a ribozyme. Another factor that makes RNA-protein interactions distinct from typical DNA-protein binding is the fact that cellular RNA is synthesized as single-stranded molecules. However, these will often contain regions of self-complementarity that allow the formation of double-stranded structures, such as hairpins, interspersed with single-stranded loops. Such structures can be very complex. Thus, RNA-binding proteins may have to recognize various folded structures, not just sites in the grooves as in DNA. In addition, the numerous base modifications found in RNA provide additional opportunities for recognition. Probably because of this great variety of potential binding modes, we do not find RNA-protein recognition involving such a limited vocabulary as DNA-protein recognition.

The example, par excellence, of protein-RNA interaction is the ribosome, the particle that is the machine for protein synthesis in all cells. The bacterial ribosome contains three RNA molecules with a total length of 4566 bases, plus 55 different proteins. The structure of this immense, complex particle has now been solved by X-ray diffraction, for which the Nobel Prize in Chemistry was awarded in 2009. Most of the proteins appear to play structural roles, complexly intertwined with the RNA, but some also have catalytic functions. The ribosome is not the only large protein-RNA complex, however. Much of the processing of messenger RNA that occurs in the eukaryotic cell nucleus involves such particles.

There does not seem to be, or perhaps we have just not yet recognized, a simple group of common protein motifs used in recognition of specific sites on RNA molecules. Rather, there are a variety of binding motifs, perhaps because there is little uniformity in RNA structure (Figure 6.13). One principle that seems to dominate RNA-protein binding is the use of a multiplicity of sites modularly. In such a case, each protein module needs recognize only a few bases, or

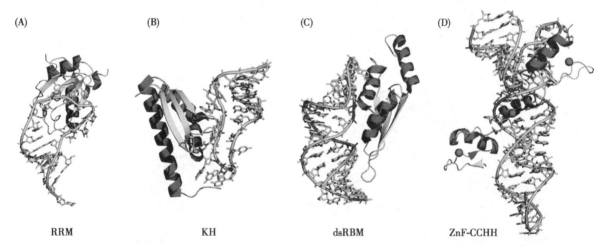

Figure 6.13 **Crystal or solution structures of some common RNA-binding domains bound to RNA.** (A–C) αβ domains; (D) zinc-finger domain (ZnF) domain of the CCHH motif type. (A) N-terminal RNA-recognition motif (RRM) of human U1A spliceosomal protein complexed with an RNA hairpin, in jade. Single-stranded bases in the RNA are recognized by the protein β-sheet and by two loops, that connect the secondary structure elements. (B) KH (K-homology) domain of Nova bound to 5′-AUCAC-3′. KH domains bind to specific sequences, in both ssDNA and RNA, through a conserved GXXG sequence located in an exposed loop. (C) Double-stranded RNA-binding domain (dsRBM) of Rnt1p RNase III bound to an RNA helix capped by an AGNN tetraloop. The protein loop interacts with the 2′-OH groups in the RNA minor groove, whereas Lys and Arg residues in the longer helix recognize the position of the phosphate atoms in the A-form helix. *In vitro*, most dsRBMs bind to dsRNA of any sequence provided it stretches for at least 12–13 base pairs and is not interrupted by too many bulges or internal loops; *in vivo*, however, proteins containing dsRBMs bind to and act on specific RNAs, and some domains even bind to other proteins. It is not clear how this non-sequence-specific domain is used to bind specific RNAs in the cell. (D) RNA-binding zinc-finger proteins: complex of three-finger peptide, fingers 4–6, from TIFIIIA and truncated 5S RNA. Recognition is primarily through residues in α-helices; in this specific case, the sequence-specific recognition is achieved through exposed bases in the loop regions of RNA. (From Chen Y, Varani G [2005] *FEBS J*, 272:2088–2097. With permission from John Wiley & Sons, Inc.)

even only one base, to achieve high specificity. How such modular proteins can function is illustrated in Figure 6.14. Usually, the length of RNA recognized, even by multiple modules, is fairly short. Thus, self-interacting proteins or proteins with multiple domains can modify the three-dimensional structures of RNA molecules in a great number of ways (Figure 6.15)

6.2.3 Studying protein–nucleic acid interactions

Throughout this chapter we have described, via boxes, ways in which protein–nucleic acid complexes could be identified and characterized. However, it has required the application of a number of sophisticated techniques to provide the details of those interactions. Obviously, the most detailed way to examine the interactions between a given protein and DNA or RNA is to prepare the complex and then use either X-ray crystallography or high-resolution NMR to reveal the molecular details of the complex. This will show exactly what the binding site is and how the protein molecule recognizes it. A multitude of examples in this book illustrate how this is becoming an ever more realizable goal. But for this to be done, it is first necessary to define, at least partially, the binding site on the polynucleotide and the protein domain involved in binding. There are a number of

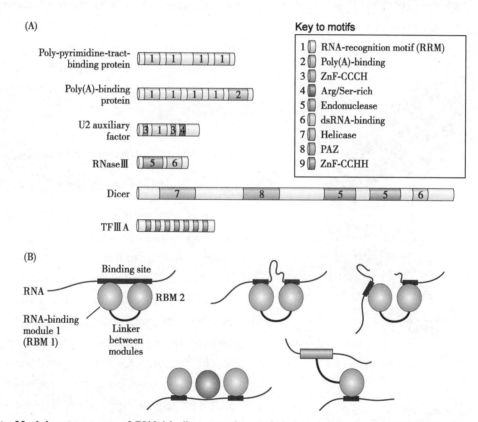

Figure 6.14 Modular structures of RNA-binding proteins and their modes of action. (A) Schematic representation of some RNA-binding protein families illustrating the arrangements of individual RNA-binding motifs along the polypeptide chain. Often, the proteins contain additional domains of varied functions. For example, Dicer and RNase III both contain an endonuclease catalytic domain followed by a double-stranded RNA-binding domain (dsRBM), so both proteins recognize dsRNA. In addition, Dicer has to specifically interact with stem-loop structures in microRNA precursors; it does this through additional domains that recognize the unique structural features of these RNAs. (B) The modular organization of RNA-binding proteins allows them to perform a variety of functional roles. The relatively low specificity and affinity of binding that is characteristic of individual domains is augmented significantly by the simultaneous binding of two or more domains. (Top) Multiple domains may combine to bind to long recognition sequences, or to sequences separated by long stretches of unrelated intervening sequences, or to sequences from different molecules. (Bottom) Two domains may function as spacers to properly position other modules; finally, the RNA-binding modules may help to define substrate specificity or regulate enzymatic activity. (Adapted from Lunde BM, Moore C, Varani G [2007] *Nat Rev Mol Cell Biol*, 8:479–490. With permission from Macmillan Publishers, Ltd.)

less sophisticated, but no less important, methods to gain this and other information about the interaction.

A technique that has allowed the rapid identification of hosts of nucleic acid-binding proteins and other proteins that bind to them is termed chromatin immunoprecipitation. The term is something of a misnomer, for while it was initially and is still applied heavily to chromatin, it is applicable to virtually any protein-nucleic acid Interactions.

Finally, we must mention site-directed mutagenesis, a method that has proved very useful for dissecting the architecture of DNA binding sites. Although not a method for identifying binding sites, it allows careful, base-by-base probing of the source of the interactions. It is possible to create mutations in DNA *in vitro*. A set of such mutated sequences can be used for *in vitro* binding studies to define the contribution of each nucleotide to the binding.

Additionally, or alternatively, these mutated sequences can be introduced into living cells to observe changes in phenotypes and thus to understand the physiological significance of the specific protein-DNA binding in which the researcher is interested.

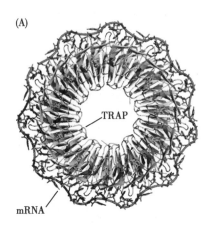

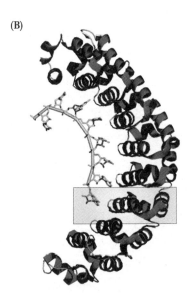

Figure 6.15 **RNA recognition.** RNA is often recognized by proteins that form heteromeric or homomeric structures through the association of multiple proteins or of the same fundamental structural motif. (A) Tryptophan RNA-binding attenuation protein (TRAP) repressor. TRAP represses tryptophan biosynthesis by regulating translation of mRNAs coding for biosynthetic enzymes. The mRNAs have multiple copies of the RNA sequence recognized by TRAP. The protein forms an oligomeric ring of 11 subunits; the RNA binds on the outside of the ring through stacking interactions and hydrogen-bonding to amino acids. (B) In the translational regulator protein Pumilio, eight copies of the same protein structural motif are used to recognize RNA. Each domain binds a single nucleotide, but the combination of multiple domains provides exquisite specificity. (Adapted from Chen Y, Varani G [2005] *FEBS J*,272:2088–2097. With permission from John Wiley & Sons, Inc.)

6.3 Protein-protein interactions

6.3.1 Protein-Protein Interactions Essentials

Proteins play central roles in different aspects of function, including transport, signal pathway, metabolism, and structural organization. Protein-protein interaction is essential for mediation functions such as environment sensing, adjusting the response of metabolic or signaling enzyme, energy convention, and cellular organization. Studying the protein-protein interaction helps us understand a protein's function and behavior, predict the biological processes that a protein of unknown function is involved in characterizing protein complexes and pathways. The protein-protein interaction networks can be used as a draft "map" or interactome mapping to add detail to biological processes and pathways. It has become one of the main scopes of current biological research, similar to the way "genome" projects were a driving force of molecular biology.

In 1876, Kuhne Wilhelm proposed the term "enzyme" and in the same article, he isolated the trypsin, which is the digestive enzyme of the pancreas. In 1906, a trypsin inhibiting activity was detected in the albumin fraction of blood serum. This interaction was analyzed in a quantitative kinetic study. The enzyme-inhibitor pair combined to form an inactive but dissociable compound. This interaction between the enzyme trypsin and its inhibitor is identified as the first protein-protein regulatory interactions.

The definition of protein-protein interactions has to consider that the interaction interface should be intentional and not accidental. For example, the result of specific selected biomolecular events/forces. And the protein-protein interactions also should be non-generic. For example, it evolved for a particular purpose distinct from totally generic functions such as protein production, degradation, and others. Therefore, the interaction is the physical contact of high specificity established between two or more proteins that occur in a cell or a living organism in a specific biomolecular context. Protein-protein interactions can modify the kinetic properties of enzymes, act as a general mechanism to allow for substrate channeling. Protein-protein interactions can also construct a new binding site for small effector molecules, inactivate or suppress a protein. Protein-protein interactions even change the specificity of a protein for its substrate through interaction with different binding partners.

Depending upon their function and properties of structure, protein-protein interaction can be classified into different groups: according to their interaction surface, can be grouped as permanent or transient; Based on their stability, it can be considered as hetero-oligomeric or homo-oligomeric and can be called as obligate or nonobligate.

6.3.2 Analysis of structural analysis of proteins

Understanding of protein structures and protein and protein interaction, often requires the knowledge of atomic details. The techniques of X-ray diffraction, Multidimensional nuclear magnetic resonance (NMR), cryo-electron microscopy, and small-angle x-ray scatting (SAXS) are used for gaining different information. X-ray crystallography can provide detailed information of each subunit. SAXS is a small-angle scattering technique by which the density differences of a solid or liquid sample can be quantified. NMR spectroscopy provides data on dynamic changes for protein conformation. Cyroelectron microscopy can provide a picture of the overall assembly of big protein complexes with multiple subunits.

(1) Max Perutz, John Kendrew, and the birth of protein crystallography

Today we know the detailed structures of an enormous number of proteins, a level of understanding unimaginable half a century ago. Most of this information has come from studies of the diffraction of X-rays from protein crystals. In 1950, the technique of X-ray diffraction had been known for decades and used widely to determine the structure of small molecules. Indeed, much of the information about amino acid configurations that Pauling used to model protein secondary structures came from such studies. Yet Pauling himself had expressed doubts that a molecule as complex as a protein could ever be elucidated by this technique.

To get a simple idea of what is involved, consider the following experiment: if a very fine-mesh wire screen is placed in the path of a collimated light beam, like that from a laser, a rectangular pattern of bright spots can be observed on the wall behind the screen. This is the diffraction pattern of the screen, produced by reinforcement of in-phase waves coming to the wall from different apertures in the screen. The use of different screens can demonstrate much about diffraction. For example, a finer screen gives more widely spaced spots; there is a reciprocal relationship between object and its diffraction pattern. If a screen with wider horizontal spacings than vertical spacings is used, the pattern will show wider vertical spacings than horizontal spacings. In fact, the spacings of patterns on the wall can be used to deduce the periodic structure of the screen.

Any crystal is a periodic structure. But there are two ways in which a protein crystal differs from the screens in our example: the spacings are very small, in the nanometer range, and the structure is three-dimensional, not two-dimensional like a screen. The small spacings necessitate the use of radiation of very short wavelength, X-rays, to get a diffraction pattern. The three-dimensional structure means that there are periodicities in three dimensions, so the crystal must be viewed from different angles. Finally, protein molecules are very complex.

To obtain details of the molecular structure, both the positions and intensities of many, sometimes hundreds, of spots in the diffraction pattern must be measured. Unfortunately, this is not enough. Because the diffraction patterns are from a three-dimensional lattice, different spots will correspond to diffracted rays in different phases. The intensities do not tell this, and it was this phase problem that held up progress for many years. For small molecules, known constraints on the structure could lead to a resolved structure, but such methods were hopeless for proteins.

Such was the conundrum faced by researchers Max Perutz and John Kendrew at Cambridge University in the early 1950s. The breakthrough came in 1953, when Perutz discovered a way through the phase problem. If one could insert a heavy metal atom at the same place in every protein molecule in a crystal, it would perturb the intensities of different spots to various extents. With a series of such isomorphous replacements, all phases could be deduced and the structure could be solved.

By 1958 the first results were published when Kendrew's group presented a low-resolution 0.6 nm structure of myoglobin (Figure 6.16). At this resolution, little detail could be resolved, but it was clear that the structure was very complex and much less regular than previous guesses had suggested. The work was pushed to higher resolution, and by 1960, a 0.25 nm structure could be published. This showed, for the first time, the α-helices predicted by Pauling in a real protein.

In the same year, Perutz and collaborators produced a 0.55 nm structure of hemoglobin. This was a much more ambitious project because each hemoglobin molecule is a noncovalent assembly of four myoglobin-like chains. The results showed, for the first time, how proteins could associate to yield quaternary structure. Perutz and Kendrew were awarded the 1962 Nobel Prize in Chemistry "for their studies of the structures of globular proteins." X-ray diffraction is the most prolific technique for determining the most detailed structures of protein. (Figure 6.17) However this method cannot capture the dynamics change of protein since it transfers protein in to static crystals.

(2) Nuclear magnetic resonance studies of biopolymers

Nuclear magnetic resonance or NMR is a second technique, in addition to X-ray crystallography, that can provide detailed information about macromolecular structure. It is based on the fact that certain atomic nuclei, such as ^1H, ^{13}C, ^{15}N, and ^{31}P, possess a feature called spin that makes them behave like tiny magnets and hence take up different preferred orien-

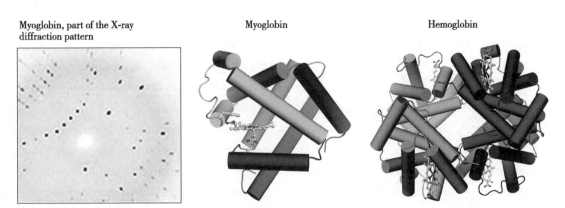

Figure 6.16 **X-ray diffraction pattern and structures of oxygen-binding protein.** The diffraction image is from Francis Crick's Thomas Splettstoesser in Max Perutz's laboratory, 1949–1950 (Left, courtesy of binding proteins).

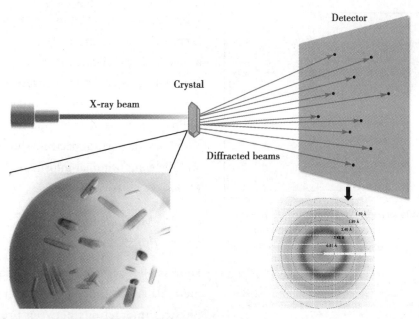

Figure 6.17 **Structural analysis of protein crystal.** The illustration of x-ray diffraction analysis. The crystal structure of AdeR DNA-binding-domain, which is a component of AdeABC efflux pump regulatory system involved in *Acinetobacter baumannii* multidrug resistance, is complexed with the intercistronic DNA and the resulting X-ray diffraction pattern. (Photographs courtesy of Yurong Wen, Xi'an Jiaotong University.)

tations in an external magnetic field. These different orientations differ slightly in energy, so the appropriate radio frequency or rf energy can manipulate the orientations. When the radio frequency is just right, the nuclei resonate and emit radio waves of their own. For example, almost every proton in a protein or nucleic acid has, by virtue of its local environment in the molecule, a different resonance frequency. Thus, scanning the rf spectrum of a sample while it is in an intense magnetic field gives a proton spectrum (Figure 6.18). This can be useful in many ways, but the real power of NMR came with the discovery of spin-spin interactions.

The spin state of one nucleus can be perturbed by another spin that is nearby. This allows mapping of internuclear distances in a folded molecule. By use of such techniques, it is now possible to deduce, with fair accuracy, the three-dimensional structures of proteins and oligonucleotides. As Figure 6.19 shows, the results are close to those from X-ray crystallography. NMR has the enormous advantage, as compared to X-ray crystallography, that it can be conducted in solution, where molecules assume their natural conformations.

Richard Ernst was awarded the 1991 Nobel Prize in Chemistry for "his contributions to the development of the methodology of high resolution nuclear magnetic resonance (NMR) spectroscopy." Kurt Wüthrich was awarded a share of the 2002 Nobel Prize in Chemistry for "his development of nuclear magnetic resonance spectroscopy for determining the three-dimensional structure of biological macromolecules in solution."

(3) NMR in medicine: MRI and MRS

Two powerful NMR variations: magnetic resonance imaging (MRI) and magnetic resonance spectroscopy (MRS), have found widespread application in medicine. The advantage of MRI is that it is completely non-invasive and does not use harmful radiation. MRI uses magnetic gradients to spatially paint the position of the water proton spins. Submillimeter resolution images can be obtained in a matter of seconds from most soft tissues. MRI can aid in tumor localization, detection of Alzheimer's disease, stroke follow up, sports injury evaluation, assessment of functional heart defects, etc. Figure 6.20 shows an example of the use of MRI to detect a tumor in the brain.

MRS uses magnetic gradients to localize one or

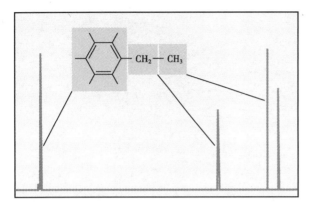

Figure 6.18 **Sample proton NMR spectrum.** The solution contains a simple organic compound, ethylbenzene. Each group of signals corresponds to protons in a different part of the molecule.

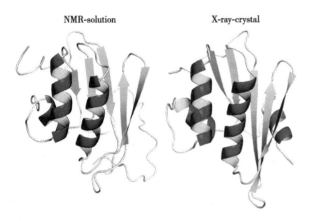

Figure 6.19 **Comparison between NMR solution structure and crystal structure.** The protein domain used here as an example is the RNase H domain of HIV-1 reverse transcriptase. Note that there are slight differences between the structures near the N- and C-termini; these may well result from the effects of packing the molecules into crystals for the X-ray study.

several regions or voxels for investigation. Still it also keeps the spectral information from the voxel to get a metabolic profile of the investigated region. The bone MRS profile shown in Figure 6.20 illustrates the connection between the known chemical structure of lipids, shown as dots of different colors connected with lines, their MRS spectrum shown at the bottom, and the compositional information obtainable with MRS.

6.3.3 Methods to investigate protein-protein interactions

Four smaller-scale techniques or so-called 'binary' methods are commonly used to investigate direct physical interactions between protein pairs. The techniques that measure physical interactions among a group of proteins are "co-complex methods", without pairwise determination of protein partners. The binary and co-complex methods could measure both direct and indirect interactions between proteins.

The most common design approach pull-down assay is based on the preselection of one "bait" protein, which is one protein tagged with a molecular marker or label, to catch or fish out "prey" proteins followed by a "pull-down" to separate them from a mix by using SDS-PAGE (see Figure 6.21A). Coimmunoprecipitation (co-IP) are similar in methodology to pull-down assay. The difference between these two approaches is that co-IP uses antibodies

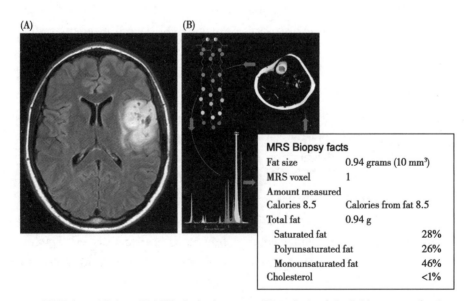

Figure 6.20 **NMR in medicine.** (A) MRI of a brain tumor; (B) analysis of the lipid content of a tissue biopsy.

to capture protein complexes (Figure 6.21B), and pull-down assay depends on the co-purification of protein groups. These two methods are easiest to perform on stable protein-protein interaction because the protein complex does not dissemble over time. The crosslinking protein interaction analysis is an approach to stabilize or freeze the components of interaction complexes during co-IP, Pull-down assay or mass spectrometry allows to study the week or transient interaction. The pull-down assay technique for the analysis of DNA-protein interactions *in vitro*.

The yeast two-hybrid system and co-IP assays are more common techniques for examine of protein-protein interactions within the living cell. The yeast two-hybrid assay required two protein domains, which will have two specific functions: DNA binding domain (DBD) that helps binding to DNA, and an activation domain (AD) responsible for activating transcription. Two domains are works together for the transcription of a reporter gene, such as *GFP* or *LacZ*. A protein of interest "bait" fused to the DNA-binding domain is screened against a library of activation domain hybrids, i.e. "prey" to select interacting partners (Figure 6.22). However, the yeast two-hybrid system involves artificial constructs and co-IP requires cell lysis for analysis, so the precise intracellular localization of protein-protein interactions cannot be determined.

To study fluorescence resonance energy transfer (FRET) or Bimolecular fluorescence complementation (BiFC) allows researchers to study protein-protein interactions *in situ*, at the precise location in the cell where they normally occur. In FRET, energy from an excited donor fluorophore e.g., CFP is transferred to an acceptor fluorophore within an approximately 10nmol/L of the excited fluorophore e.g., YFP. If FRET observed, this implies direct contact between two proteins (Figure 6.23). The BiFC is based on the association of fluorescent protein fragment e.g., GFP that is attached to components of the same molecu-

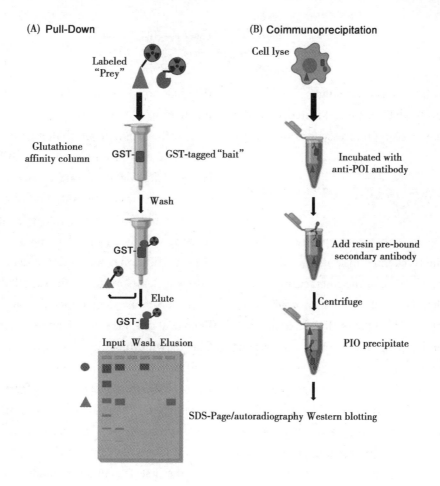

Figure 6.21 **Methods for GST pull-down assay and coimmunoprecipitation assay.**

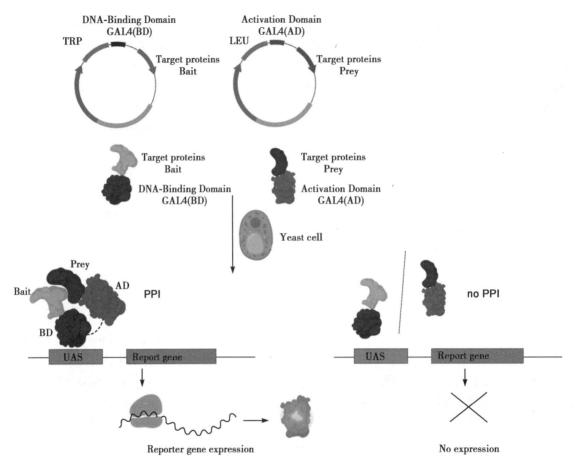

Figure 6.22 **Illustration for Yeast two-hybrid assay.**

lar complex or fluorescent exchange. The fluorescent signals can be detected and located with the cell and imaging by using fluorescence microscopy (Figure 6.24).

6.3.4 The proteome and protein interaction networks

Recent advances in high-throughput technologies have allowed unprecedented insights into the entire protein content of cells and the interactions among proteins. The library of all proteins encoded in the genome of a cell is termed the proteome. Many proteins are presently recognized only on the basis of portions of sequenced genomes; they have not yet been isolated as biochemical entities, and practically nothing is known about their possible functions. A huge amount of effort now goes into using available sequence information, at the gene and protein levels, and structural data to predict functions of proteins with closely related sequences. These predictions then have to be verified by actual biochemical experimentation.

Another significant challenge is to classify the proteome of a given species into different protein classes. Huge databases are created by the international scientific com-munity to create orderly classifications and derive biologically relevant insights into these seemingly chaotic data. Figure 6.25 illustrates the classification of the entire human proteome into different classes related to the subcellular localization of the proteins, their molecular function, and the biological processes in which the proteins participate. The known proteomes of some other model species, such as the yeast *Sac-charomyces cerevisiae*, the worm *Caenorhabditis elegans*, and the fruit fly *Drosophila melanogaster*, have also been classified into these same categories. It is important to realize that these achievements, as impressive as they are, constitute only the first steps in understanding the way the proteome functions.

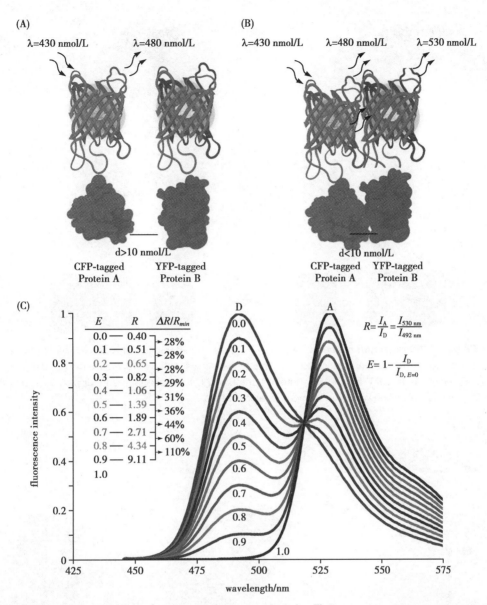

Figure 6.23 **Methods for analyzing protein-protein interactions by Fluorescence resonance energy transfer.** (A) Illustration for fluorescence resonance energy transfer (FRET), which is a special technique to gauge the distance between two chromophores, called a donor-acceptor pair. (B) For example, using a widefield fluorescence microscope with appropriate filter sets, cyan (CFP) and yellow (YFP). Adapted from Campbell RE [2009] *Anal Chem*, 81(15):5972–5979. With permission from American Chemical Society. (C) The relationship between E and R. The $\Delta R/R_{min}$ has been calculated for each increment of $\Delta E = 0.1$. Spectra are simulated for the mTFP/YFP FRET pair. The change is measured as a decrease in the intensity via the donor filter(I_D) and an increase in the intensity measured through the acceptor filter (I_A). The ratio of accepter to donor fluorescence intensity($\Delta R = I_A/I_D$) is used as a surrogate for E in the live-cell imaging. In FRET, a change in ratio($\Delta R = R_{max} - R_{min}$) is measured and maximizing this value is of the utmost concern when optimizing a biosensor.

Another state-of-the-art development in the proteome field is the creation of interactome maps. These are databases that identify, through high-throughput experimental methods, all of the possible binary interactions, that is, interactions between any two proteins, in the entire proteome. Other databases identify interactions in protein complexes, involving many protein partners. Some interactome maps focus on particular proteins or pathways. Figure 6.26 shows an example while others are more ambitious, attempting to cover large portions of proteomes. Figure 6.27 presents a recent interactome of a large portion of the yeast proteome. Figure 6.28 shows different representations of a portion of the human interactome. Clearly, the range of interactions is wonderfully complex, even with our presently limited knowledge.

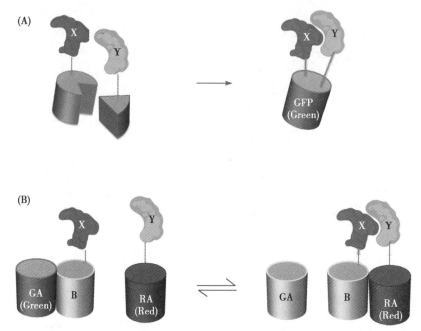

Figure 6.24 **FP-based PPi detection systems.** (A) Split FP. When the interaction of X and Y occurs, the two halves of the split FP associate form the intact and functional FP. This process is essentially irreversible, so proteins X and Y remain associated. (B) FP exchange (FPX). When X and Y interact, the B and GA protein fragments disassociate, and the fluorescent protein does not have function while the RA and B are associated.

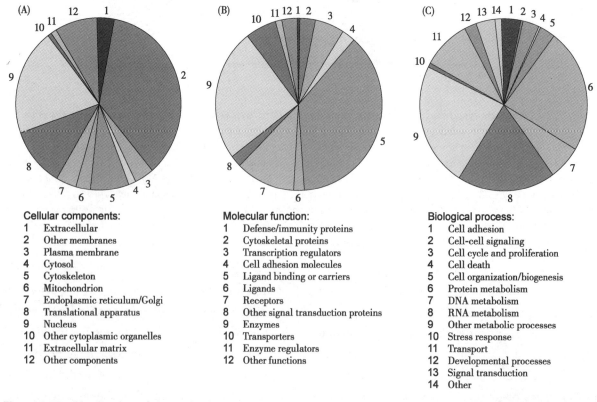

Cellular components:
1. Extracellular
2. Other membranes
3. Plasma membrane
4. Cytosol
5. Cytoskeleton
6. Mitochondrion
7. Endoplasmic reticulum/Golgi
8. Translational apparatus
9. Nucleus
10. Other cytoplasmic organelles
11. Extracellular matrix
12. Other components

Molecular function:
1. Defense/immunity proteins
2. Cytoskeletal proteins
3. Transcription regulators
4. Cell adhesion molecules
5. Ligand binding or carriers
6. Ligands
7. Receptors
8. Other signal transduction proteins
9. Enzymes
10. Transporters
11. Enzyme regulators
12. Other functions

Biological process:
1. Cell adhesion
2. Cell-cell signaling
3. Cell cycle and proliferation
4. Cell death
5. Cell organization/biogenesis
6. Protein metabolism
7. DNA metabolism
8. RNA metabolism
9. Other metabolic processes
10. Stress response
11. Transport
12. Developmental processes
13. Signal transduction
14. Other

Figure 6.25 **Distribution of the entire human proteome into three distinct large categories of proteins.** The categories, as defined by the Gene Ontology database, are (A) cellular components, (B) molecular function, and (C) biological process.

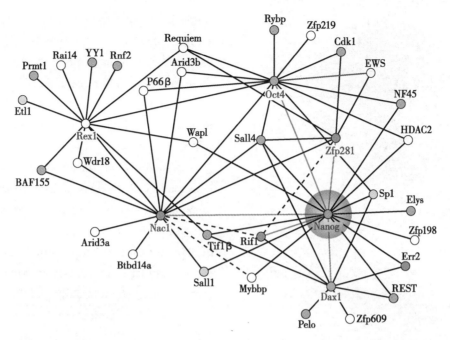

Figure 6.26 **Partial interaction network for the protein Nanog.** Nanog is known to participate in defining the pluripotent state of embryonic stem (ES) cells, that is, their ability to differentiate into numerous different cell types, depending on the applied stimuli. Affinity-purified Nanog partners were identified by mass spectrometry; in an iterative fashion, protein partners of these primary partners were then identified. The new protein partners were further verified experimentally by alternative techniques such as co-immunoprecipitation. The network is highly enriched in nuclear factors that are individually critical for the pluripotent state of the ES cell; the presence and levels of these factors are coordinated during further differentiation of the stem cells into different cell types. The tight protein network seems to function as a cellular module dedicated to pluripotency. The lines of different color that connect the proteins in the network indicate the methods by which the interaction has been verified experimentally. The circles of different colors denote proteins whose presence is essential for a particular stage of development, with the white circles designating proteins whose absence has not been correlated with developmental defects. (From Wang J, Rao S, Chu J, et al. [2006] *Nature*, 444:364–368. With permission from Macmillan Publishers. Ltd.)

These styles of representation are frequently used, but there are numerous ways to graph experimental interactome data. The estimated size of the binary interactome ranges from 130 000 to ~650 000 interactions; only ~8% of them have been identified. Ambiguity about the size of the human interactome stems from difficulties in identifying which of the possible biophysical interactions actually occur in cells. With the presently estimated number of human proteins, nearly 250 million protein pairs need to be tested to map the whole human interactome: such a huge undertaking is possible only with application of high-throughput approaches. The worm *Caenorhabditis elegans* appear to have a protein number similar to that in humans, and the sizes of the two interactomes appear to be of the same order. Then, what makes humans different from worms? We do not have the answer yet. (Courtesy of Seth Berger, Mount Sinai School of Medicine.)

Again, it must be noted that creating these beautiful-looking maps is only the first step in a process that tries to make sense of all this complexity. In fact, attempts to represent large, complex databases in a meaningful way, in both science and technology, have given rise to an entirely new discipline called visual analytics. Scientists not only seek ways to graphically represent an interactome but also try to detect "what's in there," to extract the basic principles that regulate the interactions of individual components. The insights that interactome mapping is expected to provide into the functioning of living organisms cannot be overstated.

Over the past three decades, the number of protein-protein interactions identified has increased signifi-

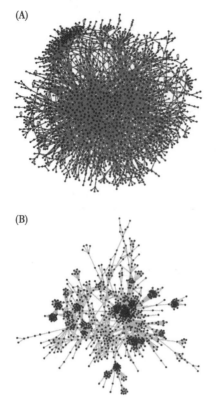

cantly. The first database was the database of interacting proteins (DIP). Since then, a large number of databases have been created to catalog and annotate these interactions, for example: DIP, IntAct molecular interaction Database (IntAct), Biological General Repository for interaction datasets (BioGRID), and so on.

It is obvious that the enormous diversity of proteins that exist in living organisms demands that every organism, in fact every cell, must have some repository of information to specify the primary structures of many proteins. The expression of this information must be under strict control as different cell types, even within a single organism, need different proteins to function, and these needs will change with growth and development. There must also be a way to preserve and transmit this information from generation to generation. That is the focus of this book.

Figure 6.27 **Yeast protein interaction maps representing binary protein-protein interactions.** Binary protein-protein interactions are those between two proteins. These maps are derived from two different high-throughput experimental methods. Individual proteins are represented by dots, and interactions are shown by lines connecting the dots. (A) This interaction map from October 2008 combines three high-quality proteome wide data sets, obtained by yeast two-hybrid screening. The map covers ~20% of the entire binary yeast interactome and contains 2930 interactions among 2018 proteins. (B) Map of binary interaction extracted from co-complex interactome maps, which are obtained by high throughput co-affinity purification followed by mass spectrometry. The map is composed of 9070 associations between 1622 proteins. The two maps constructed by use of different system biology tools are fundamentally different but complementary in nature: they reflect different connections and biological properties of the interacting partners. Certain proteins are highly connected, with numerous partners; these are termed hub proteins. The absence of such proteins leads to a number of different phenotypes. Interestingly, hub proteins are characterized by intrinsic disorder, which may form the molecular basis for their promiscuous interaction with multiple partners. Interacting with different partners in different conditions constitutes the molecular basis for pleiotropic effects, that is, different roles of the same protein in multiple biochemical pathways. (From Yu H, Braun P, Yildirim MA, et al. [2008] *Science*,322:104–110. With permission from American Association for the Advancement of Science.)

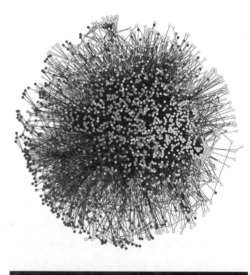

Figure 6.28 **Different graphic representations of portions of the human interactome.**

Key Concepts

- Many of the structural and functional aspects of nucleic acids *in vivo* are modulated by their interaction with proteins. Some proteins bind to specific nucleic acid sequences or binding sites; others are nondiscriminating and will bind at a very large number of sites or anywhere on the polynucleotide. Nonspecific binders tend to play structural roles, protecting the nucleic acid from degradation or maintaining a single-stranded conformation, for example. Site-specific binding is more often involved in regulatory functions, such as controlling gene transcription.

- There is a vast variety of site-specific DNA-binding proteins. These recognize short nucleotide sequences by inserting a part of the protein into the major or minor groove of the DNA, where it can recognize specific arrays of groups for interaction. Major-groove binding is more common than minor-groove binding. A limited number of molecular recognition motifs are used by a considerable fraction of major-groove-binding proteins. These include the helix-turn-helix, zinc finger, and leucine zipper motifs.

- Major-groove binding does not, in general, lead to major deformation of the DNA because the groove is wide enough to accommodate such motifs. However, if the DNA-binding protein oligomerizes and recognizes two or more sites contiguous on the DNA, bending of the DNA can result. If the binding sites are not close together, binding to a protein oligomer may result in the formation of DNA loops. Binding to the minor groove of DNA is less common and often results in bending or other deformation of the DNA. This is because the minor groove is so narrow that it must be forced open to accommodate most binding motifs.

- RNA molecules also exhibit both nonspecific and site-specific protein binding. The function here can also be either structural, such as maintaining compact RNA folds, or functional, such as promoting ribozyme activity. A common feature of protein-RNA binding is the use of multiple binding motifs, each of which recognizes only one or a few sites. The use of such multiple motifs ensures specificity of binding.

- Protein and protein interaction studies were involved in many different perspectives as biochemistry, molecular dynamics, signal transduction, and drug development.

- PPIs have been studied with many methods, and all this information enables the creation of large protein interaction networks, including metabolic, genetic/epigenetic networks. To identify the interaction of protein partners *in vitro* and *in vivo* methods often used like affinity purification, Y2H, Pull-down assay, coimmunoprecipitation assay and fluorescence-based reporter assays.

- Structural analysis of proteins can be done by methods such as X-ray crystallography, nuclear magnetic resonance spectroscopy, cryoelectron microscopy, and atomic force microscopy.

- Recent proteomic studies were beginning to unravel the very complex interactions among proteins in the cells.

Key Words

chromatin(染色质)
helix-destabilizing proteins(螺旋去稳定蛋白)
helicase(解旋酶)
helix-turn helix (HTH)(螺旋-转角-螺旋)
high-resolution NMR(高分辨率核磁共振)
recognition motifs(识别模块)
single-strand binding proteins(单链结合蛋白)
X-ray crystallography(X射线晶体学)
zinc finger(锌指结构)

Questions

1. What is the most widely used mode for BDA-Protein interactions?
2. Are the physical contacts between two domains crucial for the functioning of cellular machinery?
3. Which are the advantages of the FRET for analysis of protein interactions?
4. What is the reporter gene in the yeast two-hybrid system?
5. Could you study the protein-protein interactions using a large-scale affinity purification technique that involves

attaching fusion tags to proteins and purifying the associated protein complexes in an affinity chromatography column as an alternative approach for determination?
6. Van Holde KE, Johnson WC, Ho P-S (2006) Principles of Physical Biochemistry, 2nd ed. Pearson Prentice Hall.

References

[1] Chen Y, Varani G(2005)Protein families and RNA recognition. *FEBS J*,272:2088–2097.

[2] Kalodimos CG, Biris N, Bonvin AM, et al. (2004) Structure and flexibility adaptation in nonspecific and specific protein-DNA complexes. *Science*,305:386–389.

[3] Kozlov AG, Jezewska MJ, Bujalowski W, Lohman TM(2010) Binding specificity of *Escherichia coli* single-stranded DNA binding protein for the χ subunit of DNA pol III holoenzyme and PriA helicase. *Biochemistry*, 49:3555–3566.

[4] Lane D, Prentki P, Chandler M(1992)Use of gel retardation to analyze protein-nucleic acid interactions. *Microbiol Rev*,56:509–528.

[5] Lunde BM, Moore C, Varani G(2007) RNA-binding proteins: Modular design for efficient function. *Nat Rev Mol Cell Biol*,8:479–490.

[6] Lynch TW, Read EK, Mattis AN, et al. (2003) Integration host factor: Putting a twist on protein-DNA recognition. *J Mol Biol*,330:493–502.

[7] O'Shea EK, Klemm JD, Kim PS, Alber T(1991) X-ray structure of the GCN4 leucine zipper, a two-stranded, parallel coiled coil. *Science*,254:539–544.

[8] Privalov PL, Dragan AI, Crane-Robinson C(2009) The cost of DNA bending. *Trends Biochem Sci*,34:464–470.

[9] Theobald DL, Mitton-Fry RM, Wuttke DS(2003)Nucleic acid recognition by OB-fold proteins. *Annu Rev Biophys Biomol Struct*,32:115–133.

[10] Von Hippel PH(2007) From "simple" DNA-protein interactions to the macromolecular machines of gene expression. *Annu Rev Biophys Biomol Struct*,36:79–105.

[11] Campbell RE, Wiens MD(2018) Surveying the landscape of optogenetic methods for detection of protein-protein interactions. *Systems Biology and Medicine*,10(3): e1415.

[12] Wen YR, Ouyang ZL, Yu Y, et al. (2017) Mechanistic insight into how multidrug resistant Acinetobacter baumannii response regulator AdeR recognizes an intercistronic region. *Nucleic Acids Research*, 16: 9773–9787.

Fang Zheng

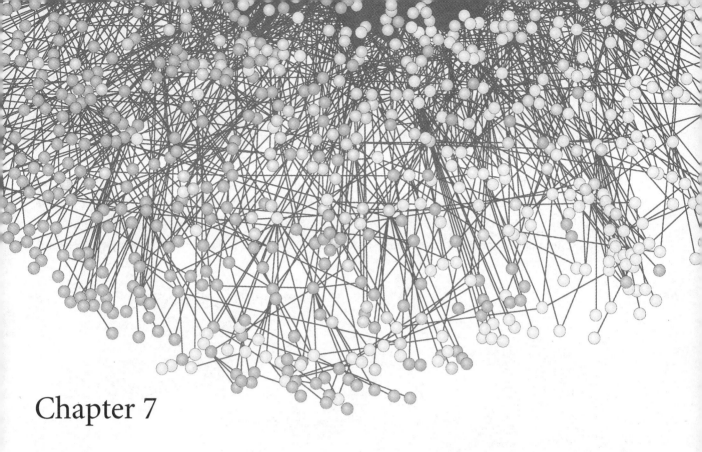

Chapter 7

Mechanism of Transcription

7.1 Introduction

Although genetic information in a cell is stored in genomic DNA, this information has to be eventually expressed in the form of specific RNA and proteins to commit the physiological functions. All these expressions initiate the synthesis of RNA molecules by the enzymes called RNA polymerases (RNAPs). Regarding the template features, RNAPs can be classified into DNA-dependent polymerases and RNA-dependent polymerases that function in most viruses. The process of RNA synthesis from a DNA template is defined as transcription, which is an essential stage of productions of both mRNA templates for protein synthesis and the special RNA molecules needed in the cell. While RNA synthesis from an RNA template is designated as RNA replication. RNA polymerases can be single-subunit mainly in viruses, or multi-subunit enzymes in all types of cells, including bacteria, archaea and eukaryotic cells.

The general transcription process is composed of three stages, including initiation, elongation, and termination, which are conserved in all known organisms. Eukaryotic transcription is more complex than that in bacteria, not only in terms of the mechanisms involved in the transcription process, but also in terms of its regulation.

In this chapter, we first provide an overview of the basic mechanisms of transcription in all organisms, and then learn the detailed pictures of how transcription occurs in both bacteria and eukaryotes. Further aspects of the regulation of transcription in both bacteria and eukaryotes will be addressed in Chapter 8.

7.2 Overview of transcription

During transcription process, RNA polymerases generate a single-stranded RNA identical in sequence (with the exception of U in place of T) to one of the strands of the double-stranded DNA template. The single DNA strand that directs the nucleotide sequence in the nascent RNA by complementary base pairing is designated as the template strand (Figure 7.1), and the generated RNA strand is called the primary transcript. The DNA template is copied in the 3′ to 5′-end direction, and the RNA transcript is synthesized in the 5′ to 3′ direction. The other unused DNA strand is called as the sense strand, coding strand, or nontemplate strand, as the transcribed RNA corresponds in sequence to this nontemplate strand.

Chemistry of RNA synthesis is that the new phosphodiester bond is formed with a magnesium-dependent catalyzed nucleophilic attack by the 3′-OH at the end of the RNA chain on the α-phosphate of the incoming nucleoside triphosphate by RNA polymerase, with elimination of pyrophosphate. The pyrophosphate (PP_i) released is quickly hydrolyzed by the abundant enzyme pyrophosphatase. This fast hydrolysis prevents the reversal of the polymerization reaction (Figure 7.2). Moving along a gene in the 5′ to 3′ direction is described as moving downstream,

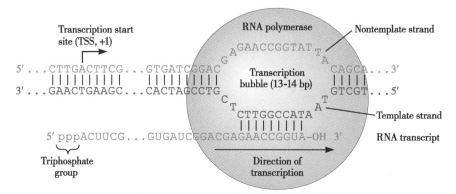

Figure 7.1 **Orientation of a gene.** A gene is transcribed from its 5′- to its 3′-end; however, the template strand is actually copied from the 3′-end. The DNA template strand and the nascent RNA are antiparallel and the bases are complementary. The nontemplate strand is also known as the coding strand or the sense strand, exactly the same as the codons in mRNA. In the process, a small, 13–14 bp portion of DNA is unwound, creating a transcription bubble. Moving along a gene in the 5′ to 3′ direction is described as moving downstream, whereas moving in the opposite 3′ to 5′ direction is known as moving upstream.

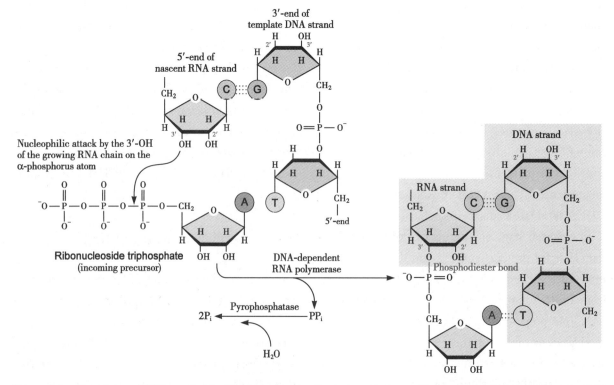

Figure 7.2 **Chemistry of RNA synthesis.** The new phosphodiester bond is formed between the 3′-OH group of the RNA and the 5′-phosphate of the incoming ribonucleoside triphosphate, with elimination of pyrophosphate.

whereas moving in the opposite 3′ to 5′ direction is known as moving upstream.

7.2.1 There are aspects of transcription common to all organisms

(1) All RNAPs initiate the synthesis of new strands in the absence of a primer, which differ from DNA polymerases.

RNAPs are able to recognize the appropriate gene to transcribe. The appropriate strand of the double-stranded DNA of a gene and the accurate start site of transcription (TSS) can be recognized by RNAPs (Figure 7.3). Promoters are specific sequences on DNA that determine where the RNA polymerase binds to and how frequently it initiates transcription. The core promoter is defined as the DNA sequence(s) around TSS bound by the polymerase to direct the transcription initiation. Motifs, such as the −10 element in bacterial or the TATA box in eukaryotes, can be the core promoter. Other regulatory sequences such as promoter-proximal elements and enhancers also affect the frequency of transcription.

(2) RNAPs melts the DNA at the promoter region, forming a transcription bubble

The process of transcription begins with the binding of RNAPs to the promoters (Figure 7.4). The first interaction of the RNAP with the promoter results in the formation of the closed complex. This is an oversimplification, as the polymerase actually binds at random sites along the DNA and then performs a one-dimensional search for the promoter; that is, it tracks the double helix until it finds the promoter sequence. This one-dimensional search is a common feature of protein-DNA interactions.

A transcription bubble is usually 13–14 base pairs long in the open complex by RNAPs. Thus, DNA in the transcription bubble exposes its bases to be used as a template for RNA synthesis. Note that DNA enters and exits the RNA polymerase as a double-stranded helix, so the bubble exists inside the polymerase itself.

(3) The formation of the first phosphodiester bond between the nucleotides of nascent RNA is rather slow.

At the beginning of transcription, the stability of the RNA-DNA duplex is very low because of the few bases involved in base pairing. This leads to frequent release of the short transcripts from the polymerase, which is dubbed abortive transcription. During this phase, the polymerase remains in its original position, without moving along the template, with the already transcribed portion of the template scrunching in its vicinity. Once the transcript overcomes the length threshold of ~15 bases, the process switches to elongation: the polymerase leaves or clears the promoter and moves into the body of the gene. This step is known as promoter escape or promoter clearance.

(4) The transcription elongation is highly processive.

Once elongation begins, the length of the DNA being transcribed may be thousands or even tens of thousands of base pairs without the enzyme dissociating from the DNA-RNA complex. The transition from initiation to elongation is accompanied by major

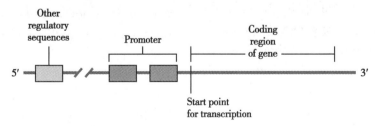

Figure 7.3 Regions of a gene. The transcribed region of a gene contains the template for an RNA synthesis, which begins at the start point. A gene also includes regions of DNA that regulate production of the encoded product, such as a promoter region. In a structural gene, the transcribed region contains the coding sequences that dictate the amino acid sequence of a polypeptide chain.

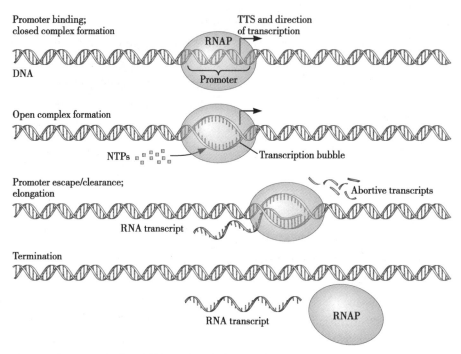

Figure 7.4 Basic steps in transcription. RNAP complex binds to the promoter region on the DNA to form a closed complex. RNAP also melts the DNA to form an open transcription bubble of 13–14 bp. The incoming ribonucleotide triphosphates (NTPs) pair with the exposed DNA bases on the template strand and the first phosphodiester bonds are formed, linking nucleotides within nascent RNA molecules. At this stage, RNA synthesis is abortive and most newly formed short transcripts are released from the polymerase. Once a transcript elongates beyond ~15 bases, the polymerase clears the promoter and enters the processive elongation phase, in which the DNA-RNA-polymerase ternary complex is very stable and the enzyme transcribes long stretches of DNA without dissociating. The characteristic high processivity of RNA polymerase distinguishes it from other DNA-tracking enzymes, such as DNA polymerases.

conformational changes in the polymerase that ensure a firm grip of the polymerase on the DNA-RNA complex.

During elongation, each nucleotide addition is accompanied by two steps of the DNA with respect to the catalytic center of the polymerase, as illustrated in Figure 7.5. The DNA template moves longitudinally to position the next template base in the catalytic center; simultaneously, the DNA rotates to make the 3'-OH group in the nascent RNA is properly aligned to attack the α-phosphorus atom of the next ribonucleoside triphosphate, thus powering the continuing chemical reaction. There are several contacts between amino acid residues in the enzyme and the incoming nucleotide to ensure that, first, it is a ribonucleotide rather than a deoxyribonucleotide, and second, it is the appropriate nucleotide to base-pair with the template base. Inappropriate nucleotides are simply rejected.

(5) The final step of transcription usually involves recognition of termination signals or termination sequences on the DNA with the help of numerous termination factors.

Termination results in dissociation of the triple or ternary DNA-RNA-polymerase complex: the transcript and the enzyme are released from the template, which regains its fully duplex character as the bubble closes. In the classical view of transcription, only one strand in any given region of the genome is transcribed. We now know that this is not always the case, as sometimes both can be transcribed in opposite directions.

7.2.2 Transcription requires the participation of many proteins

Transcription is a multi-step process, including initiation, elongation and termination. Each of these steps is quite complex and involves numerous pro-

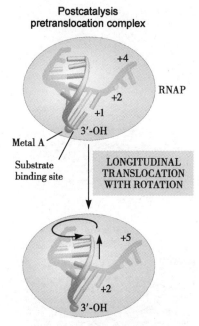

Figure 7.5 **Schematic of the DNA-RNA heteroduplex.** Based on the crystal structure of an elongating RNAP, the expected translocation movement of the DNA double helix with respect to the catalytic center of the enzyme is labeled here as metal A. Nascent RNA transcript is shown with 3'-OH. Top, postcatalysis pretranslocation complex; bottom, translocated active complex. Note that the transition involves two different kinds of motion, shown by black arrows: longitudinal translocation along the double helix and rotation, so that the 3'-OH group is aligned with the active center and the next nucleotide on the DNA template strand, at position +2, is oriented properly to determine the chemical nature of the incoming ribonucleotide.

tein factors. In addition, the RNA polymerases are multi-subunits in a majority of organisms.

For instance, all three transcription enzymes in eukaryotes are multi-subunit, varying from 12 subunits in Pol II to 17 subunits in Pol III. Ten of the subunits, including those containing the catalytic center, are shared by the three enzymes. Other subunits differ but occupy similar peripheral positions in the enzyme complex, for example, the Rpb4/7 sub-complex. Pol III contains a three-subunit complex of C82, C34, and C31 that is unique to this polymerase. Combined data from crystallographic studies and cryo-electron microscopy (cryo-EM) or imaging, together with structural modeling, produced good visualization of the respective structures, but the functions of some of the subunits and sub-complexes remain to be determined. Comparison of these structures indicates that the catalytic mechanism is same for all three polymerases except targeting to different classes of genes, as determined by peripheral proteins.

For the successful transcription in eukaryotes, the chromatin location of the transcribed DNA should be loosened or unwound, in which nucleosome-related proteins and modification enzymes are involved. Besides, numerous transcription factors, co-factors, accessory factors are needed to participate in this complicated event.

7.2.3 Transcription is rapid but often interrupted by pauses

Transcription is a rapid process, especially in bacteria, where the average *in vivo* rate has been estimated at ~45 base pairs or bp per second. In eukaryotes, it is roughly tenfold slower. The values above do not represent the actual rate of production of RNA in the cell, as many genes are being transcribed at the same time, and some of these have tandem polymerases traveling along them. The rate at which any one RNA polymerase moves along the DNA template is not uniform; the enzyme may temporarily pause on certain sequences before resuming elongation.

There are two types of pauses, long and short (Figure 7.6). Short pauses arise from sporadic alterations in the position of the 3'-OH end of RNA and/or the active site. They occur randomly once every 100 bp on average. Recovery from these short pauses is spontaneous and fast.

Long pauses are caused by backtracking, shifting the transcription bubble backward along the DNA-RNA heteroduplex, which leads to extrusion of the 3'-end of the transcript through a channel in the enzyme. The resulting drastic misalignment of the 3'-OH group with the catalytic site precludes further elongation. This pause cannot be spontaneously relieved; elongation can resume only after the extruded portion of the RNA chain is removed. This is accomplished by an endonucleolytic activity residing in the polymerase complex itself, in a reaction stimulated by specific protein elongation factors:

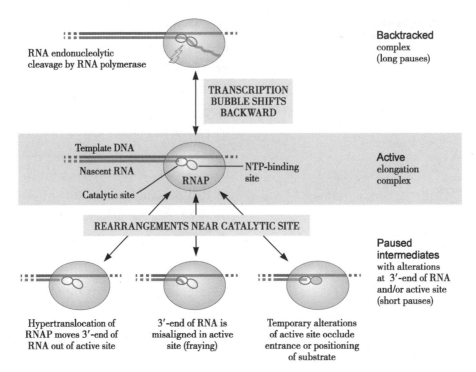

Figure 7.6 Conformational states of RNA polymerase elongation complexes. In the transcription, the 3'-end of the nascent transcript is properly aligned within the catalytic center of the enzyme for efficient elongation. Slight conformational rearrangements near the catalytic site cause displacement of the 3'-end from its proper position, slowing down transcription to lead to the frequently occurring short pauses. If the transcription bubble shifts backward along the DNA-RNA heteroduplex, known as backtracking, the 3'-end of the RNA transcript is extruded through a channel in the enzyme; the backtracked state is inactive because of long-range misalignment of the RNA end and the catalytic site.

GreA and GreB in bacteria, TFIIS in eukaryotes. This backtracking ensures the RNA transcription fidelity from DNA template, which is attributed to the proofreading ability of RNAPs, besides of base-pairing between nucleotides. Recovery from long pauses is a slow process.

7.2.4 Transcription can be visualized by electron microscopy

Visualization of transcription elongation can be observed by electron microscopy. Figure 7.7 shows the Christmas tree structure seen at sites of ribosomal RNA transcription. A spreading bacterial cell under electron microscopy showed DNA as a thin thread and polymerase complexes as small dots along the thread; RNA transcripts extend from the DNA in an almost perpendicular direction and have a knobby appearance resulting from the presence of proteins that bind to the nascent RNA chains (Figure 7.7A). The RNA chains close to the initiation site are very short, but their length gradually increases further downstream, creating structures that resemble the appearance of a tree. Interestingly, the transcription of ribosomal RNA in the eukaryotic nucleus

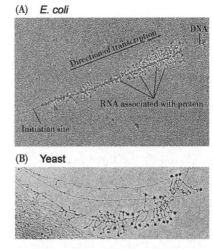

Figure 7.7 Transcription of ribosomal genes in (A) E. coli and (B) yeast. Both genes are transcribed from left to right, and the RNA products of each present the common Christmas tree appearance. The nascent rRNA chain associates with proteins that process it by nucleolytic cleavage before transcription is complete; these correspond to the knobs at the ends of transcripts.

produces very similar images (Figure 7.7B). These proteins have different functions; most of them participate in RNA processing (see Chapter 13), which begins soon after a sufficient length of RNA is synthesized and is extruded from the exit channel of the polymerase.

7.3 Transcription in bacteria

In bacteria, a single RNA polymerase produces the primary transcript precursors for all three major classes of RNA: messenger RNA (mRNA), ribosomal RNA (rRNA), and transfer RNA (tRNA). Since bacteria do not contain nuclei, ribosomes bind to mRNA while it is being transcribed, the protein synthesis occurs simultaneously with transcription.

7.3.1 Transcription initiation begins with a multi-subunit polymerase holoenzyme assembly

In bacteria, transcription initiation requires a multi-subunit polymerase complex, termed as the holoenzyme initiation. The subunit structure of the *Escherichia coli* enzyme is given in Table 7.1. Two α-subunits together with two structurally unrelated β-subunits form the core RNA polymerase that is absolutely needed for catalytic function. Sometimes the Ω subunit is included in the core, but it is not always needed. Another subunit, σ, is the initiation factor needed for promoter recognition during initiation.

Table 7.1 Subunit structure of RNA polymerase from *E. coli*. The core RNA polymerase complex comprises $α_2ββ'Ω$. β and β' are structurally unrelated. The holoenzyme consists of the core complex plus the initiation factor subunit σ.

Subunit	Function	MM, Daltons
β	active site	150 600
β'	active site	155 600
α	scaffold for assembly of other subunits, regulation platform	36 500
Ω	promotes RNAP assembly and stability	11 000
σ	initiation factor	70 300

For transcription initiation, σ first binds to the core polymerase complex to form holoenzyme that can recognize the promoter sequences to initiate transcription. The overall schematic initiation is presented in Figure 7.8. $σ^{70}$, the most abundant σ factor in bacterial and viral, or bacteriophage, can recognize a consensus sequence within a promoter. The consensus defines the base that occurs with the highest frequency at each position with respect to TSS.

Bacterial promoters contain two boxes of conserved sequences separated from each other by 16–18 nucleotides. The TSS itself is usually a purine. The consensus sequence of the box closest to the TSS is TATAAAT, and this is referred to as the −10 box. The consensus of the other box, referred to as the −35 region, is TTGACA (Figure 7.9). The relative strength of promoter depends primarily on its affinity to the appropriate σ factor. Strong promoters with a high frequency of initiation. The frequency of transcription initiation for any given gene depends on the needs of the cell. The more closely the promoter resembles the consensus sequence, the stronger the promoter: that is, the more frequently transcription initiates from that promoter. Even weak promoters still have the two boxes for recognition.

Promoter recognition resides entirely in the σ subunits, in which there are several different σ factors that recognize slightly varied promoters (Table 7.2).

Bacterial cells use slightly different σ factors to transcribe different groups of genes, and different bacterial species have different numbers of σ factors. The number seems to correlate with the complexity of the environment in which a given species lives: organisms with more varied lifestyles contain more σ subunits. Because all σ subunits employ the same core polymerase, there must be mechanisms that ensure the association of the appropriate factor under a given condition. Under certain circumstances, there is an orderly succession of alternative σ subunit usage. A common emergency encountered by bacteria is heat shock, a sudden exposure to an increase in temperature. A defensive response requires changes in protein, and hence RNA synthesis. The structure of a well-characterized $σ^{70}$ has been resolved (Figure 7.10).

σ⁷⁰ has three well-conserved domains, all of which make extensive contacts with the core enzyme. The regions of domains σ2 and σ4 recognize the DNA promoter sequences at −10 and −35 regions; when the σ subunit interacts with the promoter, it is sandwiched between the polymerase and the DNA to

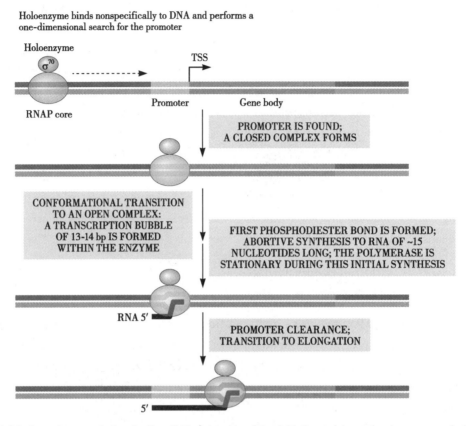

Figure 7.8 **Initiation of transcription in *E. coli*.** The transition from initiation to elongation is accompanied by conformational changes in the enzyme.

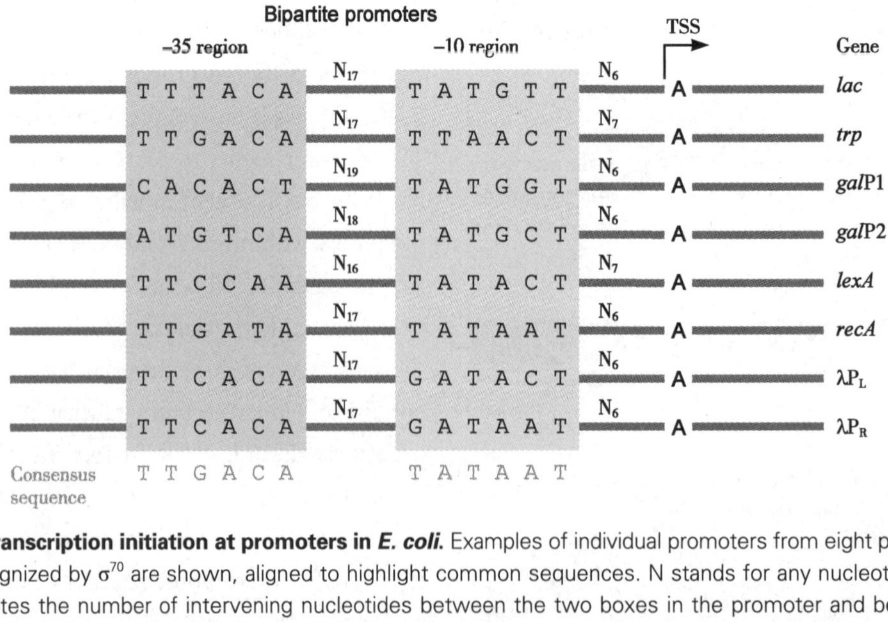

Figure 7.9 **Transcription initiation at promoters in *E. coli*.** Examples of individual promoters from eight phage and bacterial genes recognized by σ⁷⁰ are shown, aligned to highlight common sequences. N stands for any nucleotide residue; the subscript denotes the number of intervening nucleotides between the two boxes in the promoter and between the −10 region and the transcription start site. The consensus sequence was derived from a much larger database of more than 300 well-characterized promoters with the highest frequency of occurrence of a base at a specific position.

increase the affinity of polymerase to promoters and stabilizes the DNA-enzyme complex. The roles of σ are (a) to increase the affinity for promoter sequences while decreasing the affinity for nonpromoters, and (b) to stabilize the complex on promoter sequences while destabilizing it on random sequences. Promoter

Table 7.2 **σ factors in E. coli.** There are six factors in the σ^{70} family. σ^{70} and σ^S are closely related and recognize nearly identical core promoter sequences. Selectivity for specific genes may depend on protein factor binding. $\sigma^{54/N}$ constitutes a separate family on its own. N denotes any nucleotide. Housekeeping genes are defined as those that are constitutively expressed; they code for functions absolutely necessary for the life of the cell.

Subunit	Gene	Genes transcribed	Promoter sequence recognized	
			−35	−10
σ^{70}	rpoD	most housekeeping genes in growing cells	TTGACA	TATATT
$\sigma^{38/S}$	rpoS	starvation/stationary phase genes	TTGACA	TATAAT
$\sigma^{32/H}$	rpoH	genes whose expression is turned on when cells are exposed to heat	GTTGAA	CCCATNTA
$\sigma^{28/F}$	rpoF	genes involved in production of multiple flagella for rapid swimming	TAAA	GCCGATAA
$\sigma^{24/E}$	rpoE	genes involved in extreme heat-shock response	GAACCT	TCTAA
$\sigma^{19/F}$	fecI	fec (ferric citrate) gene for iron transport		
$\sigma^{54/N}$	rpoN	genes involved in nitrogen metabolism	none	CTGGCACNNNNNTTGCA

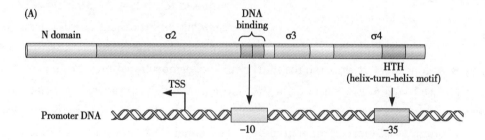

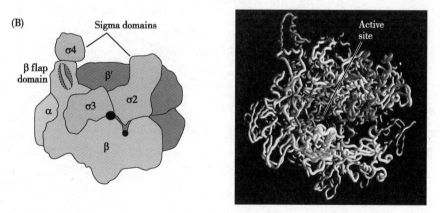

Figure 7.10 **Structure of RNAP holoenzyme from T. aquaticus.** (A) Conserved regions and functional assignment based on σ^{70}: arrows point to the regions of σ that contact conserved sequences in the promoter region. (B) Left, cartoon representation of σ domains 2, 3 and 4 of σ^{70} bound to core RNAP. Only one α subunit is visible here. The same color scheme is used in the high-resolution structure on the bottom. The pincers of the crab-claw structure formed by the β and β′ subunits create a 27-Å-wide channel containing the active site. During elongation, downstream DNA reaches the active site via this channel. Each σ domain makes extensive contacts with the RNAP core; all promoter-recognition determinants in σ are solvent-exposed until DNA is bound, with a spacing that is consistent with the predicted separation of their target promoter elements. Thus, the σ domains bridge the promoter DNA and the polymerase. Conformational changes in both partners are integral to initiation.

binding is 100 times faster than the maximum theoretical value for a diffusion-limited second-order reaction; this can be rationalized by a one-dimensional diffusion mechanism. Figure 7.11 illustrates how bacteria respond to heat shock through the consecutive use of two factors, $\sigma^{24/E}$ and $\sigma^{32/H}$. $\sigma^{24/E}$ is needed for transcription of the $\sigma^{32/H}$ gene; the $\sigma^{32/H}$ subunit then turns on an entire set of genes that participate in the heat-shock response.

In general, it must be noted that transcription is a very tightly regulated process (see Chapter 8): numerous regulatory mechanisms are at play, in addition to the mechanisms involving σ factors. In some cases, the α subunits play a role alongside σ in initiation. Each α subunit carries an extended C-terminal domain or CTD that is capable of interacting with sites upstream of the σ-sites described above. This provides another level of promoter discrimination in the initiation of bacterial transcription.

7.3.2 The initial step of elongation is frequently aborted

The initial steps of elongation are slow and the short RNA transcripts are frequently released from the ternary complex, which is known as abortive transcription. Using promoter sequences, the unusually high yields of aborted transcripts and the very slow rates of promoter escape were found. The use of one such promoter to visualize abortive transcripts on electrophoretic gels is illustrated in Figure 7.12. Note the extremely high frequency of abortion with this specific promoter: it takes the polymerase 300 trials before it can switch to stable elongation. With more conventional promoters, there are around 30–40 abortion events for each full-length transcript synthesized. The same aborted transcripts produced during the *in vitro* reaction are also observed in bacterial cells that contain introduced plasmids carrying the same template construct.

Understanding the mechanism of abortive transcription is challengeable, because two seemingly opposing events are involved. First, the polymerase moves along the DNA template to be able to synthesize a short RNA chain, a process that requires sequential reading of the sequence of bases in the template. Second, the polymerase either preserves or quickly re-forms contacts with the promoter to be able to re-initiate transcription once a short transcript is aborted. Whatever, three models are proposed to explain the occurrence of the abortive transcription: transient excursions, inch-worming, and scrunching (Figure 7.13). Current experiments demonstrated that the abortive transcription occurs through a scrunching mechanism, in which a stressed intermediate with approximately one turn of additional DNA unwinding is formed. Stress in this intermediate provides the driving force to break the existing interactions between RNAP and promoter and between RNAP and σ initiation factor, allowing promoter escape.

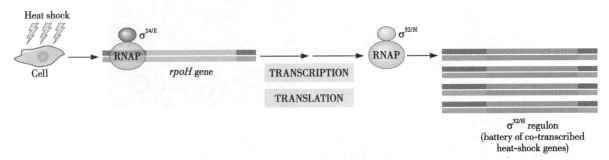

Figure 7.11 Alternative σ factors mediate *E. coli*'s global response to heat shock. The response of heat-shocked cells involves the successive use of two alternative σ factors: $\sigma^{24/E}$ is required for expression of the gene that encodes the second alternative factor $\sigma^{32/H}$, which in turn regulates the entire $\sigma^{32/H}$ regulon on a set of bacterial genes that share common regulatory elements for transcription. The $\sigma^{32/H}$ regulon is induced by excessive unfolding of proteins in the cell, caused by heat shock. The gene encoding chaperone GroEL, a protein involved in refolding misfolded or unfolded proteins, is a member of this regulon.

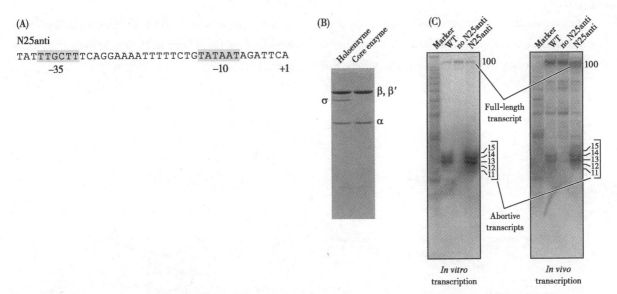

Figure 7.12 *E. coli* **RNAP and detection of abortive transcription *in vitro* and *in vivo*.** Transcription reactions were performed on DNA templates carrying the N25anti promoter, a 100-bp transcription unit, and the tR2 terminator. This entire construct is designated N25anti-100-tR2. (A) Phage T5 N25anti promoter is a classic model system for the study of abortive initiation and promoter escape, introduced in Michael Chamberlin's laboratory. The abortive/production ratio for this promoter is very high, ~300, as opposed to 40 for the closely related promoter N25. (B) SDS-polyacrylamide gel electrophoresis of purified commercially available preparations of *E. coli* RNA polymerase. The core enzyme contains α, β, and β′ subunits; the latter two co-migrate under these conditions. The holoenzyme also contains the σ subunit. This commercial RNAP preparation was used for the transcription experiments illustrated in part C. (C) Transcripts were analyzed following ethanol precipitation, electrophoresis on polyacrylamide gels containing urea, transfer to membranes, and hybridization with a ^{32}P-labeled probe complementary to the beginning of the transcribed region. The *in vivo* transcription reaction was performed on a plasmid carrying the template, which was introduced into live cells. Further experiments demonstrated that the *in vivo* products exhibited the hallmarks of abortive transcripts as defined *in vitro*: altering the strength of interactions between RNAP and promoter and between RNAP and σ factor alters the yields of 11–15-nucleotide transcripts. These interactions must not be too strong to allow ending of abortive initiation and enable promoter escape.

7.3.3 Elongation in bacteria must overcome topological problems

Unlike the transcription elongation process in eukaryotes, the process of elongation in bacteria is simpler, without nucleosome barrier to be overcome. However, when RNA polymerases of bacteria track the DNA helix, the super-helical stress will be topologically created in the DNA, which would constrain the elongation. The conventional view of transcription elongation assumed that the polymerase spiraled around the DNA as it tracked the double helix. Such a movement would entangle the transcript around the DNA. Now it is believed that in a cell, the polymerase is probably stationary and exerts both a linear pulling force and a rotary force on the DNA to thread it through its active center. The two kinds of motions—lateral translocation and rotation (as mentioned in Figure 7.5) — on the topologically constrained DNA create positive super-helical torsion ahead of the advancing polymerase and negative supercoiling in the wake of the enzyme (Figure 7.14).

As elongation proceeds along the topologically constrained loop domains in the bacterial nucleoid or in the eukaryotic nucleus, the buildup of positive supercoiling would be expected to stop transcription eventually, as the great energy cost of unwinding the overwound DNA would prohibit further elongation. Thus, a topoisomerase molecule will find the stressed region and attach to relax the super-helical stress. High levels of transcription usually reflect high frequency of initiation, which creates an array of polymerases tracking each other along the gene (as shown in Figure 7.7). Such an array will partially neutralize the positive and negative stresses that are produced by each polymerase.

The topological consequences of transcription *in vitro* transcription systems usually dissipate from the free DNA ends, or circular plasmids, where the positive and negative supercoiling propagating from opposite directions around the circle eventually cancel each other out.

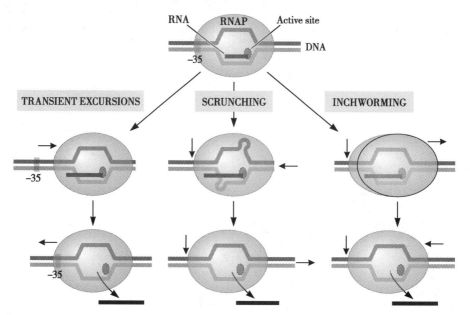

Figure 7.13 **Competing models for abortive initiation.** The models focus on the relative motions of RNAP and the DNA template, as indicated by arrows. The ability to synthesize numerous short transcripts that are released from the initially unstable transcribing complex, coupled with fast re-initiation of new transcripts, implies that the active site of the polymerase is moving along the DNA template in a forward direction but is still maintaining contacts with the promoter in order to allow re-initiation. Left, in the transient excursion model, RNAP breaks its contacts with the promoter region while transcribing the initial portion of the RNA chain but quickly returns to the promoter once the short nascent transcript is aborted. Right, in the inchworming model, the enzyme undergoes conformational changes that enlarge the footprint of the polymerase on the DNA, increasing the region over which the two molecules interact; note the slightly enlarged RNAP symbol. Thus the enzyme can continue transcribing while still maintaining its grip on the promoter. Release of the aborted transcript relaxes the RNAP to its normal dimensions. Middle, in the most recent scrunching model, the RNAP does not change shape; its effective footprint is increased by pulling in a portion of the downstream template, resulting in a stressed DNA conformation. When the short transcript is aborted, the scrunched DNA is released, and the enzyme is ready for initiation of a new chain.

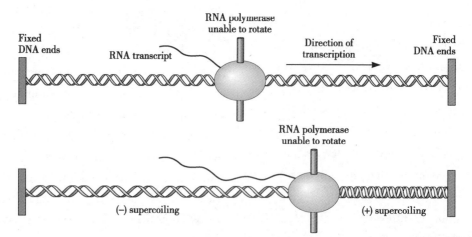

Figure 7.14 **Topological consequences of transcription elongation in a topologically constrained DNA template.** The DNA is represented as having fixed ends to satisfy the topological requirements for superhelicity. The transcription bubble does not rotate in the immobile RNA polymerase catalytic center, creating domains of positive and negative supercoiling, represented as DNA with different twist. Superhelical stress accumulates quickly because transcription of one helical turn, around 10 bp, leads to the creation of one superhelical turn. DNA structure cannot tolerate too much twist, so the stress can be relieved by the creation of writhe, which is coiling of the DNA axis.

7.3.4 There are two mechanisms for transcription termination in bacteria

Termination of transcription in bacteria requires certain sequences on genes that give rise to particular sequences in the elongating transcripts. Bacteria use two very different mechanisms to terminate transcription: one intrinsic or sequence-dependent and the other protein-factor-dependent. In *Escherichia coli*, the number of genes using one mode or the other is about equally divided.

(1) Intrinsic or sequence-dependent termination

Intrinsic termination is entirely determined by the nucleotide sequence at the 3'-end of the gene and does not require any other factors. The sequence of the template near the termination site gives rise to an RNA transcript capable of forming a hairpin. This is due to the inverted DNA repeat (Figure 7.15). In addition, the inverted repeat should be followed by a string of six or seven As. This type of DNA template leads to the formation of a hairpin in the RNA transcript, which causes the polymerase to pause and strips the RNA from the complex. The weak base pairing between the A string in the template and the complementary U string in the transcript contributes to dissociation of RNA transcript. Proteins such as NusA stabilize the RNA hairpin and helps the dissociation process.

(2) Rho-dependent termination

Rho factor is a hexameric ATP-dependent DNA-RNA helicase. The steps in Rho-dependent termination process are schematically depicted in Figure 7.16. The process begins when Rho binds to a Rho utilization site or rut site on the RNA transcript, and rut sites are very rich in C and poor in G. Once Rho binds to the rut sequence, it undergoes a conformational transition from an open ring to a closed ring that embraces the RNA. In an ATP-dependent process, Rho is propelled along the RNA transcript in a 5' to 3' direction; its DNA-RNA helicase activity pulls the RNA from the DNA in the transcription bubble, thus terminating transcription.

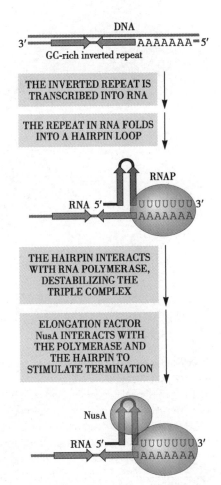

Figure 7.15 **Intrinsic termination in *E. coli*.** Transcription terminates when an inverted repeat in the RNA transcript forms a hairpin that is abutted by a string of uracils; the low stability of the RNA-DNA heteroduplex, due to A-U base pairing, contributes to the destabilization of the RNA-DNA hybrid in the transcription bubble. The RNA hairpin interacts directly with the RNA polymerase, destabilizing the triple DNA-RNA-polymerase complex. Additionally, the elongation protein factor NusA interacts with both the polymerase and the RNA hairpin structure to aid in dissociation of the complex.

In bacteria, the transcription and translation of protein-coding genes are coupled together, since they occur within a single cellular compartment. Once the transcription begins and the 5'-end of the transcribed mRNA is accessible, a ribosome attaches and initiates protein synthesis (see later chapter). When the first ribosome moves from the 5'-end of the mRNA, a second ribosome can attach.

Recent research suggests that efficient binding and movement of ribosomes along the mRNA increases the speed of RNA polymerase. In the absence of ribosomes, the polymerase slows down and waits

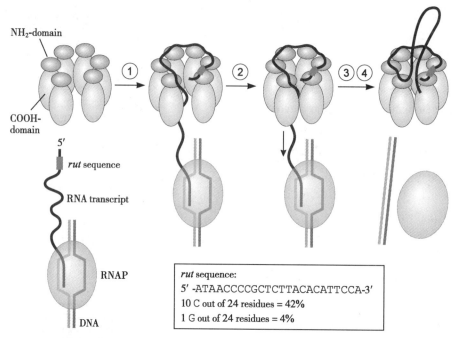

Figure 7.16 **Rho-dependent termination of transcription in *E. coli*.** Rho factor is a hexameric ATP-dependent DNA-RNA helicase that acts on DNA-RNA hybrids exclusively. Transcription termination occurs in four steps. (Step 1) The N-terminal domain of Rho (NH_2-domain) binds the Rho utilization or *rut* sequence on the transcript. (Step 2) The C-terminal domain of Rho (COOH-domain) binds the mRNA downstream of *rut*, and the ring closes. (Step 3) The C-terminus cyclically hydrolyzes ATP to propel itself along the mRNA in a 5′ to 3′ direction. (Step 4) Helicase action leads to disassembly of the transcription complex. (Inset) Sequence of Rho utilization or *rut* site of λ tR1 terminator. Note the highly skewed nucleotide content, with 42% of the residues being C and only 4% being G. This is a typical characteristic of the *rut* sequences: the efficiency of usage of these sites for termination increases with the length of the C-rich/G-poor region. *rut* sites are found at varying distances from the actual termination sites.

for the ribosome to catch up. There is evidence for direct interaction of a ribosomal protein with another protein that is bound to RNA polymerase. This interaction tethers the protein synthesis and transcription machinery together, allowing the whole process of transcription and translation to take place within minutes.

7.4 Transcription in eukaryotes

Unlike prokaryotes in which the coupling of RNA transcription and protein translation happens simultaneously due to the lack of the mature nucleus compartment, transcription in eukaryotes sequentially occurs in nucleus by three different RNA polymerases, each principally responsible for one of the major classes of RNA, and most transcripts function in cytoplasm. The primary mRNA products are trimmed and transported from nucleus to cytoplasm for the coded protein translation. The various non-coding RNAs are similarly produced and transported for the eventual functions.

A metazoan must also preferentially express different parts of the same genome at different times in its many different cell and tissue types. Of these numerous and complex regulatory mechanisms that control gene expression, regulation of transcription is considered the most important: it is mainly through effects on transcription that cells respond to developmental signals and to the environment.

7.4.1 Overview of transcriptional regulation of eukaryotes

Transcription in eukaryotes is more complex than that in bacteria, not only in terms of the cell structure and the mechanisms involved, but also in terms of the different types of RNA polymerases for various RNA classes and their regulations. Furthermore, for a typical eukaryote, the transcription must evolve to fit the needs of a host of cell types.

(1) Transcription in eukaryotes is a complex, highly regulated process

At first, the complication of eukaryote transcription is reflected both in the more structured chromatin medium in which it occurs, and in the transcriptional apparatus itself. Both the initiation and elongation phases of transcription require at least transient remodeling of chromatin structure, often including the removal of nucleosomes. The primary transcript of eukaryotic genes also needs to be modified to add "Cap" and remove the introns (splicing) before a final, mature mRNA is produced.

(2) Protein-coding gene regulatory elements

Protein-coding sequences make up only a small fraction of a typical multicellular eukaryotic genome, and they account for less than 2% of the human genome. The typical eukaryotic protein-coding gene consists of a number of distinct transcriptional regulatory elements that are located immediately 5' of the transcription start site (termed +1). The regulatory regions of unicellular eukaryotes such as yeast are usually composed only of short sequences located adjacent to the core promoter (Figure 7.17A). In contrast, the regulatory regions in multicellular eukaryotes are scattered over an average distance of 10 kb of genomic DNA, with the transcribed DNA sequence only accounting for just 2 or 3 kb (Figure 7.17B).

Protein-coding gene regulatory elements are specific *cis*-acting DNA sequences that are recognized by *trans*-acting transcription factors. *Cis*-regulatory elements in multicellular eukaryotes can be classified into two broad categories, based on how close they are to the start of transcription: promoter elements and long-range regulatory elements. Among them, enhancers and silencers are important *cis*-regulatory elements. A typical protein-coding gene is likely to contain several enhancers that act at a distance. Similar elements that repress gene activity are called silencers. These DNA sequence elements are usually 700–1000 bp or more away from the start of transcription. The hallmark of enhancers (and silencers) is that, unlike promoter elements, they can be downstream, upstream, or within an intron, and can function in either orientation relative to the promoter (Figure 7.18). A typical enhancer is around 500 bp in length and contains on the order of 10 binding sites for several different transcription factors. Enhancers increase gene promoter activity either in all tissues or in a regulated manner, i.e. they can be tissue-specific or developmental stage-specific. Silencers have the opposite effect and act to decrease gene promoter

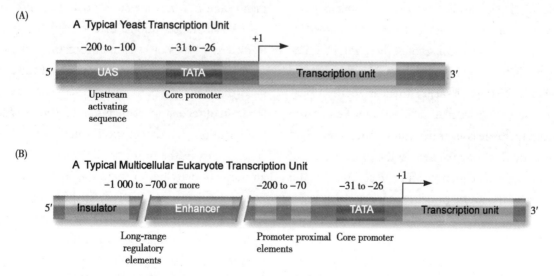

Figure 7.17 **Comparison of a simple and complex RNA Pol II transcription unit.** (A) A typical yeast transcription unit. The start of transcription (+1) of the protein-coding gene (transcription unit) is indicated by an arrow. (B) A typical multicellular eukaryote transcription unit with clusters of promoter proximal elements and long-range regulatory elements located upstream from the core promoter (TATA). Although the figure is drawn as a straight line, the binding of transcription factors to each other draws the regulatory DNA sequences into a loop.

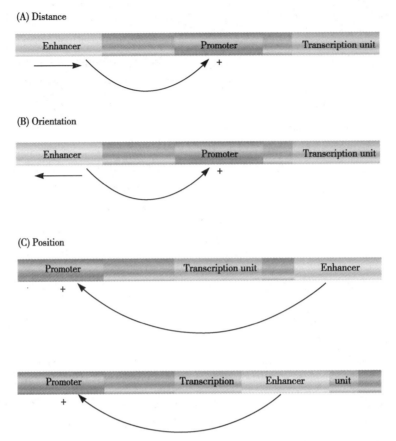

Figure 7.18 **Three key characteristics of an enhancer element.** An enhancer element can activate a promoter at a distance (A), in either orientation (B), or when positioned upstream, downstream, or within a transcription unit (C).

activity. Because enhancers and silencers can function at a vast distance from the gene promoter, they also could potentially activate or silence other nearby gene promoters inappropriately. Another long-range element called an insulator solves this problem.

When comparing the regulatory regions of two genes in multicellular eukaryotes, there will be variation in whether a particular element is present, the number of distinct elements, their orientation relative to the transcriptional start site, and the distance between them are concerned. Information content at the genetic level is expanded by the great variety of regulatory DNA sequences and the complexity and diversity of the multi-protein complexes that regulate gene expression. TATA-box, for instance, in a gene promoter region can be recognized by RNA Pol II complex (see later section).

(3) The role of specific transcription factors in gene regulation

Expression of particular protein-coding genes is essentially mediated by a group of sequence-specific DNA-binding proteins that are called transcription factors. Transcription factors interpret the information presented in gene promoters and other regulatory elements, and transmit the appropriate response to the RNA Pol II transcriptional machinery. Various genes in different types of cells in an organism share the same general transcription factors. What turns on a particular gene in a particular cell is the unique combination of regulatory elements and the gene-specific transcription factors that specifically bind to unique *cis*-elements in the promoter of specific genes. The amounts or activities of these specific transcription factors regulate a gene activity at the transcriptional level. Likewise, the genes encoding the transcription factors themselves may be transcriptionally regulated by other regulatory proteins. Transcription factors can also be activated or deactivated by proteolysis, covalent modification, or ligand binding.

Many transcription factors are members of multi-protein families. For example, nuclear receptors are members of a superfamily of related proteins, including the receptors for steroid hormones, thyroid hormone, and vitamin D. NF-κB is yet another family of proteins, and Sp1—one of the first transcription factors to be isolated—is a member of the Sp family of proteins. Within each family, the members often display closely related or essentially identical DNA-binding properties but distinct activator or repressor properties.

Transcription factors are modular proteins consisting of several domains, including three major domains: a DNA-binding domain, a transactivation domain, and a dimerization domain. In addition, transcription factors typically have a nuclear localization sequence (NLS) for entry into the nucleus, and some also have a nuclear export sequence (NES) for exit from nucleus.

Transcription factors influence the rate of transcription of specific genes either positively (activators) or negatively (repressors) by specific interactions with promoters proximal, enhancers, silencers, and other long-range regulatory elements and by their interaction with other proteins.

Transcription factors that serve as activators increase the rate of transcription by several mechanisms: (a) Stimulation of the recruitment and binding of general transcription factors and RNA Pol II to the core promoter to form a pre-initiation complex; (b) Induction of a conformational change or post-translational modification such as phosphorylation that stimulates the enzymatic activity of the general transcription machinery; (c) Interaction with chromatin remodeling and modification complexes to permit enhanced accessibility of the template DNA to general transcription factors or specific activators.

These different roles of transcription factors are achieved directly via protein-protein interaction with the general transcription machinery or via interactions with transcriptional coactivators and corepressors, which will be described in the next section.

With the help of other factors, including specific transcription factors, co-factors, accessory factors that are interacting with the particular sequences on regulatory regions of genes, different RNA polymerases bind to the matched promoters to initiate transcription for the various types of RNAs to realize the tissue or cell type specificity.

(4) Transcriptional coactivators and corepressors

Coactivators and corepressors are proteins that increase or decrease transcriptional activity, respectively, without binding to DNA sequence directly. Instead, they bind directly to transcription factors and either serve as scaffolds for the recruitment of other proteins containing enzymatic activities or have enzymatic activities themselves for altering chromatin structure. In the broadest sense, coactivators can be divided into two main categories: (a) Chromatin modification complexes: multiprotein complexes that modify histones post-translationally, in ways that allow greater access of other proteins to DNA; (b) Chromatin remodeling complexes: multiprotein complexes of the SWI/SNF family and related families that contain ATP-dependent DNA unwinding activities. Corepressors have the opposite effect on chromatin structure, making it inaccessible to the binding of transcription factors or resistant to their actions. Mechanisms for transcriptional silencing will be discussed in more detail in the next chapter.

Chromatin modification is now known to play a central role in both gene activation and repression. There are four main types of modification to histone tails in chromatin known to play a role in regulating gene expression on transcriptional level: (a) acetylation of lysines, (b) methylation of lysines and arginine, (c) ubiquitinylation of lysines, and (d) phosphorylation of serines and threonines. Other less common modifications are ADP-ribosylation of glutamic acid and sumoylation of lysine residues.

(5) Eukaryotic cells contain multiple RNA polymerases, each specific for distinct functional subsets of genes

In eukaryotic cells, there are three multi-subunit nuclear polymerases, abbreviated as Pol I, Pol II, and Pol III, each of which transcribes a certain set of genes, with the help of their transcription factors,

TF I, TF II and TF III, respectively. There are also organelle-specific enzymes found in mitochondria and chloroplasts (Table 7.3).

The organelle polymerases are monomeric, which is usually attributed to the proposed evolutionary origin of eukaryotic organelles from bacteria symbiotically living in other bacterial cells, to eventually form an integral eukaryotic cell.

7.4.2 Transcription by RNA polymerase II

RNA polymerase II is responsible for not only the protein-coding RNA transcription, but many long noncoding RNA molecules. Thus, Pol II is a very important enzyme in eukaryote transcription. We will focus on the transcription process by this enzyme as an example in detail.

(1) Structure of RNA polymerase II

Pol II structure is highly conserved across eukaryotic species, as yeast and human Pol II sequences exhibit 53% overall identity, with the conserved residues distributed over the entire structure.

Yeast Pol II is the only eukaryotic RNA polymerase whose structure has been resolved by crystallography. Lessons about how the enzyme functions can be learned from this structure, which provides a general model, as the basic mechanism of Pol II functioning seems to be shared by the other eukaryotic polymerases. The structure shows that Pol II has a positively charged active-center cleft, with the smaller subunits arranged around the periphery of the complex. At the bottom of the cleft lies a magnesium ion known as metal A that participates in the catalytic addition of each ribonucleoside monophosphate residue. The structure of the core polymerase during elongation is shown in Figure 7.19A. The mobile clamp, a domain of Rpb1, in the free enzyme has an open conformation that allows DNA binding during initiation but closes down to almost fully embrace the DNA and the RNA transcript, thus ensuring processivity of transcription. The structure also reveals the position of the bridge helix that connects the two large subunits; this helix is implicated in translocation motions of the enzyme with respect to the DNA. The conformation of the DNA-RNA hybrid at the catalytic center is intermediate between the A- and B-forms of DNA, as seen in Figure 7.19B.

The largest subunit of Pol II, Rpb1, contains a long C-terminal domain or CTD that is not present in the homologous subunits of Pol I and Pol III. CTD is characterized by the presence of numerous repeats—26 in yeast, 52 in humans—of a heptapeptide sequence, YSPTSPS. The CTD does not have a regular structure and so is not seen in crystallographic studies. The last ordered residue of Rpb1 seen in the crystal structure lies at the beginning of a linker that connects the structured body of Rpb1 to the unstructured CTD (Figure 7.20A). The unstructured CTD is probably not extended but compacts into a disordered state, close to the body of the enzyme. CTD contains

Table 7.3 Eukaryotic DNA-dependent RNA polymerases.

RNA polymerase	Number of subunits	Genes transcribed	Cellular localization	Molecules per cell
Pol I	14	pre-ribosomal RNA	nucleolus	40 000
Pol II	12	mRNA precursors; U1, U2, U4, and U5 snRNAs[a]; long noncodingRNAs	nucleoplasm	65 000
Pol III	17	5S and 7S rRNAs; tRNA; U6 snRNAa; other small RNAs	nucleoplasm	20 000
Mitochondrial polymerase	monomeric; encoded by nuclear gene; similar to T7 RNAP	mitochondrial genes	mitochondria	unknown
Chloroplast polymerase	monomeric; encoded by chloroplast gene; similar to polymerases in cyanobacteria	chloroplast genes in plants	chloroplasts	unknown

[a]Small nuclear RNAs, components of the spliceosome.

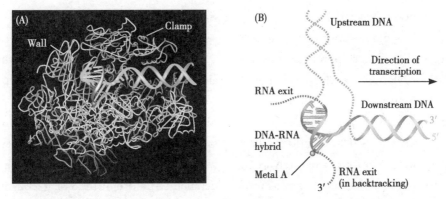

Figure 7.19 **Yeast RNA Pol II structure and the enzyme in the act of transcription.** (A) Structure of enzyme at 2.8 Å resolution. Polypeptide chains are shown in white, with the mobile clamp, a domain in Rpb1. The clamp closes on DNA and RNA in the elongation complex but has a different orientation in the free enzyme to allow DNA binding at initiation. A bridge helix connects the two largest subunits and undergoes conformational transitions during the translocation steps. The nontemplate DNA strand is shown to align with the nascent RNA transcript. DNA is surrounded by protein over ~270°. This tight grip of the polymerase on the DNA ensures high processivity of the enzyme. (B) This schematic represents the nucleic acids in the transcribing complex. The nascent RNA chain is extended from Metal A site. Solid ribbons are from the crystal structure; dashed lines indicate possible paths of the nucleic acids not revealed in the structure. Note that the entering and exiting DNA duplexes are at an angle of nearly 90° with respect to each other. The conformation of the DNA-RNA hybrid is intermediate between canonical A- and B-forms of DNA. The schematic also reveals where the nascent transcript leaves the polymerase and where it ends up if backtracking occurs.

the repeated heptapeptide sequence YSPTSPS, which can be phosphorylated. Individual amino acid residues within the repeat can undergo phosphorylation, glycosylation, and proline *cis-trans* isomerization (Figure 7.20B). Subsets of these modifications influence the function of Pol II by serving as binding platforms for specific proteins.

The phosphorylation of serine residues and possibly also other modifications change during the transcription process (Figure 7.21). The phosphorylation pattern of CTD changes during the transcription process and determines which protein factors bind to it. The serines in positions 2 and 5, as well as 7, of the YSPTSPS repeat in CTD can undergo phosphorylation in a pattern that reflects the stage of transcription. In the pre-initiation complex, CTD is not modified. The transition to elongation is accompanied by phosphorylation on Ser5 residues. Later

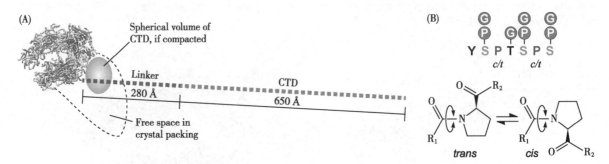

Figure 7.20 **C-terminal repeat domain of Rpb1, the largest subunit of Pol II.** (A) Rpb1 is highly homologous in all three polymerases, but only Rpb1 of Pol II contains a C-terminal repeat domain or CTD. Dashed lines represent the possible length of a fully extended linker and CTD. The absence of electron density corresponding to the linker and CTD provides evidence of disorder or motion, but even if disordered, these regions are unlikely to be in an extended conformation. The existence of free space in the crystal lattice for CTDs from four neighboring polymerases of Pol II suggests that the CTD may actually be compacted. (B) CTD contains the repeated heptapeptide sequence YSPTSPS, which can be phosphorylated (P), glycosylated (G). Finally, both prolines can undergo a *cis-trans* isomerization reaction as illustrated. R_1 and R_2 represent continuations of the polypeptide chain toward the N- and C-terminus, respectively. The number of repeats differs in different organisms, ranging from 26 in yeast to 52 in humans.

in the elongation process the CTD becomes doubly phosphorylated, carrying modifications on both Ser5 and Ser2. Finally, as the enzyme approaches the end of the gene, phosphorylation is exclusively on Ser2 residues. After termination of transcription, Pol II falls off the template and undergoes a dephosphorylation reaction, which allows it to re-initiate. It is believed that these modifications, separately or in combination, determine the binding of various protein factors to the CTD, in turn regulating transcription and RNA processing (see later section).

Sequence conservation between yeast and bacterial polymerases is less extensive, with the conserved regions clustering around the active center (Figure 7.22A), but the conservation of structure is remarkable (Figure 7.22B). The eukaryotic enzymes share a core structure, and thus a conserved catalytic mechanism, with the bacterial enzyme. The differences in structure are at the enzymes' periphery and surface, where interactions with other proteins occur. The detailed structure of the Pol II enzyme complex is especially revealing with respect to its mechanism of action during RNA elongation.

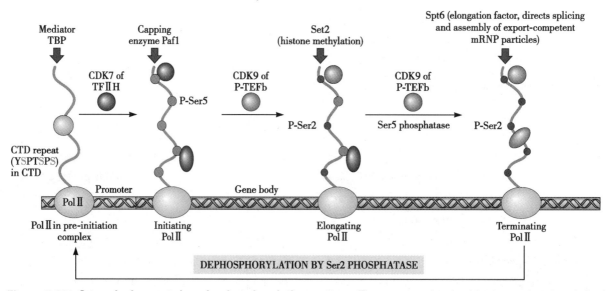

Figure 7.21 **C-terminal repeat domain phosphorylation pattern.** The enzymes involved in these phosphorylation/dephosphorylation events are indicated above and below the arrows connecting the different stage polymerases. The phosphorylation pattern of CTD determines what protein partners bind to the polymerase and control its activity. There are more than 100 different proteins that interact with CTD; some of them are depicted above the respective complex and are discussed throughout the book.

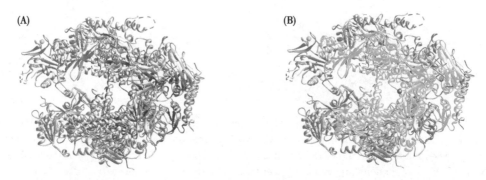

Figure 7.22 **Comparison between the two largest subunits of bacterial and eukaryotic RNA polymerases.** (A) Bacterial RNA polymerase from *Thermus aquaticus*; (B) eukaryotic RNA polymerase from *Saccharomyces cerevisiae*. The magenta sphere is the Mg^{2+} ion in the active site. The structures show considerable conservation around the active center, suggesting a conserved catalytic mechanism. Note that structural homology is much higher than sequence homology. The differences are mainly located in the peripheral and surface features, where the polymerases interact with other proteins.

(2) General transcription factors and pre-initiation complex formation

Transcription initiation in eukaryotes depends on the assembly of the pre-initiation complex (PIC) at core promoters, and the detailed information about the PIC factors is given in Table 7.4. There are two broad types of core promoters: focused or sharp-type promoters and dispersed or broad-type promoters (Figure 7.23). In vertebrates, less than a third of all promoters are focused.

The motifs constituting focused promoters are relatively well known, and the bound protein complexes are also crucial for the general and specific transcription of genes (Figure 7.24), but less is known about the characteristic sequence motifs and factors involved in PIC formation at dispersed promoters.

The RNA polymerases themselves in eukaryotes cannot recognize promoter sequences and have to seek help of the basal transcription factors (also called general transcription factors, GTFs) to bind to promoters and initiate the process, independent of the frequency at which a particular gene is transcribed.

A schematic of the minimal PIC assembled on a TATA-box promoter is presented in Figure 7.25A. The first complex to bind to the promoter *in vitro* is

Table 7.4 **Protein complexes involved in pre-initiation complex assembly.** TFIIA, previously thought of as a general transcription factor, is actually a coactivator that interacts with activators and components of the basal initiation machinery to enhance transcription.

Protein complex	Number of subunits	Function
RNA Pol II	12	catalyzes transcription of all protein-coding genes and a subset of noncoding RNA, including RNA components of the spliceosome
TFIIB	1	stabilizes TFIID binding to promoter; aids in recruiting TFIIF/Pol II to promoter; directs accurate TSS selection
TFIID	14	subunits include TBP and TAFs; nucleates PIC assembly through either TBP binding to TATA box or TAF binding to alternative promoters; coactivates transcription through TAFs/activator direct interactions
TFIIE	2	helps to recruit TFIIH to promoters and stimulates its helicase and kinase activities; binds single-stranded DNA; essential for promoter melting
TFIIF	2–3	forms a tight complex with Pol II and enhances its affinity for TBP-TFIIB-promoter complex; necessary for TFIIE/TFIIH recruitment; aids in TSS selection and promoter escape; enhances elongation efficiency
TFIIH	10	ATPase/helicase activity for promoter opening and clearance (helicase activity for DNA repair that occurs on transcribed genes); kinase for Ser5 phosphorylation of CTD; facilitates transition from initiation to elongation
Mediator	24	regulates transcription by serving as a bridge between gene-specific activators (or repressors) and basal TFs; absolutely required for basal transcription from almost all Pol II promoters (can be regarded as a general TF and a signal processor)

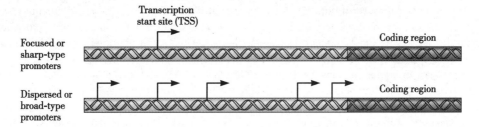

Figure 7.23 **Focused versus dispersed core promoters.** Focused or sharp-type promoters contain single or tightly clustered TSSs. They are more ancient and widespread in nature. Dispersed or broad-type promoters, on the other hand, contain multiple weak TSSs over a region of 50–100 nucleotides. They are usually found in CpG islands (see later chapter) and are more common than focused promoters in vertebrates.

transcription factor IID, TFIID, which contains the TATA-binding protein (TBP) with six core subunits, and a variable set of additional subunits, termed TATA-binding protein-associated factors (TAFs) (Figure 7.25B). TBP binding to the TATA box induces a kink in the DNA and opens the minor groove of the double helix, wherein most of its contacts occur. The other GTFs and the RNA polymerase II complex assemble around this site. The TBP-DNA complex is slightly asymmetrical, ensuring the directionality of transcription, that is, the transcription occurs on the correct DNA strand (Figure 7.25C). Most proteins recognize nucleotide sequences in duplex DNA through the major groove.

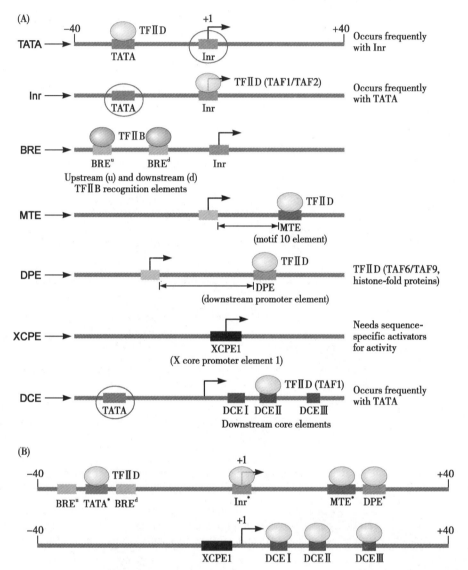

Figure 7.24 **Known core promoter motifs for Pol II transcription.** (A) Schematic representation of some individual promoters with the bound protein complexes. Each individual element occurs in only a fraction of all promoters, with the Initiator element (Inr), being the most frequent. Double-headed arrows in some promoters indicate the critical spacing between two elements. Elements that may or may not be present in a specific promoter are circled; for example, the TATA box is a secondary promoter element that is present in some Inr promoters. The synergy between motif ten element (MTE) and other promoter elements led to the creation of the super core promoter (SCP), which contains optimized versions of TATA, Inr, MTE, and downstream promoter element (DPE). These elements are marked by asterisks in part B. SCP is the strongest promoter functioning both *in vitro* and *in vivo*, with an unusually high affinity for TFIID. The green ovals represent TFIID that recognizes at least five of the seven promoters. TFIID complexes recognize different promoter elements. (B) To-scale diagram showing the mutual disposition of various promoter elements. The lower schematic illustrates elements that overlap the elements presented in the upper schematic. There is no promoter element that is universal, that is, present in all promoters.

It is believed that the six basal transcription factors TFIIB, D, E, F, H, and J are sufficient for transcription initiation (Table 7.4). Recently, another large protein complex, Mediator was discovered in yeast and mammalian cells, as the seventh basal transcription factors (Figure 7.26A). Interaction of RNAP II with Mediator is very extensive, and Mediator surrounds the face of the polymerase where TBP, TFIIB, and TFIIF presumably bind. All of these contacts are consistent with reported stabilization of the pre-initiation complex by Mediator. Mediator serves as a bridge between Pol II and the sequence-specific proteins that recognize regulatory elements surrounding the gene. Direct contact between Mediator and Pol II involves Pol II subunits Rpb4 and Rpb7 (Figure 7.26B and Figure 7.26C), which stabilizes the pre-initiation complex for efficient initiation.

The level of transcription is determined mainly by the frequency of transcription initiation and the stability of the minimal pre-initiation complex (PIC), which in turn depend on signals from regulatory proteins such as activators or repressors.

(3) Initiation of transcription

Transcription initiation in eukaryotes is exceedingly complex and depends upon assembly on the promoter of a huge multiprotein complex. The current model of transcription initiation steps is presented in Figure 7.27. (a) Mediator is recruited to a promoter through the action of activator A1. (b) The action of a second activator, A2, leads to a conformational change of Mediator that now provides a binding site for components of the basal transcriptional machinery. (c) Transcription factors TFIIB and TFIID posi-

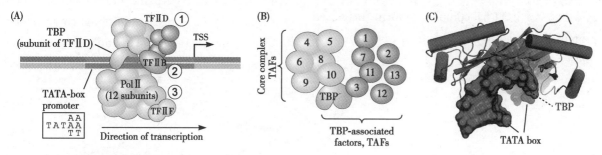

Figure 7.25 Assembly of minimal functional pre-initiation complex on the TATA box during transcription initiation. (A) Schematic of the partial PIC. TBP and the other components of TFIID bind to the TATA box and recruit all other GTFs, and the polymerase itself. Additional GTF E, H, and J are not shown. The numbers show the succession of binding *in vitro*. (B) TFIID: the core complex of six subunits, 4–6 and 8–10, depicted in light green, associates with combinations of other TAFs and TBP to form a large variety of different complexes. These different complexes bind to different types of promoters to initiate transcription. (C) Structure of TBP (in ribbon) is bound to a promoter TATA box.

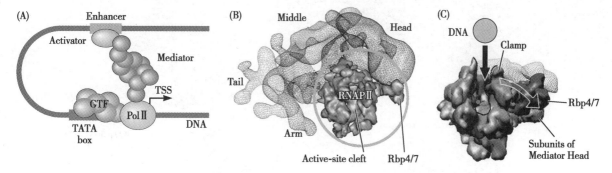

Figure 7.26 Mediator is a coactivator, co-repressor, and general transcription factor all in one. (A) Schematic showing Mediator serving as a bridge between proteins that activate or repress transcription and the transcriptional machinery assembled at the promoter. Note the direct contacts with both the activator protein and Pol II. (B) Model of the Mediator-RNAP II complex, based on cryo-EM imaging, reveals contacts between the Head module of Mediator and the RNAP II subunits Rpb4/Rpb7. (C) Interaction between the Head module of Mediator and Rpb4/Rpb7 could affect the conformation of the polymerase clamp domain and possibly facilitate opening, shown by a white open arrow, of the RNA polymerase II active-site cleft, outlined in black, to allow access of double-stranded promoter DNA to the polymerase active site.

tion themselves at the promoter and form a landing platform for the Pol II/TFIIF complex. (d) Pol II/TFIIF complex binds, or Pol II and TFIIF concomitantly bind, to the pre-formed Mediator/DNA/TBP/TFIIB complex. (e) Through the action of TFIIE and TFIIH that phosphorylates CTD of RNA Pol II, the promoter opens and the template DNA strand reaches the polymerase active site. (f) Following the abortive transcription phase, Pol II, or the Pol II/TFIIF complex, escapes the promoter, leaving behind a platform that facilitates rapid assembly of a new pre-initiation complex and re-initiation by new incoming Pol II/TFIIF complex.

Thus, once the polymerase clears the promoter to enter the processive elongation phase of transcription, a scaffold complex comprising Mediator, most of the basal TFs, and an activator protein stays behind; this scaffold can now be quickly repopulated by a free Pol II/TFIIF complex to form the closed promoter complex. This speeds up the next round of transcription initiation.

(4) Transcription elongation through nucleosomes

Transcription elongation by RNA Pol II through the coding region of the gene begins with mRNA synthesis at the promoter. Incoming (downstream) DNA is unwound before the polymerase active site and is rewound beyond it to form the exiting (upstream) duplex. In the unwound region, the DNA template strand forms a hybrid duplex with growing mRNA. The transcript synthesis passes through repetitive cycles of adding one new nucleotide to the growing RNA chain, followed by translocation steps that move the enzyme complex one base pair along the DNA (Figure 7.28). First, the DNA within the polymerase opens or denatures into a transcription bubble. The nucleoside triphosphate or NTP first binds to an open active center in the pre-insertion stage, when the selection of the appropriate NTP—that is, the one which base-pairs with the template base to be copied—takes place. Additional contacts help to discriminate between NTPs and deoxynucleoside triphosphates, dNTPs. In the second step, the correct

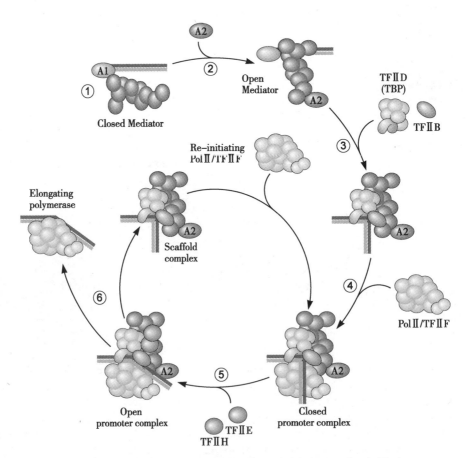

Figure 7.27 **Structural model for Mediator initiation and re-initiation.**

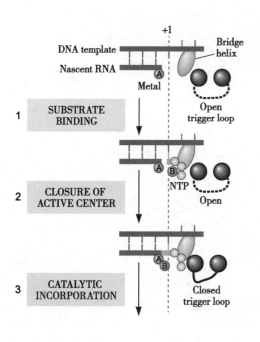
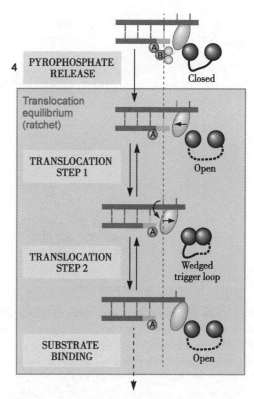

Figure 7.28 **Model of the nucleotide addition cycle.** Two functional polymerase elements: bridge helix is in green, Rpb1 residues 811–845, and trigger loop in brown, Rpb1 residues 1062–1106 and NTP substrate in orange. The two catalytic magnesium ions A and B are shown. The vertical dotted line indicates the position of DNA template base +1. The cycle is based on a Brownian ratchet model, in which the ground state of the elongation complex is an equilibrium between interconverting pre- and post-translocation states. This oscillation is temporarily stopped by substrate binding and resumes around the next template position after nucleotide addition.

NTP is delivered to the insertion site; this step is followed by the catalytic incorporation of the nucleoside monophosphate or NMP into the nascent RNA chain. The release of pyrophosphate completes this portion of the cycle.

In structural view, this cycle begins with the binding of an NTP substrate to an open active center or open trigger loop. Folding of the trigger loop leads to closure of the active center. The growing 3′-end of the RNA chain becomes elongated by catalytic addition of the nucleotide, which results in formation of a pyrophosphate ion. The release of pyrophosphate destabilizes the closed conformation, leading to trigger loop opening. In the resulting pretranslocation state, the incorporated 3′-terminal nucleotide remains in the substrate site. Translocation possibly takes place in two steps: (a) the hybrid moves to the post-translocation position, the downstream DNA translocates until the next DNA template base, +2, reaches the pre-templating position above the bridge helix; (b) the DNA template base twists by 90° to reach its templating position in the active center, +1. The altered, flipped-out conformation of the bridge helix and the transition from an open to a wedged trigger loop conformation.

After translocation of DNA and RNA, the elongation complex is in a post-translocation state, with a free substrate site for the binding of the next incoming NTP. Cycle of translocation and the post-translocation state accompanies the nucleotide selection and addition (Figure 7.29).

This one-directional translocation of the DNA through the enzyme requires an energy source. The only evident source is cleavage of the NTPs to add an NMP to the growing RNA chain and release pyrophosphate. This is an energetically highly favored reaction: the action of the polymerase is only catalytic. Thus, once NTPs are available, the process drives

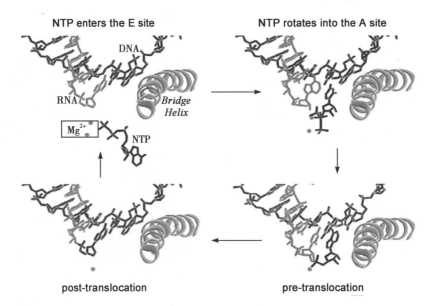

Figure 7.29 **The schematic four-step cycle of RNA transcript synthesis.** Only nucleic acids in the active center region of RNA Pol II (with template DNA and RNA transcript). The bridge helix and Mg^{2+} ions are shown. First, the incoming nucleotide binds to an entry site beneath the active center in an inverted orientation. Second, the NTP rotates into the nucleotide addition site for sampling of correct pairing with the template DNA and for discrimination from dNTPs that lack the ribose 2'-OH group. Only correctly paired NTPs can transiently bind the insertion site. Third is the pre-translocation step in which phosphodiester bond formation occurs. Fourth, translocation occurs to repeat the cycle. In the post-translocation complex, the nucleotide just added to the RNA has moved to the next position, leaving the A site open for the entry of a nucleotide. At the upstream end of the hybrid, RNA Pol II separates the nascent RNA from the DNA.

itself. The irreversibility of the process is further guaranteed by enzyme-catalyzed hydrolysis of the released pyrophosphate.

(5) Proofreading and backtracking

Unlike DNA polymerase that has proofreading capability, the fidelity of transcription, that is, avoiding incorporation of the wrong nucleotide or even a deoxyribonucleotide into the growing chain, is assured by the two-step nature of the elongation process. First, the incoming NTP is tested before the second step takes place. This testing is favored by the slow rate of transcriptional elongation in eukaryotes, which is about a tenth of the rate in bacteria. Most of the steps are highly evolutionarily conserved and seem to occur similarly in all three eukaryotic polymerases as well as in bacteria and archaea.

Second, in RNA Pol II, the growing RNA remains at a single active site that switches between RNA synthesis and cleavage. X-ray crystallographic analysis of RNA Pol II with the elongation factor TFIIS supports the idea that the polymerase has a "tunable" active site that switches between mRNA synthesis and cleavage. RNA polymerization and cleavage both require a metal ion "A" (e.g., Mg^{2+}), but differential coordination of another metal ion "B" switches RNA Pol II activity from polymerization to cleavage (Figure 7.30A).

In this model, for RNA polymerization, metal B binds with the phosphates of the substrate NTP. For cleavage during proofreading, metal B is bound by an additional unpaired nucleotide located in the active center. The tunable polymerase active site allows efficient mRNA proofreading because RNA cleavage creates a new RNA 3' end at metal A, from which polymerization can continue. Two types of proofreading reactions may occur: removal of a misincorporated nucleotide directly after its addition or cleavage of a dinucleotide after misincorporation and backtracking by one nucleotide (Figure 7.30B).

During mRNA elongation, RNA Pol II can also encounter DNA sequences that cause reverse movement, or "backtracking," of the enzyme. During backtracking, the register of the RNA-DNA base

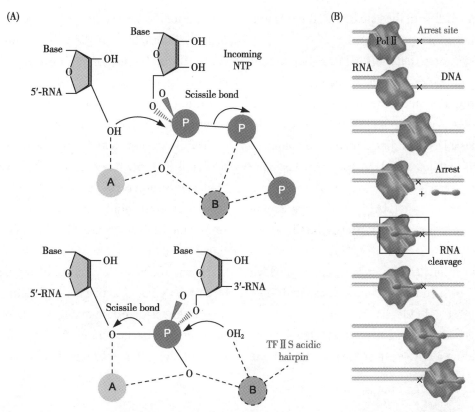

Figure 7.30 **RNA Pol II polymerization and cleavage.** (A) Model of the tunable RNA Pol II active site. (Upper panel) Proposed mechanism of nucleotide (NTP) incorporation during RNA polymerization. (Lower panel) Proposed mechanism of TFIIS-mediated RNA cleavage during backtracking. (B) A model for the rescue of arrested RNA Pol II by TFIIS-mediated mRNA cleavage. When transcribing, RNA Pol II encounters an arrest site (crossed) on DNA; it then pauses and backtracks, leading to transcriptional arrest. Cleavage of the extruded RNA is induced by TFIIS. Transcription then continues on past the arrest site.

pairing is maintained, but the 3′ end of the transcript is unpaired and extruded from the active center. This can lead to transcriptional arrest. Escape from arrest requires cleavage of the extruded RNA with the help of TFIIS. For stimulation of the weak nuclease activity of RNA Pol II, TFIIS is proposed to insert an acidic β-hairpin loop into the active center to position metal B and a nucleophilic water molecule for RNA cleavage (Figure 7.31).

(6) Transcription elongation through the nucleosomal barrier

Although chromatin is remodeled upon transcription initiation, the DNA remains packaged in nucleosomes in the coding region of transcribed genes. When RNA Pol II moves through the gene-coding region, the enzyme will encounter a nucleosome approximately every 200 bp. There are two distinct mechanisms for the progression of RNA polymerases through chromatin (Figure 7.32): (a) nucleosome mobilization or "octamer transfer" (i.e., movement of the octamer on the DNA) by RNA Pol III or SP6 RNA polymerase, and (b) H2A-H2B dimer deple-

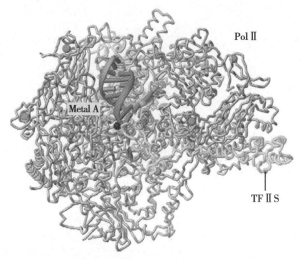

Figure 7.31 **RNA Pol II-TFIIS complex.** Cutaway view of the crystallographic model of the RNA Pol II (gray)-TFIIS (light gray) complex, showing metal ion A in the active site.

tion by RNA Pol II with the help of FACT, Elongator, TFIIS or other unidentified factors.

FACT promotes the transcription-dependent nucleosome alterations for the elongation of transcripts through nucleosome arrays without ATP hydrolysis. FACT is composed of two protein subunits, and has a DNA-binding domain and a domain that interacts with histones H2A and H2B. The large subunit also interacts with the catalytic subunit of a histone acetyltransferase (HAT), possibly explaining the extent of acetylation at transcriptionally active regions.

Elongator facilitates transcript elongation, which is composed of six major subunits, designated ELP1 to ELP6 in yeast and ELP3 has HAT activity. Like the yeast counterpart, human Elongator also has six subunits, including a HAT with specificity for histone H3 and to a lesser extent for histone H4. Elongator directly interacts with RNA Pol II and facilitates transcription by acetylating histones and opening chromatin structure in the path of the polymerase, but it does not appear to interact directly with any other elongation factors, including FACT.

The elongation factor, TFIIS, facilitates passage of RNA Pol II through regions of DNA that can cause transcription arrest. These sites include AT-rich sequences, DNA-binding proteins, or lesions in the transcribed DNA strand. TFIIS rescues transcriptional arrest by a backtracking mechanism that stimulates endonucleolytic cleavage of the nascent RNA by the RNA Pol II active center (Figure 7.31). The bottom line is that despite negotiating nucleosomes, pausing, and backtracking, RNA Pol II still moves forward along the DNA template. It catalyzes rapid

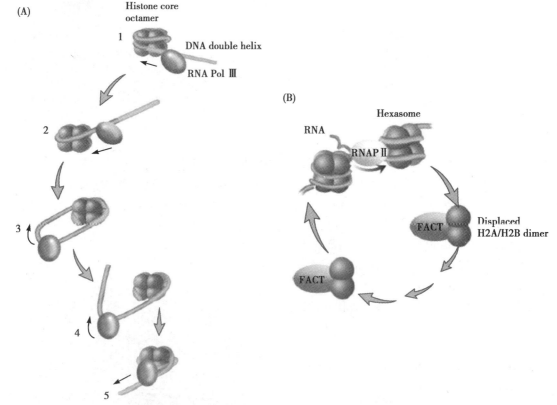

Figure 7.32 **Mechanisms of transcription through the nucleosome by RNA polymerases.** (A) Nucleosome mobilization or "octamer transfer" model for RNA Pol III. (Step 1) RNA Pol III rapidly transcribes the first ~25 bp of nucleosomal DNA, causing (Step 2) partial dissociation of the DNA from the core octamer. DNA that has not been transcribed is shown as the arrow direction. (Step 3) The DNA behind RNA Pol III (transcribed DNA) transiently binds to the exposed surface of the octamer, forming a loop. When the polymerase has moved ~60 bp into the core, the downstream portion of the DNA dissociates from the octamer (Step 4), completing octamer transfer. (Step 5) Transcription proceeds to the end of the template. (B) Histone H2A-H2B dimer depletion model for RNA Pol II. FACT mediates displacement of the H2A-H2B dimer from the core octamer, leaving a "hexasome" on the DNA. The histone chaperone activity of FACT helps to redeposit the dimer on the DNA after passage of RNA Pol II (RNAPII).

RNA chain growth at a rate of 20 to 70 nucleotides per second.

Termination of eukaryotic transcription is strongly coupled to post-transcriptional processing of the RNA transcript. When a primary transcript is synthesized by RNA polymerase II about 25–30 nucleotides in length, "Capping" of the transcript at its 5'-end occurs (Figure 7.33). The 5'- terminal, the initial nucleotide of the transcript, is a pyrimidine with three phosphate groups attached to the 5'-hydroxyl of the ribose. To form the cap, the terminal triphosphate loses one phosphate, forming a 5'- diphosphate. The beta-phosphate of the diphosphate then attacks the alpha-phosphate of GTP, liberating pyrophosphate, and forming an unusual 5'-to-5'-triphosphate linkage. A methyl group is transferred from S-adenosylmethionine (SAM), a universal methyl donor, to position 7 of the added guanine ring. Methylation also occurs on the ribose 2'-hydroxyl group in the terminal nucleotide to which the cap is attached, and sometimes the 2'-hydroxyl group of the adjacent nucleotide ribose. This cap "seals" the 5'-end of the primary transcript and decreases the rate of degradation. It also serves as a recognition site for the binding of the mature mRNA to a ribosome at the initiation of protein synthesis.

After the RNA polymerase transcribes the stop codon for protein translation, it passes a sequence called the polyadenylation signal, AAUAAA, until it reaches an unknown, and possibly nonspecific, termination signal many nucleotides later. As the primary transcript is released from the RNA polymerase elongation complex, an enzyme complex binds to the polyadenylation signal and cleaves the primary transcript approximately 10 to 20 nucleotides downstream to form the 3'-end. After this cleavage, a poly(A) tail that can be >200 nucleotides in length is added to the 3'-end (Figure 7.34). Thus, there is no poly(dT) sequence in the DNA template that corresponds to this tail; it is added posttranscriptionally. ATP serves as the precursor for the sequential addition of the adenine nucleotides. They are added one at a time, with poly(A) polymerase catalyzing each addition. The poly(A) tail is a protein binding site that protects the RNA from degradation. Another function of polyadenylation seems to be to aid in exporting RNA from the nucleus to the cytoplasm. That is why polyadenylation is not found in bacteria.

As known, eukaryotic pre-mRNAs contain regions of exons and introns. Exons appear in the mature

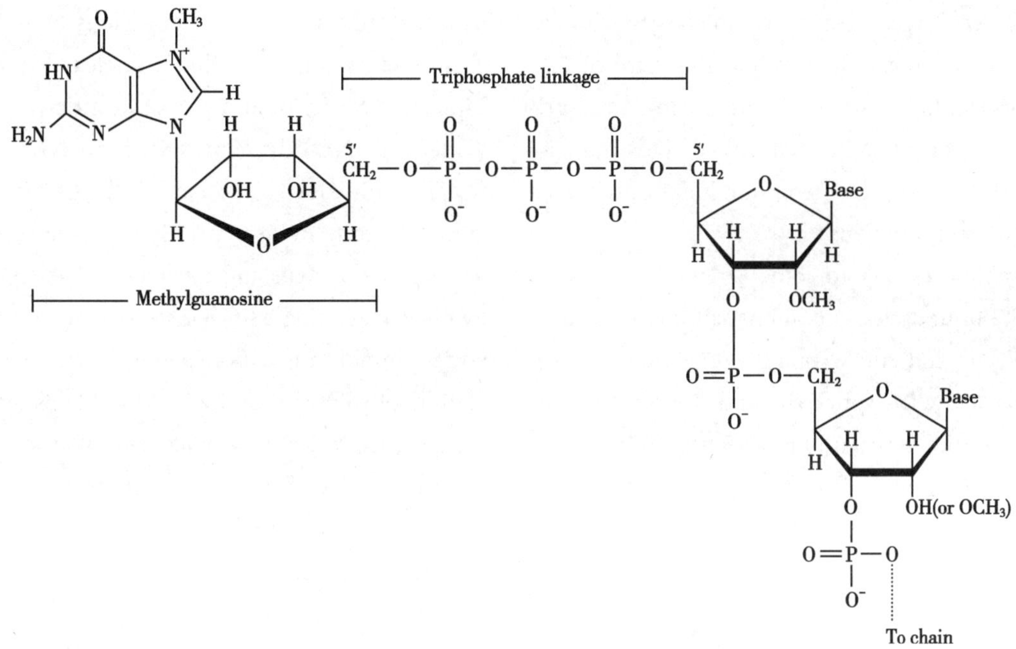

Figure 7.33 **The cap structure in eukaryotic mRNA.** The phosphates in *light gray* originated from the original RNA transcript; the phosphate in *black* comes from GTP. SAM donates the methyl groups (shown in *light gray*) required for cap synthesis. A CAP 1 structure is shown.

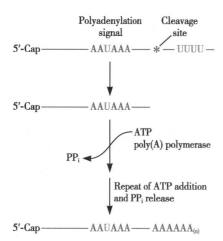

Figure 7.34 Synthesis of the poly(A) tail. As RNA polymerase continues to transcribe the DNA, enzymes cleave the transcript (hnRNA) at a point 10 to 20 nucleotides beyond an AAUAAA sequence, just before a run of Us (or Gs). Approximately 250 adenine nucleotides are then added to the 3′-end of the hnRNA, one at a time, by poly(A) polymerase.

mRNAs and introns are not found in the mature mRNAs. Once the transcription terminates, introns are immediately and carefully removed from the pre-mRNA and the exons are spliced together, resulting in the mature mRNA that will be transported out of nucleus for the appropriate protein synthesis. In general, pri-mRNAs in size varies from very small, such as a histone gene that is only 500 bp long without intron, to very large, for instance, the largest known human gene of dystrophin with a transcript of 2.5 million nucleotides in length with 79 introns, and takes over 16 hours to be transcribed. This dystrophin is usually missing or nonfunctional in the disease muscular dystrophy.

The sequences vary to some extent on the exon side of the boundaries, but almost all introns begin with a 5′-GU and end with a 3′-AG. These intron sequences at the left splice site and the right splice site are therefore invariant. Because every 5′-GU and 3′-AG combination does not result in a functional splice site, other features (still to be determined) within the exon or intron help to define the appropriate splice sites. A complex spliceosome ensures that exons are spliced together with great accuracy (Figure 7.35). Small nuclear ribonucleoproteins (snRNPs), called "snurps," are involved in formation of the spliceosome. Because snurps are rich in uracil, they are identified by numbers preceded by a U. Exons frequently code for separate functional or structural domains of proteins. Proteins with similar functional regions (e.g., ATP- or NAD-binding regions) frequently have similar domains, although their overall structures and amino acid sequences are quite different. A process known as exon shuffling has probably occurred throughout evolution, allowing new proteins to develop with similar functions.

(7) Nuclear import and export of proteins

The control of nuclear localization of transcriptional regulatory proteins represents a level of transcriptional regulation in eukaryotes that is not present in prokaryotes. Trafficking between the nucleus and the cytoplasm occurs via the nuclear pore complexes (NPCs). NPCs are large multi-protein complexes embedded in the nuclear envelope—the double membrane that surrounds the nucleus. The NPCs allow bidirectional passive diffusion of ions and small molecules. In contrast, nuclear proteins, RNAs, and ribonucleoprotein (RNP) particles larger than 9 nm in diameter, and greater than 40–60 kDa, selectively and actively enter and exit the nucleus by a signal-mediated and energy-dependent mechanism. Proteins are targeted to the nucleus by a specific amino acid sequence called a nuclear localization sequence (NLS). In some cases, a nuclear protein without an NLS dimerizes with an NLS-bearing protein and rides "piggyback" into the nucleus. In addition, some nuclear proteins shuttle repeatedly between the nucleus and cytoplasm. Their exit from the nucleus requires a nuclear export sequence (NES). Nuclear import and export pathways are mediated by a family of soluble receptors referred to as importins or exportins, collectively called karyopherins.

The largest class of soluble receptors is the karyopherin-β family, which is involved in the transport of proteins and ribonucleoprotein (RNP) cargoes. A second group of receptors, which is structurally unrelated to the karyopherin-β family, is the family of nuclear export factors (NXFs) that are involved in the export of many mRNAs. A third class

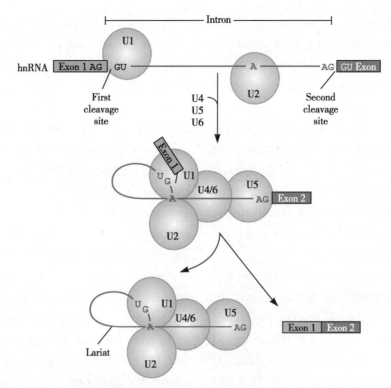

Figure 7.35 **Transcript splicing process.** Nuclear ribonucleoproteins (snurps U1 to U6) bind to the intron, causing it to form a loop. The complex is called a *spliceosome*. The U1 snurp binds near the first exon/intron junction, and U2 binds within the intron in a region containing an adenine nucleotide residue. Another group of snurp-U4, U5, and U6 binds to the complex, and the loop is formed. The phosphate attached to the G residue at the 5'-end of the intron forms a 2'- to 5'- linkage with the 2'-hydroxyl group of the adenine nucleotide residue. Cleavage occurs at the end of the first exon, between the AG residues at the 3'-end of the exon and the GU residues at the 5'-end of the intron. The complex continues to be held in place by the spliceosome. A second cleavage occurs at the 3'-end of the intron after the AG sequence. The exons are joined together. The intron, shaped like a lariat, is released and degraded to nucleotides.

is represented by the small nuclear transport factor 2 (NTF2), which imports the small GTPase Ran into the nucleus via the nuclear pore complexes (NPCs). NPCs are large multi-protein complexes embedded in the nuclear envelope—the double membrane that surrounds the nucleus. The NPCs allow bidirectional passive diffusion of ions and small molecules. In contrast, nuclear proteins, RNAs, and RNP particles larger than 9 nm in diameter, and greater than 40–60 kDa, selectively and actively enter and exit the nucleus by a signal-mediated and energy-dependent mechanism.

Unlike signal sequences targeting proteins to the endoplasmic reticulum or mitochondrion, which are generally removed from proteins during transit, nuclear proteins retain their NLSs. This may be to ensure that they can reaccumulate in the nucleus after each mitotic cell division. The best-studied NLSs are basic amino acid sequences, typically rich in lysine and arginine. There is no real consensus sequence for an NLS. Some proteins, such as nucleoplasmin, have a bipartite NLS. NLSs interact with members of the karyopherin-β family either directly or through the adapter importin-α.

(8) Regulated nuclear import and signal transduction pathways

Regulation of NLS and NES activity in a transcription factor can occur by several mechanisms, including post-translational modifications, e.g., phosphorylation and dephosphorylation that mask or unmask the NLS or NES. Where a protein is localized at "steady state" depends on the balance between import, retention, and export, and which signal is dominant.

Nuclear retention may be mediated by domains of the protein that interact with components of the nucleus, such as chromatin or the nuclear matrix. There are various ways that signals are detected by a cell and communicated to a gene. The effect may be direct, where a small molecule, such as a sugar or steroid hormone, enters the cell and binds the transcriptional regulator directly. Or the effect of the signal may be indirect where the signal induces a kinase that phosphorylates the transcriptional regulator or an associated inhibitory protein. This type of indirect signaling is an example of a signal transduction pathway. The signal-mediated nuclear import of transcription factors NF-κB and the glucocorticoid receptor are used to illustrate the wide variety of mechanisms for controlling gene activity.

NF-κB (nuclear factor of kappa light polypeptide gene enhancer in B cells) is a dimeric transcription factor that is a central mediator of the human stress response. It plays a key role in regulating cell division, apoptosis, and immune and inflammatory responses. The discovery of NF-κB attracted widespread interest because of the variety of extracellular stimuli that activate it, the diverse genes and biological responses that it controls, and the striking evolutionary conservation of structure and function among family members. NF-κB was first identified in 1986 as a transcription factor in the nuclei of mature B lymphocytes that binds to a 10-bp DNA element in the kappa (κ) immunoglobulin light-chain enhancer. Since then a family of various distinct subunits has been identified. The events leading to signal-mediated nuclear import of NF-κB involve three main stages: (a) cytoplasmic retention by I-κB; (b) a signal transduction pathway that induces phosphorylation and degradation of I-κB; and (c) I-κB degradation resulting in exposure of the NLS on NF-κB, allowing nuclear import of NF-κB (Figure 7.36). In this example of NF-κB, gene expression is induced in response to a signal received by a cell surface receptor and transmitted by a signal transduction cascade.

In contrast, another type of transcription factors, for instance, the glucocorticoid receptor that mediates a highly abbreviated signal transduction pathway: the receptor for the extracellular signal is cytoplasmic and carries the signal directly into the nucleus.

In the classical model, activation of the gluco-

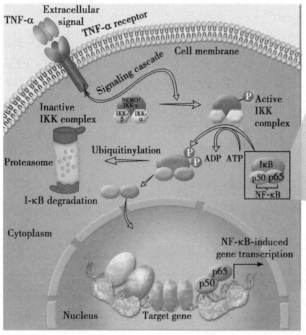
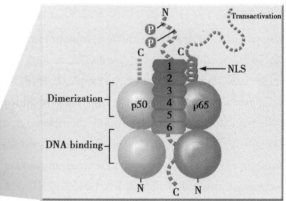

Figure 7.36 **Signal-mediated nuclear import of NF-κB.** TNF-α receptor-mediated transduction of an extracellular signal activates the I-κB kinase (IKK) complex that phosphorylates I-κB for the ubiquitinylation of I-κB and subsequent degradation by proteosome activity, releases NF-κB from I-κB and nuclear import of NF-κB to activate the target genes. Phosphorylation is indicated by a circled "P." (Inset) Schematic representation of the I-κB–NF-κB complex showing the functional domains of the p50 and p65 subunits of NF-κB. Ankyrin repeats 1 and 2 (red cylinders) of I-κB mask the p65 nuclear localization sequence (NLS).

corticoid receptor involves ligand-induced conformational changes that result in rapid dissociation of the inhibitory Hsp90 protein. Subsequently, two glucocorticoid receptors join together to form a homodimer. The NLS interacts with importins and the receptor is translocated through the nuclear pore complex into the nucleus. Once in the nucleus, the glucocorticoid receptor dimer binds DNA at a glucocorticoid-responsive element (GRE). Notice that the DNA forms a loop stabilized by protein-protein interactions, bringing long-range regulatory regions into close proximity to the core promoter. Coactivators form a bridge between GR and the general transcriptional machinery (Figure 7.37). Activation of hormone-responsive target genes leads to many diverse cellular responses, ranging from increases in blood sugar to anti-inflammatory action.

7.4.3 Transcription by RNA polymerase I and III

(1) Transcription by RNA polymerase I

As indicated in Table 7.3, Pol I transcribes the precursors of rRNA: the two large RNA molecules that participate in the formation of large and small ribosomal subunits and the small 5.8S rRNA (see later chapter). The direct product of transcription is a single 47S RNA, which is then processed to produce the above products. Thus, Pol I has only one kind of target gene. Transcription of the rRNA genes occurs in the nucleolus, where these genes are tandemly repeated multiple times (Figure 7.38). The transcriptional regulation of these genes is complex and poorly understood, as only a fraction of the genes is transcribed at any given moment, even in rapidly

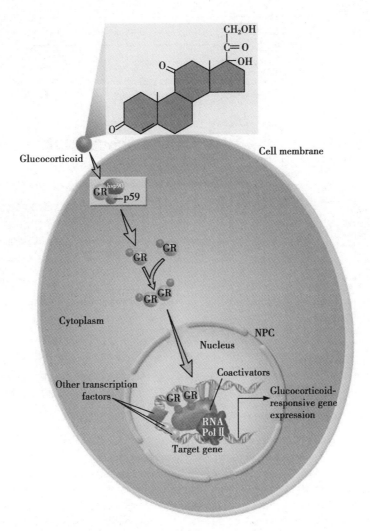

Figure 7.37 **Signal-mediated import of the glucocorticoid receptor (GR).** In the classical model, activation of the GR by glucocorticoid involves its release from the inhibitory Hsp90-p59 complex, allowing dimerization of GR and nuclear import.

proliferating cells where the demand for ribosomes, and hence rRNA, is huge.

The general transcription factors of the Pol I pre-initiation complex are schematically shown. The TBP is shared in both Pol II and Pol I transcription, although the TAFs are specific for each polymerase. The TBP complex with TAFs contains the Pol I selectivity factor 1, SL1, which interacts with numerous other factors in the PIC (Figure 7.39A, B). An important interactor is the upstream binding factor or UBF. Two UBF dimers interact with two DNA elements, the core promoter and the upstream control element or UCE, to form the highly specific architecture known as the enhanceosome (Figure 7.39C). Two adjacent enhanceosomes bring the core promoter elements and the UCE together to help to recruit TBP.

The Pol I transcription cycle consists of the following steps: pre-initiation complex assembly, Pol I recruitment, initiation, promoter escape, elongation, termination, and re-initiation (Figure 7.40). The first nucleotide addition steps are slow and the process goes through several abortive synthesis steps, during which short nascent RNA chains are released from the complex. Once the transcript reaches 10–12 nucleotides in length, the polymerase switches to a processive mode and can elongate, without dissociating from the template, for thousands of nucleotide addition cycles. Termination occurs when a specific termination factor (TTF-I) binds to the termination sequences at the 3′-end of the gene. TTF-I bends the termination site, forcing Pol I to pause. Another polymerase I and transcript release factor (PTRF)

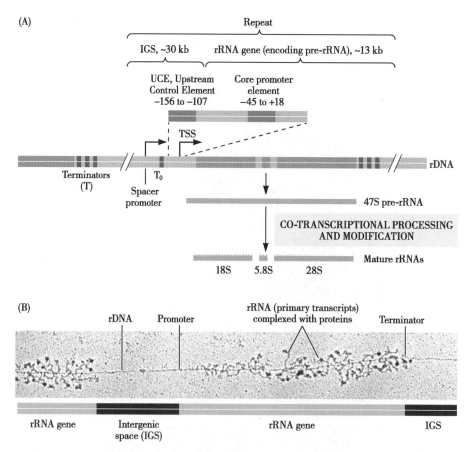

Figure 7.38 **Repetitive nature of rRNA genes in eukaryotes.** (A) Not-to-scale schematic of human rRNA repeats. The intergenic spacers, IGS, contain the transcription regulatory elements. A spacer promoter directs Pol I transcription of short-lived transcripts of yet unknown function. All rRNA sequences are transcribed as a large precursor that undergoes processing and base modifications to produce the mature RNA components of small and large ribosomal subunits, 18S and 5.8S/28S rRNA, respectively. The promoter element is bipartite and consists of an essential core and an upstream control element. The 3′-end of the transcribed region contains several transcription-termination elements. (B) Electron microscopic image of yeast nucleolar chromatin. Progressively longer RNA transcripts associated with proteins are seen to originate from the Pol I complexes, visible as small dots on DNA.

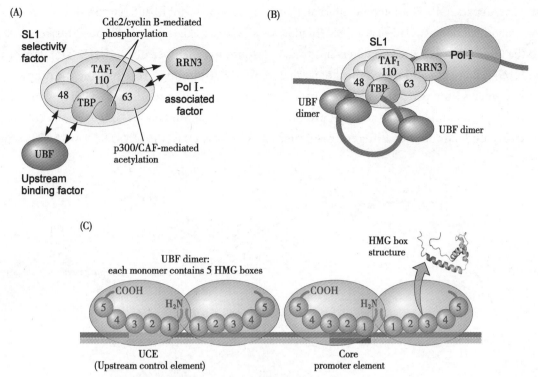

Figure 7.39 General transcription factors of eukaryotic Pol I pre-initiation complex. (A) SL1 selectivity factor is a 300 kDa complex containing the TATA-binding protein, TBP, and at least three Pol I-specific TBP-associated factors or TAFs, of 110, 63, and 48 kDa in humans. It recruits Pol I to the promoter via interaction of its TAFI63 and TAFI110 subunits with the Pol I associated factor RRN3. Some of these TAFs are known to be post-translationally modified: phosphorylated by cyclin-dependent protein kinases, which are regulators of the cell cycle, or acetylated by the abundant acetylase p300/CAF. SL1 interacts with the upstream binding factor, UBF, and with RRN3; these interactions are indicated as double-headed arrows connecting the proteins with the respective subunits of SL1. (B) Schematic reflecting recognized interactions in the Pol I PIC: SL1 selectivity factor interacts with DNA, hRRN3, and UBF; UBF interacts with rDNA and Pol I via RAF53, a homolog of Pol I A49 subunit, which is not shown. (C) Close-up of interactions of the two UBF dimers with the two DNA promoter elements: upstream control element, UCE, and core promoter element. At each of these two DNA sites, UBF binds as a dimer. Each UBF monomer has a domain structure with five HMG boxes. HMG boxes possess a characteristic L-shaped tertiary structure that is present in both sequence-specific transcription factors and nonspecific DNA binders. The box is named after the non-histone chromatin protein HMG, a high-mobility group protein, where it was first identified. Binding of HMG boxes to DNA induces DNA bending.

comes in and binds to a T-rich DNA sequence to release the polymerase and the RNA transcript. SL1 and UBF remain bound to the promoter the whole time, providing a scaffold upon which transcription can be quickly re-initiated.

An important step, termed re-initiation, allows for quick reassembly of the entire pre-initiation complex by recycling Pol I to the engaged promoter or enhanceosome structure. Recent X-ray diffraction studies of Pol I have clarified the structure and revealed elements involved in the transition from inactive to active polymerase. First, in the inactive form, the active-site cleft is held open by an expander element. This must move out of the way before the cleft can close on the substrate. Second, the inactive form can dimerize via a connector element. This must dissociate for activation. Presumably, one or both are under allosteric control.

(2) Transcription by RNA polymerase III

RNA polymerase III specializes in transcription of small genes, including the small 5S component of the ribosome (5S rRNA), tRNAs, one of the small nuclear RNAs that comprise the spliceosome and other small RNAs, often of unknown function. Each of the genes encoding these RNAs possesses specific promoter sequences. Some of them resemble Pol II promoters and are located upstream of the TSS and contain TATA boxes. Others are totally internal, that is, they are located downstream of the TSS. In addition to the three major types of promoters illustrated in Figure 7.41, there are also mixed promoters that consist of elements from two different pure types.

The functional complexity of all these promoters is poorly understood, but different promoter types do assemble different pre-initiation complexes (Figure 7.42). Type 1 promoters recruit TFIIIA, the founding member of the C2H2 zinc-finger family of DNA-binding proteins. The binding of TFIIIA then allows the binding of TFIIIC, which has five subunits in humans and *S. pombe*, six subunits in *S. cerevisiae*. Once the DNA/TFIIIA/TFIIIC complex is

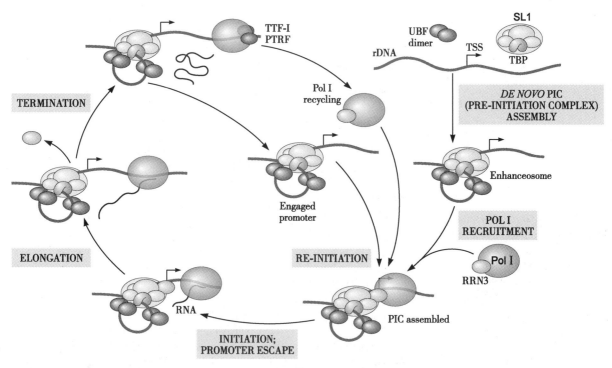

Figure 7.40 **Pol I transcription cycle.**

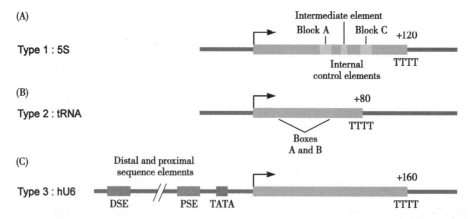

Figure 7.41 **Three types of Pol III promoters.** (A) Type 1 promoters are found in the 5S genes and consist of an internal control region, ICR, that is subdivided into A block, intermediate element, and C block. (B) Type 2 promoters are found in tRNA genes, adenovirus 2 VAI gene, and some other genes and consist of two gene-internal elements called the A and B boxes. (C) Type 3 promoters consist of a distal sequence element (DSE) which serves as an enhancer; a proximal sequence element, PSE; and a TATA box. Mixed promoters have also been described; they consist of elements from two different pure types. For example, the *S. cerevisiae* U6 gene promoter contains both a TATA box and A and B boxes; in *Schizosaccharomyces pombe*, nearly all tRNA and 5S genes contain a TATA box in addition to gene-internal elements, and the TATA box is required for transcription.

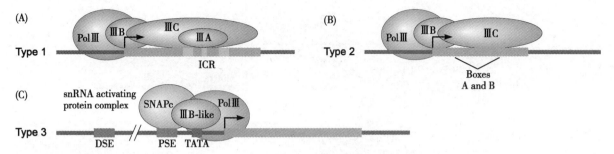

Figure 7.42 Transcription initiation complexes on Pol III promoters. pre-initiation complexes on type 1 promoters (A), type 2 promoters (B) and Metazoan-specific type 3 promoters (C).

formed, another complex Brf1-TFIIIB joins, which in turn allows the recruitment of RNA Pol III. Type 2 promoters can recruit TFIIIC without the help of TFIIIA, because TFIIIC binds directly to the A and B boxes. This then allows the binding of Brf1-TFIIIB and RNA Pol III. In yeast, once Brf1-TFIIIB has been recruited to type 1 or 2 promoters, TFIIIA and/or TFIIIC can be stripped from the DNA with high salt or heparin treatment. Brf1-TFIIIB remains bound to the DNA and is sufficient to direct multiple rounds of transcription. Metazoan-specific type 3 promoters recruit the small nuclear RNA activating protein (SNAP) complex that binds to the PSE. Type 3 promoters also recruit Brf2-TFIIIB through a combination of protein-protein contacts with SNAPc and a direct association of the TBP component of Brf2-TFIIIB with the TATA box. Then RNA Pol III joins the complex. DSE functions as an enhancer: it contains specific sequences that bind transcription factors.

7.4.4 Transcription in eukaryotes: pervasive and spatially organized

(1) Most of the eukaryotic genome is transcribed

Usually, the eukaryotic genome is viewed as a linear arrangement of genes that code either for proteins or for structural RNA molecules, such as rRNA and tRNA, which all are involved in making proteins. But exploration of the eukaryotic genome showed that there is far more DNA than is needed to account for the number of proteins in a human cell, that is, there are large amounts of junk DNA that code nothing.

With the Encyclopedia of DNA Elements or ENCODE project accomplishment by the high-throughput technologies for studying transcription on a genomewide scale, it is found that only a small percentage of the human genome codes for protein. The more evolved organisms use the smaller fraction of their genomes for transcription. Transcription in eukaryotes is pervasive: close to 75% of the sequences in the human genome give rise to some kind of primary transcript. No individual cell line shows transcription of more than ~57% of the total sequence transcribed across all cell lines. The current view presents that transcription in eukaryotes is more than just a conversion of the information encoded in DNA into mRNA molecules. Besides, the newly discovered transcripts contain two other major classes: long noncoding RNAs (lncRNAs) and short noncoding RNAs, which indicate the complexity of eukaryotic transcription (Figure 7.43).

(2) Transcription in eukaryotes is not uniform within the nucleus

As we know, chromatin in the interphase nucleus is organized in chromosome territories. Following mitosis, chromosomes decompact to each occupy a distinct portion of the nuclear volume. In addition, there is considerable preference for gene-rich chromosomes to occupy the center of the nucleus, whereas chromosomes that are relatively poor in genes tend to be at the nuclear periphery. What governs the overall nuclear architecture is unclear.

Recent sophisticated imaging techniques have uncovered another unexpected feature of genome organization in terms of its activity. Transcrip-

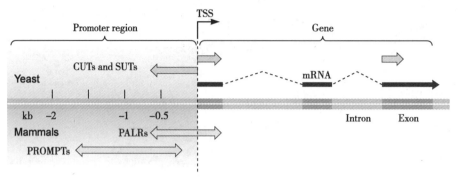

Figure 7.43 **Transcription in eukaryotes is pervasive.** In yeast, shown in the upper half of the schematic, genomic sequences around many transcription start sites, TSSs, are transcribed bidirectionally in long noncoding RNAs called cryptic unstable transcripts (CUTs) and stable unannotated transcripts (SUTs). Transcription of these noncoding RNAs is promoter-dependent, just like transcription of mRNA, and the transcripts are 5'-capped, as is mRNA. CUTs are easily degraded by RNA surveillance pathways and can be detected only in mutant cells where these pathways are disabled. They are 200–600 nt long and are usually heterogeneous at their 3'-ends. SUTs are more stable and can be detected in normal cells; the average length is ~760 nt. Both CUTs and SUTs are proposed to be broad regulators of transcription in yeast. In mammals, shown in the lower half of the schematic, the complexity of transcription is even higher. Long noncoding transcripts such as promoter-associated long RNAs (PALRs) and promoter-upstream transcripts (PROMPTs), partially overlap the same DNA sequences; PROMPTs are less stable than PALRs. PROMPTs may be transcribed from unrecognized promoters. Mammals also contain short noncoding RNAs, <200 nt.

tion does not occur evenly throughout the nuclear volume; much of it occurs in specific sites termed transcription factories. A typical eukaryotic cell contains ~10 000 such factories that harbor the full complement of molecules, such as polymerases and transcription factors for transcription (Figure 7.44). A factory is usually shared by a number of genes that may be located on different chromosomes. For such sharing to occur, genome regions that carry transcribed genes usually, but not always, move out of their respective chromosome territories to join a preexisting transcriptional factory.

It is not clear what proteins or other macromolecules contribute to the formation of this factory. Suggested proteins include the polymerase molecules themselves as well as other protein factors, such as CCCTC-binding factor (CTCF). CTCF is a ubiquitous zinc-finger protein with numerous functions; considerable evidence puts it into the enviable position of a master organizer of the genome. It can form chromatin loops in cis, that is, between DNA sites on the same chromosome, or it can bridge sequences located on different chromosomes in trans. Consistent with a role as a genome organizer, thousands of CTCF binding sites have been identified in genomewide localization studies, with considerable enrichment around transcription start sites. The 2012 ENCODE data report 55 000 CTCF sites in each of the 19 tested cell lines, including normal primary cells and immortal lines. Contrary to previous ideas, the actual CTCF occupancy of these sites varies somewhat among cell lines. This variability may reflect the presumptive involvement of CTCF in the dynamics of the transcription factories.

(3) Active and inactive genes are spatially separated in the nucleus

The 4C technology (Chromosome Conformation Capture for Contact) depicted that the active and inactive genes are spatially separated in the nucleus (Figure 7.45A). The 3C (Chromosome Conformation Capture) protocol consists of the following four steps: (Step 1) Cross-link chromatin *in vivo*; digest with restrictase, a six-cutter; and ligate cross-linked fragments. (Step 2) Reverse cross-links and digest with four-cutter to trim ligation junctions. (Step 3 and 4) Ligate to create DNA circles and amplify circles by polymerase chain reaction, PCR, using oligonucleotide primers derived from the gene region of interest (in red). Steps 1–4 are followed by microarray or sequencing analysis of interacting regions (Step 5), where the identified sequences are mapped back to the genome. This step adds a fourth dimension, the fourth C, to the assay, allowing unbiased analysis of any sequence that contacts the gene of interest. The example shown in Figure 7.45B illustrates DNA contacts of the β-globin gene in fetal liver, where the gene is actively transcribed, and in fetal brain, where the gene is not expressed. It is immediately obvious that the β-globin gene makes contacts with very different regions in the chromosome depending on its expression status. The difference in interac-

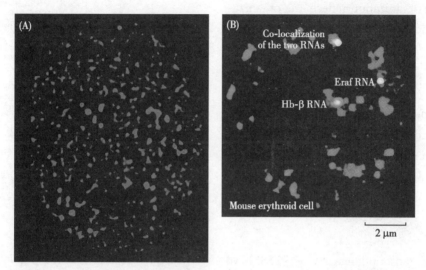

Figure 7.44 **Spatial organization of transcription in the eukaryotic nucleus: transcription factories.** Transcription in the nucleus is compartmentalized to specific, spatially confined sites as transcription factories. Each transcription center can serve several genes. (A) Transcription sites are visualized as islets by bromo-dUTP incorporation and fluorescently labeled antibody detection. (B) RNA immunofluorescence *in situ* hybridization. Hemoglobin β-chain (Hb-β) RNA and erythroid associated factor or Eraf RNA are shown, which are located on the distal third of mouse chromosome 7, separated by 25 Mb; they colocalize in a factory in one of the chromosomes. The gay islets indicate the localization of Pol II by immunofluorescence detection.

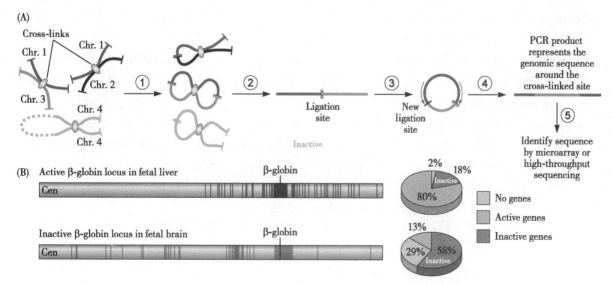

Figure 7.45 **Active and inactive genes separate spatially in the nucleus.** (A) Schematic of 4C technology used to identify DNA regions on the same or different chromosomes that interact *in vivo*. (B) Long-range intrachromosomal interactions identified for the β-globin locus in its active state in mouse fetal liver and its inactive state in fetal brain. β-globin locus is presented in Chromosome 7 as a gray horizontal bar. Vertical lines indicate the regions that interact with the β-globin locus. The pie charts characterize the distribution of interacting regions in terms of their gene content or gene activity.

tions between the active and inactive locus is that the active locus preferentially interacts with other actively transcribed regions, whereas the inactive locus predominantly contacts silent regions. The constitutively active housekeeping gene *Rad23a* contacts essentially the same regions in the two tissues, the majority of which are transcriptionally active. Presumably this organization reflects the organization of eukaryotic transcription in transcription factories.

7.4.5 Methods for Studying Transcription

Over the years, numerous methods have been introduced for the study of transcription. The majority of methods rely on radioactive or fluorescent labeling of transcripts and analysis on electrophoretic gels.

(1) Methods used to map transcription start sites or TSS

Methods to map transcription start sites (TSS) have greatly increased our understanding of transcription initiation. Mapping the exact position in a gene where transcription is initiated and important: it helps in the identification of promoter sequences and transcribed regions in the genome. Thus far, only a small portion of each sequenced genome has been annotated in terms of the genes it contains.

There are two classical methods for mapping TSSs in known genes: S1 nuclease protection and primer extension assays (see figure 4.10).

Recently, it became possible to analyze the location of TSSs across entire genomes, with the advances of human genome analysis technique. Cap analysis of gene expression (CAGE) determines TSSs genome-wide by isolating and sequencing short sequence tags originating from the 5′-end of RNA transcripts, followed by mapping the identified tags back to the genome identifies TSSs in the genome (Figure 7.46). This method is based on capturing those mRNAs that are modified or capped at their 5′-ends. mRNAs are usually labeled with biotin so they will bind to streptavidin-coated magnetic beads. Magnetic beads allow simple physical isolation of the labeled fragments. After the capped mRNAs are separated from

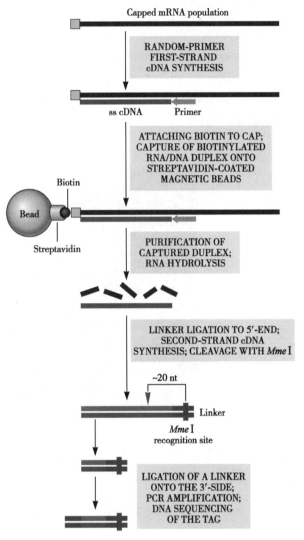

Figure 7.46 **Schematic steps of CAGE for TSS identification genomewide.**

all other RNAs, a linker containing an *Mme* I recognition site is ligated to the 5′-end and single-stranded (ss) cDNA is converted into a double-stranded cDNA copy of the mRNA. Cleavage with the type I restrictase *Mme* I cuts the DNA ~20 nucleotides downstream, producing short CAGEs that can be further PCR-amplified and sequenced.

(2) Methods used to characterize the 5′- and 3′-untranslated regions

In addition to mapping TSSs, identifying 5′- and 3′-regions of genes can significantly contribute to our overall understanding of functional elements of genomes. Methods in elucidating the structures near the 5′- and 3′-ends of genes include CAGE, as mentioned above, RACE (Rapid Amplification of cDNA Ends) to amplify the unknown cDNA sequences corresponding to the 5′- or 3′-ends of the RNA (Figure 7.47), and Pair-end di-tag analysis (PET) is also used to sequence the sequence detail in an RNA molecule (Figure 7.48).

(3) Reverse transcriptase is useful to copy RNA molecules

The procedure is based on the use of reverse transcriptase, a specialized RNA-dependent DNA polymerase that is involved in the life cycle of viruses whose genome is RNA-based. Reverse transcriptase, *in vivo*, copies the RNA genome into double-stranded DNA, which is then transcribed and translated as regular cellular DNA to produce the RNA and protein for mature viral particles. When reverse transcriptase is used in a laboratory setting to copy RNA molecules, the product is known as copy DNA or cDNA. This process is presented in detail in later chapter and illustrated with the example of the human immunodeficiency virus, HIV.

(4) High-throughput sequencing methods allow analysis of the entire transcriptome under specified conditions

Transcriptome analysis is the measurement of transcriptional activity of thousands of genes at once to create a global picture of cellular function. Recently, this analysis has been extended to cover entire sequenced genomes, including the human genome. In addition, transcriptome analysis can be used to

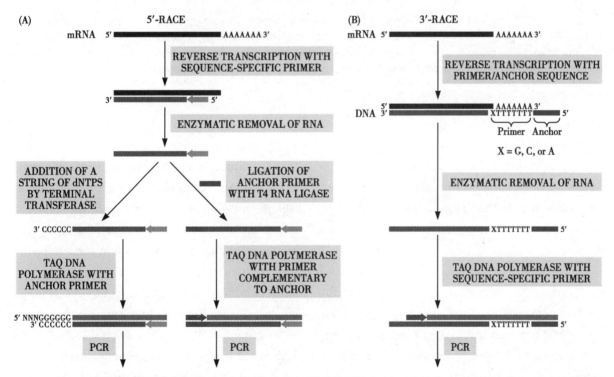

Figure 7.47 RACE: rapid amplification of cDNA ends. (A) 5′-RACE for the 5′ UTR region. The first-strand cDNA is synthesized by using an oligonucleotide complementary to a known sequence in the gene. After removal of the RNA template, terminal deoxynucleotidyl transferase adds a nucleotide tail to create an anchor site at the 3′-end of the single-stranded cDNA. An anchor primer complementary to the newly added tail is used to synthesize the second cDNA strand. An alternative strategy uses an RNA ligase to add the necessary anchor. The PCR products are sequenced to know the whole 5′ UTR region of a gene. (B) In 3′-RACE, a modified oligo (dT) sequence serves as the reverse transcription primer, which anneals to the poly(A)+tail of the mRNA, and an adaptor sequence at the 5′-end. A single G, C, or A residue at the 3′-end of the primer ensures that cDNA synthesis is initiated only when the primer/adaptor anneals immediately adjacent to the junction between the poly(A)+ tail and 3′-end of the mRNA. The PCR products are also sequenced to know the whole 3′ UTR region of a gene.

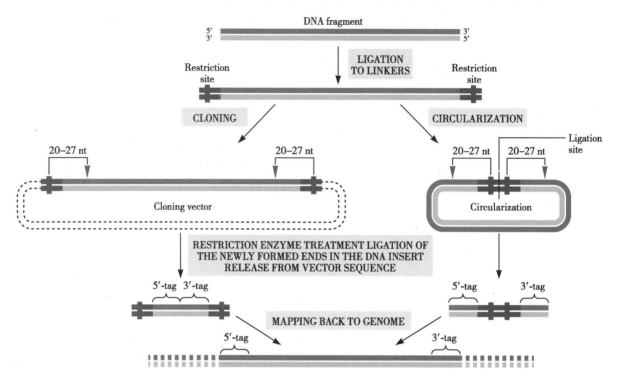

Figure 7.48 PET: pair-end di-tag analysis. Linkers containing restriction enzyme binding sites are added to each end of DNA. The DNA ends are connected by a cloning vector or by circularization. Restriction enzyme cutting 20–27 bp away from the binding site, depending on the enzyme, releases the linked PETs. They can be sequenced by all commercially available sequencing platforms, and the sequences are mapped back to the reference genome.

provide a complete map of transcriptional activity of the genome in a certain bacterial species under a certain set of conditions. This analysis can be used to monitor the transcriptional response of the cell to changing environmental conditions or to drug treatment.

cDNA libraries can then be established with reverse transcriptase and analyzed by high-throughput methods, such as microarrays, Serial analysis of gene expression (SAGE) or high-throughput DNA sequencing, to provide information on transcribed sequences genome-wide (Figure 7.49).

Microarrays are small, solid supports onto which the sequences from thousands of different genes are immobilized, or attached, at fixed locations. The supports themselves are usually glass microscope slides but can also be silicon chips or nylon membranes. The DNA, whether in the form of DNA, cDNA, or oligonucleotides, may be printed, spotted, or synthesized directly onto the support by use of robots. The RNA samples to be analyzed are fluorescently labeled and allowed to hybridize with the array, and the extent of hybridization is estimated by measuring the intensities of fluorescence in each individual spot (Figure 7.49A). SAGE is one of the genomewide techniques through high throughput DNA sequencing to measure the copy number of each gene on either genome or RNA level in a specific cell or tissue with the next-generation sequencing platform (Figure 7.49B) The more one counts—and these next-generation systems can count a lot—the better the measure of the copy number is; even rare transcripts in a population can be detected. Once either the microarray or high-throughput sequencing data are available, computer algorithms can be used to map these tags back to the reference genome.

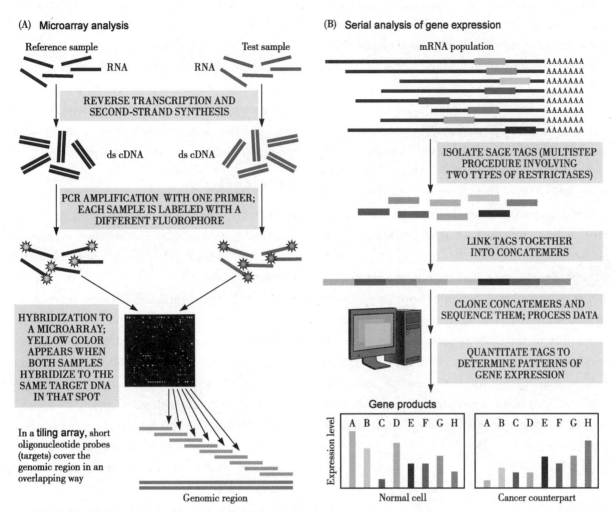

Figure 7.49 **Genome-wide transcriptome analysis.** (A) DNA microarrays with tiling arrays. The oligonucleotide sequences that cover large portions of a genome or entire genomes are spotted. For example, the Affymetrix tiling array provides 6.4 million spots, each containing oligonucleotides of length 25 nucleotides, spaced 10 base pairs apart, across the entire human genome. Such arrays allow unbiased analysis of the entire human transcriptome without any prior knowledge of genic and intergenic regions, introns and exons, etc. Microarrays provide a huge amount of noisy data. Interpretation of these data requires a combination of biological knowledge, statistics, machine learning, and the development of efficient algorithms that are able to select features above background noise. (B) high-throughput sequencing, for example, SAGE. The advent of next-generation sequencing platforms has made sequence-based expression analysis an increasingly popular, digital alternative to microarrays. Such platforms are now commercially available. In addition to providing exact sequences present in RNA populations, which are then mapped back to the genome, they give information about expression levels: one simply counts the number of times a gene is represented, or the number of hits, in the final data set.

7.5 Understanding transcription is useful in clinical practice

The toxin α-amanitin is capable of causing irreversible hepatocellular and renal dysfunction through inhibition of mammalian RNA polymerases. α-Amanitin is particularly effective at blocking the action of RNA polymerase II. Treatment was primarily supportive, with fluid and electrolyte replacement for that lost through the gastrointestinal tract. No effective antidote is available for the *Amanita phalloides'* toxin.

7.5.1 *Some common antibiotics that act by inhibiting bacterial transcription*

A deep understanding of how transcription occurs in bacteria is absolutely necessary for the development of drugs to combat bacterial infections (Figure 7.50). The introduction of Rifampicin in 1959 benefited a marked reduction of the time needed to treat tuberculosis patients from 18–24 months to 6–9 months, the effect is attributed to its ability to kill nonreplicating tuberculosis bacteria. Rifampicin acts by binding to a site within the RNAP active-center cleft, adjacent to the active center itself. It does not inhibit the binding of the polymerase to the DNA template or the catalytic step of the reaction itself. Instead, it sterically interferes with the binding of the newly formed short RNA-DNA hybrid after the addition of two or three nucleotides to the RNA chain: it blocks the channel into which the RNA-DNA hybrid must pass.

Actinomycin D is another polypeptide-containing antibiotic from a different strain of *Streptomyces*. It acts by intercalating its phenoxazone ring between neighboring base pairs in the DNA, thus affecting the ability of the DNA to serve as a template for transcription. At low concentrations, it selectively inhibits transcription without interfering with DNA replication. At higher concentrations, it inhibits the growth of all rapidly dividing cells and is used as an anti-cancer agent.

7.5.2 Drugs currently used to treat AIDS act on the viral reverse transcriptase

Drugs currently used to treat AIDS act on the viral reverse transcriptase or the protease. The non-nucleoside drugs (e.g., efavirenz) bind to reverse transcriptase and inhibit its action. The nucleoside analogs (e.g., zidovudine [ZDV]) add to the 3′-end of the growing DNA transcript produced by reverse transcriptase and prevent further elongation. The protease inhibitors (e.g., indinavir) bind to the protease and prevent it from cleaving the polyprotein.

Key Concepts

- Transcription is the process that a new RNA strand is synthesized, which is complementary to one DNA strand, and identical in sequence to the other strand.
- Transcription is catalyzed RNA polymerases or RNAPs with nucleotide triphosphates as substrates from the 5′-end toward the 3′-end. In contrast to DNA polymerases, RNAPs do not require a primer to initiate transcription, nor do they contain 3-to-5 exonuclease activity.
- The RNAP binds strongly to the promoter and allows the initiation of transcription, and the process may abort before the transcript is reached ~15 nucleotides.
- In bacteria, only one RNAP exists, which recognizes the promoter with special σ subunits com-

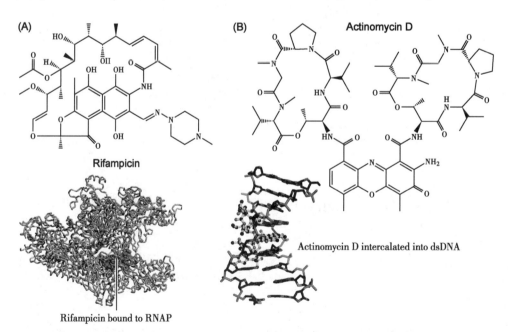

Figure 7.50 **Examples of widely used antibiotics that act by inhibiting bacterial RNAP.** (A) Rifampicin is used for the treatment of both Gram-positive and Gram-negative bacterial infections, including tuberculosis. (B) Actinomycin D inhibits transcription.

plexed with the RNAP core enzyme. There are several different σ subunits.
- Topoisomerases also play a role in relaxing the DNA superhelical stress created by the polymerase as it tracks down the template.
- Termination of transcription in bacteria occurs via two kinds of mechanisms, sequence-specific termination and a helicase enzyme-catalyzed termination.
- There are three kinds of RNA polymerases in a eukaryotic cell, Pol I, Pol II, and Pol III, which produce different kinds of RNA.
- Pol II transcribes protein-coding genes to produce mRNA, and a set of basal transcription factors, TFs, associate with promoter DNA and the polymerase to form the minimal pre-initiation complex or PIC.
- Termination of transcription in eukaryotes is closely linked to post-transcriptional processing of RNA.
- Pol I transcribes the genes for the major ribosomal RNAs. Pol III transcribes certain small RNAs involved in ribosome function and the processing of mRNAs in the nucleus.
- Other DNA sequences, such as promoter-proximal elements and enhancers, affect the rate of transcription initiation through the interactions of DNA-binding proteins with RNA polymerase and other initiation factors.
- There are some experimental approaches to be used to study gene transcription.

Key Words

-10 element(-10 顺式元件)
5′ Cap(5′ 帽子)
abortive transcription(中止性转录)
CAGE(基因表达的 CAP 分析技术)
chromatin modification(染色质修饰)
cis-element(顺式元件)
coactivator(辅助活化因子)
core enzyme(核心酶)
corepressor(辅助抑制因子)
C-terminal domain, CTD(C-末端结构域)
EMSA(电泳迁移率变动实验)
enhancer(增强子)
exon(外显子)
high-throughput DNA sequencing(高通量测序)
histone acetyltransferase(组蛋白乙酰转移酶)
holoenzyme(全酶)
intron(内含子)
microarray(微阵列技术)
mRNA(信使 RNA)
nucleus export signal, NES(细胞核输出信号)
nucleus localization signal, NLS(细胞核定位信号)
poly A tail(多聚 A 尾)
pre-initiation complex(转录起始前体复合物)
primer extension assay(引物延伸实验)
pri-RNA(前体 RNA)
promoter(启动子)
RACE(cDNA 末端快速扩增技术)
RNA polymerase(RNA 聚合酶)
S1 nuclease protection assay(S1 核酸酶保护实验)
SAGE(基因表达系列分析技术)
sigma factor(sigma 因子)
silencer(沉默子)
splicing(剪切)
TATA box(TATA 盒)
transcription(转录)
transcription factor(转录因子)
transcription start site(转录起始位点)

Questions

1. How you could use DNase footprinting to demonstrate that a sigma (σ) factor is required for specific binding of RNA polymerase to a bacterial gene promoter.
2. You suspect that a sequence upstream of a transcriptional start site is acting as an enhancer, not as a promoter. Describe an experiment you would run to test your hypothesis. Predict the results.
3. You have purified a transcription factor that has a leucine-rich region. You perform an electrophoretic mobility shift assay (EMSA) using a double-stranded oligonucleotide that you know from other studies contains the site recognized by this transcription factor *in vivo*. However, the transcription factor does not bind to the labeled oligonucleotide in your EMSA. Please explain the possible reason for this result.
4. You are studying a new class of eukaryotic promoters recognized by a novel RNA polymerase. You discover two general transcription factors that are required for the

transcription of these promoters. You suspect that one has helicase activity and that the other is required to recruit the helicase and the RNA polymerase to the promoter. Describe experiments you would perform to test your hypothesis. Provide sample results of your experiments.

5. You include an inhibitor of the protein kinase activity of TFIIH in an *in vitro* transcription assay. What step in transcription would you expect to see blocked? Describe an experiment you would run to test your hypothesis. Predict the results.

References

[1] Adelman K, Lis JT (2002) How does pol II overcome the nucleosome barrier? *Molecular Cell*,9:451–452.

[2] Agalioti T, Lomvardas S, Parekh B, et al.(2000) Ordered recruitment of chromatin modifying and general transcription factors to the IFN-β promoter. *Cell*,103:667–678.

[3] Agresti A, Scaffidi P, Riva A, et al. (2005) GR and HMGB1 interact only within chromatin and influence each other's residence time. *Molecular Cell*,18:109–121.

[4] Aravind L, Anantharaman V, Balaji S, et al. (2005) The many faces of the helix-turn-helix domain: transcription regulation and beyond. *FEMS Microbiology Reviews*, 29:231–262.

[5] Armache K J, Kettenberger H, Cramer P (2005) The dynamic machinery of mRNA elongation. *Current Opinion in Structural Biology*,15:197–203.

[6] Baumli S, Hoeppner S, Cramer P (2005) A conserved Mediator hinge revealed in the structure of the MED7/MED21 (Med7/Srb7) heterodimer. *Journal of Biological Chemistry*,280:18171–18178.

[7] Bell, AC, West AG, Felsenfeld G (2001) Insulators and boundaries. Versatile regulatory elements in the eukaryotic genome. *Science*,291:447–450.

[8] Darst SA (2001) Bacterial RNA polymerase. *Curr Opin Struct Biol*,11:155–162.

[9] Dove SL, Hochschild A (2005) How transcription initiation can be regulated in bacteria. In The Bacterial Chromosome (Higgins NP ed), pp 297–310. ASM Press.

[10] Greenleaf WJ, Woodside MT, Block SM (2007) High-resolution, single molecule measurements of biomolecular motion. *Annu Rev Biophys Biomol Struct*,36:171–190.

[11] Gruber TM, Gross CA (2003) Multiple sigma subunits and the partitioning of bacterial transcription space. *Annu Rev Microbiol*,57:441–466.

[12] Hurwitz J (2005) The discovery of RNA polymerase. *J Biol Chem*, 280:42477–42485.

[13] Kaplan DL, O'Donnell M (2003) Rho factor: transcription termination in four steps. *Curr Biol*,13:R714–R716.

[14] Lu H, Zawel L, Fisher L, et al. (1992) Human general transcription factor IIH phosphorylates the C-terminal domain of RNA polymerase II. *Nature*,358:641–645.

[15] Pal M, Ponticelli AS, Luse DS(2005)The role of the transcription bubble and TFIIB in promoter clearance by RNA polymerase II. *Molecular Cell*,19:101–110.

[16] Studitsky VM, Walter W, Kireeva M, et al. (2004) Chromatin remodeling by RNA polymerases. *Trends in Biochemical Sciences*,29:127–135.

[17] Wang G, Balamotis MA, Stevens JL, et al. (2005) Mediator requirement for both recruitment and postrecruitment steps in transcription initiation. *Molecular Cell*, 17:683–694.

[18] Yusufzai TM, Tagami H, Nakatani Y, et al. (2004) CTCF tethers an insulator to subnuclear sites, suggesting shared insulator mechanisms across species. *Molecular Cell*, 13:291–298.

Haihe Wang

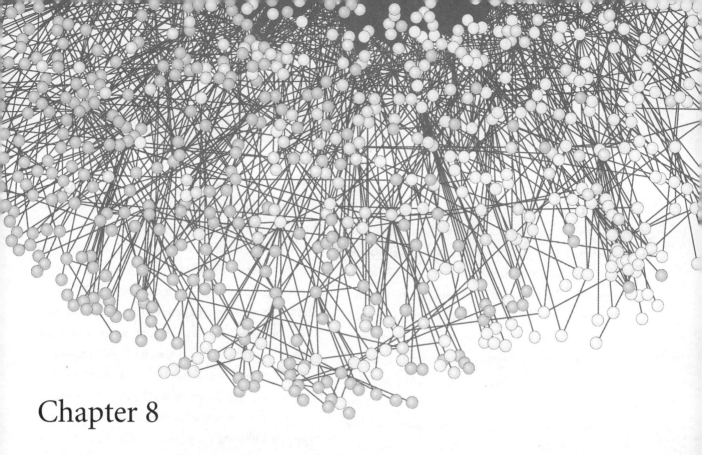

Chapter 8

Transcription Regulation

8.1 Introduction

The regulation of gene expression is a complicate event at multiple levels of gene action and regulation of transcription is the general control point. Regulation of transcriptional activation of gene is related to DNA sequence, regulatory protein and RNA polymerase activity, and so on.

Traditionally, regulation of transcription in bacteria is viewed as a way for these single-celled organisms to adjust their activity to suit their immediate environment. Some bacteria undergo major changes in lifestyle, as in sporulation. Some choose not to live alone; there are some special cases in which bacteria form colonies or biofilms, with some level of differentiation in the functions of individual members of the bacterial population.

Eukaryotic organisms must regulate their transcriptional output not only in response to environmental signals, as do bacteria, but also to adapt to programmed developmental or differentiation cues. The diversity of these responses has demanded an enormous complexity in regulatory mechanisms. Nevertheless, the basic principle of transcriptional regulation in bacteria is also valid for eukaryotes: *cis* elements in the DNA interact with *trans* factors, which can be either proteins or noncoding RNA molecules. However, two other levels of regulation were added in the evolution from bacteria to eukaryotes: regulated changes in chromatin structure or in DNA methylation provide a major part of the overall landscape of eukaryotic transcriptional regulation. In this chapter, transcription regulation in bacteria and eukaryote will be addressed in greater detail.

8.2 Regulation of transcription in bacteria

In this part, we focus on several distinct mechanisms used by the bacterial cell to adjust its metabolism to suit its immediate environment. Genes that are essential to the vital functions of any cell, termed housekeeping genes, tend to be transcribed at all times in what is called a constitutive manner, although even housekeeping genes may be expressed at different levels under different growth conditions. Most other genes are transcribed only under certain conditions; thus, their transcriptional status is highly regulated. A regulated gene can be OFF, when its product is not needed, or ON, when the cell needs the product either for some vital function, under a specific set of conditions, or to increase its efficiency in the use of resources, such as scarce nutrients. An additional level of regulation ensures that the product is available in the amounts needed. In practice, this means that a regulated gene can be expressed or transcribed to different degrees. As we see here and in later chapters, the amount of functional protein product is regulated at several different levels including transcription, processing of the primary transcript, transcript stability and translation into protein, stability of the protein, and post-translational processing and modification of the protein. Still, primary control of the amount of gene product available is through regulation of transcription, at the levels of initiation, elongation, and/or termination.

8.2.1 General models for regulation of transcription

(1) Regulation can occur via differences in promoter strength or use of alternative σ factors

The most direct way in which the transcription of different genes is controlled is from the structure of the promoter itself or from the composition of the holoenzyme complex. Supplies of both free RNA polymerase molecules and σ factors are limited in the cell, so there is intense competition between gene promoters for the holoenzyme. Because all promoters are more or less equally accessible to binding, promoter strength will be mainly defined by the comparative rates of dissociation of the holoenzyme from different promoters. The intrinsic strength of the promoter of each individual gene is a function of the extent to which the core promoter, the −10 and −35 boxes, conforms to the consensus sequences at these sites, because each base pair in the consensus sequence represents by definition the most favorable interaction with σ at that point in the sequence (see Chapter 7). The closer the −10 and −35 sequences are to the consensus, the stronger interactions will be and the stronger the promoter.

Two additional elements may be present in some promoters: the extended −10 element, or TGn element, and the upstream element, or UP element. These are schematically depicted in Figure 8.1. The extended −10 element interacts with domain 3 of σ factor, whereas the UP element interacts with the C-terminal domain of the α subunit of RNA polymerase. These additional contacts between the promoter and the holoenzyme stabilize the initiation complex, contributing to higher levels of initiation.

As mentioned in Chapter 7, another level of regulation is brought about by the use of alternative σ factors. The existence of alternative σ factors that recognize different promoters under different conditions contributes significantly to gene regulation within the bacterial cell.

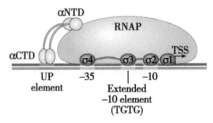

Figure 8.1 **General promoter architecture.** In addition to the −10 and −35 boxes, there are two additional gene elements: the extended −10 or TG*n* element and the upstream or UP element. The extended −10 element is a 3–4 bp motif located immediately upstream of the −10 consensus element. It is rich in TG and interacts with domain 3 of the subunit of the RNA polymerase (RNAP) holoenzyme. The UP element is an AT-rich sequence of ~20 bp that is associated with strong promoters. It is located at somewhat varied distances from the −35 box and interacts with the C-terminal domains of the RNAP α subunit, known as α CTD. The variability in the exact location of the UP element is allowed because of the highly flexible nature of the linker region that connects the CTD and the N-terminal domains or NTD of the α subunit of RNAP. All four promoter elements participate in the initial binding of RNAP to the promoter, but the relative contribution of each element differs from promoter to promoter. Thus, the UP element is characteristically present in strong promoters that dock the RNA polymerase in a relatively stable pre-initiation complex.

(2) Regulation through ligand binding to RNA polymerase is called stringent control

Bacteria have developed various highly controlled ways to respond to stress conditions such as nutrient deficiency. When the cells experience amino acid starvation, they curtail synthesis of stable RNA molecules, such as ribosomal RNA or rRNA, in favor of comprehensive restructuring of metabolic gene expression, including some amino acid biosynthesis genes. This massive, coordinated metabolic response is known as the stringent response and is mediated by a small molecule, the alarmone, ppGpp.

ppGpp binds to the β and β' subunits of RNA polymerase or RNAP, close to the active center, to inhibit the transcription of rRNA genes. Several non-mutually-exclusive models of the ppGpp-mediated negative regulation of transcription have been proposed: these include inhibiting open complex formation, reducing the stability of the open complex, inhibiting the process of promoter clearance, and increasing polymerase pausing. Although high-resolution structures of complexes of RNAP with both ppGpp and pppGpp are now available, the mechanism of polymerase inhibition is still not fully understood. The observed reduction in the stability of the open complex occurs on all promoters, but only promoters that have intrinsically unstable open complexes, such as those of the rRNA genes, are selectively inhibited.

The cell synthesizes ppGpp in response to stress by two different mechanisms (Figure 8.2). The lack of available amino acids leads to the accumulation of uncharged tRNA molecules, and these bind to the acceptor site in the ribosome, which is the site of entry of the incoming tRNA carrying its cognate amino acid (see Chapter 11). The binding of uncharged tRNAs to the ribosome signals the ribosome-associated enzyme RelA to synthesize pppGpp, which is then converted to the alarmone ppGpp. In a second, less-well-understood pathway, ppGpp is produced by the enzyme SpoT that senses most other stresses. SpoT is an enzyme of dual action, as it also hydrolyzes ppGpp. Mutant cells lacking both RelA and SpoT fail to inhibit the production of rRNA, and thus stringent control is relaxed in these mutant cells.

The second part of the stringent response is the positive regulation of a large number of genes that are induced during starvation. This positive regulation is aided by a protein factor, DskA, which is also bound to the polymerase. In addition, there is evidence that the release of the RNAP from the inhibited rRNA genes passively contributes to the up-regulation of these starvation-induced genes by raising the amount of free enzyme available.

8.2.2 Specific regulation of transcription

(1) Regulation of specific genes occurs through *cis-trans* interactions with transcription factors

It has long been recognized that gene regulation occurs through *cis-trans* interactions between *cis-*

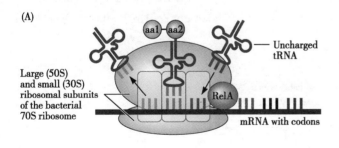

Figure 8.2 Stringent response to stress in bacteria. (A) The response is mediated by the alarmone ppGpp, whose synthesis is induced in response to nutrient stress. The absence of amino acids in the cell is sensed from large quantities of uncharged tRNA molecules. Uncharged tRNAs bind to the ribosome, signaling the ribosome-associated enzyme RelA to synthesize pppG. (B) Structure of ppGpp.

regulatory DNA elements, in the vicinity of the gene, and *trans*-acting factors, encoded elsewhere in the genome (Figure 8.3). For years, it was believed that all *trans*-acting factors are proteins; we now know that noncoding RNAs can also perform this function.

(2) Transcription factors are activators and repressors whose own activity is regulated in a number of ways

The genome of *Escherichia coli* has been predicted to contain more than 300 genes that encode proteins that bind to promoter regions to regulate transcription. Roughly 50% of these transcription factors or TFs have been studied experimentally. About 60 of these TFs each regulate only one set of genes with common functions; such a set of genes is called an operon. Regulation of a single operon is accomplished by highly sequence-specific binding of the TF to that operon's unique promoter. Other TFs may regulate several operons. Such groups of operons share identical or closely similar promoters. A small number of TFs, ~10, control very large sets of genes and have come to be known as global regulators. The existence of such classes of TFs reflects a hierarchical organization of gene regulation in bacteria.

These proteins may function either as activators, which stimulate transcription, or as repressors, which inhibit transcription. Classical examples of the regulation of three operons in *E. coli* are described below. Both activators and repressors are themselves regulated by the binding of small ligands that transmit environmental signals to the respective gene systems. Interestingly, the action of regulator proteins can be either stimulated or inhibited by the binding of specific ligands (Figure 8.4 and Figure 8.5).

In addition to being controlled by ligand binding, TF activity can be regulated by posttranslational modifications. A well-understood example is NarL, one of the global regulators, which can bind its target DNA sequences only when it is phosphorylated. The phosphorylation is carried out by two cognate sensor kinases, cell membrane proteins whose kinase activity is regulated by extracellular nitrite or nitrate ions.

Finally, the activity of some TFs in the cell is controlled by their intracellular concentration; this, in turn, is regulated at the level of their own transcription and/or translation and by their proteolytic degradation.

(3) Several transcription factors can act synergistically or in opposition to activate or repress transcription

Transcription of some genes is regulated by the action of single activators or single repressors on their cognate promoters. For most genes, several TFs act together to affect transcription activity either additively, or synergistically, or in opposition.

There are at least three major, distinct mechanisms of transcriptional activation by single activators (Figure 8.6). These differ by where in the promoter region the activator binds and how it activates transcription. In class I activation, the activator binds to a target that is located upstream of the promoter –35 element and recruits the RNAP by

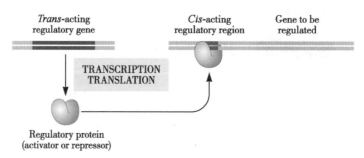

Figure 8.3 General principle of transcriptional regulation: *cis-trans* interactions. Each gene possesses one or more regulatory regions in *cis*, in the vicinity of the gene to be regulated. These are bound by regulatory protein factors, either activators or repressors, which are diffusible molecules that are encoded somewhere else in the genome; that is, they act in *trans*. We now know that the regulatory molecules can in some cases be functional RNA molecules.

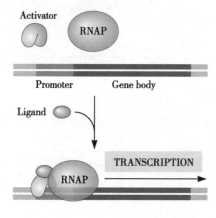

Figure 8.4 **Activity of activators can be up- or down-regulated by ligand binding.** (A) Up-regulation: the activator is active only if bound to a ligand. (B) Down-regulation: the activator dissociates from RNAP upon binding the ligand, and hence transcription is inhibited.

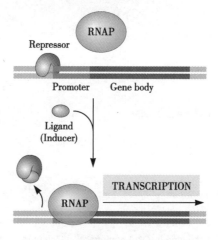

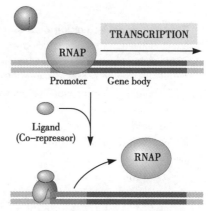

Figure 8.5 **Activity of repressors can be up-or down-regulated by ligand binding.** (A) Up-regulation: the repressor is active when not bound to a ligand, so ligand binding leads to dissociation of repressor and induction of transcription. The ligand in this case is called an inducer. (B) Down-regulation: binding of a ligand to the repressor is necessary for repressor binding and inhibition of transcription. The ligand in this case is called a co-repressor.

direct interactions with the polymerase C-terminal domain, CTD, of the α subunit. The linker that joins the CTD and the N-terminal domain, NTD, of the α subunit is flexible, so binding of the activator can occur at different distances from the −35 element. Class II activation involves RNAP recruitment via domain 4 of the σ factor. The spatial constraints of σ factor binding provide little flexibility in the positioning of the activators. In the third mechanism, the activator alters the conformation of the promoter to enable simultaneous interactions of RNAP with the −10 and −35 elements. Usually, the DNA is twisted to reorient the two elements for optimal binding of the

elements. A fourth, somewhat different mechanism depends on the modulation of repressor molecules by activators in a way that prevents the repressor from binding to the DNA.

The three mechanisms of transcriptional repression by a single repressor are steric hindrance, DNA loop formation, and modulation of an activator protein (Figure 8.7). These three mechanisms all act

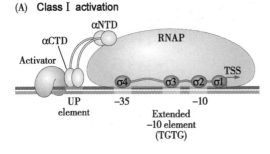

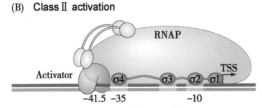

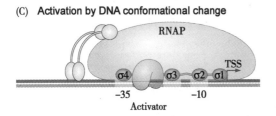

Figure 8.6 Transcriptional activation by a single activator. (A) Class I activation: the best-studied example is the action of CAP at the *lac* promoter. (B) Class II activation: the best-studied example is the activation of a phage # promoter by CI protein. (C) Transcriptional activation by conformational changes in promoter DNA caused by activator binding. (Adapted from Browning DF, Busby SJW [2004] *Nat Rev Microbiol*, 2:57–65. With permission from Macmillan Publishers, Ltd.)

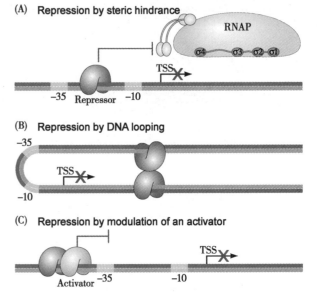

Figure 8.7 Transcriptional repression by a single repressor. (A) Repression by steric hindrance. The repressor-binding site overlaps core promoter elements and blocks promoter recognition by RNAP. The best-studied example is the action of the Lac repressor. (B) Repression by DNA looping. Repressors bind to distal sites on both sides of the promoter and oligomerize, making the intervening DNA loop out and thus preventing initiation. Examples are the Gal repressor at the *gal* promoter and the *ara* operon. (C) Repression by modulation of an activator protein: that is, the repressor functions as an anti-activator. The best studied examples are CytR-repressed promoters, which are dependent on activation by CAP. The CytR repressor and the CAP activator interact directly. (Adapted from Browning DF, Busby SJW [2004] *Nat Rev Microbiol*, 2:57–65. With permission from Macmillan Publishers, Ltd.)

at the promoter region itself or on DNA sequences in close proximity during transcription initiation. A fourth mechanism of repression by single repressors acts during elongation, when the DNA-bound repressor creates a roadblock to the advancing RNA polymerase.

Sometimes a single promoter can serve as an initiation site for a whole operon, a series of linked genes, which often code for proteins that are involved in a single metabolic pathway. In many cases, the activity of the set of genes comprising an operon is controlled at a single locus, called an operator. We now describe three such examples that are important in bacterial physiology and that have also played significant roles in developing our understanding of gene regulation.

8.2.3 Transcriptional regulation of operons important to bacterial physiology

(1) The *lac* operon is controlled by a dissociable repressor and an activator

Escherichia coli is an omnivorous microbe that is able to subsist on a number of food sources, including many sugars. Glucose is the sugar of preference but other monosaccharides, such as galactose, or disaccharides, such as lactose (Figure 8.8), can be utilized. Specific enzymes are needed for the metabolism of each sugar: in the case of lactose, both a permease, to facilitate transport into the cell, and the enzyme β-galactosidase, which cleaves the disaccharide into galactose plus glucose, are necessary (see Figure 8.8). The genes for these proteins are positioned in tandem on the *E. coli* chromosome, followed by the gene for a third enzyme, a transacetylase (Figure 8.9). Although not essential for lactose metabolism, the transacetylase serves to detoxify aberrant galactosides that are produced as by-products of the pathway. The galactosidase gene is designated *lacZ*, the permease *lacY*, and the transacetylase *lacA*; they lie in the genome in that order. These three genes comprise the *lac* operon. The *lacZ* gene is also a useful tool for assaying the success or failure of gene cloning experiments. Often, this gene is inserted along with the desired gene, and cell clones are then tested with a galactoside that produces a distinctive color when cleaved by the β-galactosidase product of the inserted *lacZ* gene.

When an *E. coli* bacterium encounters a medium containing glucose, it will use this nutrient in preference to any other. But if glucose is wholly lacking and lactose is available, the bacterium will turn on the lacZ, lacY, and lacA genes of the *lac* operon. Just how this switch occurred fascinated Jacques Monod and co-workers in the 1940s and 1950s. Their work led to the discovery of an important mode of gene regulation in bacteria. They showed, specifically, that the lac genes, in the absence of lactose, were essentially turned off by a repressor. The repressor protein binds to the DNA at operator sites near the lac promoter and can be displaced only when an inducer, such as allolactose, binds to the repressor. Allolactose is a minor product of the weak galactosidase activity that

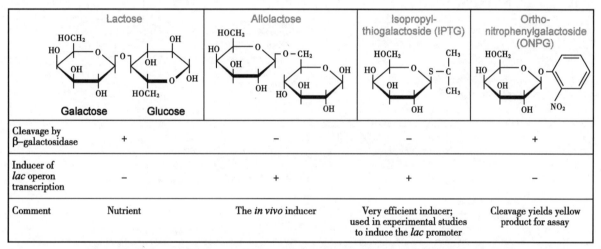

Figure 8.8 **Lactose and analogs used in studies of the *lac* operon.** Allolactose is the natural, *in vivo* inducer, but IPTG is a much more effective inducer that is used in many experimental studies. For experiments in which detection of β-galactosidase activity is important, ONPG is used as a substrate whose cleavage produces a colored product that can be observed even in cell clones.

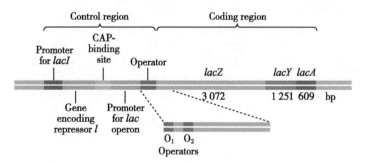

Figure 8.9 **Organization of the *lac* operon.** The three structural genes that encode the proteins involved in the overall process of lactose metabolism are *lacZ*, which codes for β-galactosidase, the enzyme that hydrolyzes the glycoside bond; *lacY*, which codes for the permease that transports the sugar from the medium through the cell membrane; and *lacA*, which codes for the transacetylase that acetylates β-galactosides that cannot be hydrolyzed, a reaction that eliminates toxic compounds from the cell. The control region contains the promoter; the operator, which binds the Lac repressor; and the gene, with its promoter, that encodes the repressor molecule. The control region also contains a binding site for the activator CAP.

is always present (see Figure 8.8).

As shown in Figure 8.10, all three genes, *lacZ*, *lacY*, and *lacA*, are transcribed together, with the polymerase starting from the promoter site just upstream from the operator sites. The two operator sites bind two dimers of the repressor protein, which effectively blocks polymerase binding. The gene for the repressor protein itself, *lacI*, is upstream from the *lac* control region and has its own promoter; the gene is designated *lacI* because the protein was at first thought to be an inducer. Between the *lacI* gene and the promoter for the *lac* operon lies another control element, the **catabolite activator protein** or **CAP**-binding site, which is involved in an additional, broader level of control. CAP is also known as the catabolite regulatory protein, CRP. Monod's initial observation was that lactose metabolism would begin only when glucose in the medium had mostly been consumed. When glucose levels in a cell are low, the production of ATP is limited and cyclic adenosine

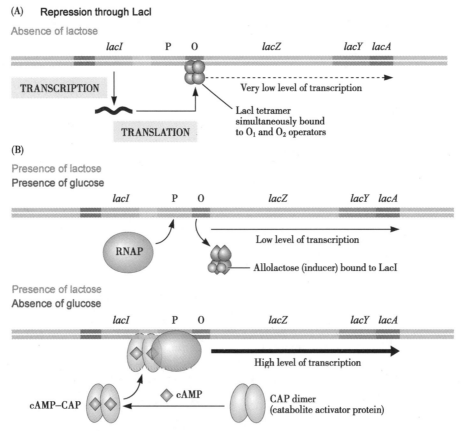

Figure 8.10 **Transcriptional activity of the lactose operon is regulated by two sugars.** When glucose is present in the medium, independent of the presence or absence of lactose, the *lac* operon is repressed, as glucose is the preferred carbon source. The presence of glucose leads to catabolic repression, which affects the activity of all operons that encode enzymes that catabolize alternative sugars. For the *lac* operon to be active, two conditions must be met: lactose should be present but glucose should not. (A) In the absence of lactose, the *lac* operon is repressed: the LacI repressor is bound to a site that partially overlaps the *lac* promoter, preventing the binding of RNAP. To be specific, LacI binds as a tetramer to two operator sites, O_1 and O_2, the latter of which is within the coding region of *lacZ*, and the intervening DNA between O_1 and O_2 loops out. (B) Behavior of the operon in the presence of lactose. In the presence of glucose, in the upper part, the genes are only slightly expressed. A secondary lactose metabolite, 1, 6-allolactose, binds to the repressor, changing its conformation and reducing its affinity for the operator. Allolactose acts as an inducer. For full activation to take place, there should be no glucose in the medium The mechanism of activation in the absence of glucose, in the lower part, involves an activator—catabolite activator protein or CAP, also known as cyclic AMP regulatory protein or CRP—which binds to a regulatory site upstream of the promoter only if cAMP is bound to it. The contact between CAP and RNAP is thought to be through the α subunit of the polymerase. The absence of glucose is sensed by the cell through the intracellular concentration of cAMP: when glucose is high, cAMP levels are low, but when glucose levels are low, there is a high concentration of cAMP. cAMP then binds to CAP to change its conformation and facilitate its binding to DNA.

monophosphate, cAMP, accumulates. Cyclic AMP will bind to the CAP protein, favoring a conformational change that facilitates binding of the protein to the CAP-binding site. This, in turn, recruits RNA polymerase to the promoter by direct interaction with the CAP dimer bound to the DNA.

The effects of various combinations of available nutrients on the expression of the *lac* operon can now be readily explained (see Figure 8.10). To summarize:

- In the absence of lactose, levels of inducer will be very low, and the repressor will bind strongly; the *lac* operon will be silent.
- In the presence of both lactose and glucose, RNAP can bind, albeit weakly, and the *lac* genes will be transcribed at a low level.
- When lactose is present but glucose is absent, the inducer will be present and the repressor will dissociate from the operator. Furthermore, cAMP will be present in high concentrations, favoring the binding of CAP and thus RNAP recruitment. A high level of *lac* operon transcription will result.

The structures of the Lac repressor and CAP are shown in Figure 8.11. The repressor functions as a tetramer of identical dimers, each dimer binding to one of the operator sites. CAP also binds to the DNA as a dimer. Both repressor and CAP bind to their specific sites via helix-turn-helix or HTH motifs and produce bending in DNA (see Chapter 6). The repressor dimer, in the absence of inducer, can also bind more weakly and nonspecifically to DNA, allowing it to search along the DNA for its specific site. In this condition, the protein does not produce DNA bending. The presence of bound inducer produces a conformational change in which the HTH motifs become disorganized, and binding to DNA is weakened.

Both the Lac repressor and the CAP protein provide excellent examples of allosteric regulation (see Box 2.1). In fact, it was the discovery of the repressor and its control by an inducer that suggested the very idea of allostery to Jacob and Monod. The binding of two small molecules, inducer to repressor and cAMP to CAP, produces opposite effects: the inducer weakens the DNA binding of the repressor, whereas cAMP strengthens CAP's affinity for DNA. Likewise, the allostery of interacting proteins within a heterodimer can be either positive or negative (see Box 2.1). Within the *lac* operon, interaction between repressor dimers to produce the tetramer strengthens their binding at each of the pair of operator sites,

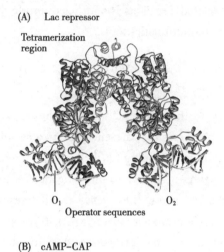

(A) Lac repressor

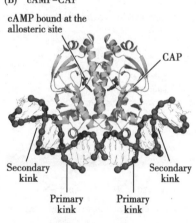

(B) cAMP–CAP

Figure 8.11 Structure of the Lac repressor and the cAMP-CAP dimer bound to DNA. (A) Structure of the Lac repressor tetramer bound to O_1 and O_2. Each dimer, one shown in red/blue and the other in green/orange, binds to one operator site through helix-turn-helix (HTH) motifs interacting with the major groove. The dimers are coupled at the tetramerization region; thus the tetramer binds O_1 and O_2 simultaneously. (B) Structure of the cAMP-CAP dimer bound to DNA. The crystallization DNA fragment contained a single-phosphate gap between positions 9 and 10 of each DNA half-site. The conformational change conferred by cAMP binding brings the two HTH motifs of each monomer together at an appropriate distance for interaction with neighboring major grooves. CAP binding causes 90° bending in DNA. (A, courtesy of Daniel Parente, University of Kansas Medical Center. B, from Lawson CL, Swigon D, Murakami KS, et al. [2004] *Curr Opin Struct Biol*, 14:10–20. With permission from Elsevier.)

and interaction between CAP and the polymerase enhances the DNA binding of both.

Currently, the binding sites of TFs across the genome can be visualized by the recently developed bioinformatic tool, termed as **sequence logos** that are a quantitative and visually effective way of presenting sequence conservation information for a family of closely related TF-binding sites or for a family of closely related TFs. This is another example of the power of bioinformatics tools.

(2) Control of the *trp* operon involves both repression and attenuation

Synthesis of the amino acid tryptophan in bacteria involves enzymes that are encoded by a set of five contiguous genes, coding for the tryptophan biosynthesis enzymes themselves plus a leader sequence. As in the *lac* operon, these genes are all transcribed from a single promoter (Figure 8.12). Tryptophan is not used very much in proteins, so to synthesize a lot of tryptophan or even to transcribe much messenger when the amino acid is abundant would be wasteful of cellular resources and energy. Therefore, the entire operon is regulated, mostly by allosteric feedback control of the kind mentioned in Box 2.1. As Figure 8.12 shows, there is a repressor protein that can bind to an operator site, blocking the binding of RNAP. In analogy with the Lac repressor, this is coded by a separate gene. The Trp repressor, however, operates in a way that is opposite to the Lac repressor. Binding of the effector molecule, in this case tryptophan itself, promotes binding of the repressor to the operator element, blocking transcription initiation. Thus, tryptophan accumulation nearly shuts down transcription of the genes necessary for its synthesis. Transcription at high tryptophan levels is only about 0.1% of that when no tryptophan is present.

However, there is evidence that not all of this effect is due to binding of the repressor. There exist mutants of the *trp* operon that have wholly effective repressor and operator, yet they transcribe at a rate tenfold higher than wild type under the same

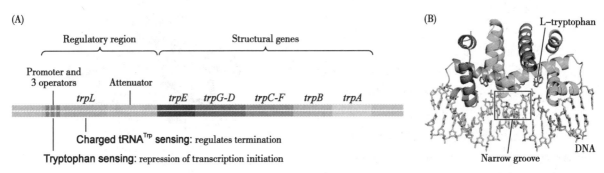

Figure 8.12 Organization and regulation of the *trp* operon in *E. coli*. The genes required for tryptophan biosynthesis are organized and regulated as a single transcriptional unit or operon. (A) Seven genetic elements, *trpEGDCFBA*, are under the control of a regulatory region located at the beginning of the operon. Two pairs of genes are fused: *trpG* with *trpD*, and *trpC* with *trpF*; separate polypeptide segments of the products of these fused genes catalyze different reactions in the tryptophan biosynthesis pathway. The regulatory region senses and responds to two different signals through two different mechanisms. The first signal is the concentration of tryptophan in the cell: the response to tryptophan is exerted through regulation of binding of the Trp repressor to the operator elements. The second signal is the proportion of tRNA^{Trp} that is attached to tryptophan, or charged, which depends on the concentration of free tryptophan in the cell; the response to this signal is exerted through the attenuation mechanism (see Figure 8.13). (B) Trp repressor bound to one of the operator sites. Trp repressor is a homodimer; one subunit is colored green and the other orange. The recognition helices of the DNA-binding HTH motifs insert themselves into the major groove of the DNA binding site. In the structure of the protein dimer itself, the distance between the DNA-interacting helices on the two monomers is 26.5 Å; upon binding of tryptophan, the dimer undergoes a conformational change that increases the distance between the DNA-interacting helices to 32.7 Å. This change is instrumental in securing a good fit between the Trp repressor and a *trp* operator site; in the protein-DNA complex the distance is 32.1 Å. Binding of repressor to the operator site interferes with the ability of the DNA to bind RNAP, preventing transcription initiation. Simultaneous binding of repressor dimers to adjacent operator sites in the promoter-operator region increases the stability of the repressor-DNA complex and thus the effectiveness of repression. (A, adapted from Yanofsky C [2007] *RNA*, 13:1141–1154. With permission from The RNA Society.)

tryptophan-poor conditions. This led to discover a whole other level of regulation, overlaid upon repression; this is called **attenuation**, and it operates in an entirely different manner.

The mechanism of attenuation, as it occurs in the *trp* operon, is illustrated in Figure 8.13. This method of regulation acts on the termination of transcription. It depends upon the fact that in bacteria even the earliest stages of transcription are accompanied by attachment of ribosomes to the mRNA: transcription and translation are coupled. Also important is the precise sequence structure near the 5'-end of the operon. Following the start codon is a leader sequence, followed by three sequences in which the transcribed RNA has the potential to form a hairpin. These sequences have the potential to form hairpins in two different places: between segments 2 and 3 or between segments 3 and 4.

Even in the repressed state, there is some initiation of transcription on the *trp* operon. When tryptophan levels are high, the polymerase will transcribe right through the leader sequence, with a ribosome following along the new RNA to effect translation. This ribosome physically blocks the formation of hairpin 2–3, with the consequence that hairpin 3–4 can form as soon as the polymerase has passed (see

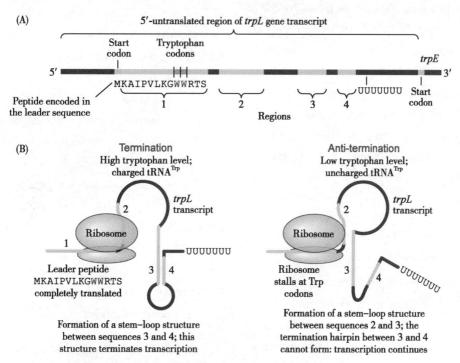

Figure 8.13 **Attenuation as a regulating mechanism in the *trp* operon.** Functioning of the attenuation mechanism depends on the two-dimensional structure of the leader transcript and on translation, or protein synthesis, in that region, which is coupled to transcription. (A) The first 141 nucleotides of the *trp* operon transcript can fold into three alternative RNA hairpin structures involving regions 1, 2, 3, and 4, bracketed below the transcript. (B) Left, formation of a stem-loop structure between regions 3 and 4 leads to termination of transcription through the typical intrinsic termination process. Right, the competing formation of a stem-loop structure between regions 2 and 3 causes anti-termination; that is, it allows transcription to continue because of the absence of the terminator 3–4 hairpin. The formation of terminator versus anti-terminator structures is controlled through the coupled translation process. Note that the 5'-untranslated region, 5'-UTR, of the *trp* operon transcript contains a small nucleotide sequence that can be translated into a short leader peptide that contains two adjacent tryptophan residues, WW. This peptide sequence is shown alongside the corresponding region of the operon. As soon as the 5'-UTR is transcribed, a ribosome is attached to the ribosome-binding site immediately preceding the start codon and initiates translation. Whenever the level of charged tRNA^{Trp} is adequate for rapid translation of the two Trp codons, translation of the leader peptide is completed, the ribosome dissociates, and hairpin structures 1–2 and 3–4 form: transcription is terminated. When there is not enough tryptophan to charge all tRNA^{Trp}, the ribosome stalls at one of the two Trp codons, allowing for formation of the 2–3 hairpin; this, in turn, prevents the formation of the terminator 3–4 structure and the operon is actively transcribed. The regulation of the entire operon is thus subject to a two-level feedback response to the presence of tryptophan, the final product of the synthetic pathway encoded in the operon. (Adapted, courtesy of Michael King, Indiana University.)

Figure 8.13). Hairpin 3–4 is followed immediately by a stretch of seven adenine nucleotides, A, in the DNA, which transcribes into seven uracil nucleotides, U, in the RNA. But this is the exact condition, a hairpin followed by several U residues, that we encountered in Chapter 7 as the intrinsic termination signal. Thus, under these conditions, the polymerase encounters a termination signal and is released, and transcription of the operon is aborted.

Now consider what happens when tryptophan is in short supply and needs to be made. Attenuation must somehow be overridden. Within the leader sequence is a pair of adjacent Trp codons. If tryptophan is scarce, it will be difficult for the ribosome to find enough charged tRNA^{Trp} to allow it to translate the transcript beyond these Trp codons. Accordingly, it stalls within the leader sequence while the polymerase proceeds on its way. The stalled ribosome allows the 2–3 hairpin to form, which disrupts the formation of the 3–4 hairpin (see Figure 8.13). Consequently, the termination signal is not formed, and the polymerase can proceed to transcribe the rest of the *trp* operon.

(3) The same protein can serve as an activator or a repressor: the *ara* operon

Bacteria can utilize the sugar arabinose, metabolizing it into xylulose 5-phosphate, which can enter the pentose phosphate pathway. The *ara* operon codes for three enzymes involved in this pathway, AraB, AraA, and AraD, plus the regulatory protein AraC. The *araB*, *araA*, and *araD* genes have a separate promoter; *araC* is transcribed in the opposite direction (Figure 8.14A). There are four AraC binding sites: $araO_1$ and $araO_2$ are located upstream from the *araC* promoter, and $araI_1$ and $araI_2$ are located between the two promoters.

AraC can bind arabinose, and its choice of DNA binding sites depends upon whether or not arabinose is present. In the absence of arabinose, AraC binds as a dimer between $araO_2$ and $araI_1$. As shown in Figure 8.14B, this forces a loop in the DNA, which has the effect of blocking transcription from the *araBAD* promoter, repressing the operon. This loop is critical, which was demonstrated in an elegant way by inserting an extra five base pairs between $araO_1$ and $araO_2$. This insertion produced an extra half-twist in the DNA, rotating the $araO_1$-$araO_2$ binding sites by 180° with respect to each other and thus hindering the interaction between AraC and these two binding sites. The effect was to relieve repression of *araBAD*. Adding or subtracting a full 10-base-pair turn had no effect, serving as an elegant control.

In the presence of arabinose, AraC no longer causes loop formation or acts as a repressor. Instead, as shown in Figure 8.14B, it binds as a dimer to the $araI_1$ and $araI_2$ sites and promotes adjacent binding of CAP protein. These effects stimulate transcription of the *araBAD* genes. Thus, AraC can act as either a repressor or an activator, depending on the concentration of arabinose.

An overabundance of AraC is avoided by a simple autoregulatory mechanism. $araO_1$ is just downstream, and in the *araC* gene orientation, from the *araC* promoter. High concentrations of AraC will tend to bind at this site, repressing its own gene via a roadblock mechanism (Figure 8.14C). Note the contrast between *trp* and *ara* operon regulation. In the *trp* operon, the end product of a metabolic pathway regulates transcription of the operon; the *ara* operon is regulated by the first substrate of the metabolic pathway. There are many different modes of transcriptional control.

8.2.4 Other modes of gene regulation in bacteria

(1) DNA supercoiling is involved in both global and local regulation of transcription

As seen in Chapter 5, the bacterial chromosome is organized in supercoiled loops. The degree of superhelical stress in individual supercoiled domains is finely regulated by topoisomerases (see Chapter 2) and by structural proteins that twist, bend, or loop the DNA upon binding. In addition, enzymes that translocate along the DNA, such as RNA polymerase and helicases, create temporary waves of positive supercoiling in front and negative supercoiling in their wake.

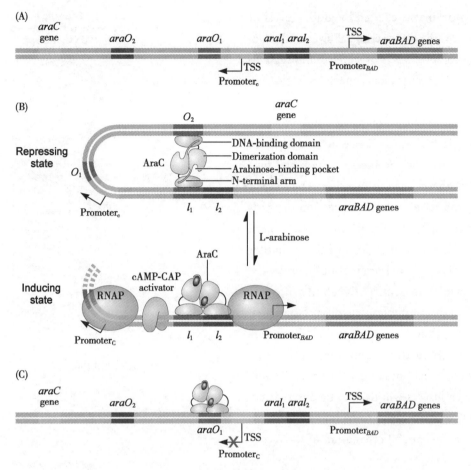

Figure 8.14 **Transcriptional control of the *ara* operon in *E. coli*.** (A) Map of the arabinose control region. Four binding sites for the transcriptional factor AraC (shown in brown) are located just upstream of the promoter for the arabinose structural genes, *araB*, *araA*, and *araD*. The promoter for the *araC* gene drives transcription from the opposite DNA strand. (B) Negative control of transcription, in the absence of arabinose, and positive control of transcription, in the presence of arabinose, through the action of AraC, which acts as a dimer. Each monomer has a C-terminal DNA-binding domain connected through a flexible linker to the N-terminal dimerization domain, which also contains the arabinose-binding pocket. In the absence of arabinose, the two monomers interact through their N-terminal tails, so that the two C-terminal domains bind to the *araO₂* and *araI₁* regulatory elements, thus forcing the intervening DNA to loop out. The promoter is inaccessible to polymerase binding in this configuration. When arabinose becomes available, it binds to AraC, changing the conformation of the dimer such that it now binds to the *araI₁* and *araI₂* sequences; this opens up the *araBAD* promoter to bind the polymerase. For full transcriptional activation, glucose should be absent from the medium. This condition is sensed by the levels of cAMP in the cell, which increase when glucose is absent, and thus the cAMP-bound positive regulator CAP binds to the promoter region of the *araC* gene to up-regulate its transcription. The response of the cell to lack of glucose is exactly the same as in the case of *lac* operon regulation. (C) Autoregulation of the *araC* gene. This occurs through utilization of the *araO₁* element, which does not participate in regulation of the *araBAD* genes; it instead allows AraC to regulate its own synthesis. As levels of AraC increase, it binds to *araO₁*, thus preventing polymerase binding to the *araC* promoter and inhibiting the transcription of its own gene from the bottom DNA strand. (B, adapted from Weldon JE, Rodgers ME, Larkin C, et al. [2007] *Proteins*, 66:646–654. With permission from John Wiley & Sons, Inc.)

The level of global supercoiling is different under different environmental conditions. Gyrase, the enzyme that pumps negative superhelical stress into DNA, utilizes ATP to do so. Thus, the overall cellular energetics play a role in determining the global level of supercoiling. It is well known that the ratio [ATP + 1/2ADP]/[ATP + ADP + AMP], known as energy charge, drops when *E. coli* cells are shifted from aerobic to anaerobic conditions. This shift is accompanied by a decrease in the overall level of DNA supercoiling. Under optimal growth conditions, the energy charge is high, ~0.85, and the superhelical density σ equals −0.05. Under anaerobic conditions, σ drops to ~−0.038. As is not difficult to imagine, any control on transcription that is exerted by global levels of supercoiling must be rather crude. However, the factors that control local levels of supercoiling serve to fine-tune the nonspecific global effects.

How can local supercoiling affect transcription? The first mechanism involves changes in the twist of the double-helical DNA. A change in twist changes the base-pair spacing in the affected region and thus the physical distance between two sites. The level of supercoiling that gives optimal expression depends on the number of base pairs in the spacer region between the −35 and −10 regions of the promoter (Figure 8.15). The optimal orientation between the two regions, which maximizes the ability of σ^{70} to locate and bind to a promoter, depends on the actual physical length of the −35 to −10 spacer. Promoters with 17-bp spacers have been optimized during evolution for the normal physiological levels of supercoiling. Longer spacers will have to be overtwisted, and shorter spacers will have to be undertwisted, in order to create the optimal orientation of the recognition elements for σ^{70} binding.

DNA supercoiling can also regulate transcription through changes in writhe: short regions can loop out, and long regions can form plectonemes. Both these structures can bring sequence elements into close proximity so that they can interact, either directly or through the proteins bound to them. An example of a loop structure that can inhibit RNAP binding is presented by the regulation of the *ara* operon in the absence of arabinose (see Figure 8.14).

Finally, transcription can be regulated through the formation, under negative stress, of **alternative DNA structures** such as locally denatured regions, cruciforms, Z-DNA, or triplex DNA. If these form downstream of the promoter, transcription elongation will be inhibited. Alternative structures in regulatory regions can also serve as recognition elements for binding of TFs or the polymerase itself.

(2) DNA methylation can provide specific regulation

DNA methylation in bacteria is widespread and can occur on both cytosines and adenines. The three recognized base modifications are 5-methylcytosine,

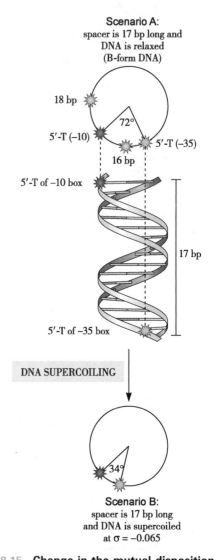

Figure 8.15 Change in the mutual disposition of the −35 and −10 promoter boxes due to changes in helical twist. The boxes are represented by their 5′-T residues, shown as green and red starbursts. The projections of the σ70 promoter, represented as circles, are perpendicular to the helical axis. In scenario A, the spacer is 17 bp long and the DNA is relaxed; that is, the average rotation angle is 34.5° per base pair, as in B-DNA. In this case, the angle between the two 5′-Ts in the projection is 72°. For reference, the positions of the −10 box Ts if the spacer were 16 or 18 bp long are also shown, as pink starbursts. In scenario B, the DNA is now negatively supercoiled or untwisted so that the superhelical density σ = −0.065, and the angle between the 5′-Ts of the two promoter boxes decreases to 34°, bringing them much closer to each other in the projection plane. In other words, the two boxes are now much closer to being on the same side of the helix. This scenario presumes that 2/3 of the change in linking number is absorbed by a change in twist and 1/3 by a change in writhe. As can be seen from comparison of the two projections, the two boxes in the supercoiled DNA molecule that are spaced 17 bp apart will be located similarly with respect to each other as would the two boxes spaced by 16 bp in the relaxed molecule.

m⁵C; N^6-methyladenine, m⁶A; and N^4-methylcytosine, m⁴C (Figure 8.16). The first two modifications occur in bacteria, protists, and fungi, whereas the third is bacteria-specific. These marks can be propagated from generation to generation because of the capacity of the respective enzymes to recognize and act upon hemimethylated DNA. This is the state of the DNA immediately following replication: the parental strand keeps its methyl groups, whereas for a while the newly synthesized DNA strand remains unmethylated. Methyltransferases then add a new methyl group to the new DNA strand. The specific enzymes that introduce these modifications are DNA cytosine methyltransferase or Dcm for m⁵C, DNA adenine methyltransferase or Dam for m⁶A, and cell-cycle-regulated methyltransferase or CcrM for m⁶A; different bacterial taxa may contain any one of these. m⁴C is modified by members of restriction/modification enzyme systems. Thus, DNA methylation is a true epigenetic mark. DNA methylation is used by numerous cellular processes, including mismatch repair (see Chapter 13). As far as transcription regulation is concerned, the methylation of adenine in GATC sequences in *cis*-regulatory elements can either increase or decrease transcription by affecting regulatory protein binding. An interesting example of how the expression of pathogenic genes in uropathogenic *E. coli* is regulated through DNA methylation is presented in Box 8.1.

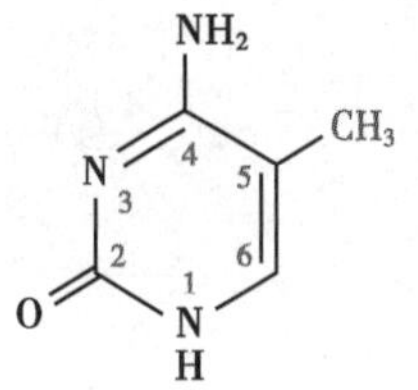

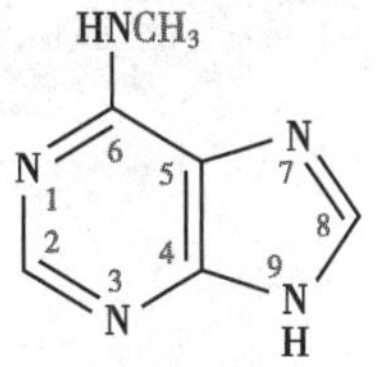

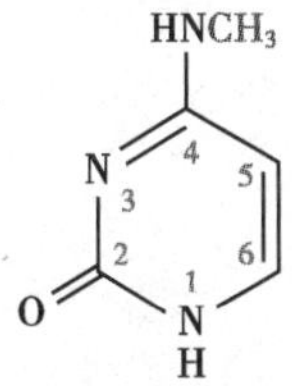

8.2.5 Coordination of gene expression in bacteria

Regulation of gene expression in bacteria is not restricted to the individual behavior of single genes or even operons. Rather, there are multiple levels

Figure 8.16 DNA methylation in bacteria. Bacterial DNA can be methylated at cytosine and adenine nucleotides, as shown in the structures.

of coordination. The DNA-binding proteins that regulate specific operons are usually highly specific to the regulatory regions of the respective operon and are present in relatively small amounts, sometimes only a few molecules per cell. A higher level of control is seen when sets of operons form regulons, whose expression is regionally controlled. The operons constituting a regulon participate in a common function, such as utilization of nitrogen or carbon, and share common regulators, either activators or

---

Box 8.1 Regulation of bacterial pathogenicity by phase variation

Phase variation is defined as the reversible generation of variant bacteria that differ in the presence or absence of specific surface antigens. We illustrate the regulation of phase variation with the example of synthesis of pyelonephritis-associated pili or Pap in uropathogenic *E. coli*. Synthesis of pili is either turned ON or OFF, producing two bacterial populations: one with pili that are capable of adhering to the mucosa in the urinary tract, which are pathogenic, and the other without pili, which are nonpathogenic. The switch in Pap expression is controlled at the transcriptional level by a mechanism involving DNA methylation by Dam methyltransferase (see Figure 8.16) and the leucine-responsive regulatory protein Lrp (Figure 1).

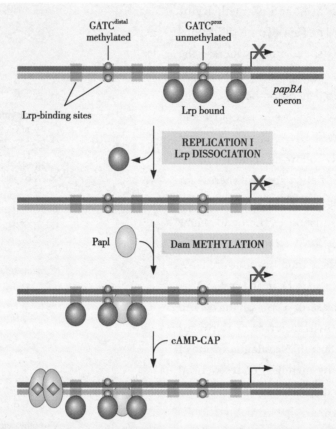

Figure 1 **Switching the *pap* operon from OFF to ON phase.** The upstream region of the operon contains six binding sites for leucine-responsive regulatory protein Lrp; sites 2 and 5 contain methylatable adenines in a GATC context. When the *papBA* operon is OFF, Lrp is bound to sites 1–3, closest to the transcription start site or TSS, to prevent methylation of GATCprox following passage of the replication fork (see Chapter 12). Lrp binding to sites 1–3 reduces the affinity of Lrp for sites 4–6 and RNA polymerase binding. Every round of replication offers a window of opportunity to switch the transcriptional status from OFF to ON, the necessary condition being the presence of PapI. The GATC^{distal} sequence at Lrp binding site 5 is not protected from methylation because Lrp is not bound, and thus it is methylated in the OFF state. The OFF state is perpetuated by the high affinity of Lrp for unmethylated GATC and its inability to bind to its methylated distal counterpart. The switch to the ON state is initiated by the activity of the ancillary protein PapI, which tends to move Lrp to the distal GATC motif; this allows the methylation of Lrp binding site 2 and protects the distal GATC motif from methylation. Thus the ON state is defined by passive demethylation, during replication, of GATC^{distal} and methylation of GATC^{prox}. Activation is dependent on the presence of cAMP-CAP, as already illustrated with the *lac* operon activation example. One of the *pap* products, PapB, activates *papI* transcription, thus providing a positive-feedback loop that propagates the ON state. Note that switching in both directions requires DNA replication because DNA methylation on both strands and passive DNA demethylation occur in two consecutive hemi-steps.

repressors, that recognize DNA sequences common to all member operons and respond to nutrient or environmental conditions. The regulon regulators are more abundant than those of individual operons and bind to multiple targets. Regulation of the heat-shock regulon is an example (see Figure 8.11). This regulation involves the use of alternative σ factors in succession, with $\sigma^{32/H}$ being the factor recognizing the promoters of heat-shock operons.

Multiple regulons may also be controlled in coordination; sets of such regulons have been termed stimulons or modulons. Operons in a modulon may be under individual controls as well as under the control of common, pleiotropic regulatory proteins. For example, the CAP modulon contains all regulons and operons, including the *lac* operon and *ara* regulon, that are regulated by cAMP-bound CAP; each operon has other regulators as well. Finally, there are global controls of overall expression patterns (Figure 8.17). The global level of DNA supercoiling represents one such global control.

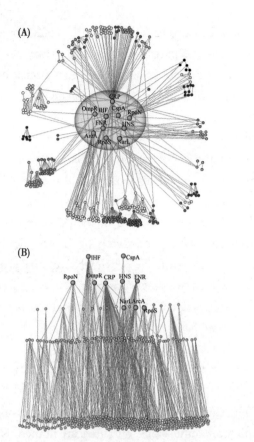

Figure 8.17 Transcriptional regulatory network of *E. coli*. (A) Operons can be organized into modules; different modules are shown in different colors. The 10 global regulators shown inside the oval form the core part of the network. The peripheral modules are connected mainly through the global regulators. (B) A different representation of the hierarchical regulation structure, in which all of the regulatory links are downward. Nodes in the graph are operons. Links show transcriptional regulatory relationships. Global regulators are labeled at the top. (From Ma HW, Buer J, Zeng AP [2004] *BMC Bioinformatics*,5:199. With permission from BioMed Central.)

Networks of transcription factors form the basis of coordinated gene expression

The different levels of transcriptional regulation in bacteria must rely on highly interactive networks of transcription factors, TFs. Often two or more TFs control the transcription of a single gene; on the other hand, sets of target genes in regulons are co-regulated by the same set of TFs. It is as if the organism were conducting a sophisticated computer search using a Boolean algorithm with keywords: transcribe if x AND y but NOT z, and so on. By use of bioinformatics tools, it is now possible to unravel transcription regulatory networks, which depict the complex interplay of individual TFs in the overall regulation of transcriptional activity of the bacterial genome.

In *E. coli*, these networks are dominated by about 10 global TFs (see Figure 8.17). Many of these global regulators represent nucleoid-associated factors, such as IHF and H-NS, that control the organization of the bacterial chromosome (see Chapter 5). CAP, the activator of the *lac* and *ara* operons described above, is also a global TF: it regulates numerous operons that form the CAP-regulated modulon. Local TFs usually act in concert with global TFs. Most of the local TFs tend to be encoded in DNA neighboring the regulated gene or operon. This proximity is explained by the fact that small amounts of local factors must find their way to the regulated gene without being diluted in the large cellular volume. Often, this search is along the genome, not through space. As a rule, global TFs do not regulate each other but regulate more local TFs. In addition, global and local TFs act in distinct ways; global factors and DNA supercoiling induce continuous changes in transcription rates, whereas local TFs control ON/OFF switches in expression.

8.3 Regulation of transcription in eukaryotes

After an overview of general features of bacterial gene transcription, we mainly discuss general principles of eukaryotic gene control and try to provide the background to and an overview of eukaryotic gene transcription in detail.

8.3.1 Regulation of transcription initiation: regulatory regions and transcription factors

The demands for exquisite gene regulation in eukaryotes have led to the appearance of complex arrays of regulatory elements, both close/proximal to and sometimes far/distal from the core promoter region. In considering any aspect of transcription in eukaryotes, it must be kept in mind that transcription occurs within the context of chromatin, so the accessibility of any of these regulatory elements is modulated by the local chromatin structure. We dis-

(1) Core and proximal promoters are needed for basal and regulated transcription

The promoter regions of eukaryotic genes are located immediately upstream of transcriptional start sites, TSSs, and are usually several hundred base pairs in length. Frequently, the promoter region is found to be nucleosome-free (see Chapter 9 and this chapter). The entire promoter can be subdivided into core promoter region and proximal promoter region. The core promoter is the docking site for the basal transcriptional machinery, including the polymerase and any associated factors needed to establish the pre-initiation complex, as shown in Chapter 9. The first characterized eukaryotic promoter contained a TATA box, but there are a variety of core promoters that bind to different protein factors: see Figure 7.24 for Pol II promoters and Figure 7.41 for Pol III promoters. Recent evidence shows that only ~20% of eukaryotic promoters carry TATA boxes.

Proximal promoter regions, which generally lie just upstream of the core promoter, serve to regulate the activity of the core promoter. These promoter regions can also be involved in the regulation of distant genes, as part of complex regulatory networks. Proximal promoters are highly gene-specific as they bind a variety of factors that regulate the expression of specific genes. For example, in the human *Hsp70* gene (Figure 8.18) both the core promoter and the proximal promoter region are complex, each with numerous sequence elements and protein-binding sites. Regulatory elements often act synergistically; their synergism results from physical interactions between transcription factors or TFs that are bound to them. In yeast, the proximal promoter elements are replaced by a single upstream activating sequence or UAS; a single UAS can control multiple regulatory elements.

(2) Enhancers, silencers, insulators, and locus control regions are all distal regulatory elements

The regulation of eukaryotic gene expression has still more dimensions to it: there are numerous distal regulatory elements outside the promoter that act in response to signaling pathways to either enhance or inhibit the transcription of a particular gene.

Enhancers and silencers, although they have opposite effects on gene activity, seem to share some

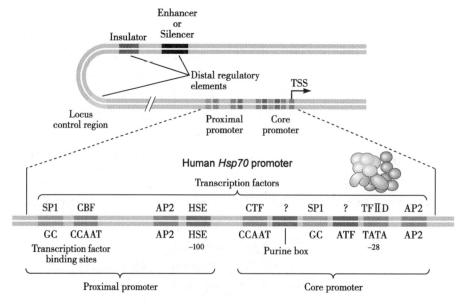

Figure 8.18 **Typical gene regulatory region of a eukaryotic proteincoding gene.** The promoter itself rarely exceeds 1 kb. The distal regulatory elements may include enhancers, silencers, insulator elements, and locus control regions. The expanded region represents proximal and core promoter regions for the human *Hsp70* gene. Multiple control elements are bound by multiple transcription factors to ensure precise regulation of the amount of pre-mRNA synthesized. Many of these factors act in collaboration with each other, responding to the same stimuli.

basic features: (a) they are capable of acting on a gene from long distances, up to several kilobase pairs away; (b) they can be located on either side of the gene or even within an intron; and (c) they act in an orientation-independent manner, that is, these elements would work even if they were flipped over with respect to the sequence they regulate. Each enhancer may be capable of binding a number of transcription factors.

The mechanisms that allow enhancers and silencers to act from a distance are still not entirely clear. Enhancers are better understood, and two models prevail. In the looping model of enhancer or silencer action, transcription factors that bind to specific sequences in the enhancer or silencer interact with protein factors that are bound to the promoter, with the intervening DNA looping out. The alternative scanning or tracking mechanisms of enhancer action posit that activators originally bound to enhancers move continuously along DNA until they encounter other proteins that are already bound to the promoter. If, as expected, the tracking mechanism involves actually moving along the backbone of the double-helical DNA, then it may affect the superhelical tension in the DNA template and the integrity of nucleosomes. It is now known that many enhancers of developmentally regulated genes can be more than a megabase away from the target gene, sometimes with one or two active genes in between, making the scanning mechanism highly unlikely in such cases. The ENCODE project has revealed some fascinating, previously unrecognized features of enhancers, which are discussed in Chapter 9.

Insulators are regulatory regions that insulate sections of the genome from the spread of either activating or repressing influences from nearby genomic regions. Two major types of insulators have been described: those that possess enhancer-blocking activity, that is, they block communications between enhancers and promoters, and those that physically prevent the spread of repressive heterochromatin structures in what is known as barrier activity. Figure 8.19 illustrates the enhancer-blocking activity of an insulator that controls mutually exclusive gene transcription at a well-studied imprinted gene locus.

The expression patterns of the two alleles of an imprinted gene depend on the origin of the allele: whether it is inherited from the mother or the father. The two alleles are characterized by distinct DNA methylation patterns at specific DNA sequences, and this distinct pattern of methylation determines whether certain proteins can bind to the sequence in question. Thus, the DNA methylation pattern is the inherited imprint that distinguishes the two alleles. During gametogenesis, specific parental patterns of DNA methylation, either maternal or paternal, are established at imprinted loci. The binding of proteins, or the lack thereof, to the differentially methylated sequences determines the alternative formation of DNA loops; some loops may separate a promoter from an enhancer, thus inhibiting transcription.

Insulators that have barrier activity are even more frequent. A good example is the insulator in the locus control region presented in Figure 8.20. In this case, one of the DNase I hypersensitive sites, sites in chromatin that show higher than average sensitivity to DNase I cleavage of DNA, in the region serves as a barrier to the spread of highly compacted heterochromatin from the region containing olfactory genes into the region containing globin genes. This may happen because the hypersensitive site with its bound transcription factors represents a gap in the sequence of nucleosomes (see Chapter 9).

The locus control regions or LCRs, are probably the most complex of all distal control elements. These regions control the developmentally regulated expression of individual genes within a gene cluster. One of the best-studied examples is the LCR of the mouse β-globin locus. The locus contains four globin genes that are expressed at different stages of erythroid cell differentiation: ε^γ and $\beta h1$ are expressed in primitive red blood cells during early embryonic development, while β^{maj} and β^{min} are expressed in definitive red blood cells, later during embryogenesis and in the adult organism. The globin genes are embedded in the midst of numerous olfactory receptor genes that are transcriptionally inactive in erythroid cells and possess highly compacted chro-

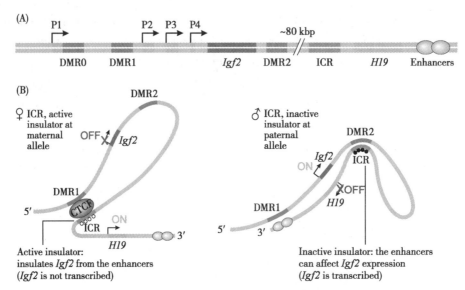

Figure 8.19 Action of enhancerblocking insulator in control of mutually exclusive gene expression at imprinted gene loci. Differential loop formation in the maternal and paternal alleles of the *Igf2-H19* imprinted locus is responsible for the differing transcriptional status of the *Igf2* and *H19* genes in the two alleles. (A) Map of the region shows *Igf2* and *H19* genes as well as sequences involved in regulation of the locus: differentially methylated DNA regions DMR0, DMR1, and DMR2 and the imprinting control region ICR. Arrows above the blue boxes designate the start sites of alternative promoters. (B) Schematic demonstrating differential gene expression in the maternal and paternal alleles. In the maternal allele, loop formation is mediated by binding of the transcription factor CTCF to the unmethylated ICR and to another differentially methylated region, DMR1, upstream of the *Igf2* gene. This conformation creates an active insulator region that precludes utilization by the *Igf2* promoter of the two adjacent enhancers that are located downstream of the *H19* gene: hence the *Igf2* gene is inactive. In the paternal allele, the ICR is methylated and cannot bind CTCF. As a result, a different spatial conformation is formed that allows the enhancers to affect the expression of *Igf2*, and the enhancer-blocking activity of ICR is lost. The alternative spatial organization of the region in the maternal and paternal alleles regulates the expression of the *Igf2* gene; what regulates the expression of *H19* is unclear.

matin structure. Two DNase I hypersensitive sites, HS5 and 3′ HS1, flank the entire β-globin locus (see Figure 8.20). Figure 8.20 also shows the changes that occur in the organization of this locus during erythroid cell development. These changes are initiated through the action of transcription factors such as EKLF and GATA-1 and are mediated by sequence-specific binding of the transcription factor CTCF to four sites. Interestingly, this figure also shows two of these sites are upstream of the globin genes, in intergenic regions between olfactory receptor genes, while the other two fall within the DNase I hypersensitive sites that flank the locus.

The regulation of the β-globin locus also introduces a new concept in molecular biology, the chromatin hub, a dynamic spatial organization or clustering of genes and chromatin regions that affect the transcriptional activity of the resident genes. In erythroid progenitors, which express globin at basal levels, a poised chromatin hub is formed through interactions between CTCF-binding sites and DNase I hypersensitive sites (see Figure 8.20). During erythroid differentiation, the pairs of specific globin genes that become highly activated and the LCR stably interact with the poised chromatin hub to form a functional active chromatin hub or ACH; high β-globin gene expression is LCR-dependent. Clustering of binding sites for transcription factors in the ACH causes local accumulation of cognate proteins and associated positive chromatin modifiers, which are required to drive efficient transcription of the globin genes. During hub reorganization, inactive globin and olfactory receptor genes loop out.

Thus, in summary, an important feature of chromatin hubs is the presence of chromatin loops, formed through protein-protein interactions between proteins bound at different sites. Gene activation is accompanied by dynamic reorganization of poised hubs, leading to the formation of active chromatin hubs.

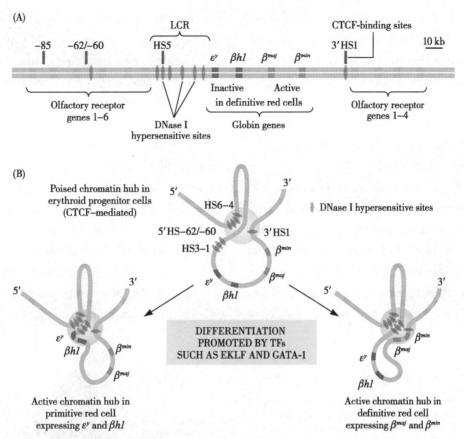

Figure 8.20 Locus control region in the mouse β-globin locus. (A) The locus contains four globin genes that are expressed at different stages of erythroid cell differentiation: ε^y and $\beta h1$, shown as orange blocks, are expressed in primitive red blood cells during early embryonic development, and β^{maj} and β^{min}, shown as green blocks, are expressed in definitive red blood cells, later during embryogenesis and in the adult organism. Developmental regulation of these genes occurs through the action of the LCR. Two DNase I hypersensitive sites, HS5 and 3'HS1, flank the entire β-globin locus. The domain is embedded in the midst of numerous olfactory receptor genes that are transcriptionally inactive and possess highly compacted chromatin structure. The transcription factor CTCF binds to four sites, shown by red bars; two of these sites are upstream of the globin genes in intergenic regions between olfactory receptor genes. Upon binding of CTCF, the CTCF-binding sites contact each other in both progenitor and erythroid cells to form the chromatin hub. (B) In erythroid progenitors, which express globin at basal levels, a hub is formed through interactions between the upstream 5' HS-60/-62, the downstream 3' HS1, and hypersensitive sites HS4–HS6 within the 5' side of the LCR. During erythroid differentiation, which occurs through the action of transcription factors such as EKLF and GATA-1, the β-globin genes that become highly activated and the rest of the LCR stably interact with this substructure to form a functional active chromatin hub or ACH, shown as a green shaded circle; high β-globin gene expression is LCR-dependent. Clustering of binding sites for transcription factors in the ACH causes local accumulation of cognate proteins and associated positive chromatin modifiers, which are required to drive efficient transcription of the globin genes. Inactive globin and olfactory receptor genes, shown in blue, loop out. (Adapted, courtesy of Wouter de Laat, Hubrecht Institute.)

(3) Some eukaryotic transcription factors are activators, others are repressors, and still others can be either, depending on context

Transcription factors belong to two general categories, activators and repressors of transcription. Activators are modular proteins that have distinct DNA-binding domains or DBDs and transcriptional activation domains or ADs. Activation domains are the actual functional parts of the TFs. Contrary to some earlier beliefs, ADs do not interact directly with DNA but with protein components of the transcriptional machinery. Other domains, such as those involved in signal responses or protein-protein interactions, may also be present. In some cases, these domains are part of the same polypeptide chain, but in other instances they are located on separate subunits of multiprotein complexes; Figure 8.21 shows some well-studied examples. The multisubunit organization is beneficial in terms of combinatorial control. However, not all activators are modular in structure: in some proteins, residues that participate in one function are interdigitated with residues performing another function. Representative activators belonging to this class are MyoD, the key developmental regulator of expression of muscle-specific genes, and the glucocorticoid receptor (see Figure 8.21C).

Although the detailed structure of protein domains that interact with DNA, known as DBDs was shown in Chapter 6, much less is known about the structure and function of the activation domains, ADs. These domains are functionally defined as regions of proteins that stimulate transcription when attached to a DBD which is heterologous, that is, from a different protein. They interact either directly or indirectly, through co-activators, with components of the general transcriptional machinery. According to some experts, the activator is misnamed, as it does not directly activate transcription through binding to DNA but rather merely recruits the transcriptional machinery to the gene promoter. This view forms the basis of the recruitment model for transcriptional activation, which seems to be supported by much experimental evidence (Figure 8.22). As very little is known about the exact mechanisms involved, the classification of ADs is mainly based on their amino acid composition: acidic, glutamine-rich, and so on.

The majority of activators contain more than one AD. In the examples presented in Figure 8.21,

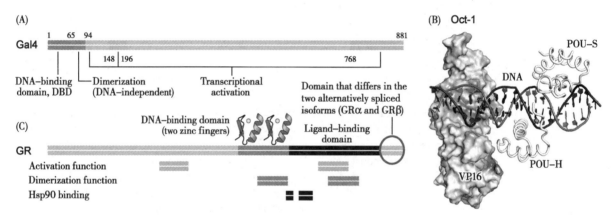

Figure 8.21 **Versatility of eukaryotic transcription factors.** (A) Gal4 is a positive protein regulator of the expression of a whole battery of galactose-induced genes; these genes encode enzymes that convert galactose to glucose. Gal4 recognizes a 17-base-pair sequence in the upstream activating sequence or UAS of these genes. The DNA-binding domain or DBD is often used as a tool for identification of genome sequences that contain activation domains. For that purpose, the sequence encoding the tested presumptive activation domain is fused, through recombinant DNA technology, to the DNA sequence encoding the DBD of Gal4. If transcription of a DNA template containing the binding sequence for the DBD is stimulated by the resulting fusion protein, *trans*activating activity of the tested activation domain is indicated. Note that there are three activating regions in Gal4; the region between 94 and 100 is nonfunctional in yeast *in vivo*. (B) Oct-1 is an example of a composite transcription factor, which harbors individual domains on different subunits of a multisubunit complex. In mammals, Oct-1 is a ubiquitous DNA-binding protein that contains a homeodomain and a separate POU domain. The POU domain contains two separate DNA-binding domains, POU-specific, or POU-S, and POU-homeo, or POU-H, joined by a flexible linker. The two domains bind to opposite faces of the DNA. On its own, Oct-1 binds weakly to 8-bp sites or octamers, but this binding is significantly enhanced by interaction with protein partners such as OCA-B or VP16, a herpes simplex virus factor involved in determining the outcome of viral infection. In the model shown, VP16 is bound to a TAATGARAT element that also interacts with the Oct-1 POU domain; VP16 serves as a bridge between Oct-1 and general transcription factors, thus providing Oct-1 with an activator domain. (C) Glucocorticoid receptor, GR, is an example of a transcription factor in which the individual domains are not discernible as such but overlap to different degrees: the same portion of the polypeptide chain can participate in multiple functions. The N-terminal region of GR encompasses a hormone-independent activation domain that is important in recruiting positive or negative regulatory factors. The DNA-binding domain contains two zinc fingers, which bind to the GR elements in the DNA and also play roles in nuclear transport and in dimerization functions. The ligand-binding domain binds the steroid hormone; in addition, it binds Hsp90, which is instrumental in opening the hormone-binding cleft. Hsp90 dissociates once the steroid binds: this occurs in the cytoplasm. The binding of the hormone induces a conformational change in the hormone-dependent activation domain, allowing interaction with co-activators.

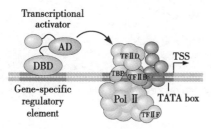

Figure 8.22 Model of transcription activation by recruitment of the basal transcriptional machinery to the promoter. The example shows a TATA-box promoter in a protein-coding gene. The typical activator contains a DNA-binding domain, DBD, and an activation domain, AD, in addition to other domains of various functions. The activation domain acts by recruiting core promoter-binding factors such as TFIID, and hence the polymerase Pol II, to the promoter sequence.

Gal4 has three identifiable ADs, whereas the glucocorticoid receptor has two. These domains function additively or synergistically to increase the activation efficiency of each individual domain. Multiple domains increase the probability that any individual domain will interact with its target in the basal transcriptional machinery. In addition, different domains may interact with different targets. In terms of structure, activation domains are mainly α-helical; in some cases, the α-helices are clearly amphipathic (see Chapter 6 and Figure 6.12).

It must also be noted that the same TF can serve as an activator or a repressor, depending on the specific context. Such situations are reminiscent of the regulation of the bacterial *ara* operon, where the protein AraC acts as a repressor in the absence of arabinose or as an activator in the presence of arabinose.

Finally, it is clear that combinatorial interactions among TFs are critical for directing tissue-specific gene expression patterns. Transcription factors often interact with each other physically, forming homo- and heterodimers or larger complexes. As pointed out in Chapter 6, such interactions can function in cooperation, enhancing interactions with targets. It has been estimated that ~75% of all metazoan transcription factors heterodimerize with other factors, allowing enormous complexity and subtlety in gene regulation. Contemporary high-throughput methods allow for sensitive and precise detection of such interactions, and this information can be used for the construction of combinatorial transcription regulation atlases. Complete TF interaction networks for both humans and mice already exist. From such atlases, it becomes apparent that, in general, highly connected TFs are broadly expressed across tissues, and these TFs are highly conserved between humans and mice. On the other hand, TFs that have few interactions tend to be expressed in tissue-specific patterns.

(4) Regulation can use alternative components of the basal transcriptional machinery

It was long believed that gene-specific transcriptional regulation required gene-specific activators and/or repressors but that the remainder of the factors that participated in formation of the pre-initiation complex—components of the basal transcriptional machinery—were invariant and common to all genes (see Figure 7.26). Clues that this is probably not the case came from the discoveries of different core-promoter architectures that were found to recruit unique combinations of TATA-binding protein, TBP, and various TBP-associated factors or TAFs (see Figure 7.25), even in the same cell. Thus, variant basal complexes could regulate different sets of genes. The picture became even more complicated when cell-type-specific basal factors were described, including cell-specific TAFs and specific TBP-related factors. Moreover, convincing evidence emerged that pointed to major dynamic changes in the basal transcriptional machinery during processes involved in cell differentiation. We illustrate this new aspect of transcriptional regulation with the example of changes that occur during muscle differentiation (Figure 8.23).

(5) Mutations in gene regulatory regions and in transcriptional machinery components lead to human diseases

The complexity of *cis-trans* regulation of transcription in eukaryotes creates special problems if one or more of the DNA sequences and/or their protein partners undergo mutational changes. Often such mutations are involved in various diseases; some prominent examples are presented in Table 8.1. Knowledge

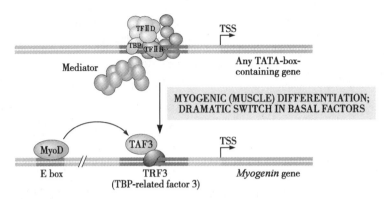

Figure 8.23 Global changes in gene expression during cell differentiation are carried out by significant changes in the general transcriptional machinery. This figure illustrates the example of the myogenin gene promoter. (Top) In myoblasts, the precursor cells to mature multinuclear muscle cells or myotubes, one sees the usual players involved in formation of the pre-initiation complex on TATA-box-containing promoters: TBP, the TAFs that bind to it as a part of TFIID, and Mediator. The process of differentiation is accompanied by drastic changes in the availability of these factors: there is a near-wholesale elimination of the prototypic TFIID, including TBP, and disappearance of Mediator. (Bottom) Active transcription of muscle-specific genes such as myogenin, which encodes the key regulator of myogenic differentiation, is directed by a much pared-down complex consisting of only TAF3, which is a subunit of TFIID, and TBP-related factor 3 or TRF3. TRF3 is nearly identical to TBP in its C-terminal region, which includes the TATA-box-binding and TFIIA- and TFIIB-binding domains, but it differs considerably in the N-terminal region. The myogenin gene is regulated by a classical cell-type-specific transcriptional activator, MyoD, that directly interacts with TAF3.

Table 8.1 Transcriptional regulatory elements and transcriptional machinery components involved in some human diseases. (Adapted from Maston GA, Evans SK, Green MR [2006] *Annu Rev Genomics Hum Genet*,7:29–59.)

Regulatory element	Disease	Affected gene[a]	Mutations (bound factor)
core promoter	β-thalassemia	β-globin	TATA box, CACCC box (EKLF), DCE
proximal promoter	hemophilia hereditary persistence of fetal hemoglobin δ-thalassemia	factor IX Aγ-globin δ-globin	CCAAT (C/EBP) ~175 bp upstream of TSS (Oct-1, GATA-1) 77 bp upstream of TSS (GATA-1)
enhancer	X-linked deafness	*POU3F4*	microdeletions 900 kb
silencer	asthma and allergies	TGF-β	509 bp upstream of TSS (YY1)
insulator	Beckwith–Wiedemann syndrome	*H19–Igf*	CCCTC-binding factor (CTCF)
LCR	α-thalassemia	α globin	~62 kb deletion upstream of gene cluster
	β-thalassemia	β-globin	~30 kb deletion removing 5′HS2–5
Transcriptional component	**Disease**		**Mutated factor**
general transcription factors	xeroderma pigmentosum, Cockayne syndrome, trichothiodystrophy		TFIIH
activators	congenital heart disease Down syndrome with acute megakaryoblastic leukemia prostate cancer X-linked deafness		Nkx2-5 GATA-1 ATBF1 POU3F4
repressors	X-linked autoimmunity–allergic deregulation syndrome		FOXP3
co-activators	Parkinson's disease type II diabetes mellitus		DJ1 PGC-1
chromatin remodeling factors	cancer retinal degeneration Rett syndrome		BRG1/BRM ataxin-7 MeCP2

[a]POU3F4 = POU domain, class 3, transcription factor 4; TGF-β = transforming growth factor β; Igf = insulin-like growth factor.

about such disease connections would lead to better diagnostic tools and treatment options in the future.

8.3.2 Regulation of transcriptional elongation

(1) The polymerase may stall close to the promoter

Not only formation of the initiation complex but also the process of transcription elongation itself is under complex regulation. It has long been recognized that some genes, like the HS heat-shock genes in *Drosophila*, exhibit RNA polymerase stalling. That is, RNA polymerase that has just begun transcribing the body of the gene may become stalled at a position very close to the promoter region. The existence of this promoter-proximal stalling was once considered an idiosyncratic feature of HS genes, but genome-wide studies have revealed that promoter-proximal stalling may be widespread, affecting ~10%–15% of all *Drosophila* genes, for example. Parallel studies in human cells indicate a similar widespread occurrence of promoter-proximal stalling. This phenomenon is now considered an important aspect of the regulation of inducible genes, as it allows for a very fast induction of transcription when an appropriate signal appears; the stalled transcriptional machinery is ready to go at the signal. The stalling itself requires negative transcription factors, which need to be displaced for the stalled polymerase to resume transcription. This kind of stalling must be distinguished from the transient, unprogrammed pausing that occurs frequently during transcription elongation of any gene (see Chapter 7 and this chapter).

(2) Transcription elongation rate can be regulated by elongation factors

The rate of transcription elongation is not uniform along the gene body. The polymerases often enter pauses of various durations (see Figure 7.6); the exit from some of these pauses requires the action of specific transcription elongation factors. In bacteria, these elongation factors are GreA and GreB; in eukaryotes, the best-studied factor is TFIIS. TFIIS acts to overcome transcriptional pausing caused by backtracking of the polymerase enzyme. Backtracking results in the extrusion of the 3′-end of the nascent transcript through a channel in the enzyme, thus misaligning the 3′-OH group and the catalytic site (see Figure 7.6). TFIIS acts by engaging transcribing Pol II and stimulating a cryptic, nascent RNA cleavage activity that is intrinsic to the polymerase. Structural studies have indicated that TFIIS extends from the polymerase surface through a pore to the internal active site, spanning a distance of 100 Å (Figure 8.24). Two essential and invariant acidic amino acid residues in TFIIS complement the polymerase active site, acting to properly position a metal ion and a

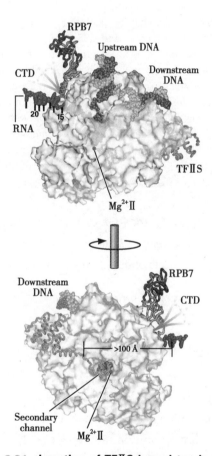

Figure 8.24 Location of TFIIS bound to elongating yeast RNAP II. Portions of RNAP II are rendered semi-transparent to reveal RNA, shown in red, DNA, shown in green, and TFIIS, shown in orange, in the internal channels of the enzyme. The approximate length of exiting RNA and the distance between the exiting RNA and TFIIS are indicated. The C-terminal domain or CTD emerging from the enzyme is shown in magenta, the RPB7 subunit of the polymerase is shown in blue and gray, and the weakly bound Mg^{2+} ion required for transcript cleavage is represented by a sphere. (From Palangat M, Renner DB, Price DH, et al. [2005] *Proc Natl Acad Sci USA*, 102:15036–15041. With permission from National Academy of Sciences.)

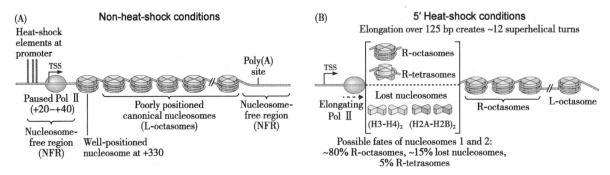

Figure 8.25 **Transcription elongation affects nucleosome structure at a distance.** This distance effect occurs through the positive supercoiling stress or torque created by the polymerase advancing along the DNA template. (A) Chromatin structure of the *Hsp70Ab* locus in *Drosophila* under normal, non-heat-shock conditions. The heat-shock elements in the promoter provide binding sites for the heat-shock transcription factor HSF, whose binding induces the genes. A number of other proteins including TBP, GAGA, Spt5, PARP-1, and the negative elongation factor NELF are present on the gene. Importantly, the gene harbors a paused RNAP II between positions +20 and +40. The chromatin structure over the *Hsp70Ab* gene is characterized by two nucleosome-free regions or NFRs and a well-positioned first nucleosome; note that nucleosomes within the gene body gradually show less and less defined positioning as one proceeds downstream from the transcription start site, TSS. (B) Changes in chromatin structure of the locus upon heat shock. As a first step, the paused polymerase is released into productive elongation through the action of the positive transcription factor P-TEFb, a kinase that phosphorylates Ser2 on the C-terminal domain of Pol II and NELF. Phosphorylated NELF leaves the complex, and other proteins join and travel with the polymerase along the gene. During active transcription, new unphosphorylated Pol II molecules enter the promoter; these molecules also pause at the promoter-proximal site, but the duration of the stall is dramatically decreased in comparison with that of the inactive gene. Importantly, the positive superhelical stress created as a result of polymerase movement along the DNA template disrupts downstream nucleosomes. In ~5s, Pol II advances 125 bp, which creates ~12 positive superhelical turns; this is enough to convert six L-nucleosomes into R-octasomes (see Figure 7.14) at a rate far exceeding that of Pol II transcription. (Adapted from Zlatanova J, Victor J-M [2009] *HFSP J*,3:373–378. With permission from Taylor and Francis Group.)

water molecule for the RNA cleavage to occur. TFIIS also induces extensive structural changes in Pol II that realign nucleic acids with the active center.

8.3.3 Transcription regulation chromatin structure

What happens to nucleosomes during transcription?

The question of what happens to nucleosomes during transcription has plagued researchers for almost 30 years, but we still do not know exactly how the transcription process deals with the presence of nucleosomes. One thing is clear: the nucleosomes must go, at least partially/temporarily, because when the double-helical DNA is wrapped around the histone core, it cannot open to provide a template DNA strand for the polymerase to copy. Different mechanisms for overcoming this problem may operate at promoter regions and within gene bodies, but in the majority of cases, promoter regions are devoid of nucleosomes altogether, forming nucleosome-free regions or NFRs. These are defined primarily by specific nucleotide sequences that are resistant to bending and thus cannot wrap easily around the nucleosome histone core. Such regions will remain nucleosome-free regardless of the transcriptional state of the linked gene. In cases where inactive promoter regions contain nucleosomes, some kind of remodeling will be needed for activation.

Figure 8.25 illustrates the chromatin structure at the heat-shock protein 70 locus, Hsp70Ab, in *Drosophila* and the changes that occur in this structure upon heat-shock treatment. The heat-shock genes provide a useful system in which to study the regulation of transcription of inducible eukaryotic genes. Their expression is induced rapidly, within seconds, and robustly, with ~500-fold increase in mRNA levels, under a variety of stress conditions including heat shock. This rapid response is favored by the presence of a stalled polymerase in the NFR (see Figure 8.25A). The genes undergo numerous highly synchronous changes in chromatin structure upon induction: these changes are recognized cytologically as the appearance of decondensed puffs in polytene chromosomes (see also Figure 1.7 and Figure 1.8).

Some classical remodeling, performed by chromatin remodelers, may also occur in the transcribed regions, but most proposed models of how elongation proceeds through nucleosomal regions presume that physical invasion of the nucleosome by the progressing polymerase removes the nucleosomal barrier. Whether the histone octamer is removed as a whole in a single step or is sequentially dismantled and reassembled during passage of the polymerases is not known.

Nucleosome disassembly may be driven from a distance, the driving force being the creation of positive superhelical stress in front of the translocating polymerase. Transcription elongation requires relative rotation of the enzyme around the DNA. In the twin supercoil domain model, the polymerase is assumed to be fixed, anchored to some nuclear structure such as a transcription factory or the nuclear matrix, so that the DNA must instead rotate in a screwlike fashion inside it. This also suppresses the entangling of the RNA transcript around the DNA that would occur if the polymerase and the transcript it carries were to rotate around the DNA, following the helix backbone. Solving the entangling problem, however, leads to another topological problem: transcription on a fixed, topologically constrained template creates positive supercoiling in the downstream portion of the template and negative supercoiling in the upstream portion. Transcription-coupled supercoiling has been observed *in vitro* and *in vivo*, including in genes transcribed by Pol II. One model of the effect of transcription-induced supercoiling on the integrity of the nucleosome is presented in terms of the *Hsp70* gene in Figure 8.25B. The basic proposition of this model is that the supercoiling induced by the polymerase advance results in a reversal of chirality in downstream nucleosomes. Nucleosomes exist as a family of particles of different histone composition and/or of different chirality; that is, the DNA may be wrapped in either a left-handed or a right-handed sense around the histone core. Even if most of the nucleosomes do not immediately fall apart but just change the chirality of the DNA superhelix around their histone octamers, these right-handed particles are highly unstable and can be easily disrupted by independent forces, possibly including the invasion of the polymerase.

Other changes in the nucleosome particle that can occur under positive superhelical tension have also been described: for example, nucleosomes may split into two halves, presumably giving rise to the well-known elevated DNase I sensitivity of transcriptionally active regions.

A number of other models have been proposed to explain the fate of nucleosomes during transcriptional elongation. Most agree that, at a minimum, H2A-H2B dimmers must be displaced and then rebind. In one model, the DNA released from the nucleosome by H2A-H2B removal forms a loop that can be occupied by the polymerase as it moves around the nucleosome. As yet, no one model is supported by conclusive evidence.

In any event, the disruption of nucleosomes during transcription may require specific protein factors in addition to the polymerase. Among the proteins that have been shown *in vitro* to aid elongation is the FACT or facilitates chromatin transcription complex, which interacts with DNA and with histones H2A and H2B. The interaction with H2A-H2B facilitates their release from nucleosome particles in front of the polymerase and their subsequent rebinding after the polymerase has passed. In fact, FACT appears to act as a chaperone for the H2A-H2B dimer.

8.3.4 *Regulation of transcription by histone modification variants*

(1) Modification of histones provides epigenetic control of transcription

It is now clear that the mechanisms controlling gene transcription are not encoded directly in the nucleotide sequence of the DNA. Rather, they reside outside that sequence, in the form of proteins and their modifications, as well as in postreplicational modifications of the DNA molecule itself. As an everyday analogy of this situation, suppose there were a vast document, which is mostly unimportant to a reader but contains important, readable passages buried within it. Some of these need to be read, others do not. To save the searcher the work of hunting through the vast amount of irrelevant material, one could go in and mark these passages, maybe by underlining the important ones, after the whole document had been prepared. These would be like *epi* marks, from the Greek for on, upon, over: not there in the original but put in later. All those molecular mechanisms that convey regulatory information beyond and above the informational content in the DNA sequence are now grouped under the term epigenetic regulation. However, there is a problem with this designation.

The term epigenetics was originally introduced by geneticists in the 1930s to describe heritable cellular events that could not be explained by genetic principles. The term was subsequently used to define heritable controls in gene expression that are not based on changes or mutations in DNA. Nowadays, the term has acquired a broader meaning: it is used to designate all postsynthetic modifications, to proteins, DNA, and noncoding RNAs, that affect gene expression and other processes, without taking into account the inheritance of the respective state. This broader definition deviates from the original meaning of the terms *epi* and *genetics*, which focus on inheritance. Under this new, broader definition, all post-translational modifications or PTMs of histones are described as epigenetic. However, since these modifications are very dynamic and are not inherited through mitosis, we prefer not to use the term epigenetic in reference to them. The term is, however, appropriate with respect to DNA methylation, because DNA methylation patterns that regulate transcription are inherited; they pass from mother cell to daughter cells during mitosis.

In terms of gene regulation, several classes of postsynthetic modifications form part of the regulatory mechanisms. We discuss the post-translational modification of histones and DNA methylation separately. In addition, we discuss other players and mechanisms that contribute to regulation, such as chromatin remodelers and long noncoding RNAs or lncRNAs.

(2) Gene expression is often regulated by histone post-translational modifications

As mentioned in Chapter 5, histones can undergo multiple chemical modifications that affect their properties. Modifications that are especially important in transcriptional regulation are acetylation, methylation, phosphorylation, ubiquitylation, and poly(ADP) ribosylation. For the most part, these modifications occur on the N- or C-terminal tails of the histones, which project from the nucleosome and are thus accessible for interactions with other proteins. It turns out that most of these modifications can communicate with each other in terms of stimulating or inhibiting further modifications of other residues on the same or a different histone molecule. Several mechanisms could be responsible for communication among modifications: the initial histone modification may trigger increased or decreased activity of a histonemodifying enzyme, or alternatively, different histone-modifying enzymes may be present in a single protein complex, thus coordinating the simultaneous occurrence of several modification or demodification reactions needed to achieve the desired transcriptional outcome.

When a modification of one amino acid residue affects the modification of another amino acid residue in the same histone molecule, this is known as cross-talk in cis (Figure 8.26). A modified amino acid residue in one histone molecule may also affect the modification pattern of a different histone molecule, residing either in the same nucleosome particle or in a different nucleosome within the chromatin fiber. Such interactions are known as cross-talk in trans (Figure 8.27). These kinds of interactions ensure the coordination of modifications that together lead to a unique transcriptional outcome.

(3) Readout of histone post-translational modification marks involves specialized protein molecules

Three different classes of proteins interact with nucleo-

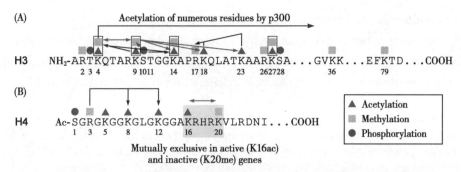

Figure 8.26 Cross-talk in cis among PTMs occurring on histones H3 and H4. (A) Histone H3; (B) histone H4. Inhibition of modification is represented by a red arrow; the direction of the arrow shows the directionality of inhibition. For example, H3K9 methylation inhibits acetylation of H3K14, H3K18, and H3K23 and methylation of H3K4. Promotion of modification is marked by a green arrow. Thus, for example, H3K4 methylation promotes acetylation of numerous other residues by the acetyltransferase p300. For methylation to occur on an acetylated residue, the acetyl group has to be removed, shown by boxed red and blue symbols.

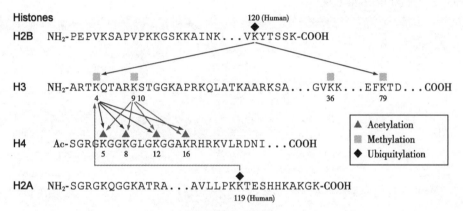

Figure 8.27 Cross-talk in *trans* among PTMs occurring on different histone molecules. Inhibition of modification is marked by a red arrow; the direction of the arrow shows the directionality of inhibition. Promotion of modification is marked by a green arrow. Ubiquitylation of H2AK119 inhibits di- and trimethylation of H3K4 but not monomethylation. H2A is ubiquitylated only in metazoans.

somal histones to regulate transcription. First, there are writers, enzymes that catalyze various modification reactions. Then, there are readers, proteins that recognize specific histone marks. Finally, there are erasers, enzymes that remove the PTM marks from histones.

There are many writer enzymes, often with very well-defined specificities. We cannot describe them all, but some information about representative enzymes is provided in Table 8.2. Note that different enzymes have very distinct specificities and modify specific residues on specific histones. The example shown for acetylation is typical in that most modifications are on histone termini that lie on or near the nucleosome surface (Figure 8.28).

A remaining uncertainty in the field is this: what determines which sites on which proteins on which nucleosomes receive which modification marks? To some extent, the question has been answered by the observation that certain modifications favor or disfavor certain other modifications. But it still seems that there is a chicken-and-egg problem here.

In order for histone modifications to have an effect on transcription, they must affect nucleosome structure, chromatin fiber structure, or the interaction of chromatin with other factors. Each of these consequences is probably modulated by the effect of modifications on the interactions of chromatin with other proteins or protein complexes. The proteins that recognize specific histone marks are called readers (Figure 8.29). In general, protein modules that recognize acetylated lysines belong to bromodomain families; the bromodomain modules are the only known protein recognition motifs that selectively target ε-N-acetylation of lysine residues. They were first recognized as domains in a protein encoded by the *Drosophila brm* gene, hence the name bromodomains. The human proteome contains more than 40 proteins with more than 60 diverse bromodomains. These domains are characterized by low sequence

Table 8.2 **A list of major histone modification enzymes.** The selection of enzymes presented in the table is based on well-known enzymes that are present, in slightly different forms, in both human and yeast. The new nomenclature considered existing close relationships in primary sequence and domain structure/organization: the comparable domain structures may not extend across the entire protein but are clearly recognizable as evolutionarily related. The second consideration, if the domain structure is not recognizable, is sequence homology in the catalytic domains and in substrate specificity. Related enzymes from a single species have been given the same name with a capital letter as a distinguishing suffix, for example, A or B. Related enzymes from different species have been given an identical name but with a different prefix to denote the species of origin: for example, h for human, d for *Drosophila*, Sc for *Saccharomyces cerevisiae*, or Sp for *Schizosaccharomyces* pombe. As an example, not shown in the table, the human demethylase LSD1/BHC110 becomes hKDM1, the *Drosophila* equivalent Su(var)3-3 becomes dKDM1, and the fission yeast equivalent SpLsd1/Swm1/Saf110 becomes SpKDM1. (Adapted from Allis CD, Berger SL, Cote J, et al. [2007] *Cell*,131:633–636.)

New name	Human	S. cerevisiae	Substrate specificity	Function
K-Methyltransferases, KMTs; Formerly Lysine Methyltransferases				
KMT2		Set1	H3K4	transcription activation
KMT2A	MLL		H3K4	transcription activation
KMT3		Set2	H3K36	transcription activation
KMT3A	SET2		H3K36	transcription activation
KMT4	DOT1L	Dot1	H3K79	transcription activation
KMT6	EZH2		H3K27	polycomb silencing
K-Acetyltransferases, KATs; Formerly Acetyltransferases				
KAT1	HAT1	Hat1	H4K5/12	histone deposition, DNA repair
KAT2		Gcn5	H3K9/14/18/23/36; H2B	transcription activation, DNA repair
KAT2A	hGCN5		H3K9/14/18; H2B	transcription activation
KAT4	TAF1	Taf1	H3>H4	transcription activation
KAT5	TIP60	Esa1	H4K5/8/12/16; H2A	transcription activation, DNA repair
KAT8	HMOF/MYST1	Sas2	H4K16	chromatin boundaries, dosage compensation, DNA repair
KAT9	ELP3	Elp3	H3	

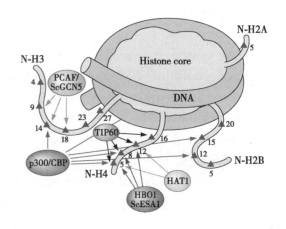

Figure 8.28 **Specificity of action of human acetyltransferases.** The schematic shows the nucleosome with four protruding histone tails, in which lysine residues that are modifiable by acetylation are depicted as red triangles. As denoted by arrows of different colors, different acetyltransferases modify lysines within a specific sequence context. Some enzymes, such as p300/CBP, have rather broad specificities, modifying numerous residues on several tails. Others modify several residues on a specific histone tail; for example, HBO1 acetylates only the tail of H4, whereas PCAF is specific for the tail of H3. In some cases, the human enzyme has a close ortholog in yeast, denoted by *Sc* in front of the name. Only one residue-specific histone deacetylase has been identified, SIRT2 or *Sc*Sir2 in yeast: it deacetylates H4K16, not shown here.

identity, but all share a conserved fold comprising a bundle of four α-helices (see Figure 8.29A).

Protein modules that recognize methylated residues form families of chromodomain or chromatin organization modifier motifs. Recall from Chapter 5 that lysine and arginine residues can accommodate one, two, or three methyl groups, yielding considerable diversity in the physicochemical properties of the modified proteins. The need to recognize these diverse methylation states led to the existence of numerous variants of the chromodomain. Proteins that read the methylation marks belong to a large superfamily, the Royal superfamily, which is subdivided into proteins that read higher and lower methylation states. The basic fold of chromodomains consists of incomplete β-barrels, with the connecting

Figure 8.29 **Protein domains known as readers recognize and bind posttranslationally modified histones.** (A) Bromodomain modules bind to acetylated lysines. Bromodomains are present in several histone acetyltransferases, nucleosome remodelers, and TAFs. The overall topology of the fold consists of four α-helices; the loops connecting the helices participate in formation of the acetyllysine or Kac reader pocket. The pocket is essentially hydrophobic and neutral with significant hydrogen (H)-bonding capacity to form direct or water-mediated H-bonds. The particular structure shown is that of the bromodomain of Gcn5p, a histone acetyltransferase or HAT. The existence of HAT activity in an acetyl mark reader suggests a potential mechanism for the spreading of acetylation marks in chromatin. (B) Chromodomain modules bind to methylated residues. Recall from Chapter 5 that lysine and arginine residues can accommodate one, two, or three methyl groups, yielding considerable diversity in the physicochemical properties of the modified proteins. The need to recognize these diverse methylation states led to the existence of numerous variants of a basic binding module, the chromodomain. (Top) The basic fold of chromodomains consists of incomplete β-barrels, with the connecting loops forming aromatic cages around the modified residue. (Bottom) In the structure given, that of the chromodomain of heterochromatin protein 1 or HP1, the domain has three core strands, strands 2–4, and one orphaned β-strand, strand 5. Upon complex formation, the histone peptide completes the β-barrel by introducing an extra β-strand, strand 1', sandwiched between strands 2 and 5 and shown in yellow. (C) Protein 14-3-3 reads phosphoserine in the H3Ser10 context. Phosphorylation adds a bulky, negatively charged phosphate moiety to the -OH group of amino acid side chains, substantially expanding the ion-pairing and H-bonding capacities of the modified residues. The mammalian 14-3-3 protein family that recognizes the phosphorylation mark plays roles in signal transduction, chromosome condensation, and apoptosis. The phosphate-bearing histone peptide from the N-terminal tail of H3, H3S10ph, is buried in the V-shaped 14-3-3, an all-α-helical protein; the phosphate group forms multiple contacts and its charge is neutralized by the basic side chains of two arginine residues in 14-3-3. (Adapted from Taverna SD, Li H, Ruthenburg AJ, et al. [2007] *Nat Struct Mol Biol*, 14:1025–1040. With permission from Macmillan Publishers, Ltd.)

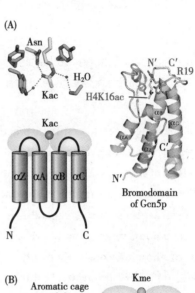

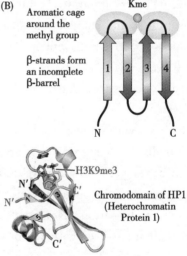

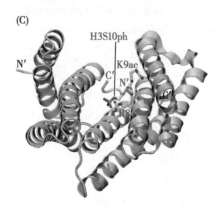

loops forming aromatic cages around the modified residue (see Figure 8.29B).

Finally, phosphoserine marks have their own designated readers. Phosphorylation adds a bulky, negatively charged phosphate moiety to the -OH group of serine side chains, substantially expanding the ion-pairing and hydrogen-bonding capacities of the modified residues. Here, we present only the structure of the protein 14-3-3 reader domain (see Figure 8.29C).

(4) Post-translational histone marks distinguish transcriptionally active and inactive chromatin regions

The major markings used to distinguish active from inactive genes include acetylation, methylation, and ubiquitylation of histones. Histone phosphorylation is also observed during the transcriptional activation of some specific genes, but its general role in transcription regulation remains controversial because high levels of phosphorylation, especially in histones H1 and H3, are observed during chromosome compaction in mitosis, and compaction is considered incompatible with active transcription. The roles of two other PTMs, poly(ADP) ribosylation and ubiquitylation, are discussed separately because of their peculiarities.

It has been known for about half a century that transcriptionally active chromatin is enriched in acetylated histones, yet even today we do not fully understand the structural consequences of this or other histone modifications. Nevertheless, much more information has been gathered by studying individual gene systems. As a result, a clearer picture has emerged about the presence or absence of at least some modifications in active and inactive genes (Figure 8.30). It has also become obvious that practically all modifications are distributed in gradients along eukaryotic genes (Figure 8.31). The case for the different methylation levels at residue K4 of histone H3 is particularly striking: trimethylation is prevalent at the beginning of the transcribed DNA region and falls off quickly and dramatically toward the end of the region. At the same time, H3K4me2 and

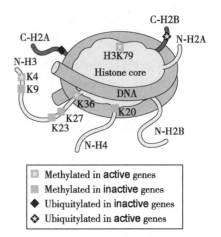

Figure 8.30 **Histone marks that are commonly present in transcriptionally active or inactive genes.** Histone acetylation at lysine residues is a characteristic feature of active genes. These post-translational modifications may act by changing the stability of the nucleosome particle and/or the interactions between nucleosomes in the context of chromatin fiber structure or by interacting with reader proteins that contain modification-recognition modules.

H3K4me1 show different gradual changes along the gene. We still do not understand why these gradients exist: are they needed for regulation of transcription, or do they occur as a consequence of the elongation process? Nevertheless, the very fact that they are created, by different enzymes, is amazing. Understanding the complexity of the regulatory system that controls the activity of these enzymes remains a challenge for the future.

(5) Some genes are specifically silenced by post-translational modification in some cell lines

It is clear that histone PTMs also play crucial roles in processes that lead to gene silencing. Gene silencing is defined as a permanent, irreversible loss of the ability of a gene to be transcribed. Cytologically, silenced genes are organized as dense constitutive heterochromatin. Constitutive heterochromatin is distinguished from a less dense form known as facultative heterochromatin. Genes packed as facultative heterochromatin are not expressed but can revert to active expression. These types of heterochromatin differ significantly from each other and from transcriptionally active euchromatin in the nature and extent of their histone modifications; they are also characterized by the presence or absence of specific

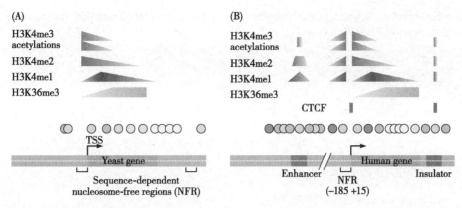

Figure 8.31 Partial chromatin maps of typical transcriptionally active genes in yeast and humans. In both yeast and humans, the TSS is embedded in a nucleosome-free region or NFR. The existence of such regions is determined by the underlying DNA sequence; for example, it may contain poly(dA/dT) tracts. In addition, the NFR is flanked on both sides by nucleosomes containing the histone H2A replacement variant H2A.Z, depicted by yellow circles. The stringency of nucleosome positioning is depicted with increasing intensity of blue color. Note that nucleosomes close to the beginning of the gene are well-positioned but nucleosomes further down the gene lose their strict positioning. It is believed that the positioning signals are determined by the presence of the H2A.Z-containing nucleosome at the beginning of the gene; this nucleosome may serve as a boundary element for the positioning of both downstream and upstream nucleosomes. (A) Typical yeast gene, showing the distribution of some histone marks associated with transcription. Note that the different histone modifications appear as characteristic gradients along the gene. Thus, for example, H3K4 trimethylation is maximal at the start of the gene and diminishes to zero toward the middle of the gene. By contrast, H3K4 monomethylation is relatively weak at the beginning of the gene, then picks up and decreases again. (B) Typical human gene, with an enhancer and an insulator linked to it. Note again the gradients in the distribution of histone marks that exist along the gene. Note also the existence of clear differences in the modification status of enhancers and insulators. Both enhancers and insulators contain H2A.Z nucleosomes and are marked by H3K4 monomethylation but not by trimethylation. In addition, the enhancers are enriched in the acetyltransferase p300, which is absent from promoters and not shown here. Thus the chromatin signature of enhancers and promoters is different enough to allow prediction of the location and function of novel enhancers. Another notable feature of both the NFR and of insulators is significant enrichment in transcription factors, such as CCCTCbinding factor, CTCF. Both enhancers and insulators contain DNase I hypersensitive sites, which are generally attributed to more relaxed chromatin structure, eventually the absence of nucleosomes, and binding of TFs. (Adapted from Rando OJ, Chang HY [2009] *Annu Rev Biochem*,78:245–271. With permission from Annual Reviews.)

trans-acting factors and other chromatin components. Finally, they differ in the presence of bound RNA components and in their DNA methylation status. We have summarized these characteristic features in a concise format in Table 8.3.

Next we discuss a few well-understood examples of gene silencing. Note the involvement, in each specific case, of a particular histone modification. We must state, though, that this knowledge is still at the descriptive stage; we do not understand the exact molecular mechanisms that underlie these phenomena.

(6) Polycomb protein complexes silence genes through H3K27 trimethylation and H2AK119 ubiquitylation

Polycomb complexes bring about silencing of hundreds of genes that encode crucial developmental regulators in a wide variety of plants and animals. Classic examples are the homeobox genes, which were first discovered in *Drosophila* and then shown to regulate organ development and body form in many eukaryotic organisms. These genes are active in early development but are silenced when they are no longer needed later in development. Two critical repressive complexes, polycomb repressive complexes (PRC1 and PRC2), are recruited to genes to be silenced. Both complexes contain subunits that have specific histone-modifying activities (Figure 8.32). PRC2 recognizes specific sequence elements in the DNA, whereas the recruitment of PRC1 usually occurs through interactions with H3K27me3, a modification introduced by PRC2. It must be noted that PRC1 and PRC2 act through a variety of mecha-

Table 8.3 Molecular characteristics of euchromatin, constitutive heterochromatin, and facultative heterochromatin. (Adapted from Trojer P, Reinberg D [2007] *Mol Cell*, 28:1–13.)

		Histone modifications	Chromatin components and *trans*-acting factors	DNA methylation	RNA component
		Euchromatin			
	hyperacetylation	H3K4me2/3 H3K36me3	ATP-dependent chromatin remodelers; H3.3, H2A.Z, H2ABbd	–	–
		Constitutive Heterochromatin			
	hypoacetylation	H3K9me3 H4K20me3		+	+
		Facultative Heterochromatin			
local gene silencing	hypoacetylation	H3K9me2 H4K20me1 H2AK119ub1	PRC1, PRC2, and other PcG proteins; HP1γ, MBT proteins[a]	?	?
long-range silencing, such as of *Hox* gene clusters	hypoacetylation	H3K27me2/3 H4K20me3 H2AK119ub1	PRC1, PRC2, and other PcG proteins	+	+
autosomal imprinted genomic loci	hypoacetylation	H3K9me2/3 H3K27me3 H4K20me3	PRC2; macroH2A; CTCF	+[b]	+
inactive X chromosome, Xi	hypoacetylation	H3K9me2 H3K27me3 H4K20me1 H2AK119ub1	PRC1, PRC2, and other PcG proteins; macroH2A; CULLIN3/SPOP[c]	+	+

[a] MBT, malignant brain tumor-domain proteins are low-methylation-state-specific readers. The MBT-domain-containing protein specifically interacts with histone H1.4 methylated at lysine 26, compacting chromatin.
[b] Associated with the inactive allele.
[c] CULLIN3/SPOP is a ubiquitin U3 ligase that ubiquitylates macroH2A on the inactive X chromosome; this modification of macroH2A is important for association with Xi.

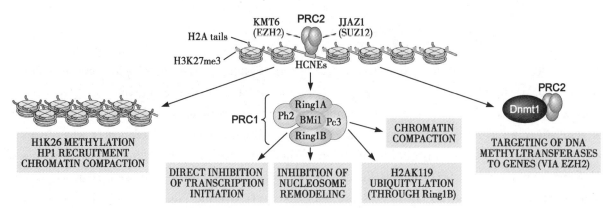

Figure 8.32 Epigenetic gene silencing by polycomb protein complexes. Binding of polycomb repressive complex 2, PRC2, to polycomb-group or PcG target genes induces H3K27 trimethylation through the activity of the KMT6 enzyme. KMT6 was previously known as enhancer of zeste homolog 2 or EZH2. H3K27me3 is recognized by a second polycomb complex, PRC1, through the chromodomain of Pc3. The recruitment of PRC1 might act through different mechanisms, all leading to gene silencing. Here, HCNEs stands for highly conserved noncoding DNA sequence elements in genes silenced by polycomb complexes. One such mechanism involves the ubiquitylation of H2AK119; the downstream consequences of this modification are not yet understood. Protein subunits of both PRC2, subunits EZH2 and JJAZ1/SUZ12, and of PRC1, subunit BMi1, are highly overexpressed in certain human cancers; high levels of EZH2 are associated with poor prognosis and indicative of the metastatic stage of the disease.

nisms, only some of which involve their histone-modifying activities.

(7) Heterochromatin formation at telomeres in yeast silences genes through H4K16 deacetylation

Another well-understood example of silencing mechanisms that involve histone PTMs is the formation of heterochromatic structures at yeast telomeres (Figure 8.33). Three proteins are involved. The initial step depends on the action of an unusual enzyme, Sir2 or silent information regulator 2, the founding member of a family of NAD dependent histone deacetylases. During the reaction, nicotinamide adenine dinucleotide, NAD, is hydrolyzed to O-acetyl-ADP-ribose, which promotes the binding of multiple Sir3 molecules to the Sir2-Sir4 complex (see Figure 8.33). This facilitates heterochromatin spreading. As a curious note, Sir2 is considered a longevity factor in both yeast and humans, and as such it attracts much attention.

(8) HP1-mediated gene repression in the majority of eukaryotic organisms involves H3K9 methylation

Organisms other than budding yeast have a different mechanism for gene silencing. The process involves the action of HP1 or heterochromatin protein 1, first discovered in *Drosophila*. The domain structure of HP1 is characterized by the presence of two closely related domains—a chromodomain, CD, and a chromo-shadow domain, CSD (Figure 8.34)—that perform different functions. CD specifically recognizes methylated lysines in H3. CSD, on the other hand, interacts with other proteins, including the methylation enzyme SUV3–9 in *Drosophila* and its orthologs in mouse and human. This domain also interacts with histones H3, H4, and H1 to strengthen the interaction with chromatin; with DNA methyltransferases Dnmt1 and Dnmt3a; with MeCP2; with histone deacetylases or HDAC; and with other closely related isoforms of HP1. CSD is also the interaction interface with transcriptional activators and components of the basal transcriptional machinery, such as TFIID. HP1 participates in the formation of heterochromatin by recognizing mono-, di-, or trimethyl-

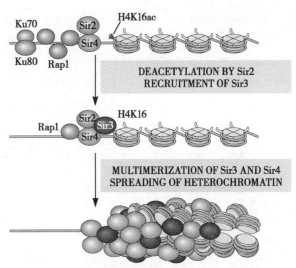

Figure 8.33 Stepwise model for assembly of heterochromatin at telomeres in budding yeast. The initial step involves recruitment of Sir2-Sir4 complexes to DNA through interactions between Sir4 and the telomere-bound proteins Ku70/Ku80 and Rap1, repressor activator protein, a sequence-specific protein that can function either as a repressor or as an activator of transcription, depending on the binding-site context. A series of 16–20 Rap1-binding sites within telomeres ensures Rap1 binding and thus recruitment of Sir2-Sir4. In the second step, the acetyl group at H4K16 on the adjacent nucleosome is removed by Sir2, the founding member of an interesting family of nicotinamide adenine dinucleotide- or NAD-dependent histone deacetylases. Sir3 is recruited via interactions with Rap1, Sir4, and the deacetylated histone tails. Multimerization of Sir3 and Sir4 leads to spreading of the Sir complex along nucleosomes to create compact chromatin structures at telomeres. Slight overexpression of Sir2 in yeast extends the life span by about 30%; the human homolog SIRT1 is also implicated in longevity.

ated H3K9. Unexpectedly, it can also repress or activate genes in euchromatin through the mechanisms depicted in Figure 8.34. Thus the protein is stuck with a name that does not accurately reflect its seemingly opposing functions in gene regulation. It is not alone: many other proteins have been described that can act as either activators or repressors, depending on the specific context.

(9) Poly(ADP) ribosylation of proteins is involved in transcriptional regulation

Poly(ADP) ribosylation, or PARylation, is a very distinct, bulky post-translational modification of proteins that significantly affects the transcriptional status of eukaryotic genes. The modification consists of adding one or more adenosine diphosphate-ribose or ADP-ribose units from donor NAD+ molecules onto target proteins (Figure 8.35A). This reaction is catalyzed by the enzyme poly(ADP-ribose) polymerase 1, known as PARP-1, and its homologs. The enzyme has a modular structure comprising three domains (Figure 8.35B) and can modify itself in a reaction known as automodification. Modification of other acceptor proteins is known as heteromodification. The enzyme is inactive for ADP-ribosylation reactions until an appropriate stimulus appears. Automodification then activates the enzyme for further PARylation reactions. A unique aspect of PARylation is that, in addition to the standard covalent modification, many proteins, including histones, can undergo noncovalent modification (see Figure 8.35B).

PARP-1 is extremely abundant, with ~1–2 million molecules per cell, second only to histones. It participates in numerous biological processes, such as DNA damage detection and repair, DNA methylation and gene imprinting, insulator activity, and chromosome organization. These apparently disparate roles of PARP-1 can be divided into two major categories: emergency responses and housekeeping. The emergency function occurs after DNA damage and involves the numerous PARP-1 molecules normally present in the nucleus in inactive, non-PARylated form; these non-PARylated PARP-1 molecules rapidly become auto-poly(ADP) ribosylated upon DNA injury. The housekeeping role is played under normal unstressed conditions and involves only the few PARP molecules that are PARylated under such conditions. The main housekeeping role is transcriptional regulation, through a wide variety of mechanisms (Figure 8.36).

The first role for PARylation in the regulation of transcription to be recognized was the modulation of chromatin structure. Chromatin structure is modulated either by direct binding of PARP-1 to chromatin or by PARylation of a number of chromatin proteins, some of which are released when modified; more details are given in the legend to Figure 8.36. In addition, PARP-1 can serve in traditional activator or co-regulator roles. A fourth mechanism is involved in insulator function, through PARylation of the insulator CTCF protein (see Figure 8.19). In all these scenarios, it remains unclear how poly(ADP) ribosylation is targeted to specific genes and how such a

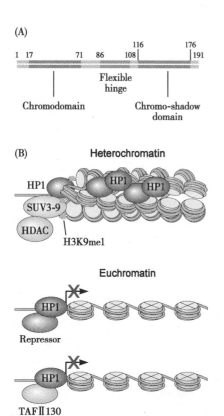

Figure 8.34 **HP1-mediated repression of transcription in organisms from *S. pombe* (A) to humans.** (A) Domain structure of heterochromatin protein 1, HP1, with the evolutionarily conserved chromodomain, CD, and chromo-shadow domain, CSD. The CD recognizes and binds to H3K9me/me2/me3. The flexible hinge shown in green interacts with RNA, DNA, and chromatin and contains a number of regulatory phosphorylation sites. The CSD is the site of dimerization and protein binding. (B) Various mechanisms of transcriptional repression in heterochromatin and euchromatin. Heterochromatization involves the formation of highly compacted chromatin structures through numerous HP1 molecules. In addition, heterochromatin contains deacetylated and methylated histones; the involved writer enzymes are depicted. In euchromatin, HP1 repression can occur through the formation of similar highly compacted chromatin structures, not shown here, or by short-range action over very short stretches of chromatin fibers, sometimes even at the level of a single nucleosome. Repression in euchromatin can also occur via repressive interactions with components of the basal transcriptional machinery, such as TAFII130 or TFIID, at promoter regions. HP1 can also activate transcription, usually via interactions with transcriptional activators.

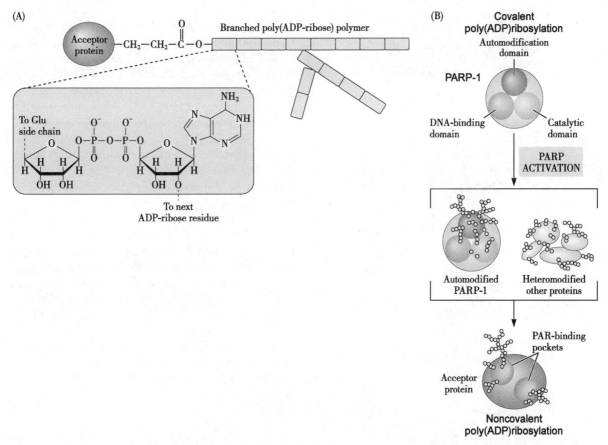

Figure 8.35 **ADP-ribose polymerization reaction.** (A) Structure of poly(ADP-ribose) or PAR polymer attached to an acceptor protein. The polymer chain is attached to glutamate residues within the acceptor protein. The polymers are usually long, up to 200 residues, and highly branched. Mono (ADP-ribose) adducts also exist. (B) Covalent and noncovalent poly(ADP) ribosylation or PARylation. Covalent PARylation begins with activating the catalytic activity of poly(ADP-ribose) polymerase 1 or PARP-1, usually in response to nicks in the DNA. The enzyme has a modular structure comprising multiple independently folded domains: the three most prominent domains are an N-terminal DNA-binding domain or DBD, a central automodification domain, and a C-terminal catalytic domain. The enzyme builds numerous long and branched poly(ADP-ribose) polymers onto its automodification domain; it also modifies numerous other proteins in a process known as heteromodification. A number of additional acceptor proteins can be modified by noncovalent attachment of PAR polymers to specific binding pockets. The interacting polymers could be still hosted on the automodified PARP-1 or on heteromodified proteins; alternatively, noncovalent modification could occur through protein binding of free polymers that are transiently available in the cell because of the activity of enzymes that cleave the polymer, PAR glycosylases. PARP-1 is the founding member of a family of at least 17 members that share homology with its catalytic domain.

bulky and poorly defined modification can have so many specific effects.

(10) Histone variants H2A.Z, H3.3, and H2A.Bbd are present in active chromatin

Several of the nonallelic replacement variants described in Chapter 5 are thought to be involved in the regulation of transcription, or at least transcribability. The role of histone H2A.Z in this context (Figure 8.37) is particularly intriguing for several reasons: (a) H2A.Z is significantly enriched in promoter regions genomewide; (b) there are two H2A.Z-containing nucleosomes flanking the NFR at the transcription start site in at least two-thirds of all genes; (c) H2A.Z levels anticorrelate with transcription in yeast, whereas in humans they actually correlate with transcription; and (d) H2A.Zcontaining nucleosomes have very high turnover rates, making them among the most popular nucleosomes known. The mechanisms of H2A.Z action remain to be elucidated.

The other replacement histone variant that has

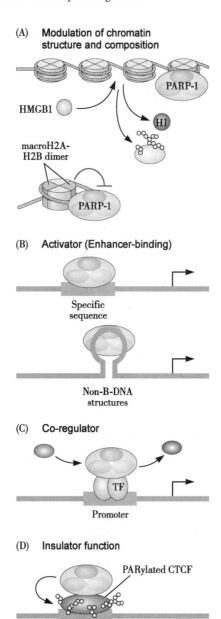

Figure 8.36 **Poly(ADP-ribose) polymerase regulates transcription through multiple mechanisms.** (A) (Top) PARP-1 modulates chromatin structure and composition by directly competing with H1, or by removing H1 by PARylation from its binding site on the nucleosome, or by stimulating the binding of HMGB1, a chromatin non-histone protein that stimulates transcription, or by releasing other proteins after modifying them. PARylated proteins acquire a highly negative charge that is incompatible with chromatin or DNA binding. (Bottom) PARP-1 also interacts with nucleosomes that contain the histone variant macroH2A; the interaction leads to inhibition of the enzymatic activity of PARP-1, with consequences for chromatin structure and function in the specific regions containing macroH2A. (B) PARP-1 can serve a typical activator function by binding to specific sequences in enhancers or to non-B-DNA structures such as hairpins, cruciforms, cross-overs, or double-strand breaks. (C) PARP-1 can be a co-activator or co-repressor, depending on the context. As a co-regulator, it may function as an exchange factor, promoting the release of some factors and the recruitment of other factors, in a promoter-specific way. (D) PARP-1 is involved in the function of insulator elements by PARylating the insulator factor CTCF.

been associated with transcriptional activity is histone H3.3. Recall from Chapter 5 that replacement variants can be synthesized and incorporated into chromatin throughout the cell cycle, in contrast to the canonical histones H2A and H3.3, which are produced only in S phase. H3.3 marks actively transcribed regions, where nucleosomes are constantly disassembled and reassembled (see Figure 8.37). Genomewide, the two nucleosomes flanking the NFR in promoters contain H3.3 as well as H2A.Z. The exact properties and functions of these hybrid nucleosomes remain to be established. In addition, H3.3-containing nucleosomes are enriched over *cis*-regulatory boundary elements in the *Drosophila* genome, suggesting that chromatin structure in these regions is in constant flux, probably as part of a mechanism to keep cis elements exposed to factor binding. Thus, it remains unclear whether the presence of H2A.Z and H3.3 variants is just a reflection of high nucleosome turnover or whether H2A.Z and H3.3 endow certain structural characteristics upon the nucleosomes or chromatin fibers that directly affect their behavior. Recent *in vitro* results suggest that the answer may actually be different for the two variants.

There is another very interesting histone variant, H2A.Bbd or Barr body-deficient histone H2A, that is largely excluded from the inactive X chromosome or Barr body of mammals. Moreover, its genome deposition pattern overlaps with regions of histone H4 acetylation, suggesting that it is associated with transcriptionally active euchromatic regions. The polypeptide is relatively short, lacking the C-terminal tail and a portion of the docking domain of canonical H2A; the docking domain is the portion of H2A

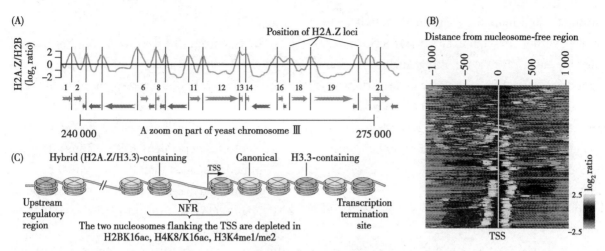

Figure 8.37 **Genomic localization of histone variants H3.3 and H2A.Z.** (A) Genomewide location of H2A.Z on ~40 kbp of yeast chromosome III. Genes occupying the region are represented by blue and red arrows that denote the direction of transcription. The centers of the H2A.Z loci are marked by vertical lines. Note the significant enrichment in promoter regions: there is a H2A.Z locus upstream of every open reading frame or ORF. Intergenic regions that do not contain promoters, such as regions between convergent genes, are not enriched in H2A.Z, and intergenic regions between divergently transcribed genes contain two separable H2A.Z loci. (B) High-resolution mapping of H2A.Z-containing nucleosomes in 2000-bp regions surrounding the nucleosome-free region or NFR in each promoter. Genes present in the database are aligned by their NFRs, with each row representing a single promoter region. Yellow indicates H2A.Z enrichment; blue denotes H2A.Z depletion. Note that the two nucleosomes surrounding the TSS contain H2A.Z. (C) Schematic of genomic localization of the two variants, H2A.Z and H3.3, believed to be associated with actively transcribed regions. Note that H3.3 is enriched not only in the gene body but also in the upstream regulatory regions. Repressed genes also contain H3.3 in these regulatory regions and in the promoter region: thus H3.3 enrichment may be a characteristic feature of genomic regions that are in flux, with nucleosomes constantly assembling and disassembling, and may not be a specific marker of active transcription. (A, adapted from Guillemette B, Bataille AT, Gévry N, et al. [2005] *PLoS Biol*,3: e384. With permission from Luc Gaudreau, Université de Sherbrooke. B, from Raisner RM, Hartley PD, Meneghini MD, et al. [2005] *Cell*,123:233–248. With permission from Elsevier.)

that interacts with H3, thus stabilizing the nucleosome. H2A.Bbd is considered the most specialized of all of the histone variants known to date because of its very low sequence identity with H2A, only 48%. The protein is very rapidly evolving.

Nucleosomes that contain H2A.Bbd show high turnover rates, in agreement with the view that the presence of H2A.Bbd destabilizes nucleosomes. Indeed, histone octamers containing H2A.Bbd organize only 120–130 bp of DNA in a nucleosome, leaving ~10 bp at each end free from interactions with histones. Such nucleosomes are organized in more relaxed fiber structures.

All of these properties would seem to underlie a function in transcriptionally active chromatin. *In vitro* studies using reconstituted arrays of H2A.Bbd-containing nucleosomes do show more efficient transcription than arrays of canonical nucleosomes. It is still unclear, however, whether the variant facilitates transcription *in vivo*. It may have other roles, such as an involvement in mammalian spermatogenesis, as part of the mechanism of replacement of histones by protamines. This replacement also requires histone acetylation, suggesting coordinated action of histone variants and histone PTMs.

(11) MacroH2A is a histone variant prevalent in inactive chromatin

MacroH2A is another recently discovered histone H2A variant that is present only in vertebrates. It occurs abundantly on inactive X chromosomes, and, in general, its presence in chromatin leads to transcriptional silencing. On average, one out of every 30 nucleosomes contains macroH2A. MacroH2A contains a long C-terminal non-histone region termed the macrodomain. The macrodomain recruits PARP-1 to chromatin and inhibits its enzymatic activity, with many ensuing consequences for transcription. In addition, macroH2A-containing nucleosomes are more stable than their canonical counterparts, which may also be a factor in their inhibition of transcription. Additional mechanisms of gene silencing may include interference with TF binding and chromatin remodeling.

(12) Problems caused by chromatin structure can be fixed by remodeling

Given that chromatin structure and its modification can have major effects on transcription, it is not surprising that mechanisms exist to remodel chromatin. Chromatin remodeling is an active process in which remodeling complexes use the energy of ATP hydrolysis to introduce changes in the structure of the nucleosomal particle and/or the location of the particle with respect to the underlying DNA sequence. Although the mutual disposition or spacing of nucleosomes along DNA has the potential to affect folding of the chromatin fiber, most efforts have been directed toward understanding remodeler action at the level of individual nucleosomes, especially in promoter regions. Thus, chromatin remodelers are also termed nucleosome remodelers.

Chromatin remodelers form a large superfamily containing at least three or four subfamilies; each subfamily contains multiple complexes in each species. In humans, four different subfamilies have been recognized. Although the subfamilies differ considerably, each is characterized by the shared ATPase subunit (Figure 8.38):

- SWI/SNF or switch/sucrose nonfermenting: the name originates from the yeast genes identified by mutations that affect the ability of yeast to switch mating type or to use sucrose as a carbon source. Compositions of human and prototypic yeast SWI/SNF complexes are presented in Figure 8.39; note that the two species share some subunits, especially the subunits that carry out ATP hydrolysis, but also contain distinctive subunits.
- ISWI or imitation switch subfamily: contains four different complexes, RSF, hACF/WCFR, hCHRAC, and WICH.
- CHD or chromo-helicase-DNA-binding protein subfamily: at least nine members of the subfamily, CHD1–9, have been recognized in humans. CHD3 is a component of the Mi2/NuRD complex that contains both histone deacetylase and nucleosome-dependent ATPase subunits. Current models suggest that the Mi2/NuRD complex functions primarily in repression of transcription.
- INO80/SRCAP or inositol-requiring/SNF-2-related CREB-binding activator protein.

Remodelers can act in a variety of ways (Figure 8.40); in fact, each subfamily is characterized by the specific outcome of the remodeling activity. Figure 8.41 illustrates the proposed mode of action of the best understood remodeler, ISWI, which causes sliding of the histone octamer with respect to the underlying DNA sequence. Other remodelers change the internal nucleosome structure, destabilizing the particle, which may lead to histone replacement or dissociation. Such destabilization may be crucial in allowing access to nucleosomal DNA by the machineries that utilize DNA as their substrate.

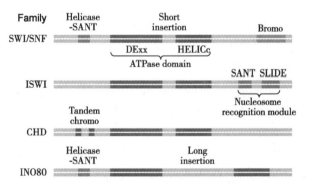

Figure 8.38 **Defining ATPase subunits of remodeler families.** The conserved ATPase domain is split into two parts characterized by specific amino acid sequences: DExx and HELICc. Each family has distinct unique domain(s) residing on one or both sides of the ATPase domain, which define the unique functions of the respective families. The unique domains recognize differently modified histones: bromodomains recognize acetylated histones, whereas chromodomains bind to methylated histones. The SANT-SLIDE domain in ISWI binds unmodified histone tails and DNA. The helicase-SANT domain interacts with actin and actin-related proteins, which are subunits of some remodeling complexes of undefined function.

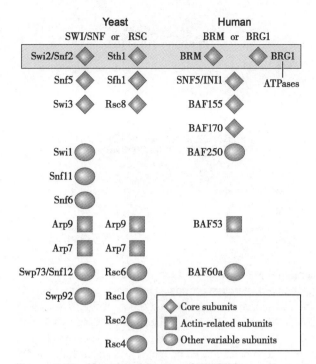

Figure 8.39 Yeast and human SWI/SNF complexes. The columns show the subunit composition of each complex; the rows indicate the homologs and/or orthologs of each individual subunit. Note that each complex possesses core subunits, one of which is the subunit with ATPase activity, as well as variable subunits. The core subunits are necessary for full activity *in vitro*. The role of the variable subunits is not well understood; they may participate in directing the specificity of the complex through protein-protein interactions. Curiously, some subunits carry sequence similarity to the contractile muscle protein actin; actin is also involved in the formation of the cytoskeleton and of nuclear skeletal structures. The amino acid sequences of the ATPase subunits BRM and BRG1 are 75% identical, and both are widely expressed. Nonetheless, the presence of these subunits in a complex is mutually exclusive.

Finally, there is a well-documented link between mutations in genes coding for chromatin remodeling factors and disease, especially cancer. Thus, mutations in the hSNF5/INI1 subunit of the human SWI/SNF complex are characteristically present in aggressive pediatric tumors as well as in some acute leukemias. Strong support for considering *hSNF5* a tumor-suppressor gene comes from the observation that heterozygous knockout mice develop tumors. Mutations in BRG1, the ATPase subunit of human SWI/SNF, have been identified in several cancer cell lines; *BRG1+/–* mice are also predisposed to tumors, again confirming the role of BRG1 as a tumor suppressor.

(13) Endogenous metabolites can exert rheostat control of transcription

Small-molecule metabolites can regulate transcription as well as translation, either directly or by affecting chromatin structure or remodeling. Because the level of activity will vary in a continuous manner with metabolite concentration, this is sometimes called rheostat control.

Direct regulation occurs when metabolites affect the activity of activators or repressors. Thus, for example, estrogen receptors activate gene transcription only when steroid hormones are bound to them. Similarly, interactions of the C-terminal binding protein or CtBP with activators or repressors are

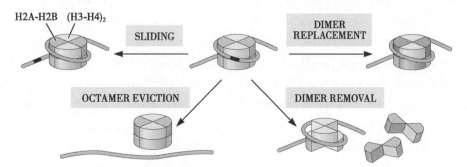

Figure 8.40 Models of nucleosome remodeling. Remodelers can act in a variety of ways, depending on the biological context. They can slide the histone octamer with respect to the underlying DNA sequence, so as to allow accessibility of factor-binding sites. They can totally evict the histone octamer, an event that frequently occurs in promoters of genes undergoing transcriptional activation. They can remove one or both H2A-H2B dimers during transcriptional elongation. This activity is not unique to remodelers and can sometimes be performed by unrelated protein complexes that contain histone chaperones: a well-studied example is FACT, facilitates chromatin transcription complex. Finally, remodelers can replace existing H2A-H2B dimers with dimmers containing variant histones, such as H2A.Z.

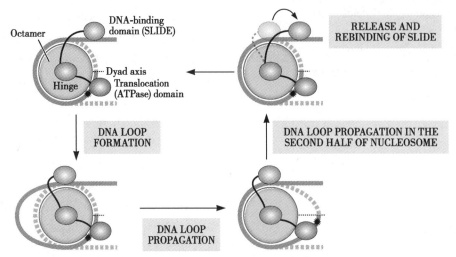

Figure 8.41 **Loop or bulge model for the sliding remodeling action of ISWI.** The second gyre of DNA is presented as a broken line to reinforce the perspective of DNA wrapping. The black starburst provides a reference point on the DNA to facilitate visualization of the translocation of DNA along the octamer surface. Concerted action of the DNA-binding SLIDE domain, bound to the linker region, and the translocation or ATPase domain, bound close to the dyad axis, generates a small bulge that propagates on the nucleosomal surface. The loop is initially created by the DNAbinding domain, which pushes DNA into the nucleosome. Generation of the loop is accompanied by conformational changes in the SLIDE domain. The directional propagation of the loop is then conducted by the ATPase domain, which remains anchored at its position on the nucleosome: it draws the DNA from the linker and pumps it toward the dyad and then further in the second half of the particle. Histone-DNA contacts are being broken at the leading edge of the loop and re-formed at its lagging edge. (Adapted from Clapier CR, Cairns BR [2009] *Annu Rev Biochem*,78:273–304.With permission from Annual Reviews.)

controlled by NAD⁺/NADH. Chromatin-mediated control by metabolites depends on the presence or concentration of metabolites that are used to add or remove marks on histones and other chromatin components. Thus, for example, both histone and DNA methylation require *S*-adenosylmethionine as a donor of methyl groups; acetyl-CoA is needed for the acetylation of lysine residues on histones; and the Sir2 deacetylase needed for heterochromatin silencing in yeast is dependent on the presence of NAD. NAD is, of course, the precursor needed to build poly(ADP-ribose) polymers onto proteins. Such metabolite-induced alterations in protein function and chromatin structure may allow fine-tuning of gene expression in response to changes in the levels of cellular metabolites and, in a broader sense, to the environment.

8.3.5 DNA methylation

DNA methylation by DNA methyltransferase (DNMT) enzymes is the major epigenetic modification of DNA. In eukaryotes, it is achieved through the transfer of methyl groups from *S*-adenosylmethionine to cytosine, converting it into 5-methylcytosine, m^5C (Figure 8.42). The modification preferentially occurs on CpG dinucleotides. As cytosine methylation does not affect base pairing, the deposition of this postreplicational mark can regulate processes without affecting the genetic information in DNA. In addition, mechanisms exist that propagate this modification throughout DNA replication (Table 8.4 and Figure 8.42B), thereby passing the modification pattern from mother to daughter cells during mitosis. In this sense, and in contrast to histone PTMs, this mark can be considered truly epigenetic: it carries regulatory information that is not encoded in the DNA sequence and is inherited from cell generation to cell generation.

DNA methylation patterns are not directly inherited during sexual reproduction in many organisms; they are erased during early development and later established *de novo* (Figure 8.43). The mechanisms that instruct the somatic cells in the new organism to establish patterns are not clear. It is clear, though,

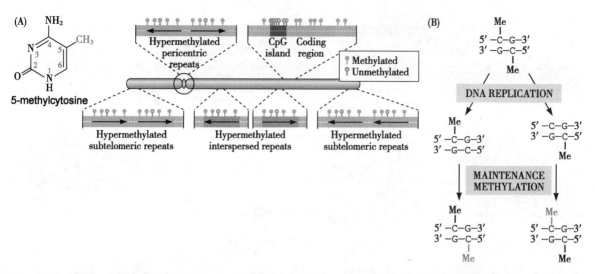

Figure 8.42 **DNA methylation.** (A) Schematic of a chromosome with representative regions that are rich in CpG dinucleotides, shown as lollipops, whose methylated or unmethylated pattern is required for normal cellular functions. The structure of methylated cytosine is shown to the left. (B) Maintenance of DNA methylation of CpG sites following DNA replication. The newly synthesized DNA strand is unmethylated for a while; then methyl groups are added to it by use of the opposite strand as a template. (A, adapted, courtesy of Paola Caiafa, University La Sapienza.)

Table 8.4 **DNA methyltransferases in mammals.**

Mammalian enzyme	Function	Phenotype of homozygous deletion mutants
Dnmt1	maintenance, functions after replication on hemimethylated DNA to restore existing methylation patterns; associates with histone deacetylase	$Dnmt1^{-/-a}$ die *in utero*; DNA in these embryos is hypomethylated and imprinted genes are biallelically expressed; embryonic stem (ES) cells from $Dnmt1^{-/-}$ mice are viable and capable of *de novo* methylation
Dnmt3a	*de novo* methylase	$Dnmt3a^{-/-}$ mice die at 4 weeks; $Dnmt3a^{-/-}$ embryos and ES cells exhibit demethylation of centromeric satellite repeats; ES cells from $Dnmt3a^{-/-}$ mice are viable and capable of *de novo* methylation
Dnmt3b[b]	*de novo* methylase	$Dnmt3b^{-/-}$ mice die *in utero*; ES cells from $Dnmt3b^{-/-}$ mice are viable and capable of *de novo* methylation; $Dnmt3b^{-/-}$ embryos and ES cells do not exhibit demethylation of centromeric satellite repeats

[a]The designation $^{-/-}$ refers to lack of both alleles or homozygosity of the respective gene.
[b]Heterozygous mutations affecting the catalytic domain of DNMT3b have been identified in humans with ICF syndrome, which stands for immunodeficiency, centromere instability, and facial anomalies. It is characterized by variable reduction in serum immunoglobulin levels, which causes severe susceptibility to infectious diseases in childhood. Facial abnormalities include hypertelorism, increased distance between the eyes; low-set ears; Mongolian eye folds; and macroglossia, unusual enlargement of the tongue.

that the DNA methylation patterns are established in totally defined ways; thus there must be mechanisms, albeit indirect and unknown, to ensure the preservation of these patterns in the new organism. Although in vertebrates m^{5}C has occasionally been found on cytosines followed by a C, an A, or a T, the best substrate for the addition of methyl groups is cytosine in the CpG or C-phosphate-G dinucleotide.

(1) DNA methylation patterns in genomic DNA may participate in regulation of transcription

The first complete map of the human DNA methylome, published in 2009, examined the methylation state of 27 million CpG locations on the 23 pairs of human chromosomes. It also looked for C methylation in the context of the other dinucleotides. The effort relied on the use of bisulfate treat-

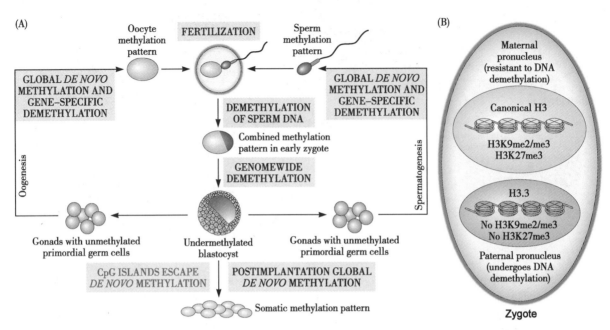

Figure 8.43 Dynamics of DNA methylation patterns during mouse development. DNA methylation patterns are heritably propagated in somatic lineage cells, but during the development of an organism, methylation is a very dynamic process. (A) The sperm genome is rapidly demethylated following fertilization; this creates the combined methylation pattern in the early zygote. The global loss of DNA methylation in the sperm-derived pronucleus occurs independently of DNA replication and thus involves active DNA demethylation. During the first two to three cleavage divisions, the level of methylated cytosines decreases further and stays low through the blastula state; the loss of methylation in the maternal pronucleus is a passive, replication-dependent event. Postimplantation, the embryo genome undergoes *de novo* methylation; the CpG islands remain mostly unmethylated. The primordial germ cells also remain unmethylated; during gametogenesis, specific parental (maternal or paternal) patterns of DNA methylation are established at imprinted loci. (B) Correlation between DNA demethylation and chromatin structure in the early zygote. The paternal and maternal pronuclei behave very differently in terms of DNA demethylation: the former undergoes the process, while the latter is resistant to DNA demethylation. They also differ in several chromatin-structure characteristics. Maternal chromatin contains inactive chromatin marks and canonical H3, whereas paternal chromatin is devoid of these inactive chromatin marks and contains the replacement histone variant H3.3. As embryonic transcription of many genes starts only after the first cell division, the differences in chromatin structure between maternal and paternal pronuclei are probably not transcription-related but serve to allow access of the DNA demethylation machinery to DNA.

ment and high-throughput sequencing; treatment of methylated DNA with sodium bisulfate converts the nonmethylated cytosines to uracil while leaving the methylated cytosines untouched. Because of limitations in the method, the sequencing effort amounted to sequencing the equivalent of the entire genome 57 times. This effort was worthwhile because it produced some rather surprising results. First, there were dramatic differences between pluripotent embryonic cells and differentiated cell types such as Fibroblasts in terms of which dinucleotides are methylated: 99.98% of all methylation occurs at CpG dinucleotides in fibroblasts, whereas ~25% of the methylation in stem cells does not occur in the CpG context. Second, transcriptionally active genes contain undermethylated CpGs. Importantly, if terminally differentiated cells are engineered to revert to pluripotent stem cells, they regain the unusual modifications at sites different from CpGs. The question that still needs to be addressed is whether the observed differences are consequences of differential gene activity or are actively involved in the regulation of transcription.

Differentiated cells are characterized by an uneven distribution of methylated cytosines along chromosomes. Of particular interest are CpG islands, which despite being unusually rich in this dinucleotide somehow evade being heavily methylated (Figure 8.44).

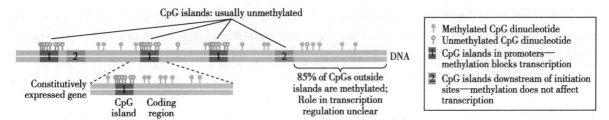

Figure 8.44 **Distribution of methylated CpGs in genomic DNA and their effect on transcription.** Approximately 70%–80% of all CpG dinucleotides are methylated in vertebrate genomes. mCpG, shown as blue lollipops, are randomly distributed throughout the genome but are excluded from regions that have unusually high CpG density, known as CpG islands. Most of the CpG islands are associated with gene promoters and are maintained unmethylated, shown as light gray lollipops, in all types of somatic cells. Aberrant methylation of CpG islands occurs in cancer cells and leads to the silencing of tumor suppressors and other essential genes.

Furthermore, the presence of methylated DNA in promoters is directly linked to silencing of the downstream genes. Numerous mechanisms suggesting how promoter DNA methylation may affect gene expression have been proposed, including interplay between DNA methylation and histone modifications or chromatin remodeling. DNA methylation could also directly affect chromatin structure by compacting the chromatin fiber. Indeed, it has been convincingly demonstrated that chromatin regions that contain methylated cytosines are more compact than their unmodified counterparts. However, DNA methylation alone is not sufficient: the compaction also seems to require the presence of bound linker histones.

(2) Carcinogenesis alters the pattern of CpG methylation

Importantly, carcinogenesis alters the pattern of CpG methylation in two opposing ways: the genome as a whole becomes hypomethylated whereas CpG islands, which regulate the expression of housekeeping genes, including a variety of tumorsuppressor genes, cell-cycle-related genes, DNA mismatch-repair genes, hormonereceptor genes, and others, become hypermethylated. Hypermethylated CpG islands lead to silencing of the linked genes, which has catastrophic consequences: the cells become malignantly transformed. DNA hypermethylation is now the best-characterized epigenetic change in tumors, and it is found in virtually every type of human neoplasm. Silencing of tumor-suppressor genes by promoter hypermethylation is at least as common as silencing caused by genetic mutations. Numerous genes have been shown to undergo promoter hypermethylation in cancer, but a unique profile of hypermethylated CpG islands defines each neoplasia. Thus, hypermethylation in a limited number of gene markers can be used for cancer diagnostics; attempts are also being made to use these patterns as a predictor of response to treatment. Finally, drugs are being developed that should lead to the demethylation of hypermethylated promoters and thus to the reactivation of silenced tumor-suppressor genes. Unfortunately, these drugs suffer from high cytotoxicity levels because they are not selective for the cancer-causing genes but affect numerous other genes, leading to undesirable side effects.

(3) DNA methylation changes during embryonic development

As we have said, DNA methylation patterns are stably inherited during mitosis. During early embryogenesis in many vertebrate species, however, DNA methylation is a highly dynamic process. Two successive waves of demethylation occur in the fertilized egg or zygote: First in the male pronucleus, immediately after fertilization, and then during the transition to the blastocyst stage of development (see Figure 8.45). It is noteworthy that the paternal and maternal pronuclei which undergo differential changes in DNA methylation are also characterized by clear-cut differences in their histone methylation patterns, especially at K9 and K27 of histone H3. They also differ in their content of the active histone variant H3.3, which is present only in the paternal pronucleus. DNA methylation levels are restored by de novo methylation following implantation. Gametogenesis is also characterized by dynamic and highly selective changes in methylation patterns. Global demethylation and remethylation is, however, not obligatory in all vertebrates, with zebrafish being a notable exception. A more limited drop in methylation occurs in *Xenopus laevis*. How and why all these changes occur is one of the unresolved problems in development. Another unresolved question concerns a purely structural issue: DNA methylation and de-methylation must occur in the context of chromatin. Thus there must be challenges in terms of which CpG would be accessible in the context of the nucleosome particle and/or chromatin fiber structure and compaction. These issues have not been addressed yet.

(4) DNA methylation is governed by complex enzymatic machinery

There are two classes of enzymes that convert cytosine to methylcytosine in the context of CpGs. One enzyme, DNA methyltransferase 1 or Dnmt1, recognizes and methylates the unmodified C in hemimethylated sites generated during DNA replication, thus preserving the methylation pattern of the genome on both strands. In higher eukaryotes, this process occurs within a minute or two after replication. Two other enzymes, Dnmt3a and Dnmt3b, are responsible for introducing new methyl groups onto DNA at sites in which neither strand was previously methylated: these are the enzymes involved in global de novo methylation during development (see Figure 8.43). For more information on the important roles of all three Dnmts *in vivo*, see Table 8.4. In terms of structure, all Dnmts contain a conserved catalytic domain and additional domains, which are mainly responsible for protein-protein interactions (Figure 8.45).

The enzymes that put the methylation mark on CpGs are well understood, but persistent efforts to identify the enzyme(s) that actively demethylates DNA have created what is probably the most contentious area of research in the past decade. Several enzymes were described that reportedly performed this function, only to be quickly rejected. The long-sought-after, mysterious pathway may have finally been found: in 2010, it was reported that the active removal of methyl groups from cytosine actually occurs in steps, in a rather indirect, convoluted way (Figure 8.46). The first step involves deamination of cytosine, with the production of thymidine; in a second step, the resulting mismatched T-G base pair is recognized by the base excision repair pathway (see

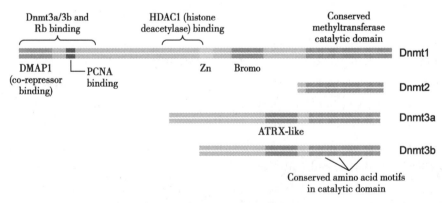

Figure 8.45 **Domain structure of mammalian DNA methyltransferases.** All known DNA methyltransferases contain a conserved catalytic domain with conserved amino acid motifs. Each DNA methyltransferase interacts with a number of regulatory proteins; some of them modulate enzymatic activity, while others participate in transcriptional repression mechanisms. When DNA methyltransferase 1 represses transcription, it interacts with histone deacetylases such as HDAC1 either directly or through other proteins, such as co-repressor DMAP1 or retinoblastoma protein Rb.

Chapter 13), and the mismatched T is replaced by C. In this way, the cell avoids the challenge of having to break a very strong chemical bond, C–CH3.

(5) There are proteins that read the DNA methylation mark

As is customary in molecular biology, if there is a specific mark on any molecule that is added following its synthesis, there are players and mechanisms that recognize the mark. These molecules perform effector functions: they translate the signal provided by the mark into biochemical outcomes. DNA methylation is no exception. Methylated regions of the genome are recognized by a variety of proteins, dubbed methyl-CpG-binding proteins that share a common methyl-CpG-binding domain, MBD. Some MBD proteins are listed in Table 8.5, which provides information on structures of the respective proteins, characteristic features of the DNA binding site, specific effects on transcription, and, Finally, *in vivo* expression and localization of each protein. One of these proteins, MeCP2, whose mutations contribute to the neurodevelopmental disorder known as Rett syndrome. Undoubtedly, more MBD proteins will be discovered in the future.

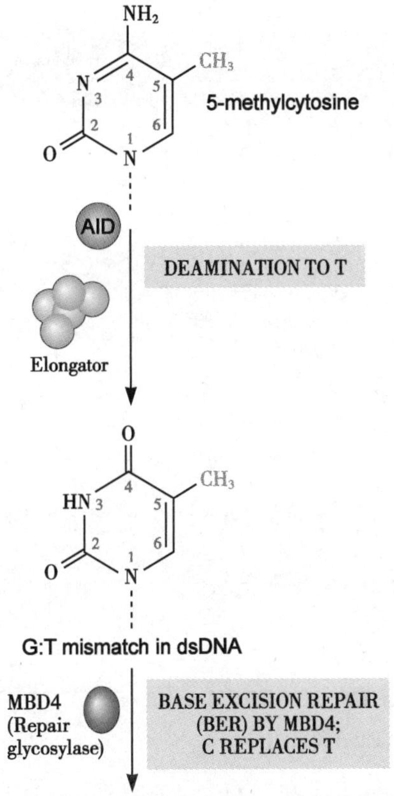

Figure 8.46 Active DNA demethylation in mammals. Active demethylation occurs in a number of biological contexts, including development and gene activation. DNA demethylation occurs in several steps: first, activation-induced cytidine deaminase or AID deaminates methylcytosine, with the production of thymidine. Additional proteins such as the Elongator complex, which is implicated in transcription elongation, may be required. Then the resulting mismatched T is thought to be replaced by cytosine via the base excision repair or BER pathway, which involves the action of MBD4, a thymine-specific glycosylase. This mechanism is active in primordial germ cells to effect the global genome demethylation observed there. AID is also implicated in the rapid removal of methyl groups from methylated promoters of genes that undergo transcriptional activation.

Table 8.5 **Methyl-CpG-binding proteins.**

Protein	Features of protein structure and DNA binding	Effect on transcription; other properties	*In vivo* expression and localization
MBD1	contains MBD domain and CxxCxxC motifs; has several splice variants	represses transcription from a methylated promoter *in vitro* and *in vivo*	expressed in somatic tissues but not embryonic stem (ES) cells
MBD2a	contains MBD domain and (Gly-Arg)$_{11}$ [(GR)$_{11}$] domain	represses transcription; component of Mi2/NuRD deacetylase complex	co-localizes with heavily methylated satellite DNA in mouse cells
MBD2b	truncated version of MBD2a that lacks the (GR)$_{11}$ domain; translation starts at a second methionine codon in MBD2a	represses transcription; component of Mi2/NuRD deacetylase complex	expressed in somatic tissues but not ES cells

Protein	Features of protein structure and DNA binding	Effect on transcription; other properties	*In vivo* expression and localization
MBD3	contains MBD domain and a C-terminal stretch of 12 Glu residues; mammalian MBD3 does not bind methylated DNA *in vivo* or *in vitro*	component of Mi2/NuRD and SMRT/HDAC5-7 deacetylase complexes	expressed in somatic tissues and ES cells
MBD4	contains MBD domain and a repair domain, T-G mismatch glycosylase	thymine glycosylase that binds to deamination product at methylated CpG sites	co-localizes with heavily methylated satellite DNA in mouse cells; expressed in somatic tissues and ES cells
MeCP2	contains MBD domain and a transcriptional repression domain; binds to a single symmetrically methylated CpG	represses transcription from a methylated promoter *in vitro* and *in vivo*; participates in several co-repressor complexes, Sin3a/HDAC1-2, NCoR/Ski, and Rest/CoRest; activates transcription from both methylated and unmethylated promoters	co-localizes with heavily methylated satellite DNA in mouse cells; expressed in somatic tissues and ES cells

(Continue)

8.4 Long noncoding RNAs in transcriptional regulation

8.4.1 Noncoding RNAs play surprising roles in regulating transcription

As discussed in Chapter 7, the pervasive transcriptional activity throughout the genome is existed at the genomewide level. For instance, there are approximately 180 000 mouse cDNAs identified, whereas a mere ~20 200 protein-coding genes are known. Less than 3% of the human genome encodes proteins, leaving many of the remaining transcribed RNA sequences with unknown functions. Analysis of the mouse and human transcriptomes thus reveals a huge number of transcripts that do not encode proteins, hence their name noncoding RNAs or ncRNAs. The functional importance of ncRNA transcription may lie in the process of transcription itself rather than in the products. Transcription of ncRNAs helps to maintain a transcriptionally competent state for the transcription of functional RNA- and protein-coding genes. In the known few cases, the function of these molecules seems remarkably diverse (Figure 8.47).

8.4.2 The sizes and genomic locations of noncoding transcripts are remarkably diverse

Noncoding RNAs may be relatively short or relatively long, with the arbitrary dividing line set at ~200 nucleotides. There are several well-defined classes of short ncRNAs; we describe their biogenesis and the roles of others in Chapter 10. Here we discuss the properties and transcription regulatory functions of some long noncoding RNAs or lncRNAs.

Many ncRNAs are transcribed away from the 5'- or 3'-ends of protein-coding genes; for the lncRNAs, there is significant concentration near promoters, initial exons, and initial introns. At least some of these RNA molecules are both capped and polyadenylated, as are regular mRNA molecules. Some also contain short, stable stem-loop structures that are deemed important in protein binding. The proteins bound could be either basal or gene-specific transcription factors, elongation factors, and/or chromatin remodelers. Figure 8.47 presents several well-understood lncRNAs that function as regulators of transcription. The mechanisms involved are incredibly varied and show no common feature. Figure 8.48 illustrates another example of how one lncRNA mediates the apoptotic response of cells to stress, by repressing the transcription of pro-survival genes. The mechanisms of ncRNAs on transcriptional regulation is diverse and the lncRNAs and their functions are need to be further investigated. More insights gleaned from the ENCODE project will be of help to systematically analyze the entire human genome and transcriptome (see Chapter 9).

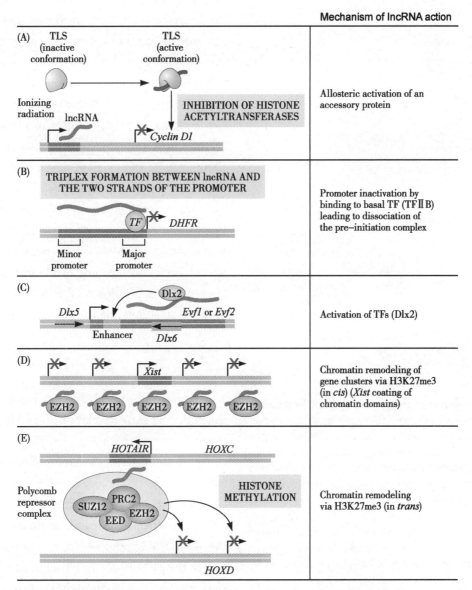

Figure 8.47 **Mechanisms of lncRNA-mediated regulation of transcription in eukaryotes.** (A) In response to stress, the long noncoding RNA (lncRNAs) transcribed upstream of the *Cyclin D1* gene interacts with the RNA-binding protein TLS, translocated in liposarcoma, and induces a conformational change in the protein. The activated TLS inhibits histone acetyltransferase activities and thus inhibits *Cyclin D1* transcription. (B) The promoter of the *DHFR* or dihydrofolate reductase gene is inhibited by binding of the lncRNA to the basal transcription factor TFIIB, preventing initiation. The mechanism also involves the formation of a stable triplex complex, purine-purine-pyrimidine, between the single-stranded lncRNA and the double-stranded *DHFR* promoter. (C) The lncRNA *Evf1-Evf2* interacts with the homeodomain protein Dlx2 to activate the enhancer serving *Dlx5* and *Dlx6*, two genes that are involved in neuronal differentiation and migration. Expression of *Evf2* is highly regulated in the developing mouse brain. (D) Epigenetic silencing of gene clusters in *cis* by lncRNAs: the example shown is *Xist*, the ncRNA crucial for X-chromosome inactivation in mammals. *Xist* coats the X-chromosome that is to be stably inactivated. It is believed that *Xist* coating establishes a specialized nuclear compartment devoid of Pol II; it also interacts with EZH2, the subunit of PRC2 that introduces repressive methylation marks on H3K27. (E) Epigenetic repression of genes in *trans*. The *HOTAIR* lncRNA is transcribed within the human *HOXC* gene cluster; it then directly targets epigenetic modifiers, such as PRC2, to the *HOXD* gene cluster. Other histone-modifying complexes, such as G9a methyltransferase, may also be targeted.

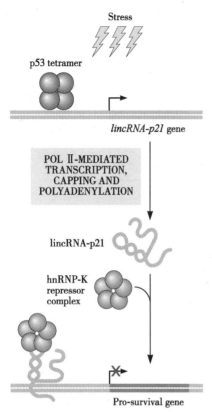

Figure 8.48 Participation of lincRNA in p53-mediated transcriptional repression. *p53* is an important tumor-suppressor gene mutated in around 50% of human cancers. The p53 protein binds to DNA as a tetramer; it becomes stabilized in response to DNA damage and triggers a complex transcriptional response, activating or repressing the expression of numerous genes. The transcriptional response leads either to cell-cycle arrest or to apoptosis, programmed cell death. The p53-mediated gene repression that is part of apoptosis occurs through a mechanism involving a long intergenic noncoding RNA, lincRNA-p21, so named for its physical proximity to the p53 target gene *p21*, although *p21* is not involved in the lincRNA pathway. The sequence of events is as follows: p53 induces the transcription of lincRNA-p21, which then interacts with a repressive RNA-protein complex, hnRNP-K; the interaction is required for proper localization of hnRNP-K on responsive genes. The repressor complex contains the linker histone variant H1.2 and blocks histone acetylation by the histone acetyltransferase p300. The changes in chromatin organization are the probable cause for transcriptional repression of the linked pro-survival genes, leading to cell death.

8.5 X chromosome inactivation

In many animal species, males and females have unbalanced genetic content regarding the sex chromosomes. In mammals, females have two X chromosomes and males have an X and a Y chromosome. If the X-linked genes performed equally well in each sex, women would express twice as much in each gene product as men. In 1961, Mary Lyon was firstly put forward the theory of X chromosome inactivation. She concluded that there must be either of the two X chromosomes randomly inactivated in each cell of female mice, by studying coat-color variegation in mice. Lyon's view is now known as the Lyon hypothesis or "dose compensation". Different organisms deal with different ways of dose compensation. In *Drosophila melanogaster*, the level of gene expression on male X chromosome is twice as high as that on female X chromosome, while in *C. elegans*, the level of gene expression on female X chromosome is only by half comparing with male X chromosome.

8.5.1 Random X chromosome inactivation in mammals

X chromosome inactivation is randomly established during early development, and it is inherited throughout the following somatic divisions. Both maternal and paternal X chromosomes are active in fertilized eggs. In mice, X chromosome inactivation occurs in two waves during early development. A first wave of X chromosome inactivation starts between the 4–8-cell stage and results in imprinted inactivation of the paternal X chromosome. At the blastocyst stage, the paternal X chromosome is reactivated and random X chromosome inactivation occurs, implying that either the maternal X chromosome or the paternal X chromosome is silenced. After this choice is made, the inactive state is inherited throughout cell division. In human, instead, both X chromosomes are maintained in an active state throughout pre-implantation development. It is not known exactly when this choice is made in human development, but based on mouse studies it is hypothesized to happen shortly after the embryo implants. The same X is inactive in all offspring of a given cell. This leads to cell chimerism in adult tissues. The inactive X is dark-stained heterochromatin, which forms highly concentrated substances in the interphase nuclei known as "Barr bodies".

8.5.2 Molecular mechanisms for stable maintenance of X chromosome inactivation

X inactivation is controlled by the X inactivation center, a regulatory domain. A key role in the function of X-inactivation centers is the X-inactivation-specific transcript XIST (pronounced "exist"), which is transcribed from the XIST gene located at the X-inactivation centers. XIST is a 17–19 kb long RNA with complex repeat patterns, which was initially discovered in humans by Carolyn Brown and Hunt Willard. This lncRNA as a critical factor for X chromosome inactivation based on its location within the X-inactivation centre and its unique expression pattern that is completely female-specific in adult somatic cells. It is expressed at low levels from both X chromosomes, together with an overlapping antisense transcript, Tsix. When X inactivation is initiated during differentiation, the transcription level of XIST is up-regulated from chromosome to inactivated X, while the expression of Tsix is down regulated.

The activity of active X requires continuous expression of Tsix in cis. XIST is negatively regulated on active X, while positively regulated on active X by Jpx transcripts. Although the mechanism by which inactive X chromosomes initiation of silence is not fully understood, the coating of X chromosomes destined to inactivate XIST RNA is known to be the core of this process. Also, several factors were reported to involve in XIST coating of the inactive X-chromosome, such as YY1 and HnrnpU.

In addition, inactivated X is characterized by a series of epigenetic chromatin modifications, which include histone H3 methylation, histone H4 hypoacetylation, macroH2A enrichment and DNA methylation. One of the functions of XIST is to recruit Polycomb group proteins directly or indirectly, which include histone modified enzymes catalysing the deposition of H2AK119 ubiquitination and H3K27 trimethylation. Follow-up maintenance of inactive state is independent of XIST. X chromosome inactivation is accompanied by the loss of DNA methylation specifically in the CpG islands of affected genes, and its extent varies in different cell lines, ranging from only a handful of genes to almost the entire inactive X chromosome. How XIST RNA and other epigenetic modifications are located along inactive X chromosomes and how inactivation is spread over 155 Mb X chromosomes are still poorly understood and need to further explore.

8.6 Methods for measuring the activity of transcriptional regulatory elements

Almost all our definitions of regulatory elements in the eukaryotic genome depend on methods that use recombinant DNA technology to incorporate the DNA element of interest into some sort of construct and then to measure the transcriptional activity of the construct after its introduction into living cells. Although it took years to come up with such methods, at present we have at our disposal a whole battery of various strategies to study all kinds of regulator elements (Box 8.2). Nevertheless, a big question still remains to be answered: do the elements identified in these reporter assays really function as expected *in vivo*?

Box 8.2 **Measuring the activity of transcriptional regulatory elements *in vivo***

Assays that measure the activity of transcriptional regulatory elements are based on the use of reporter gene constructs, whose transcriptional activity can be easily monitored (Figure 1). The most frequently used reporter genes are chloramphenicol acetyltransferase or CAT, β-galactosidase, or luciferase genes. CAT detoxifies the antibiotic chloramphenicol by acetylation, preventing its binding to the ribosome. Cells that express CAT can grow on medium containing the antibiotic. β-Galactosidase catalyzes the hydrolysis of β-galactosidase into monosaccharides. X-gal, a colorless modified galactose sugar, is metabolized by the enzyme to an insoluble product, 5-bromo-4-chloroindole, which is bright blue and thus functions as an indicator of enzymatic activity. Finally, luciferase, an enzyme derived from the firefly, is responsible for oxidation of luciferin pigment, a reaction that is accompanied by the production of light or bioluminescence.

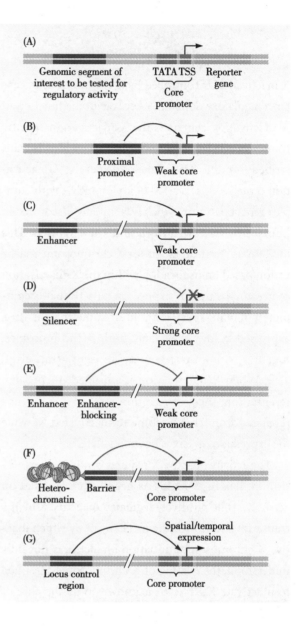

Figure 1 Functional *in vivo* assays to measure the activity of transcriptional regulatory elements. (A) Regions to be tested for regulatory activity are cloned into a plasmid bearing the reporter gene; the constructs are transfected, either transiently or stably, into cultured cells; and the activity of the reporter gene is monitored. If the segment is tested for core promoter activity, then it is placed immediately upstream of the reporter gene lacking its endogenous promoter, not shown here. (B) Testing for proximal promoter elements: an increase in transcription is expected. (C, D) Testing for enhancers or silencers would require the use of appropriate strength promoters. (E, F) Testing for two different types of insulator activities. A potential enhancer blocking element should be active when it is cloned between an existing enhancer and the gene. A barrier element should block the spreading of heterochromatin structures from neighboring regions. This assay requires stable integration into the genome. A barrier element should insulate the gene construct from position effects, that is, the gene should always be active wherever in the genome the integration occurred. (G) Definitive identification of a locus control region, LCR, also requires stable integration. The LCR should confer regulated expression of the linked gene, independently of where the construct integrates. (Adapted from Maston GA, Evans SK, Green MR [2006] *Annu Rev Genomics Hum Genet*, 7:29–59. With permission from Annual Reviews.)

Key Concepts

- Within a bacterial cell, the expression of some genes is regulated to respond to environmental changes. Other constitutive or housekeeping genes are transcribed uniformly to maintain cellular functions.

- Regulation may occur in many ways: the most direct is through differences in promoter strength or the use of alternative σ factors.

- A general type of regulation in bacteria is the stringent response. Major changes in the expression of many genes are induced when general starvation for critical nutrients occurs.

- A more precise method of controlling transcription is the activation or repression of specific operons, groups of genes with related functions and a common promoter, in response to specific environmental changes. This is frequently effected by the presence of certain transcription factors that can bind to the DNA, usually in the promoter region.

- Examples of control by transcription factors include the lactose operon, which has both repressor and activator factors; the arabinose operon, which has one factor that can act as either activator or repressor through a mechanism that involves DNA looping; and the tryptophan operon, which also involves the linkage of translation and transcription that occurs in bacteria.

- Transcription factors are usually allosteric proteins, in which one or another conformational state with specific DNA-binding capabilities is favored by the absence or presence of a small effector molecule.
- Bacterial transcription can also be regulated by the control of DNA supercoiling, on either a global or local scale.
- DNA methylation can also affect the transcription of particular genes.
- Regulation of transcription in bacteria is hierarchical; there are several distinguishable levels of control. The most specific control occurs at the operon level. However, sets of operons with related functions, called regulons, may have common control elements, and several functionally related regulons may be grouped into a few modulons.
- Because of the demands of development and of responding to environmental factors, the regulation of transcription is much more complex in eukaryotes than in viruses or bacteria.
- Promoter regions in eukaryotes are usually split into core and proximal elements; the core contains the polymerase docking site, while the proximal promoter region has regulatory functions.
- Distal DNA regulatory elements include enhancers, silencers, insulators, and locus control regions. Each of these elements acts through its own mechanisms.
- Eukaryotic transcription factors are molecular complexes that bind to DNA regulatory elements. Typically, they contain both DNA-binding and activation domains that act to bind to specific DNA sequences in specific genes and then activate transcription through recruitment of the basal transcriptional machinery.
- Mutations in gene regulatory regions or in the associated protein machinery can give rise to disease states, including cancer.
- Regulation also occurs at the level of transcriptional elongation. Sometimes polymerases are poised near the promoter until a signal releases them to continue transcription. Some elongation factors help the polymerase to overcome temporary transcriptional pausing as it travels along the gene body.
- There are also protein factors that help polymerases to pass through nucleosomes. The mechanisms for transcription of DNA within nucleosomes are still not well understood.
- Transcription in eukaryotes is markedly influenced by post-translational modification or marking of histones on nucleosomes. These marks include acetylation, methylation, phosphorylation, ubiquitylation, and poly(ADP) ribosylation events. Specific enzymes exist for each of these modifications: the enzymes are highly specific for individual amino acid residues on individual histone molecules. There is also cross-talk between such modifications.
- Specific readers of these markings exist that, upon recognizing the marks, initiate the modification of gene expression. Certain modifications, like lysine acetylation, correlate strongly with gene activity. Others are associated with gene silencing.
- Histone replacement variants also play a role in gene regulation. For example, H2A.Z is often found in nucleosomes that flank nucleosome-free regions, which occur around transcription start sites. H3.3 and H2A.Bbd are often associated with active transcription.
- Chromatin structure must sometimes be remodeled to permit transcription. This can be accomplished by a battery of ATP-dependent chromatin remodelers. These may slide nucleosomes on DNA, modify their structure, or partly dissociate the nucleosomal particle.
- Genes may also be regulated by methylation of cytosine residues, especially at CpG sites, to form 5-methylcytosine. This is considered a true epigenetic modification because it is carried through mitosis.
- Major changes in DNA methylation occur during development, upon differentiation of pluripotent stem cells, and in carcinogenesis.
- There are enzymes for methylating DNA sites and for reading or removing those DNA methylation marks.

- In addition to all the other mechanisms described above, it is now becoming clear that long noncoding RNAs can carry out a remarkable number of regulatory functions.

Key Words

activator（激活蛋白）
attenuation（衰减）
catabolite activator protein（分解代谢物激活蛋白）
cis-trans interactions（顺反交互）
CpG islands（CpG 岛）
chromatin remodelers（染色质重塑）
DNA methylation（DNA 甲基化）
DNA methyltransferase（DNA 甲基转移酶）
epigenetic regulation（表观遗传调控）
epigenetics（表观遗传学）
enhancer（增强子）
gene silencing（基因沉默）
gene expression（基因表达）
histone post-translational modification（组蛋白翻译后修饰）
histone variants（组蛋白变体）
housekeeping genes（管家基因）
insulator（绝缘子）
imprinted gene（印记基因）
mutation（突变）
nucleosome（核小体）
noncoding RNA（非编码 RNA）
operon（操纵子）
operator（操纵元件）
promoter（启动子）
regulation of transcription（转录调节）
repressor（阻遏蛋白）
silencer（沉默子）
stringent response（严谨反应）
trans-acting factor（反式作用因子）
transcriptional elongation（转录延长）

Questions

1. Describe briefly the organization of the *lac* operon.
2. How to regulate the transcriptional activity of the lactose operon by RNAP and CAP?
3. What happens to nucleosomes during transcription?
4. What is the concept of epigenetics?
5. Describe briefly the concept and function of enhancer, silencer and insulator in Eukaryotic transcription regulation?
6. List the common types of histone post-translational modifications.

References

[1] Allis CD, Berger SL, Cote J, et al. (2007) New nomenclature for chromatin-modifying enzymes. *Cell*,131:633–636.
[2] Browning DF, Busby SJ (2004) The regulation of bacterial transcription initiation. *Nature reviews. Microbiology*,2:57–65.
[3] Clapier CR, Cairns BR (2009) The biology of chromatin remodeling complexes. *Annual review of biochemistry*, 78:273–304.
[4] Lawson CL, Swigon D, Murakami KS, et al. (2004) Catabolite activator protein: DNA binding and transcription activation. *Current opinion in structural biology*, 14:10–20.
[5] Ma HW, Buer J, Zeng AP(2004)Hierarchical structure and modules in the *Escherichia coli* transcriptional regulatory network revealed by a new top-down approach. *BMC bioinformatics*,5:199.
[6] Maston GA, Evans SK, Green MR(2006)Transcriptional regulatory elements in the human genome. *Annual review of genomics and human genetics*,7:29–59.
[7] Palangat M, Renner DB, Price DH, et al.(2005) A negative elongation factor for human RNA polymerase II inhibits the anti-arrest transcript-cleavage factor TFIIS. *Proceedings of the National Academy of Sciences of the United States of America*,102:15036–15041.
[8] Rando OJ, Chang HY(2009) Genome-wide views of chromatin structure. *Annual review of biochemistry*,78:245–271.
[9] Raisner RM, Hartley PD, Meneghini MD, et al. (2005) Histone variant H2A.Z marks the 5′ ends of both active and inactive genes in euchromatin. *Cell*,123:233–248.
[10] Taverna SD, Li H, Ruthenburg AJ, et al.(2007) How chromatin-binding modules interpret histone modifications: lessons from professional pocket pickers. *Nature structural & molecular biology*,14:1025–1040.
[11] Weldon JE, Rodgers ME, Larkin C, et al.(2007)Structure and properties of a truely apo form of AraC dimerization domain. *Proteins*,66:646–654.
[12] Yanofsky C(2007) RNA-based regulation of genes of tryptophan synthesis and degradation, in bacteria. *RNA (New York, N.Y.)*,13:1141–1154.
[13] Zlatanova J, Victor JM(2009) How are nucleosomes disrupted during transcription elongation? *HFSP journal*, 3:373–378.

Yang Yang

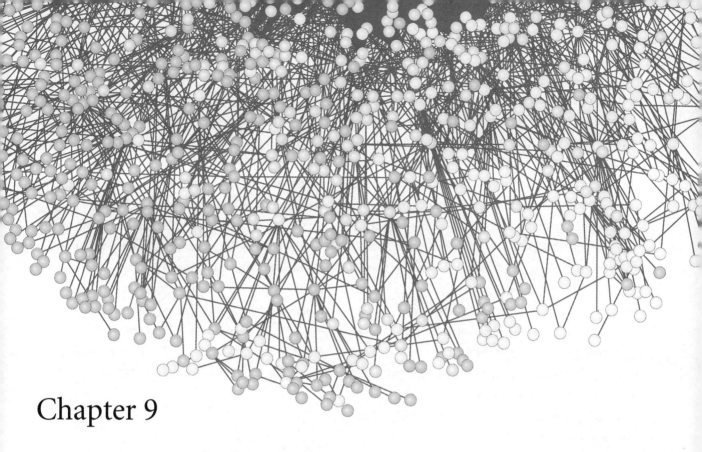

Chapter 9

Transcription Regulation in the Human Genome

9.1 Introduction

The year 2012 saw a revolution in molecular biology when the full results of the ENCODE or Encyclopedia of DNA Elements project were released. This project originated shortly after the completion of the Human Genome Project. ENCODE aimed at nothing less than to mine the entire human genome sequence for an encyclopedia of elements characteristic of various genomic functions, known as functional elements. These elements have immense diversity and include, among many others, transcription start sites or TSSs, promoters, enhancers, nucleosome locations, and methylation sites. The first output from ENCODE, published in 2007, was an analysis of only a selected 1% of the genome. The development of powerful new methods provided remarkable progress, allowing 100% of the genome to be analyzed in the next few years. In total, 147 different human cell types were analyzed. This required the collaboration of hundreds of scientists around the world, each centering on special aspects of the larger problem.

In Chapter 8, we outlined the knowledge about transcription regulation that had been obtained through the use of traditional reductionist methodologies. In these approaches, when genetic elements of functional interest were identified, they were systematically altered, for example, truncated or mutated at predefined positions, and then introduced into living cells, where their function was investigated by traditional methods. As we have seen, these approaches have yielded a wealth of information on the functions of specific genomic regions of limited lengths and the factors that interact with them.

9.2 Rapid full-genome sequencing allows deep analysis

Researchers gradually recognized that functional

elements were characterized by common biochemical or biophysical features, called signatures. This realization led to the development of genomewide methods, both experimental and computational, that look for these specific biochemical signatures in order to scan whole genomes for the occurrence and distribution of individual functional elements. The biochemical signature strategy became the cornerstone of the ENCODE project's efforts to identify all functional elements in the human genome that are specified by the genomic sequence. In this chapter, we present those major results of the second, production stage of the ENCODE project that are relevant to transcription. Some of the findings were expected on the basis of knowledge accumulated from studies of gene-specific systems, whereas others revealed a complex picture of transcription regulation and provided a wealth of new insights. Excitingly, the mine of information has not been exhausted; in fact, its full potential has hardly been touched. We are concerned here only with what has been learned to date about transcription and its regulation.

The material and methods covered in this chapter will become more and more important in the future development of molecular biology, but this content is necessarily more advanced. We are at the frontier right at the moment, looking into unknown territory. ENCODE results are mentioned in several other chapters, but those who wish to more fully understand the cutting edge of molecular biology should focus on this chapter and the references herein.

9.3 Basic concepts of ENCODE

9.3.1 ENCODE depends on high-throughput, massively processive sequencing and sophisticated computer algorithms for analysis

Locating thousands or millions of functional elements of a particular type requires that the genome be fragmented in some way. Then, fragments containing the sequence elements must be separated and sequenced so as to locate them in the known sequence of the whole genome. The techniques for isolation are varied and too detailed to recount here; we give an example in Box 9.1. Then there is the problem of sequencing thousands or millions of fragments. However, recent decades have produced a host of high-throughput methods that can provide rapid, automated sequencing. Most of these are also massively parallel, meaning that they can analyze thousands of samples simultaneously. Again, to center on any one of the currently used methods would be pointless, for improvements and new methods are appearing all the time. Just to give an idea of current capabilities, there are techniques that can simultaneously read 50 000 samples, each of length up to several thousand base pairs, in less than an hour, with better than 99.9% accuracy. The cost is less than one dollar per one million bases.

9.3.2 The ENCODE project integrates diverse data relevant to transcription in the human genome

The ENCODE project has systematically mapped features in the genome that relate to transcription: transcribed regions, transcription factor or TF binding sites, chromatin structure and histone modifications, and DNA methylation. In addition, the project addressed issues concerned with both evolutionary conservation of regulatory elements and sequence variations in these elements. The results are of unprecedented magnitude and are impossible to cover in a textbook. Still, we present here at least some of the data and the conclusions of these studies to give the reader an appreciation of the direction in which molecular biology is heading, and to expand on the more familiar facts in Chapter 8.

The project analyzed numerous aspects of functional genome organization, often providing overlapping data sets centering around a specific genomic element. For example, the analyses of promoters and TSSs provide information on many related features of the chromatin organization in these elements—nucleosome positioning, histone modification patterns, and so on—while at the same time investigating the connectivity of these elements with other recognized regulatory elements. We follow this pattern of presenting the most significant findings of

Box 9.1 **A Closer Look: FAIRE, a procedure for isolating regulatory elements genomewide**

Traditionally, open chromatin regions that occur in regulatory elements are identified by their hypersensitivity to DNase I. A more selective, simple, and highly efficient technique has recently been introduced for the isolation and analysis of active regulatory elements from eukaryotic genomes. The technique, termed FAIRE or *formaldehyde-assisted isolation of regulatory* elements, consists of protein-DNA covalent crosslinking with formaldehyde, fragmentation of DNA by shearing, and phenol/chloroform extraction of the non-cross-linked DNA fragments (Figure 1A). The DNA that is organized within nucleosomes has greater formaldehyde crosslinking efficiency than the open chromatin regions that contain DNA-binding proteins. Thus, the non-cross-linked DNA, which is recovered in the aqueous phase, originates from nucleosome-depleted regions (Figure 1B). These regions are also coincident with the location of DNase I hypersensitive sites as well as TSSs, active promoters, enhancers, and insulators (Figure 2).

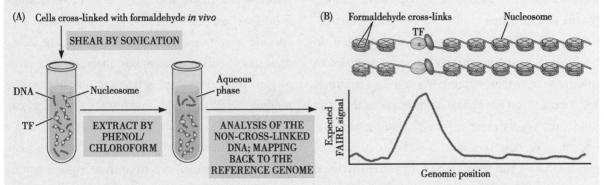

Figure 1 **FAIRE-seq depends on formaldehyde cross-linking efficiency.** (A) Overview of FAIRE, or formaldehyde-assisted isolation of regulatory elements: DNA recovered in the aqueous phase following phenol/chloroform extraction of the cross-linked and sheared sample is subjected to high-throughput sequencing or analyzed in microarrays. TF, transcription factor. (B) Cross-linking captures only a portion of potential protein-DNA interactions. Given that histone-DNA interactions constitute the majority of cross-linkable interactions in the genome, these interactions are captured preferentially; interactions between other DNA-binding proteins and DNA will be only occasionally captured by cross-linking. The two rows represent two cells in a population. (Adapted from Giresi PG, Lieb JD [2009] *Methods*,48:233–239. With permission from Elsevier.)

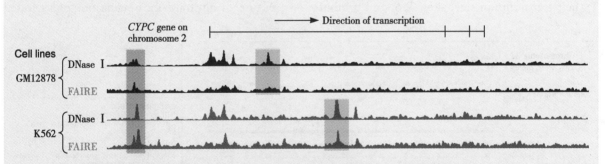

Figure 2 **Coincidence of open chromatin regions as identified by DNase I hypersensitivity and FAIRE-seq in two cell lines.** The *CYPC* gene is the only gene in this ~90 kb subsection of human chromosome 2. -The right two shading shows open chromatin regions identified by both methods in just one of the cell lines; note that open regions are also present in the body of the gene. The left shading shows open regions identified by both methods in both cell types. (Adapted from Song L, Zhang Z, Grasfeder LL, et al. [2011] *Genome Res*,21:1757–1767. With permission from Linyung Song.)

the ENCODE project, focusing on specific elements and whatever properties were interrogated. We begin with a general overview of the genome organization in terms of distinguishable elements that are relevant to transcription. In addition to the extreme sequencing demands mentioned above, interpretation of the enormous mass of raw data, especially when interrelationships between functional elements are addressed, has necessitated the construction of a host of new computer algorithms.

9.4 Regulatory DNA sequence elements

Seven classes of regulatory DNA sequence elements make up the transcriptional landscape

The project discovered numerous candidate regulatory elements that are physically associated with one another and have potential for regulating gene expression. Regulatory elements are DNA sequences characterized by three related features: (a) they bind sequence-specific TFs, often in a cooperative manner; (b) the segments that bind TFs are devoid of nucleosomes; and (c) they are hypersensitive to digestion by DNase I. This last property was first identified in specific gene systems in the 1980s and is so common and robust that it has been used throughout the years to identify regulatory elements, including promoters, enhancers, silencers, insulators, and locus control regions. As we will see, DNase I hypersensitivity is one of the major characteristics of regulatory elements interrogated systematically by the ENCODE project.

In general, seven genome segmentations, or major classes of genome states, were agreed upon as characterizing segments of the genome with regard to the role they play in transcription: transcribed genes, T; their transcription start sites, TSS, and promoters, PF; two classes of predicted enhancers, strong enhancers E and weak enhancers WE; CTCF-binding sites, CTCF; and transcriptionally repressed regions, R (Figure 9.1). The established notion that active promoters and transcribed genes go together with TSS was confirmed. In addition, three active distal states were recognized. Two of these were labeled as predicted enhancers and predicted weak enhancers because they occur in regions of open, DNase I-sensitive chromatin with high levels of the histone modification H3K4me1. A high content of this specific histone modification has been seen in the majority of enhancer elements analyzed on a gene-to-gene basis. The third active state has high CTCF binding. CTCF is a multifunctional protein that can serve as a transcription factor, an insulator, and a master organizer of the genome. In all of these roles, CTCF is intimately involved in transcription regulation. Finally, the repressed state includes sequences belonging to actively repressed or inactive, permanently repressed quiescent chromatin. The enhancer and transcribed gene states are highly cell-specific and undoubtedly reflect the long-recognized fact that cell types differ in the specific portion of the genome they express.

Figure 9.1 illustrates the organization of a selected

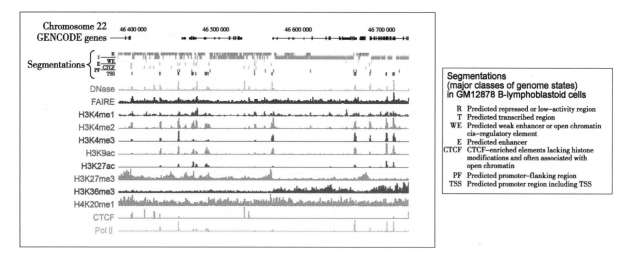

Figure 9.1 Major classes of genome transcriptional states. Seven segmentation classes of genome states and their characteristic features of human chromosome 22. FAIRE, or formaldehyde-assisted isolation of regulatory elements (see Box 9.1), isolates nucleosome-depleted genomic regions by exploiting the difference in cross-linking efficiency between nucleosomes and sequence-specific regulatory factors. The CTCF-enriched element is composed of CTCF-binding sites that lack histone modifications. They are often associated with open chromatin but may also have other functions, for example, as insulators. (From The ENCODE Project Consortium [2012] *Nature*, 489:57–74. With permission from Macmillan Publishers, Ltd.)

region of human chromosome 22, with its characteristic DNase sensitivity patterns, histone post-translational modifications, and patterns of CTCF and Pol II occupancy. This figure also shows nucleosome-depleted segments, as determined by the formaldehyde-assisted isolation of regulatory elements or FAIRE technique (see Box 9.1).

9.5 Specific findings concerning chromatin structure from ENCODE

9.5.1 Millions of DNase I hypersensitive sites mark regions of accessible chromatin

DNase I hypersensitivity is a common feature of all regulatory elements, reflecting their open chromatin structure. The long-held view that DNase I hypersensitivity is linked to the binding of TFs has been confirmed beyond doubt by the genomewide analysis of DNase I hypersensitivity included in the ENCODE project. This analysis, performed by DNase-seq on 125 cell types, identified 2.89 million unique, non-overlapping DNase I hypersensitive sites or DHSs, the overwhelming majority of which lie far from TSSs. Figure 9.2 provides an illustrative example of the coincidence between DHSs and transcription factor occupancy for a selected region of human chromosome 19. This figure also shows the distribution of DHSs over the recognized gene annotations in the GENCODE gene set, and it points to the high degree of cell specificity of DHSs. It is significant that only 0.1% of all DHSs are ubiquitously present in all human cell types, underlining the involvement of DHSs in cell differentiation.

9.5.2 DNase I signatures at promoters are asymmetric and stereotypic

The recognized involvement of promoters in gene regulation was the impetus behind the considerable effort that has gone into the genomewide characterization of these elements. The focus within the ENCODE project was on the hypothesis that RNA output can be predicted from patterns of chromatin structure and/or TF binding in and around promoters. In general, and consistent with earlier-held views, two distinct types of promoters were recognized: broad, CG-rich TATA-less promoters and narrow, TATA-box-containing promoters.

The discovery of new promoters was driven by biochemical knowledge obtained by DNase I sensitivity assays, mapping of 5′-ends of transcripts, and

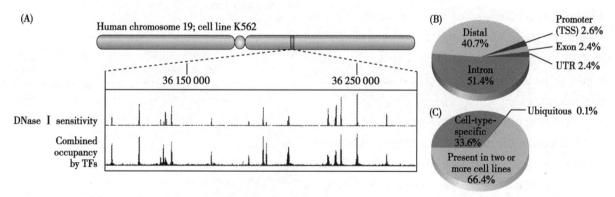

Figure 9.2 **General features of accessible chromatin landscape as determined by DNase I hypersensitivity or DNase-seq.** (A) Chromatin accessibility is driven by TF binding. Data for a 175-kb region of chromosome 19 obtained from cell line K562 show the coincidence of regions hypersensitive to DNase I and those bound by TFs; the lower graph is the cumulative sum of 45 TFs tested by ChIP-seq. (B) Distribution of 2 890 742 DNase I hypersensitive sites or DHSs with respect to GENCODE gene annotations. Promoter DHSs are defined as those located within 1 kb upstream of a TSS. Note that promoter DHSs are a small portion of all DHSs, with practically the same percentage present in exons and untranslated regions, UTRs. The majority sites are about evenly distributed between intronic and intergenic, or distal, regions. (C) Cell specificity of DNase I hypersensitive sites. The majority of sites are shared among two or more cell types, but a large proportion are cell-type-specific, and only a very small minority are present in all cell types. (Adapted from Thurman RE, Rynes E, Humbert R, et al. [2012] *Nature*, 489:75–82. With permission from Macmillan Publishers, Ltd.)

the recognition that promoters are highly enriched in H3K4me3. Well-annotated promoters were systematically studied for these properties: DNase I cleavage profiles were plotted against ChIP-seq data for H3K4me3 for 56 cell types, revealing that these features have highly stereotypical, asymmetric patterns (Figure 9.3A). The promoter regions immediately preceding TSSs were found to be highly sensitive to DNase I cleavage, whereas the region that codes for the transcript contains a wave of high H3K4me3 content, which dwindles as the distance from the TSS increases. This genomewide pattern was then used in computational approaches to scan the entire genome for the possible presence of such patterns. In total, 113 622 distinct putative promoters were identified, 39.5% of which were new. The distribution of all promoters, both previously annotated and newly recognized, in the genome is presented in Figure 9.3B. We find a novel feature: a significant portion of new promoters is found within gene bodies and 3′-regions of already annotated genes. These are oriented either sense or antisense to the annotated direction of transcription.

The location and orientation of the newly discovered promoters are highly reminiscent of the situation already observed in the mouse genome, where pervasive transcription was first recognized. Although the functions of these numerous transcripts interleaved with the mRNA portions of genes remain to be elucidated, it is clear that pervasive transcription is a common property, at least of highly complex metazoans.

Another important feature recognized by the ENCODE project is the highly stereotypic structural motif in the immediate upstream vicinity of TSSs. There is a robust 50 bp footprint region in the center of the region of high DNase sensitivity that typifies promoters (Figure 9.4). Genomic DNase I footprinting performed on 41 cell types identified 8.4 million distinct footprints. *De novo* motif discovery methods recovered ~90% of known TF-binding motifs, together with hundreds of novel motifs; many of these display high cell selectivity, suggesting their involvement as regulators of differentiation.

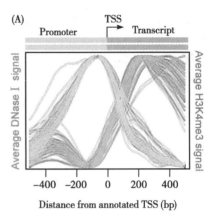

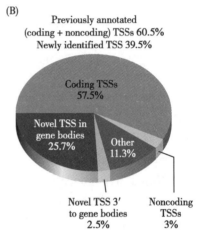

Figure 9.3 **Invariant directional signature of two features of promoter chromatin structure.** The features shown are DNase I sensitivity and presence of H3K4me3. (A) The same biological samples from 56 cell types were analyzed by ChIP-seq to show the presence of H3K4me3 and for DNase I sensitivity. Averaged H3K4me3 signal, shown in the right curve, is plotted against averaged DNase sensitivity signal, shown in the left curve, across 10 000 randomly selected TSSs oriented 5′→3′. Each curve is for a different cell type. Note that the patterning of these two chromatin features is highly directional, or asymmetric, and has a precise relationship to the TSS. The pattern is consistent with a rigidly positioned nucleosome immediately downstream from the promoter DHS. (B) A pattern-matching search across the genome for the pattern shown in part A was used to identify novel promoters. The overall distribution of TSSs in the genome is presented in the pie chart. It may be highly significant that ~60% of the promoters that were newly identified in annotated gene bodies or 3′ to them are oriented antisense to the annotated direction of transcription. (A, from Thurman RE, Rynes E, Humbert R, et al. [2012] *Nature*, 489:75–82. With permission from Macmillan Publishers, Ltd.)

9.5.3 Nucleosome positioning at promoters and around TF-binding sites is highly heterogeneous

To understand nucleosome positioning at promoter sites genomewide, the ENCODE researchers generated data by digestion of chromatin DNA with micrococcal nuclease, MNase, followed by DNA sequencing for two cell lines. In addition to analyzing regions around TSSs in these cell lines, they also related the nucleosome positioning signals to the binding sites of 119 DNA-binding proteins across a large number of cell lines.

Traditionally, data quantifying relationships among genomic signals are presented in the aggregation plots. In these plots, the signal of interest is averaged for each position within a predefined window around the center of an anchor site; the anchor sites are all aligned at the location of a feature that they share in common. In the specific example we consider, TSSs or the location of TF-binding sites serve as the aligned features of the anchor sites, while the nucleosome positions are the signal whose shape, magnitude, and asymmetry is being analyzed. This method, despite its popularity and effectiveness, suffers from a major disadvantage. The averaging over all anchor sites of a particular type, such as all TSSs, produces a misleading aggregate that could obscure underlying heterogeneities that may be of major biological significance. To avoid this drawback, a novel methodology, termed clustered aggregation tool or CAGT, was introduced.

The application of CAGT to the analysis of nucleosome positioning around TSSs revealed 17 clusters of distinct patterns. Figure 9.5 shows the 11 clusters that each contained >2% of the TSSs. Broadly, the clusters fall into two categories in terms of their regularity of nucleosome positioning with respect to the TSS: either upstream or downstream. Surprisingly, no cluster had equally strong positioning on both sides of the TSSs, suggesting that the aggregate plot (see top-left plot in Figure 9.5) is an averaging artifact. Attempts to relate these diverse patterns to the level of transcriptional activity of each cluster showed no

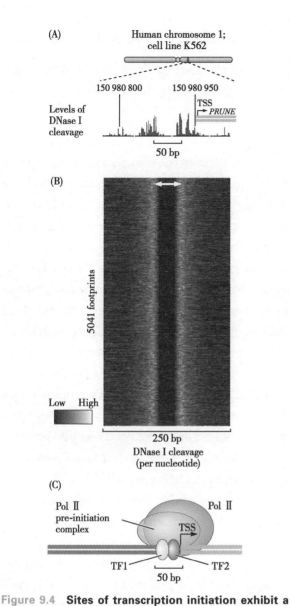

Figure 9.4 **Sites of transcription initiation exhibit a highly stereotypic chromatin structural motif of ~80 bp.** The motif consists of a ~50 bp central DNase I footprint flanked symmetrically by ~15 bp regions of uniformly elevated levels of DNase I cleavage. (A) Promoter region of the gene *PRUNE* on chromosome 1 is presented as an example. Note the tight spatial coordination of the motif with the TSS of the gene. (B) Heat map of the per-nucleotide DNase I cleavage patterns over 5041 instances of this stereotypical signature. (C) Schematic presentation of the molecular structure underlying the signature. This interpretation was derived from the finding that there are two distinct peaks of evolutionary conservation within the central footprint, not shown here, compatible with binding sites for paired canonical sequence-specific TFs. The TSS is localized precisely within the footprint. (Adapted from Neph S, Vierstra J, Stergachis AB, et al. [2012] *Nature*, 489:83–90. With permission from Macmillan Publishers, Ltd.)

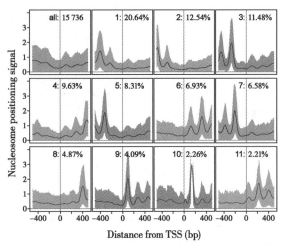

Figure 9.5 **Nucleosome positioning patterns around TSSs in the K562 cell line.** The top-left plot, labeled all, is a traditional aggregation plot of the nucleosome positioning signal in a 1001-bp window centered on each of the 15 736 TSSs annotated by GENCODE. The rest of the plots show the patterns of nucleosome positioning uncovered by CAGT, ordered by the percentage of TSSs that follow each pattern. All TSSs are reoriented so that the direction of transcription is from left to right. (Adapted from Kundaje A, Kyriazopoulou-Panagiotopoulou S, Libbrecht M, et al. [2012] *Genome Res*, 22:1735–1747. With permission from Anshul Kundaje.)

consistent positioning features. A general trend can be discerned between high expression levels and the particularly pronounced nucleosome positioning peaks, either upstream or immediately downstream of the TSSs. Much remains to be learned about how or whether this structural heterogeneity plays a role in transcriptional regulation.

9.6 ENCODE insights into gene regulation

9.6.1 Distal control elements are connected to promoters in a complex network

The ENCODE project performed experiments to compare the dynamics of the appearance of DNase sensitivity of known cell-selective enhancers with that of the promoters of their target genes, when these genes are activated. The researchers found that in many cases both elements became hypersensitive at the same time following gene activation, an observation that prompted them to analyze the possible connection between promoters and enhancers genomewide. They studied the patterning of 1 454 901 distal DHSs, where distal means a DHS separated from a TSS by at least one other DHS. They used 79 diverse cell types and correlated the DNase I signal at each distal position with the signal at all promoters within ±500 kb. A total of 578 905 DHSs were highly correlated with at least one promoter. These data provided an extensive map of candidate enhancers controlling specific genes. Experimental validation of this connectivity was provided by identifying chromatin interactions by the chromosome conformation capture carbon copy or 5C technique (Box 9.2). The connectivity derived from both DNase I hypersensitivity studies and 5C is presented in Figure 9.6A which uses the example of the phenylalanine hydroxylase gene, *PAH*. Figure 9.6B shows the genomewide pat-

---

Box 9.2 **A Closer Look: 5C methodology, a massively parallel solution for mapping interactions between genomic elements genomewide**

The 5C technology, chromosome conformation capture carbon copy, was developed as an expansion of the 3C method, chromosome conformation capture, introduced in 2002 to detect physical interactions between genomic sequences. 3C uses formaldehyde crosslinking to trap covalently interacting segments in the genome, followed by restriction endonuclease treatment and ligation of the interaction segments. The ligated products are then quantified individually by polymerase chain reaction, PCR. 3C is particularly suited for relatively small-scale analysis of interactions between a set of candidate sequences. PCR detection is, however, not suitable for large-scale mapping of previously unidentified chromatin interactions. To perform such mapping, the 5C method was introduced, in which highly multiplexed ligation mediated amplification, LMA, is used to first copy and then amplify parts of the complex 3C libraries produced as the end product of the 3C methodology. LMA is widely used to detect and amplify specific target sequences by use of primer pairs that anneal next to each other on the same DNA strand; only primers annealed next to each other can be ligated. Inclusion of universal tails that contain strong promoter sequences at the end of these primers allows subsequent amplification. LMA-based approaches can be performed at high levels of multiplexing, using thousands of primers in a single reaction. The amplified libraries are analyzed by microarrays or DNA sequencing (Figure 1).

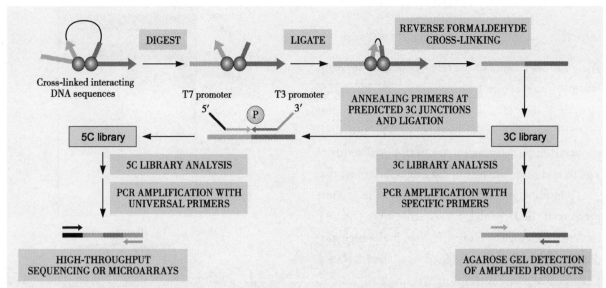

Figure 1 **5C detects ligation products in 3C libraries.** A 3C library is generated by conventional methods and then converted into a 5C library by annealing and ligating 5C oligonucleotides in a multiplex setting. The new 5C libraries are then analyzed by sequencing or in microarrays. 5C libraries are produced by annealing 5C primers at predicted 3C junctions in a multiplex setting, followed by specific ligation of annealed primers with a NAD-dependent DNA ligase. The universal tails of 5C primers are illustrated as black and green lines and contain T7 and T3 promoter sequences, respectively. These promoters are used to amplify libraries in a single PCR step. (Adapted from Dostie J, Richmond TA, Arnaout RA, et al. [2006] *Genome Res*,16:1299–1309. With permission from Josee Dostie.)

tern of connectivity among these control elements. Most promoters were connected to more than one distal DHS, and vice versa, most distal DHSs interacted with more than one promoter. The number of distal DHSs connected with a particular promoter provided, for the first time, a quantitative measure of the overall regulatory complexity of that particular gene.

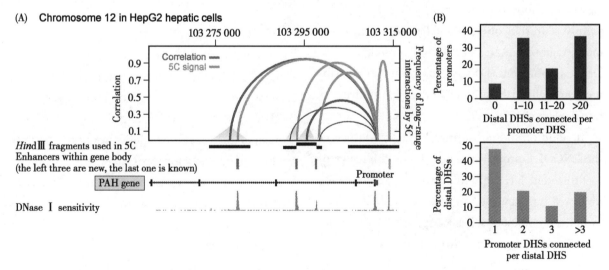

Figure 9.6 **Promoter-to-distal DHS connectivity for the *PAH* gene.** (A) Cross-cell-type correlation, shown as dark grey arcs as measured on the left *y*-axis, of distal DHSs and *PAH* promoter closely parallels chromatin interactions measured by 5C, shown as light grey arcs as measured on the right *y*-axis. Black bars indicate *Hind*III fragments used in 5C assays. Known and new enhancers are shown as indicated vertical bars, respectively. (B) Proportions of connected elements. (Top) Each promoter is connected to the number of DHSs. Note that more than half of the promoters are connected to more than 11 DHSs. (Bottom) Each distal DHS is connected to 1, 2, 3, or >3 promoters. (Adapted from Thurman RE, Rynes E, Humbert R, et al. [2012] *Nature*,489:75–82. With permission from Macmillan Publishers, Ltd.)

9.6.2 Transcription factor binding defines the structure and function of regulatory regions

We are already aware of the role TFs play in the regulation of eukaryotic transcription. The ENCODE analysis of the entire human genome confirmed this role and revealed novel features of TFs and their cooperation in defining the transcriptional status of a gene, and more broadly in shaping cellular identity.

TF binding to regulatory elements in the genome protects the underlying DNA sequence from DNase I cleavage, creating the footprints seen in the high-resolution electrophoretic gels used to analyze DNase I cleavage patterns. Footprinting was originally used to study known gene-specific *cis*-regulatory sequences and led, among other things, to the discovery of the first human sequence-specific TF, SP1.

It is now possible to perform genomewide mapping of DNase I footprints. Efficient footprinting in large genomes requires focused analysis of a small fraction, 1%–3%, of the genome that is characterized by a substantial concentration of DNase I cleavage sites. Analyses of sequences that are enriched for DNase I cleavage across 41 cell types identified an average of ~1.1 million footprints, 6–40 bp long, per cell type. The majority of DHSs (99.8%) contained at least one footprint, indicating that DHSs do not simply represent open or nucleosome-free chromatin regions but are constitutively populated with DNase I footprints. DNase I footprints are distributed throughout the genome in the quantitative pattern depicted in Figure 9.7.

In addition to interrogating DNase I footprints, the ENCODE project mapped the binding sites of 119 different DNA-binding proteins genomewide in 72 cell types by use of ChIP-seq methods. These included canonical or sequence-specific TFs, histone-modification enzymes, chromatin remodelers, a number of components of RNA polymerases II and III, and their associated basal TFs. Overall, ENCODE identified 636 336 binding regions across all studied cell types, covering 8.1% of the genome. In addition, the project looked for known and new DNA-binding motifs. Both high- and low-affinity binding sites

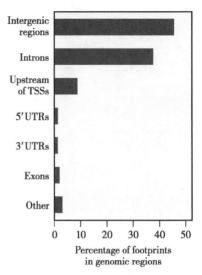

Figure 9.7 Distribution of DNase I footprints among different elements of the genome. There is an enrichment of footprints in recognized control regions located in intergenic portions of the genome and upstream of TSSs, commensurate with the DNase I cleavage densities observed in these regions. A significant portion of DNase I footprints occurs in introns. Of interest, and unexpectedly, 2% of footprints were localized within exons. The functional significance of the footprints in introns and exons is unclear.

were recognized; moreover, it was possible to identify TF target regions that were indirectly bound by TFs through interactions with other factors.

9.6.3 TF-binding sites and TF structure co-evolve

As discussed above, the application of genomic DNase I footprinting across regulatory genomic sequences defines the recognition landscape of hundreds of DNA-binding proteins. Mining footprint sequences for recognition motifs has nearly doubled the human *cis*-regulatory lexicon, identifying numerous new *cis*-regulatory elements with characteristic structural and functional features. Of importance, the *cis*-regulatory compartment that comprises all *cis*-acting regulatory sequences has co-evolved with TFs, so that the structures of the DNA elements closely fit the structures of the TFs that bind to them. Figure 9.8 illustrates this point with the specific example of USF1, upstream stimulatory factor 1. It also points out that all recognized sequences that bind the factor *in vivo* share a common nucleotide-level DNase I signature. The DNase cleavage pattern closely parallels the topology of the protein-DNA interface, including a marked depression in DNase I cleavage at nucleotides directly involved in protein-DNA contacts and increased cleavage at exposed nucleotides. Thus, the high-resolution aggregated DNase I cleavage signature reflects fundamental features of the protein-DNA interaction interface.

Interestingly, the conservation of DNA residues in vertebrates closely correlates with the DNase I cleavage pattern, implying that regulatory DNA sequences have evolved to fit the morphology of the TF-DNA binding interface.

9.7 ENCODE overview

9.7.1 What have we learned from ENCODE, and where is it leading?

Aside from the many specific revelations detailed above, what are the main lessons obtained to date from ENCODE? It seems to us that they are twofold. First, it is now clear that human cells utilize, in some fashion, much more of the genomic information encoded in their DNA than was hitherto expected. Perhaps it will turn out to be 100%, when all cell types have been interrogated. At any rate, junk DNA is no longer a useful term. To be sure, we do not know the functional importance of even a fraction of the transcripts that have been identified, and unless new screening methods are devised, we will not know for a very long time. But if there is any practical lesson from the history of molecular biology, it is that powerful new methods often appear quickly when needed.

A second surprise is the finding that the vast majority of the newly detected functional entities and interactions among them are cell-type-specific. Perhaps this revelation should not have been surprising, given the remarkable number of very different varieties of cells present in the adult human. A complicating aspect of this is that the function of a given gene or regulatory element may vary, depending on the cellular milieu in which it exists. This will make the unscrambling of interconnected regulatory pathways even more complex but additionally rewarding. Perhaps the major impact of this line of research will be in the areas of developmental biology and evolution. There are already hints that the fundamental reason for the difference in size between mammalian and invertebrate genomes lies in this difference in cell-type diversity. Now it can be analyzed and quantified. It may be that a new age in biology has been born.

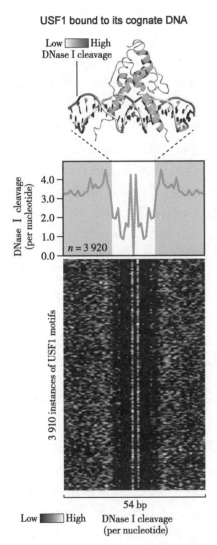

Figure 9.8 **DNase I footprint structure parallels the structure of the cognate TF.** The co-crystal structure of a TF, upstream stimulatory factor or USF1, bound to its specific binding site is positioned above the average nucleotide-level DNase I cleavage patterns. Nucleotides that are sensitive to DNase I cleavage are black in the co-crystal structure. The heat map at the bottom shows the DNase I signature at each individual USF1 binding motif; these individual signatures have been averaged to produce the black line profile shown. (Adapted from Neph S, Vierstra J, Stergachis AB, et al. [2012] *Nature*, 489:83–90. With permission from Macmillan Publishers, Ltd.)

9.7.2 Certain methods are essential to ENCODE project studies

For a monumental project such as ENCODE to be meaningful and successful, it is important to develop and standardize experimental and computational methods to be used by the involved research community. We have already covered the more specific methods at the appropriate places throughout the chapter. We will now list and briefly introduce the more common methods, following the format used in the master publication of the second, production phase of the ENCODE project. To fully describe all these techniques would in itself require a very large book.

(1) RNA-seq

Isolation of different RNA subpopulations, often combining different purification techniques for different RNA fractions, followed by high-throughput sequencing.

(2) CAGE

Capture of the 5′-caps on mRNAs, followed by high-throughput sequencing from a small tag adjacent to the 5′-methylated caps.

(3) RNA-PET

Simultaneous capture of the 5′-caps and poly(A) tails on mRNAs, indicative of a full-length RNA, followed by high-throughput sequencing from each end.

(4) ChIP-seq

Cross-linking of DNA to bound proteins in chromatin *in vivo*, fragmentation of the DNA, and selection of fragments bound to a specific protein by use of antibodies specific to that protein. The enriched sample is subjected to high-throughput sequencing. In addition to identifying bound proteins, the technique can be used to identify features such as specific histone post-translational modifications or DNA methylation. In the latter case, the antibodies directly recognize methylated CpGs or specific methyl DNA-binding proteins.

(5) DNase-seq

Cleavage of chromatin by DNase I occurs preferentially at sites that are exposed and is thus used as a probe for open chromatin. Openness is defined by the absence of nucleosomes and the binding of TFs. The cut segments are then subjected to high-throughput sequencing to determine the location of hypersensitive sites on the known genomic sequence.

(6) FAIRE-seq, formaldehyde-assisted isolation of regulatory elements

Isolation of nucleosome-depleted genomic regions by exploiting the difference in cross-linking efficiency between nucleosomal histones and DNA, which is highly efficient, and sequence-specific regulatory factors and DNA, which is characterized by low efficiency. Cross-linking is followed by phenol extraction and sequencing of DNA fragments in the aqueous phase (see Box 9.1).

(7) CAGT

A novel methodology for relating functional elements, such as TF-binding sites or TSSs, and their associated signals, such as histone modifications or nucleosome positioning, and for discovering meaningful and robust signal patterns around these loci.

(8) 5C, chromosome conformation capture carbon copy

A modification of the 3C method introduced in Chapter 7, designed to convert physical chromatin interactions into specific ligation products that can be quantified by PCR-based methods. The method has been expanded for large-scale parallel detection of chromatin interactions. (see Box 9.2)

Key Concepts

- The ENCODE project has made use of the total human genome sequence and sophisticated data analysis algorithms to probe transcriptional functional elements and their regulation in depth.
- The project defines seven major classes of elements: transcribed genes, their transcription start sites

and promoters, two classes of predicted enhancers (strong and weak enhancers), CTCF-binding sites, and transcriptionally repressed regions.
- DNase hypersensitive sites or DHSs indicate more open chromatin structure and correlate strongly with transcribed genes, transcription start sites, promoters, and strong enhancers.
- Nucleosomes are often precisely positioned about TF-binding sites near TSSs.
- Footprints in the DHS regions correspond to the binding of specific TFs. The topographies of these DHS regions match the TF structures, indicating co-evolution.
- Although most of the genome is transcribed, the pattern of transcribed genes and their control is highly dependent on cell type.

Key Words

cis-regulatory lexicon（顺式调控元件）
DNase I footprints（DNase I 足迹法）
DNase I hypersensitivity（DNase I 超敏位点）
ENCyclopedia Of DNA Elements project, ENCODE project（DNA 元件的百科全书计划）
formaldehyde-assisted isolation of regulatory elements, FAIRE（甲醛辅助法分离调控元件）
genome segmentations or major classes of genome states（基因组元件的片段切分/基因组形态元件的主要类别）
regulatory elements（调控元件）
5C technology（5C 技术）

Questions

1. What's the aim of ENCODE? Please describe the specific findings from ENCODE.
2. What's DNase I hypersensitivity sites?
3. Which new technologies are being used in ENCODE?

References

[1] Chanock S (2012) Toward mapping the biology of the genome. *Genome Res*,22:1612–1615.
[2] Dostie J, Richmond TA, Arnaout RA, et al. (2006) Chromosome Conformation Capture Carbon Copy (5C): A massively parallel solution for mapping interactions between genomic elements. *Genome Res*,16:1299–1309.
[3] Ecker JR, Bickmore WA, Barroso I, et al. (2012) Genomics: ENCODE explained. *Nature*,489:52–55.
[4] Frazer KA (2012) Decoding the human genome. *Genome Res*,22:1599–1601.
[5] Giresi PG, Lieb JD (2009) Isolation of active regulatory elements from eukaryotic chromatin using FAIRE (Formaldehyde Assisted Isolation of Regulatory Elements). *Methods*,48:233–239.
[6] Harrow J, Frankish A, Gonzalez JM, et al. (2012) GENCODE: The reference human genome annotation for The ENCODE Project. *Genome Res*,22:1760–1774.
[7] Kundaje A, Kyriazopoulou-Panagiotopoulou S, Libbrecht M, et al. (2012) Ubiquitous heterogeneity and asymmetry of the chromatin environment at regulatory elements. *Genome Res*,22:1735–1747.
[8] Neph S, Vierstra J, Stergachis AB, et al. (2012) An expansive human regulatory lexicon encoded in transcription factor footprints. *Nature*,489:83–90.
[9] Song L, Zhang Z, Grasfeder LL, et al. (2011) Open chromatin defined by DNase I and FAIRE identifies regulatory elements that shape cell-type identity. *Genome Res*,21:1757–1767.
[10] Stamatoyannopoulos JA(2012)What does our genome encode? *Genome Res*,22:1602–1611.
[11] The ENCODE Project Consortium(2012) An integrated encyclopedia of DNA elements in the human genome. *Nature*,489:57–74.
[12] Thurman RE, Rynes E, Humbert R, et al. (2012) The accessible chromatin landscape of the human genome. *Nature*,489:75–82.

Lin Hou

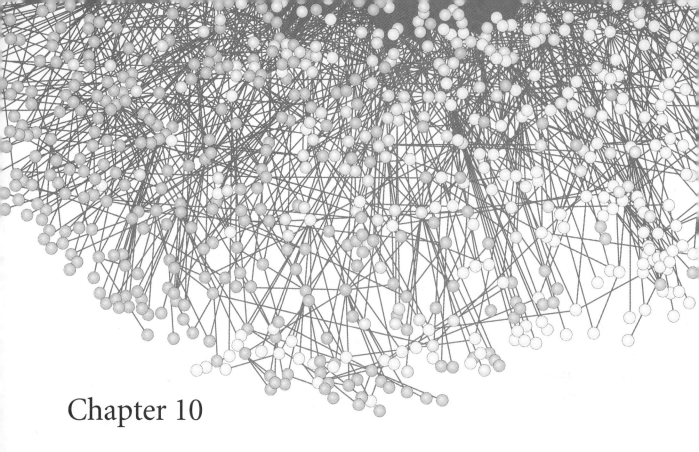

Chapter 10

RNA Processing

10.1 Introduction

In earlier chapters, we have seen that large portions of the genome can be accurately copied into RNA. In most cases, however, these primary transcripts are still not appropriate to the cell's needs and must be modified. In other word, most RNA molecules undergo post-transcriptional processing. The transformation of primary transcripts into mature functional RNA molecules is collectively called RNA processing. Additional processes occur to degrade RNA molecules that have reached the end of their physiologically relevant lifetimes or to degrade RNA molecules that are defective or misfolded. Virtually every cellular RNA undergoes one or, more often, several forms of processing.

There are four general categories of processing. The first category involves removal of nucleotides from the primary transcript; this may involve substantial portions of the transcript, as, for example, in the removal of introns from eukaryotic messenger RNA, mRNA. The second category adds nucleotides to the 3′- and 5′-termini of RNA chains in a template-independent manner. Third, RNAs can be edited by removal or insertion of nucleotides within the sequence originally prescribed by the DNA. Finally, bases can be modified covalently by a number of different enzyme-catalyzed reactions.

Eukaryotic RNAs exhibit much more processing than bacterial RNAs. This chapter covers the best-characterized processing reactions. Firstly, the main focus will be on eukaryotic encoding RNA, mRNA, because bacterial RNAs exhibit few post-transcriptional modifications. Bacterial protein-coding mRNAs are produced in the cytoplasm and can immediately attach to ribosomes and begin to be translated, without undergoing any processing. Eukaryotic mRNA, on the other hand, is produced in the nucleus and must be transported to the cytoplasm for translation. Furthermore, the directly transcribed message in eukaryotes almost invariably contains introns, which must be spliced out to produce the mature, translatable forms of mRNA. Two other modifications are common: capping of the

5'-end and addition of a 3'-polyadenylate or poly(A) tail. In describing eukaryotic mRNA processing, we first consider the modifications at the 5'- and 3'-ends, and then we will describe the complex reactions that carry out and regulate splicing, and mRNA methylation, m^6A. Secondly, the main focus will be on the processing of constitutive noncoding RNA, transfer RNA (tRNA) and ribosomal RNA (rRNA) molecules. The modifications that do occur in bacteria mainly involve cutting up long tandem transcripts to produce functional tRNA and rRNA, molecules. Furthermore, the biogenesis, function and processing of regulatory noncoding RNA, including small noncoding RNAs, siRNA, miRNA and piRNA and long noncoding RNAs, lncRNAs and circRNAs, will be discussed. Finally, the main focus will be on the turnover of RNA in nucleus and cytoplasm.

10.2 Processing of eukaryotic mRNA: end modifications

Processing of the eukaryotic precursors to mRNA usually involves three distinct steps: 5'-end capping; splicing, or removal of introns and splicing together of exons; and polyadenylation, or addition of poly(A) tails to the 3'-end of the transcript (Figure 10.1). We first discuss the modifications that occur at the 5'- and 3'-ends before considering the complexities of splicing. Not all mRNAs experience all types of processing; thus, for example, the histone genes do not contain introns, so no splicing occurs to produce the mature mRNA. Most histone mRNAs also do not undergo polyadenylation; the creation of their mature 3'-end involves other specialized pathways.

10.2.1 *Eukaryotic mRNA capping is co-transcriptional*

Capping is the addition of a methylated guanosine monophosphate cap at the 5'-end of the mRNA precursor via an unusual 5'-5' triphosphate linkage (Figure 10.2). Capping occurs in the nucleus. The cap performs multiple roles: it protects mRNA from exonucleolytic degradation from the 5'-end, it creates the appropriate substrate for splicing, and it serves as the site for initiation factor attachment during translation initiation. The enzymatic activities that perform the capping reaction are attached to the C-terminal domain of Pol II, as are many of the other proteins that couple transcription to processing. Once the cap structure is in place, a dimeric Cap-binding complex, CBC, attaches to it, and stays bound until the mRNA is exported to the cytoplasm. CBC serves to recruit the transcription-export complex TREX to the mRNA early in the process.

10.2.2 *Polyadenylation at the 3'-end serves a number of functions*

Addition of a string of consecutive adenylate or A residue to the 3'-end of mRNA, known as polyadenylation, is an important step in eukaryotic mRNA processing. Figure 10.3 depicts the process and the major proteins involved. The discovery that the actual site of poly(A) addition is situated 5' to the transcription termination site was unexpected and led to the discovery of a poly(A) signal in the DNA that is located in what will be transcribed into the 3'-untranslated region or UTR of the mRNA. The actual poly(A) signal in the primary transcript, AAUAAA, acts in conjunction with a poly(U) sequence

Figure 10.1 **Structure of a typical eukaryotic mRNA following processing.** The coding region and two flanking untranslated regions or UTRs are included. The coding region contains both exons and introns in the primary transcripts, but only exons are present in the mature mRNA, which is ready to be translated on the ribosome. Note the cap structure at the 5'-end and the poly(A) tail at the 3'-end of the molecule. Not all mRNAs are polyadenylated; metazoan histone mRNAs are a notable example of mRNAs that lack a poly(A) tail.

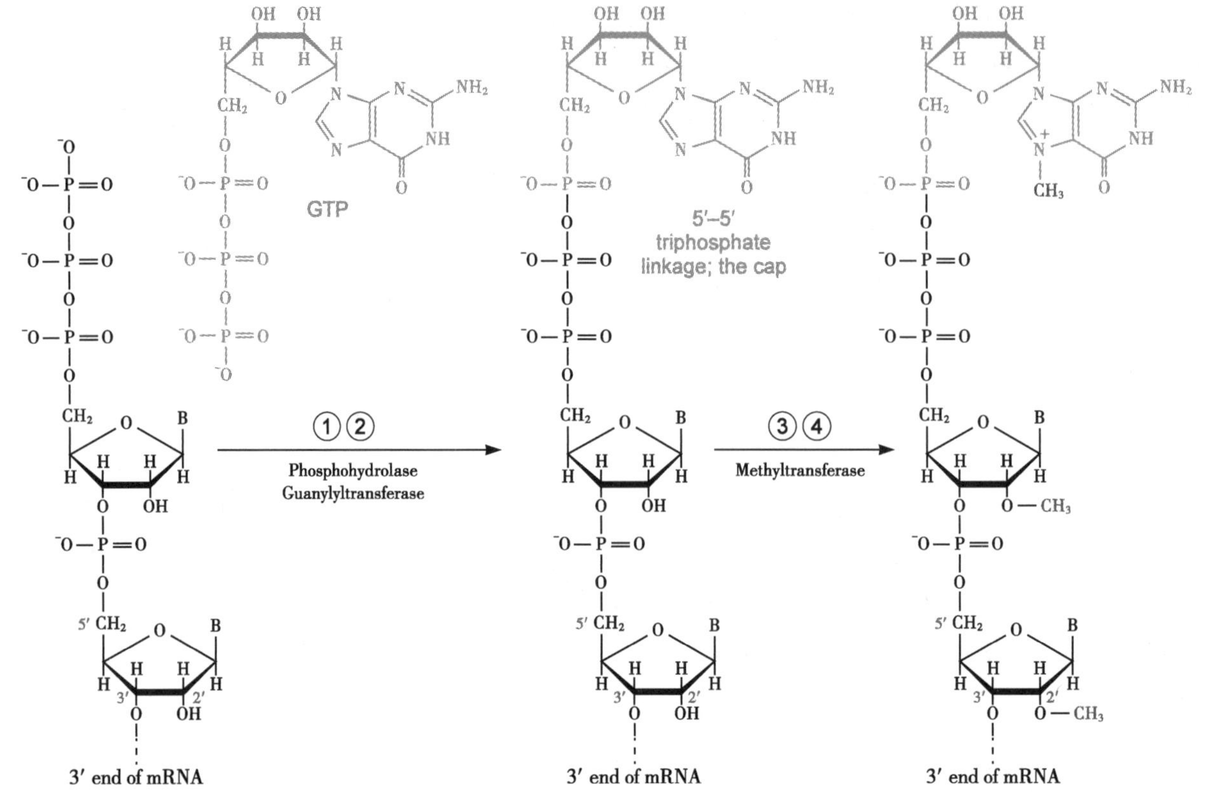

Figure 10.2 **Formation of a cap at the 5'-end of a eukaryotic mRNA precursor.** The reaction proceeds in several steps: (Step 1) A phosphohydrolase catalyzes removal of the phosphate group at the 5'-end of the precursor. (Step 2) The 5'-end then receives a GMP group from GTP with release of pyrophosphate, in a reaction catalyzed by guanylyltransferase. (Step 3) The base of the guanylate group is methylated at N-7. (Step 4) The 2'-hydroxyl groups of the terminal and penultimate ribose groups of the precursor may also be methylated.

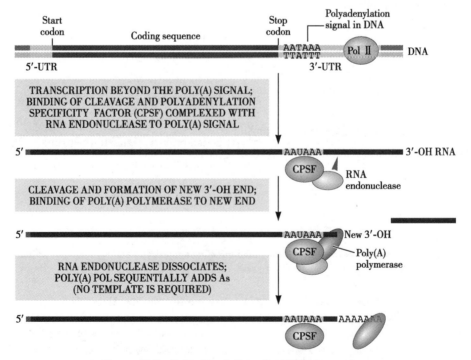

Figure 10.3 **Polyadenylation of mRNA precursors.**

downstream to form the active poly(A) signal. The poly(A) signal in the message interacts with a protein factor dubbed CPSF, cleavage and polyadenylation specificity factor, which is complexed with an RNA endonuclease. This endonuclease cleaves the RNA to form the 3'-OH group needed for the polymerization activity of the special poly(A) polymerase PAP, which adds the A residue in a template-independent process.

Poly(A) tails are highly heterogeneous in length, ranging from a few adenylate residues to more than 200–300. The length varies with biological species, developmental stage, and type of RNA. An important role of the poly(A) tail is believed to be stabilization of the message against exonucleolytic degradation: mRNAs that need to be around for a long time contain long poly(A) tails. Generally, mRNAs have short half-lives, measured in minutes.

The poly(A) tail shortens progressively during the lifetime of the mRNA; the shortening is executed by exonucleases and starts even before the mRNA leaves the nucleus. Where such a shortening of an mRNA is undesirable, there is a cytoplasmic mechanism that restores the length of the poly(A) tail. This is achieved through the action of a cytoplasmic polyadenylation element that is present in the 3'-UTR of some mRNAs.

How does the poly(A) tail stabilize the message? In the nucleus, stabilization is achieved through the binding of a poly(A)-binding or PAB protein to the tail that protects it from exonucleolytic degradation. PAB also helps the polymerase to synthesize long tails. Once a short poly(A) tail has been synthesized, PAB binds to it and forms a quaternary complex with CPSF, PAP, and the substrate RNA. This complex transiently stabilizes PAP binding, thus supporting rapid processive catalysis. Another class of PABs is cytoplasmic and in some cases required for poly(A) removal.

Additional functions for the poly(A) tail have been recognized. In bacteria and during eukaryotic nuclear surveillance, the addition of a poly(A) tail initiates mRNA decay. The tail binds to a protein, Nab2, which positions the 3'-end of the export-competent mRNA in close proximity to the nuclear pore channel. Thus the tail plays a crucial role in mRNA export. In addition, it enhances translation initiation through an incompletely understood mechanism.

10.3 Processing of eukaryotic mRNA: splicing

All ribonucleic acids studied are transcribed from their respective genes, and (especially in eukaryotes) require further processing to mature and function. Interrupted genes have been found in all kinds of eukaryotes. They represent a small fraction of genes in unicellular eukaryotes, but most genes in the genome of multicellular eukaryotes. Genes vary greatly due to the number and length of introns, but a typical mammalian gene has 7 to 8 exons distributed over 16 kb. Exon is relatively short (−100 to 200 bp), intron is relatively long (> 1 kb). The difference between interrupted tissues of genes and uninterrupted tissues of their mRNA requires the processing of primary transcripts. The primary transcript has the same tissue as the gene and is called a pre-gene. Introns were removed from the front gene, leaving ribonucleic acid molecules with an average length of −2.2 kb. Intron removal is the main part of all eukaryotic ribonucleic acid processing. The process of removing introns is called RNA splicing.

10.3.1 The splicing process is complex and requires great precision

Most eukaryotic genes contain noncoding sequences, known as introns, that need to be removed from the primary transcript. The flanking exons must then be spliced together. This must be done with great precision so as not to cause frameshifts. To meet these strict requirements, a complex and highly regulated nuclear machine, named the spliceosome, has evolved. For the spliceosome to work properly, it has to recognize precisely the boundaries between exons and introns. To facilitate this recognition, conserved sequences have evolved that mark the boundaries on both the exon and the intron sides (Figure 10.4). Consensus sequences have been derived for the boundary sequences.

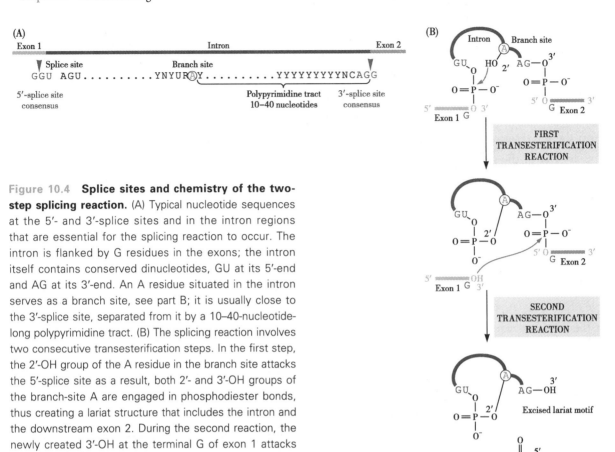

Figure 10.4 **Splice sites and chemistry of the two-step splicing reaction.** (A) Typical nucleotide sequences at the 5'- and 3'-splice sites and in the intron regions that are essential for the splicing reaction to occur. The intron is flanked by G residues in the exons; the intron itself contains conserved dinucleotides, GU at its 5'-end and AG at its 3'-end. An A residue situated in the intron serves as a branch site, see part B; it is usually close to the 3'-splice site, separated from it by a 10–40-nucleotide-long polypyrimidine tract. (B) The splicing reaction involves two consecutive transesterification steps. In the first step, the 2'-OH group of the A residue in the branch site attacks the 5'-splice site as a result, both 2'- and 3'-OH groups of the branch-site A are engaged in phosphodiester bonds, thus creating a lariat structure that includes the intron and the downstream exon 2. During the second reaction, the newly created 3'-OH at the terminal G of exon 1 attacks the 3'-splice site; as a result, the two exons are linked together and the intron is released as a shortened lariat, which is then destroyed.

However, these sequences are very short and do not bind the spliceosome very tightly, which can allow alternative splicing to occur. Spliceosome binding is enhanced by the existence of the branch site, which consists of an A residue embedded in a somewhat conserved internal intron sequence. The branch site is located relatively close to the 3'-splice site, separated from it by a polypyrimidine tract of 10–40 nucleotides that serves as a binding site for some of the spliceosome proteins. The splicing reaction occurs through two consecutive transesterification reactions, as detailed in Figure 10.4 and Figure 10.5.

10.3.2 Splicing is carried out by spliceosomes

The spliceosome is a huge macromolecular machine that contains five small nuclear RNAs or snRNAs, known as U1, U2, U4, U5 and U6, complexed with 41 different proteins that constitute ~45% of the mass of the particle. The U snRNAs have been named for uracil, the base prominent in all five members of the family. There is extensive intrachain base-pairing in all U snRNAs (Figure 10.6A), and many of the bases are post-transcriptionally modified. U snRNAs are very abundant, present at around 100 000 copies per nucleus. This number is not surprising considering the level of splicing that almost every precursor mRNA undergoes and the large average number of introns in each pre-mRNA. A curious, and still unexplained, fact is that four of the five U snRNAs are transcribed by Pol II; the exception is U6, which is synthesized by Pol III. Overall, more than 300 proteins have been recognized that bind to the spliceosome; some of those proteins act as splicing factors, whereas the function of most of these proteins is yet to be identified.

Most studies on splicing have been performed in *in vitro* systems, usually using simple synthetic mRNA containing a single intron and two flanking

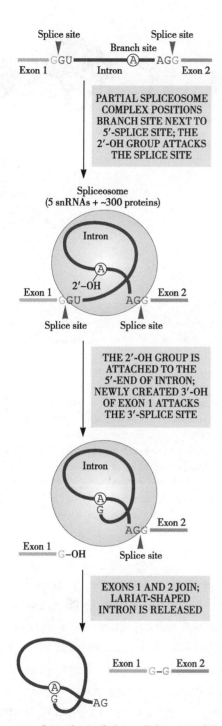

Figure 10.5 **Overview of the splicing reaction.** Intron removal in mRNA precursors is performed by a specialized ribonucleoprotein or RNP machine, termed the spliceosome. The schematic also indicates the main chemical reactions, depicted in Figure 10.4B, catalyzed by the spliceosome. (Adapted from Sharp PA [1987] *Science*,235:766–771. With permission from American Association for the Advancement of Science.)

exons as substrate. These investigations led to the view that the spliceosome complex is assembled, during the process of transcription, in a stepwise manner (Figure 10.7). The binding of U2 small nuclear ribonucleoprotein, snRNP, to the branch site is assisted by the protein dimer U2AF, which recognizes and binds to the conserved polypyrimidine tract.

Extensive *in vitro* studies also helped to identify the roles of the individual spliceosome components and led to acquisition of electron microscopy (EM) images of individual A, B, and C complexes (Figure 10.6B). Nevertheless, experiments in which native spliceosome particles were isolated in a form bound to nascent mRNA during processing produced micrographs that present a somewhat different picture of the structure. Supraspliceosome particles that contain four closely packed monomeric spliceosomes have been recognized by EM (Figure 10.6C). Further cryo-EM imaging and reconstitutions led to a relatively high-resolution structure of these native monomeric spliceosomes. These studies allowed the creation of a model that possibly explains how four splicing events can take place on a long pre-mRNA simultaneously.

10.3.3 Splicing can produce alternative mRNAs

Alternative splicing is a process that uses some splice sites while neglecting others; it can also use cryptic sites that are located internally in exons or introns. Thus, alternative splicing can be viewed as the suppression of optimal splice sites and/or the use of those that are suboptimal or cryptic. The outcome of such selective splice-site usage is the creation of alternative mRNA forms that deviate in sequence and/or structure from the mature mRNA that is produced through conventional splicing. In principle, this should also lead to the generation of alternative variants of protein sequence. As we shall see, it sometimes does. The alternative forms may differ from the conventionally spliced forms in a variety of different ways.

Alternative splicing was discovered in individual gene systems some years ago, but only recently, with

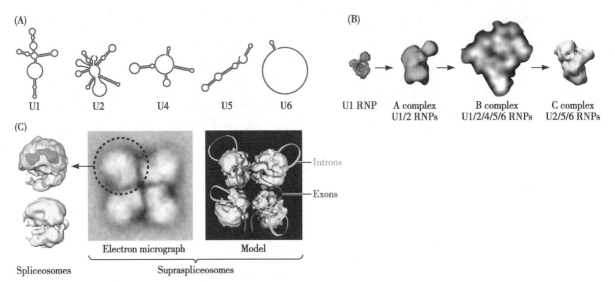

Figure 10.6 **The spliceosome: a structural view.** (A) Secondary structures of the five snRNAs involved in formation of the spliceosome. (B) Reconstructions based on EM images of the structures of individual RNP complexes participating in the stepwise assembly and splicing cycle, as observed *in vitro*. (C) Structure of the native spliceosome, isolated from a nuclear fraction that contained 85% of all nascent Pol II transcripts. The particles associated with the transcripts were large tetrameric structures, as visualized by EM in the middle: these were termed supraspliceosomes. The structure of the monomeric spliceosome, as reconstructed from cryo-EM images, is presented on the left. Two distinct subunits are distinguishable, interconnected with a tunnel running between them; the tunnel is large enough to allow the pre-mRNA to pass through. The dark grey region represents the position of the five spliceosomal snRNAs. On the right is a model of the supraspliceosome. The supraspliceosome presents a platform onto which the exons can be aligned and splice junctions can be checked before splicing occurs. When a pre-mRNA is not yet processed, it is folded and protected within the cavities of the spliceosome. When splicing occurs, the RNA is proposed to unfold and loop out. The existence of the supraspliceosome allows for simultaneous splicing of four exons in the pre-mRNA. Exons are shown in red; introns are shown in blue. An alternative exon is depicted in red in the upper-left corner of the model. (A, adapted, courtesy of Wellcome Trust Sanger Institute. B, from Jurica MS [2008] *Curr Opin Struct Biol*, 18:315–320. With permission from Elsevier. C, from Sperling J, Azubel M, Sperling R [2008] *Structure*, 16:1605–1615. With permission from Elsevier.)

the advent of genomewide methods, have we come to appreciate its pervasiveness. Indeed, more than 90% of human pre-mRNAs undergo alternative splicing events, leading to the existence of families of mature mRNA molecules and, eventually, of closely related protein isoforms. Most studies have addressed the heterogeneity of the mRNA molecules that result from alternative splicing of a single mRNA precursor. Technically, this is achieved by creating cDNA libraries from the mRNA populations in a single cell and then cloning and sequencing the cDNA clones. In few cases, however, have the actual protein isoforms been identified and their structure or function determined. The characterization of protein isoforms remains an enormous challenge because of the lack of appropriate high-throughput methods.

The computational tools used to date suggest that very few of the potential isoforms are functional, since alternative splicing often leads to deleterious changes in protein structure. Thus, the accepted view that alternative splicing creates a large diversity of proteins—of altered or eventually even of different functions, known as neofunctionalization—is based on studies of a limited number of genes and may require a reevaluation. Despite this uncertainty, it is clear that alternative splicing itself is a frequent occurrence that is regulated according to the physiological needs of the cell. According to the most recent data from the ENCODE project in 2012, each human gene produces, on average, 6.3 differently spliced transcripts. The fact that alternative splicing can explain the presence of multiple forms of a protein found in higher organisms may indicate an important role for this phenomenon in evolution.

Alternative splicing can occur through a variety of mechanisms; the major types are outlined in

Figure 10.7 **Stepwise spliceosome assembly during transcription.** As soon as the 5'-splice site exits the transcription complex, U1 snRNP binds to it through base-pairing between GGU and U1 RNA; the resulting complex is known as complex E, for early. In the next step, U2 snRNP binds to the branch site in an ATP-dependent process to form complex A. The complex also contains the protein factor U2AF, not shown here, which helps to recruit the U2 RNP. When the 3'-splice site emerges from the transcription complex, U4, U5, and U6 bind as a triple RNP complex to form complex B. Following release of U1 and U4 snRNPs and conformational transitions in the U2-U5-U6 complex, the catalytically active complex C is formed. Both of the chemical reactions that are involved in splicing occur on complex C. Finally, the U2-U5-U6 complex dissociates from the lariat structure and the individual RNP complexes can be reused; the lariat structure is destroyed.

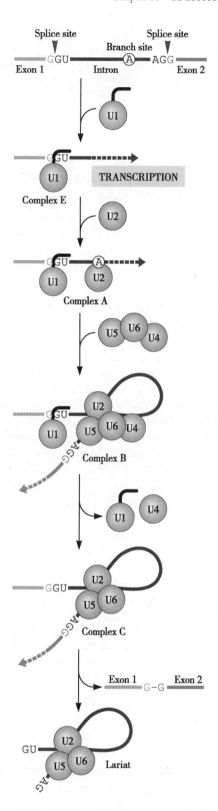

Figure 10.8. In addition to these four mechanisms, alternative mRNAs can arise through the use of alternative promoters or alternative polyadenylation sites (Figure 10.9). Examples of specific mechanisms as they occur in specific systems are presented in Box 10.1.

There is a well-established connection between alternative splicing and certain types of cancer (Table 10.1).

10.3.4 Tandem chimerism links exons from separate genes

The mechanisms of alternative splicing described and illustrated thus far concern the classical case in which one pre-mRNA, the product of transcription of one gene, gives rise to alternative mRNA products. Recently, researchers have come to realize that alternative splicing may involve exons from two or more neighboring genes, or from annotated genes and 5'-exons from previously unannotated genes that lie very far away. The 5'-ends of 399 genes were mapped from the region of the human genome selected in the initial phase of the ENCODE project. This analysis revealed that many genes use alternative 5'-ends that lie tens or hundreds of kilobase pairs away from the annotated 5'-ends. Often, other genes are located between the annotated genes and these distal, previously unannotated 5'-ends. As a consequence of long-range transcription, multiple exons from separate protein-coding genes can be spliced together, creating intergenic splicing products. The creation of these kinds of intergenic products has been termed tandem chimerism.

Two examples, one involving two neighboring genes and the other affecting widely separated exons,

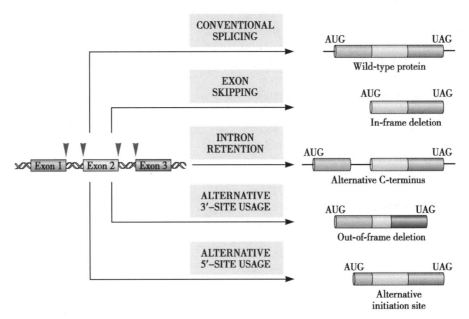

Figure 10.8 **Alternative splicing can occur through a variety of mechanisms that lead to different protein products.** Exon skipping, also known as the exon cassette mode, is the most common known alternative splicing mechanism in which one or more exons are excluded from the final gene transcript, leading to shortened mRNA variants. In intron retention, also known as the intron retaining mode, an intron is retained in the final transcript. Intron retention can lead, depending on the length of the intron, to a frameshift and the creation of an altered amino acid sequence downstream of the intron. In humans, ~15% of a set of 20 687 known genes have been reported to retain at least one intron. Finally, the utilization of alternative 3'- and/or 5'-splice sites can lead to the creation of a great variety of proteins. In this splicing mechanism, two or more alternative 5'-splice sites compete for joining to two or more alternative 3'-splice sites.

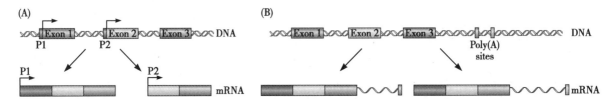

Figure 10.9 **Alternative selection of promoters and cleavage-polyadenylation sites.** (A) The production of isoforms of human p53 is an example of the use of three alternative promoters: the resultant protein isoforms are either full-length or truncated at their N-termini. (B) Some genes, such as that encoding tropomyosin, have two alternative polyadenylation sites, which can be used to produce mRNAs that differ in their 3'-termini.

Box 10.1 **A closer look: examples of mechanisms and outcomes of alternative splicing**

The high incidence of alternative splicing produces numerous alternative mRNA molecules and possibly some functional protein isoforms. Here we present some well-studied examples that illustrate the complexity of the processes in multicellular eukaryotes. Figure 1 depicts a case where the utilization of alternative 5'- and 3'-splice sites bordering constitutive and alternative exons creates more than 500 different mRNA molecules from a single primary transcript. Figure 2 illustrates the diversity and tissue specificity of protein isoforms produced via alternative splicing mechanisms.

In addition to the common form of alternative splicing, exon skipping, the production of some isoforms involves the utilization of alternative polyadenylation sites (top and bottom variants). (Adapted from Breitbart RE, Andreadis A, Nadal-Ginard B [1987] *Annu Rev Biochem*, 56:467–495. With permission from Annual Reviews.)

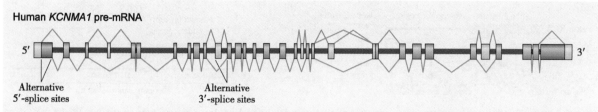

Figure 1 Alternative splicing through the use of alternative 5'- and 3'-splice sites. The human *KCNMA1* pre-mRNA provides the example shown. Blue boxes represent constitutive exons; other colors represent alternative exons. Possible splicing patterns are indicated as blue connecting lines. The alternative splicing mechanism makes use of multiple alternative 5'- and 3'-splice sites as marked. Alternative splicing creates >500 mRNA isoforms from a single pre-mRNA molecule. (Adapted from Nilsen TW, Graveley BR [2010] *Nature*,463:457–463. With permission from Macmillan Publishers, Ltd.)

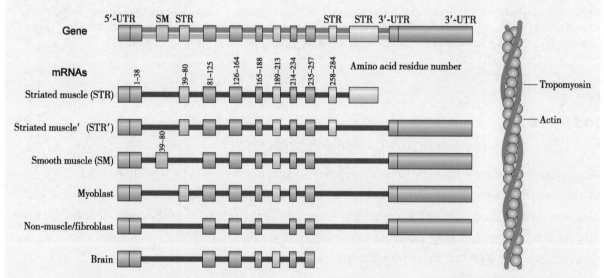

Figure 2 Alternative splicing of the tropomyosin gene. Tropomyosin is an actin-binding protein that regulates the binding of myosin to actin; this interaction is important in muscle contraction and other actomyosin functions. Distinct tropomyosin variants are expressed in different tissues, with some exons selectively present or absent in some tissue-specific isoforms.

Table 10.1 Alternative splicing and cancer.

Disease	Gene	Mutated sequence	Consequences of mutation on alternative splicing
hepatocellular carcinoma	CDH17, LI cadherin	intron 6 A35G exon 6 codon 651	exon 7 skipping exon 7 skipping
prostate cancer	KLF6, tumor suppressor	intron 1 G27A, IVSDA allele	generation of ISE[a], binding site for SRp40; increased splice variant production; novel splice variants functionally antagonize wtKLF6's[a] growth suppression properties
breast and ovarian cancer	BRCA1, tumor suppressor	exon 18 G5199T or E1694X multiple other mutations	disruption of ESE[a] leading to exon 18 skipping effects on splicing enhancers and silencers

[a]ISE, intronic splicing enhancer; ESE, exonic splicing enhancer; wt, wild type.

are presented in Figure 10.10. Thus, exons that have been considered until recently as discrete modules of a specific gene should now be viewed as more general functional modules that can be joined together in multiple RNA molecules. About 65% of the genes tested in ENCODE are involved in the formation of chimeric RNAs. These may, in turn, lead to the formation of chimeric proteins. Indeed, much of the modular domain structure we see in proteins may have had its evolutionary origin in such splicing.

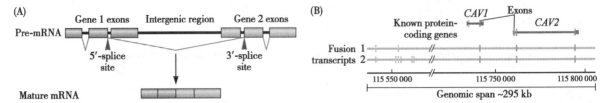

Figure 10.10 Tandem chimerism, long-range splicing events based on long-range transcription. (A) In this case, the transcribed region spans two consecutive genes and the intergenic region between them. Splicing of the pre-mRNA involves a 5'-splice site in the upstream gene and a 3'-splice site in the downstream gene; thus, the intergenic region is removed from the mature mRNA. (B) Two different fusion transcripts combine selected exons from the caveolin *CAV1* and *CAV2* genes with novel, unannotated 5'-exons that lie far from the two known genes. The creation of such fusion transcripts is very common in both human and *Drosophila* genomes. In the pilot phase of the ENCODE project, 65% of the genes tested are involved in the formation of such chimeric RNAs. Exons are shown as vertical bars and introns as horizontal lines; slanted lines indicate a gap of ~200 kbp. (A, adapted from Akiva P, Toporik A, Edelheit S, et al. [2006] *Genome Res*, 16:30–36. With permission from Cold Spring Harbor Laboratory Press. B, adapted from Kapranov P, Willingham AT, Gingeras TR [2007] *Nat Rev Genet*, 8:413–423. With permission from Macmillan Publishers, Ltd.)

10.3.5 Trans-splicing combines exons residing in the two complementary DNA strands

Most frequently, chimeric RNAs originate from genes on the same chromosome that are transcribed in the same direction. In these cases, the underlying molecular mechanism involves long-range transcription that creates a very long transcript containing these exons. Then alternative splicing comes into play to create new combinations of exons. However, alternative splicing does not rely on this mechanism alone. We now know of cases where the spliced exons originate from the two complementary DNA strands of a single gene. Some exons in the final mRNA product come from the gene transcribed in the sense direction, whereas other exons originate from antisense transcripts of the same gene. This phenomenon is known as *trans*-splicing. The repertoire of known *trans*-splicing events constantly increases: separate and remotely located genes on the same chromosome or even on a different chromosome may contribute some of their exons to a new product. The latter occurrences exclude the mechanism of initial transcription of very long primary transcripts, followed by alternative splicing, as a plausible explanation for the formation of these *trans*-spliced mRNAs. The molecular mechanism of *trans*-splicing remains to be elucidated.

10.3.6 Regulation of splicing and alternative splicing

Splicing is subject to fine-tuned regulation, which occurs through a number of different mechanisms. Since the mechanisms that regulate splicing are also those that participate in alternative splicing decisions, we will treat them together.

(1) Splice sites differ in strength

Because the utilization of different splice sites depends on relatively weak protein-RNA and protein-protein interactions, recognition of specific sites can be easily and subtly modified by control of these interactions. When the sequence of a splice site deviates somewhat from the consensus shown in Figure 10.4A, the site is used with reduced efficiency because these weaker sites have less affinity for their spliceosomal protein partners. The selective utilization of these weaker sites constitutes the main mechanism through which alternative splicing occurs and is regulated.

(2) Exon-intron architecture affects splice-site usage

Unexpectedly, the length of the exons and introns has emerged as a factor in determining splice-site usage. The recognition of splice sites is most efficient when introns and exons are small. Genomewide, this is an important regulatory feature, especially in view of the broad distribution of intron lengths in the human genome. In the intron-definition model, assembly of the spliceosome occurs at the 5′- and 3′-splice sites of a given intron. The contrasting exon-definition model posits that the initial recognition of splice sites occurs across an exon, at splice sites flanking the exon. The two models differ in the possible orientations of interactions between the U1 and U2 small nuclear ribonucleoproteins, or snRNPs, across the intron or across the exon. In the latter case, U1 bound to the 5′-splice site of the downstream intron interacts with U2 bound to the 3′-splice site of the upstream intron. The exon-definition model was recognized when scientists moved from *in vitro* investigations of simple artificial splicing substrates, exon-intron-exon, to multi-intron substrates.

Figure 10.11 illustrates the notion behind intron and exon definition. Kinetic experiments demonstrated that splice-site preference across introns is lost when intron sites are larger than 200–250 bp; beyond this length, splice sites are recognized across exons. Intron definition is much more efficient than exon definition, and so intron size profoundly influences the probability of exon inclusion or skipping in the final mRNA generated during alternative splicing of exons with weak splice sites. Finally, both experimental and computational approaches showed that the length of the upstream intron is more important in alternative splicing than the length of the downstream intron. Thus, exon-intron architecture defines the very mechanism of splice-site recognition and affects the frequency of alternative splicing events.

(3) *Cis-trans* interactions may stimulate or inhibit splicing

In addition to the sequences of the splice sites, other sequences can affect the efficiency of splicing by forming *cis-trans* interactions. These sequences, found in both introns and exons, can exert either a stimulatory or an inhibitory effect on splicing. Depending on their location and effect, they are termed exonic splicing enhancers, ESE; exonic splicing silencers, ESS; intronic splicing enhancers, ISE; or intronic splicing silencers, ISS (Figure 10.12). The enhancers and silencers are relatively short conserved sequences of ~10 bp, found in isolation or in clusters, that affect the use of weak splice sites. These regulatory elements are operationally identified if

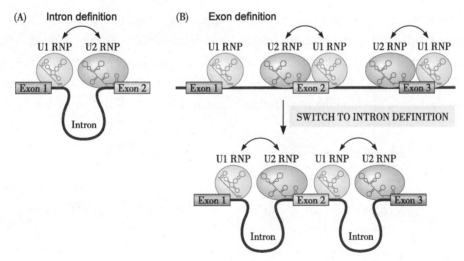

Figure 10.11 **Intron and exon definition in spliceosome assembly.** (A) Intron definition: in a pre-mRNA molecule containing one intron and two exons, interaction between U1 RNP and U2 RNP, as well as auxiliary factors, occurs across the intron. (B) Exon definition: in a pre-mRNA containing multiple exons and introns, the initial interactions between factors bound at the splice sites occur across the exon. The two transesterification reactions of splicing occur across an intron, that is, in an intron-defined complex. Hence, once an exon-defined complex is formed, it needs to switch to an intron-defined complex for the reactions to take place. At least one factor that is involved in the switch has been identified: the polypyrimidine-tract-binding protein PTB. (Adapted from Schellenberg MJ, Ritchie DB, MacMillan AM [2008] *Trends Biochem Sci*,33:243–246. With permission from Elsevier.)

mutations in their sequences lead to enhancement of splicing, for silencers, or to inhibition of splicing, for enhancers. As in all cases of *cis-trans* regulation, the cis elements are recognized by and interact with proteins encoded somewhere else in the genome, known as *trans*-factors.

There are two general categories of regulatory proteins that affect splicing. The first class comprises Ser-Arg or SR proteins, which are usually activators of splicing, although under certain circumstances they can also serve as repressors. Members of this protein class share similar organization: they possess one or two RNA-binding domains at their N-terminus and a variable-length domain at the C-terminus that contains repeats of RS or Arg-Ser dipeptides. RS domains serve the activation function and are extensively phosphorylated. Activation of splicing occurs by enhancing the recruitment of spliceosome components to the 5'- and 3'-splice sites (Figure 10.13A).

In addition, SR proteins participate in numerous other aspects of mRNA metabolism, such as nonsense-mediated mRNA decay, nuclear export, and translation. The second class of *trans*-regulators acts through binding to silencers; usually they inhibit splicing, although they can sometimes act as activators too. These proteins are members of a large, structurally diverse group of RNA-binding proteins, usually complexed with small RNA molecules, hence their name heterogeneous nuclear ribonucleoproteins or hnRNPs. hnRNPs may act through a number of mechanisms, including blocking the recruitment of spliceosome snRNPs, looping out exons, or multimerization along exons; the first two mechanisms are illustrated in Figure 10.13B.

Cis-trans interactions also constitute a major part of alternative splicing regulation, since the concentrations of splicing activators or repressors, as well as those of spliceosomal components, can be regulated in physiologically meaningful ways and can simultaneously modify splicing at many loci. Thus, these concentrations may differ between different terminally differentiated cell types, may change during differentiation and development programs, or may fluctuate during the cell cycle. As a result of these changes, exon inclusion or exclusion may be differentially regulated to reflect the needs of the cell.

(4) RNA secondary structure can regulate alternative splicing

We often depict the structure of mRNA or its precursors as a straight or a wavy line. This is, of course, an oversimplification, as we now know that significant portions of any pre-mRNA and its mature counterpart are double-stranded helices, forming the stems of hairpin structures of various lengths. The kinetic stability of such secondary structures will determine their half-lives. Structures that persist may modulate splice-site recognition and usage. For example, local secondary structures may interfere with splicing if they conceal splice sites or enhancer-binding sites from their binding partners. The opposite effect will occur if local RNA structures mask splicing repres-

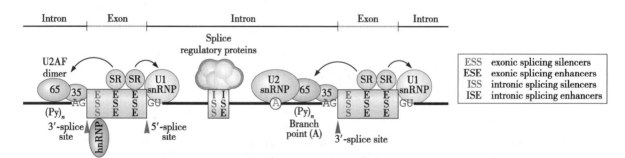

Figure 10.12 **Overview of *cis-trans* regulation of splicing.** The removal of introns is guided by specific protein factors that interact with specific nucleotide sequences in *cis-trans* interactions. Such sequences are present in both exons, shown in green, and introns, shown in black. Splice regulatory proteins include SR or Ser-Arg proteins, which act by binding exonic splicing enhancers, ESE, and heterogeneous nuclear ribonucleoproteins or hnRNPs, which bind to the exonic, ESS, or intronic, ISS, splicing silencers to repress splicing. (Adapted from Schwerk C, Schulze-Osthoff K [2005] *Mol Cell*,19:1–13. With permission from Elsevier.)

sors. Thus, the existence and stability of secondary RNA structures adds another level of complexity to the already extremely complex regulation of splicing.

(5) Sometimes alternative splicing regulation needs no auxiliary regulators

A recently discovered regulatory mechanism does not involve auxiliary splicing factors of the SR or hnRNP classes; sequences close to splice sites can affect splicing by changing the configuration of the U1 snRNP that interacts with the 5′-splice site. The *in vitro* experiments that led to the discovery of this kind of regulation are illustrated in Figure 10.14. Thus, mutations in sites proximal to splice sites may lead to different patterns of alternative splicing *in vivo*.

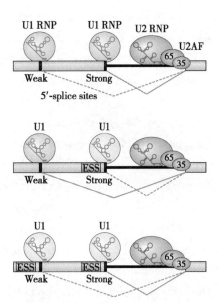

Figure 10.14 **Regulation of alternative splicing without regulators.** Experiments performed in T. Nilsen's laboratory uncovered a new type of regulation of alternative splicing that occurs without the involvement of auxiliary splicing factors of the SR or hnRNP classes. The authors designed a synthetic splicing substrate containing two competing 5′-splice sites, one weak and one strong, and a single 3′-splice site. *In vitro*, the strong downstream 5′-splice site was predominantly used in the absence of a splicing silencer, despite the fact that U1 snRNP binds to both 5′-splice sites. A splicing silencer, ESS, was created by inserting randomized sequences and selecting for predominant use of the weak site. Importantly, the newly created ESS does not prevent U1 snRNP from binding to the strong 5′-splice site, though the configuration in which U1 snRNP interacts is altered, depicted here as a change in color for the U1 snRNP particles. When the ESS is present near both 5′-splice sites, it alters the association of U1 snRNP with both 5′-splice sites in the same way. As a result, efficient splicing to the strong downstream 5′-splice site is restored. These observations suggested that the newly created silencers do not impair the ability of U1 snRNP to recognize and bind to the 5′-splice sites but rather alter the efficiency with which the U1 snRNP-5′-splice site complex engages in splicing; the efficiency is altered because of subtle changes in the way U1 snRNP interacts with the splice site. Thus, the sequences flanking the 5′-splice site themselves regulate alternative splicing events without the need for auxiliary protein factors. (Adapted from Graveley BR [2009] *Nat Struct Mol Biol*,16:13–15. With permission from Macmillan Publishers, Ltd.)

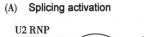
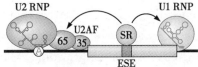
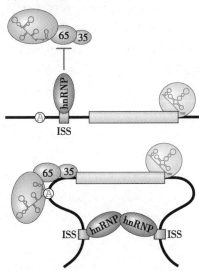

Figure 10.13 **Proposed models for the most common mechanisms of traditional splicing regulation.** (A) SR or Ser-Arg proteins bind to exon enhancers, ESEs, to stimulate binding of U1 snRNP to the 5′-splice site and binding of the U2 auxiliary factor U2AF, a heterodimer of 65 and 35 kDa proteins, to the polypyrimidine tract and the conserved AG at the 3′-splice site. U2AF in turn guides the U2 snRNP to the branch-point A nucleotide. (B) Two alternative, non-mutually-exclusive models for inhibition of splicing by hnRNPs, which bind to silencer elements occurring in both introns and exons. hnRNP binding can interfere with the binding of U2AF to the 3′-splice site, as shown in the top drawing. Alternatively, hnRNPs can bind to ISSs in the introns flanking an exon and then interact with each other, which leads to looping out of the intervening exon. This exon is excluded from the final mature mRNA. (Adapted from Graveley BR [2009] *Nat Struct Mol Biol*,16:13–15. With permission from Macmillan Publishers, Ltd.)

(6) The rate of transcription and chromatin structure may help regulate splicing

It has been known for years that splicing is a co-transcriptional event. The fact that these two processes occur at the same time does not necessarily mean that they are mechanistically connected, but the coincidence of the two processes at least sets the stage for mechanistic coupling. Indeed, such a coupling has been indicated in many experimental studies, and it has numerous possible connections to chromatin structure.

Two models exist to account for the coupling of transcription and splicing. In one of these models, the C-terminal domain or CTD of the advancing polymerase serves as a landing pad to recruit various splicing factors (Figure 10.15). Hence, this model came to be known as the recruitment model. In fact, recruitment of splicing factors may also occur through direct interactions with chromatin components. The second model of coupling is known as the kinetic model. It states that the rate of movement of the polymerase along the gene is important for recognition of weak splice sites: the more slowly the polymerase progresses, or if it pauses, as we know it often does, the greater the opportunity for weak splice sites to be recognized and used by the spliceosome.

Since the rate of transcription is dependent on the chromatin structure of the underlying template, the coupling between transcription and splicing may also involve chromatin organization. In the past decade, numerous scattered studies have supported this idea. Both bona fide ATP-dependent chromatin remodelers and histone modification enzymes have been shown to affect splicing regulation. Recent advances in bioinformatics have allowed comparisons between different genomewide sets of experimental data obtained by independent methods. One data set identified sequences that are organized as nucleosomes. The other data set was obtained by chromatin immunoprecipitation using antibodies against various histone modifications, followed by sequencing of the immunoprecipitated DNA.

Importantly, it was established that there is an enrichment, ~1.5-fold, of nucleosome occupancy in exons as compared to introns. On the surface, this level of enrichment may seem small, and thus probably insignificant, but simple calculations indicate that it could be important. For example, a stretch of DNA 4800 bp in length would accommodate 30 nucleosomes of repeat length ~160 bp, which is the shortest nucleosome repeat length described *in vivo*. If the number of nucleosomes on the same stretch of DNA is reduced 1.5-fold, to 20, then the nucleosome repeat length would jump to 240. It may be just a

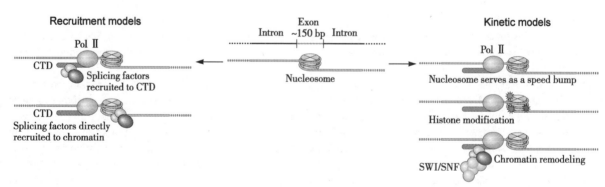

Figure 10.15 **Multiple connections between splicing, transcription, and chromatin structure.** The center schematic illustrates increased nucleosome occupancy at exons. Note that because of the relatively short sequences in exons, most frequently a single nucleosome organizes an entire exon. (Left) The recruitment models state that Pol II, via its C-terminal domain or CTD, helps to bring the splicing machinery to the mRNA; alternatively, recruitment may occur through direct binding to chromatin components. (Right) Possible scenarios of kinetic models. The mere presence of a nucleosome can slow down the progression of the polymerase, serving as a speed bump; histone modifications can affect the rate of polymerase movement; and finally, remodeling of chromatin structure by remodelers such as the SWI/SNF complex can also affect the rate of transcription.

coincidence that this nucleosome repeat length is the largest found in nature. In any case, the difference between the two cases would be in linker length of 80 bp, which would have profound effects on both the structure of the chromatin fiber and the accessibility of the resident DNA to protein binding. It may be important that introns are not densely populated by nucleosomes; according to current ideas about the participation of introns in gene regulation, a looser chromatin structure may help this function. It is also intriguing that the average exon length in metazoans is ~140–150 bp, exactly the length of DNA organized in a nucleosome. So the majority of exons contain a single nucleosome, which may carry specific histone modifications. Indeed, there are differences in the histone modification patterns of exonic and intronic nucleosomes, even after nucleosome occupancy is accounted for.

10.3.7 Self-splicing: introns and ribozymes

(1) A fraction of introns is excised by self-splicing RNA

Most pre-mRNAs use spliceosomes for splicing, but some employ self-splicing mechanisms catalyzed by the very intron that needs to be excised. RNase P and self-splicing introns were the first ribozymes to be discovered in the early 1980s. Ribozymes are so termed because they possess enzymatic activity in the RNA molecule itself. They are often bound to proteins that serve to stabilize the complex folded RNA structures that are needed for catalysis. The initial small group of known ribozymes, which were considered exceptional in structure and biochemical activities, quickly grew to include other molecules, such as hammerhead, hairpin, and hepatitis delta virus ribozymes. Moreover, the enzymatic activity that creates the peptide bond during protein synthesis also turned out to be a ribozyme.

Self-splicing introns occur in many organisms, but most are not essential for viability, unlike RNase P. They may be viewed as selfish genetic elements that found a way to propagate themselves by persisting in the genome as normal components of intron-containing genes but splicing themselves out at the RNA level, so that they do not bring about the destruction of their hosts. This view may be changing, however, as most of the DNA sequences that were previously considered junk may actually participate in gene regulation.

(2) There are two classes of self-splicing introns

There are two major classes of self-splicing introns: group I and group II. The two groups differ mainly in their co-factor requirements: group I introns use a guanosine molecule as a co-factor, whereas group II generally use an internal adenosine, similar to the A in the branch site that is involved in spliceosome-catalyzed splicing. The second distinguishing feature is their quite different structures.

Group I introns require an external G cofactor for splicing

Group I introns are very abundant, with more than 2000 members described, mainly in bacteria and lower eukaryotes; they are rare in animals, and to date none have been found in archaea. They catalyze their own excision from mRNA, tRNA, and rRNA precursors. Their structure is formed by a specific arrangement of nine double-helical elements, called paired regions or P, capped by loops and connected by junctions; a tenth helix is formed during the reaction itself (Figure 10.16 and Figure 10.17). The catalytic G-binding site is located in helix P7. Some group I introns require a protein for activity; well-studied examples come from *Neurospora crassa* and *Saccharomyces cerevisiae*. The proteins do not, however, directly participate in catalysis; they are involved in stabilizing the catalytic core by reinforcing long-range interactions between individual P elements.

Group II introns require an internal bulged A for splicing

Group II introns are phylogenetically unrelated to group I introns; they are mainly found in the mRNA, tRNA, and rRNA of organelles in fungi, protists, and plants and in the mRNA of bacteria. Group II introns possess an unusually diverse repertoire of chemical activities, including catalysis of

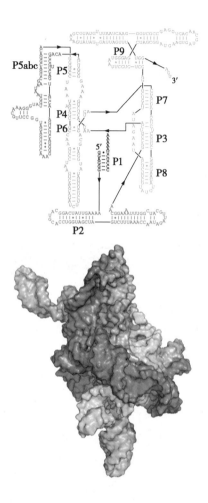

Figure 10.16 **Representative secondary and tertiary structures of a group I intron.** This intron is from the ciliate *Tetrahymena thermophila*. The color scheme in the secondary structure corresponds to that used in the three-dimensional structure. The intron is highly structured, containing nine double-stranded helices or paired regions, numbered P1–P9. The structure is highly evolutionarily conserved: the crystallographically resolved structures from *Tetrahymena*, the purple bacterium *Azoarcus*, and the *Staphylococcus aureus* phage Twort are almost superimposable. This conservation of the structure is impressive when one bears in mind the very poor sequence conservation, apart from a few crucial nucleotides located at the active or G site. The intron misfolds substantially *in vitro* and needs chaperone proteins to help proper folding *in vivo*. (Adapted from Jarmoskaite I, Russell R [2011] *Wiley Interdiscip Rev: RNA*, 2:135–152. With permission from John Wiley & Sons, Inc.)

2′–5′ phosphodiester bond formation and the ability to reinsert themselves into DNA with the help of intron-encoded proteins. The latter process is known as retrotransposition and is discussed in detail in Chapter 14. Structurally, group II introns possess six helical domains (Figure 10.18).

The most conserved domain, DV, consists of a 30–34-nucleotide stem-loop structure and contains a highly conserved catalytic triad of nucleotides, located exactly five base pairs away from the two-nucleotide bulge in the helix. This structural arrangement is strikingly similar to that of U6 spliceosomal RNA. This similarity, together with other similarities that occur during the self-splicing process, such as the use of an internal A, rather than an external cofactor as in the case of group I introns, and the formation of a lariat structure, led to the hypothesis that group II introns and the spliceosome machinery may be evolutionarily connected. More in-depth analysis is needed to prove or disprove this hypothesis. Finally, we note that group II introns are not, in fact, true catalysts, as the entire intron is degraded once self-splicing occurs.

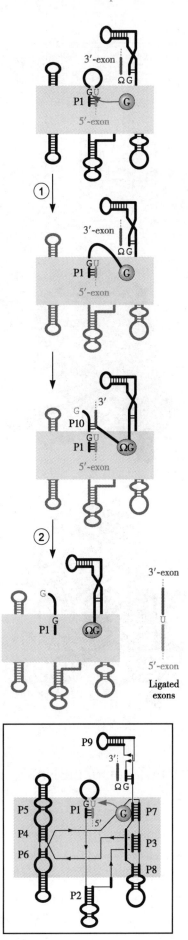

Figure 10.17 Mechanism of group I intron self-splicing. The example shown is the *Azoarcus* intron. The shaded box delimits the catalytic core of a group I intron. This is shown in more detail in the box at the bottom, which depicts the conserved secondary structure, with the nine helices forming the catalytic core and the branching peripheral elements. Introns are shown in black. Only the peripheral elements that branch out of the catalytic core are shown for simplicity. The 5′-splice site contains a conserved GU pair, where G is in the intron and U is in the exon; a conserved guanine, termed ΩG, is at the 3′-terminal position of the intron. The splicing reaction occurs in two catalytic steps: (Step 1) The 3′-OH group of a guanosine co-factor binds to the intron at the G-binding site, in P7, and attacks the 5′-splice site. After the reaction occurs, the G is covalently linked to the 5′-end of the intron. (Step 2) A conformational change occurs: another helix, P10, is formed, which involves base pairing between the intron and the 3′-exon. Recognition of the 3′-splice site is achieved, in part, by ΩG, which displaces the guanosine from the G-binding site. Then the 3′-OH group of ΩG attacks the 3′-splice site. The two exons are ligated and the intron is released. (Adapted from Vicens Q, Cech TR [2006] *Trends Biochem Sci*, 31:41–51. With permission from Elsevier.)

294 Chapter 10 RNA Processing

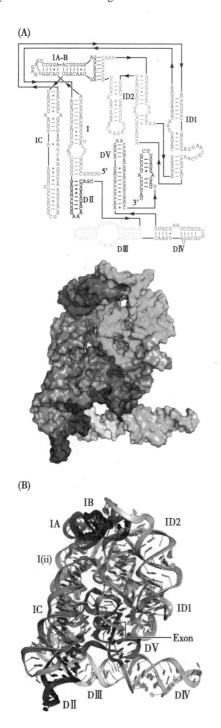

Figure 10.18 **Secondary and tertiary structures of a group II intron.** The example shown is from the halophilic and alkaliphilic eubacterium *Oceanobacillus iheyensis*. This bacterium was isolated from seabed mud at a depth of 1 km off the coast of Japan. A crucial step in getting the intron in its native, catalytically active form was to isolate it immediately after it had gone through both steps of splicing, thus ensuring that it is properly folded. This procedure is a departure from the routine purification of RNAs for structural analysis, which involves denaturing polyacrylamide gel electrophoresis followed by renaturation steps. (A) The color scheme in the secondary structure corresponds to that used in the three-dimensional (3D) structure. The conserved DV domain involved in catalysis is depicted in red. (B) The same structure is presented as a ribbon diagram to show the RNA helices within DV, shown in red, forming the catalytic core and the bound exon, shown in purple, with the rest of the domains encapsulating this active site. Note that different colors are used to indicate domain I A-B in the two 3D structures. (A, adapted from Jarmoskaite I, Russell R [2011] *Wiley Interdiscip Rev: RNA*,2:135–152. With permission from John Wiley & Sons, Inc. B, from Toor N, Keating KS, Pyle AM [2009] *Curr Opin Struct Biol*,19:260–266. With permission from Elsevier.)

10.4 Methylation of mRNA: N^6-methyladenosine

Abundant noncoding RNAs such as rRNAs and tRNAs are extensively modified, while mRNA modifications are thought to be relatively low in frequency apart from the common terminal modifications, m^7G cap and poly(A) tail. Recent developments have unveiled that at least some of the modifications, such as methylation, are considerably abundant and widely conserved.

10.4.1 The m^6A methylation and demethylation reaction of mRNA

The most common internal modifications of mRNA include N^6-adenosine methylation (m^6A), N^1-adenosine methylation (m^1A), and cytosine hydroxylation (m^5C). The m^6A methylation is the most abundant internal modification on mRNA, discovered in the 1970s. Because the mutation of specific m^6A sites did not affect mRNA abundance and processing and frequently modified viral m^6A was rarely detected in cellular mRNAs examined, this led to the idea that adenosine methylation may occur in a limited subset of viral and cellular mRNAs. Forty years later, with developments of new sequencing techniques and discoveries from genetic and biochemical studies, the interest in m^6A has been renewed recently and a new understanding of it have been updated.

Methyltransferase hetero complexes, containing METTL3, METTL14, WTAP, KIAA1492 and other factors, catalyze m⁶A methylation of mRNA *in vivo* and *in vitro*. These Writers have a consensus motif. METTL3 and METTL4 have a SAM-binding domain required for m⁶A methylation, whereas WTAP contains no characteristic domain. The demethylases, eraser proteins, FTO (Fat mass and obesity-associated protein) and ALKBH5, catalyze oxidative demethylation of m⁶A of mRNA. FTO and ALKBH5 demethylases have an AlkB domain in common, mediating m⁶A demethylation modification. Compared to ALKBH5, FTO has an additional C-terminal domain. YTHDF family members, generally containing a YTH RNA binding domain, are effector proteins, which Identify methylation modification information and participate in downstream translation, degradation and other processes of mRNA. As shown in Figure 10.19, YTHDF2 participates in the mRNA degradation process. (Figure 10.19, Table 10.2).

10.4.2 The biological functions of mRNA m⁶A methylation

Like the modifications of DNA and protein, RNAs undergo chemical modifications that can affect their activity, localization, and stability. As early as the 1960s, Cohn et al. had found a large number of base site modifications on RNA in addition to the traditional four base types of ACGU. Holley et al. first identified more than a dozen different RNA modifications, including pseudouridine, in yeast tRNA in 1965. It is known that in most eukaryotes, mRNA is methylated at 5′Cap, and its functions include maintaining mRNA stability, mRNA precursor shearing, polyadenylation, mRNA transport and translation initiation, etc. The modification of 3′polyA contributes to nuclear export, initiation of translation, and maintenance of structural stability of mRNA with polyA binding proteins. M⁶A methylation may affect the export and splicing of mRNAs in the nucleus. M⁶A can accelerate the processing time of mRNA precursors, and accelerate the transport speed and the speed of nuclear export of mRNA in cells (Figure 10.20).

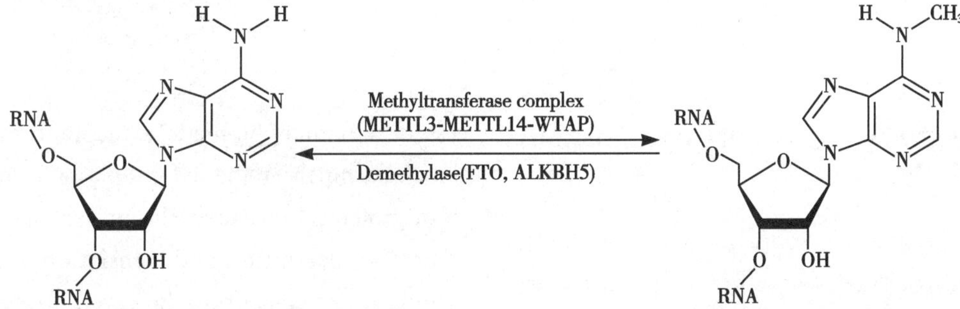

Figure 10.19 N⁶-methyladenosine methylation and demethylation reactions of mRNA. Methyltransferase complex containing METTL3, METTL14, and WTAP catalyzes m⁶A methylation of mRNA, whereas FTO and ALKBH5, the demethylases, catalyze oxidative demethylation of m⁶A of mRNA.

Table 10.2 The functions of writers, erasers and readers in mRNA m⁶A methylation.

Category	Gene	Function
writers	METTL3, METTL14, WTAP, KIAA1492	METTL3 and METTL14 form complexes that catalyze m⁶A methylation of RNA *in vivo* and *in vitro*. WTAP and KIAA1492 and other factors are also important components of the complex. These Writers have a consensus motif.
erasers	FTO, ALKBH5 and other homology	Demethylase mediating m⁶A demethylation modification
readers	YTHDF1, YTHDF2, YTHDF3 and other homology	Identify methylation modification information and participate in downstream translation, degradation and other processes of RNA. These and the proteins of YTHDF family generally contain the YTH domain.

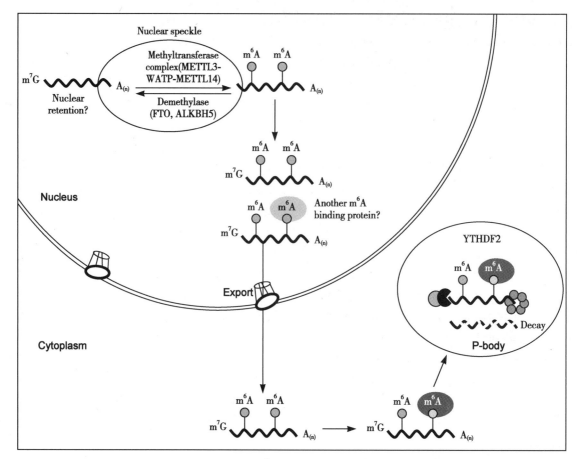

Figure 10.20 **The biological functions of mRNA m⁶A methylation.** Reversible methylation/demethylation occurs in nuclear speckles where the related enzymes are concentrated. Methylation may affect the export and splicing of mRNAs in the nucleus. Exported methylated mRNAs are recognized by YTHDF2 in the cytoplasm and then localize to P bodies to decay.

10.5 Overview: the history of an mRNA molecule

10.5.1 Proceeding from the primary transcript to a functioning mRNA requires a number of steps

We have described many ways in which mRNA molecules are processed; it is useful here to stand back and look at the overall process. The history of a typical pre-mRNA from synthesis to export to the cytoplasm is presented in Figure 10.21. It is clear that the cell has evolved a set of highly complex mRNA processing events that are intimately connected to transcription. Practically all individual steps are subject to regulation, the result being the production of the right type of mRNA in the right amount, at the right place, at the right time.

After mRNA molecules have been properly processed, they must be made available to the translational machinery in the cell cytoplasm for the synthesis of proteins. In bacteria, this presents no problems, as both transcription and translation occur in the cytoplasm. In eukaryotes, however, mRNA that has been transcribed and processed in the nucleus must be transported to the cytoplasm through the membranes of the nuclear envelope. There exists a whole molecular machine to accomplish and regulate this transport.

Finally, all mRNA molecules must eventually be degraded. There are two quite different reasons for this. First, even completed and processed RNA molecules may contain errors in sequence, processing, or packaging that would impede proper cell function. These must be removed. Second, it is not desirable for the cell to maintain mRNA molecules active in protein synthesis continually. Proper control of cellular function requires the shutting off of protein synthesis that is no longer needed. Recall that the clue that led Jacob and Monod to postulate the existence of mRNA was the evidence of a short-lived intermediate in gene expression. Many small RNA molecules that play regulatory roles are also short-lived. The only long-lived RNAs in most cells are those involved in the mechanisms of protein synthesis, such as rRNA and tRNA. Therefore, there must be a selective mechanism for the degradation of even competent mRNAs in the cytoplasm.

10.5.2 mRNA is exported from the nucleus to the cytoplasm through nuclear pore complexes

Both export and import of macromolecules or their complexes to and from the nucleus occur through the nuclear pore complexes or NPCs. Despite unique features characteristic of specific trafficking pathways, all these processes share common features and mechanisms. All cargo passes through the channel of the nuclear pore with the help of soluble carrier proteins in a three-step process: (a) generation of a cargo-carrier complex in the donor compartment of the cell (b) passage of the complex through the NPC, and (c) release of the cargo in the target compartment, followed by recycling of the carrier back to the donor compartment. The passage itself is a Brownian or random-walk process facilitated by the FG nucleoporins that line the NPC channel.

The majority of nuclear trafficking pathways that transport proteins or small RNA molecules employ members of the β-karyopherin superfamily of proteins as carriers. mRNA, however, uses the Mex67-Mtr2 heterodimer (Figure 10.22). We shall refer to the RNA-carrier complex as messenger ribonucleoprotein or mRNP. There are also mechanisms that allow recognition of the donor and target cell compartments, thereby ensuring that appropriate assembly of the cargo- carrier complex or release of cargo occurs. When β-karyopherins are used as carriers, compartment recognition is achieved through the

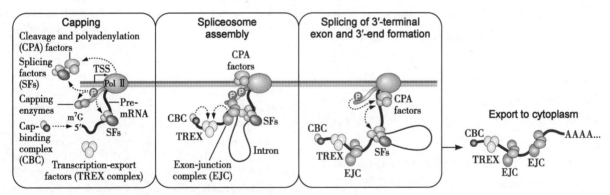

Figure 10.21 Overview of co-transcriptional RNA processing. Pre-mRNA is represented by a thicker line at the exon portions and a thinner line at the introns. The three adjacent boxes represent the composition of protein complexes bound to either the polymerase, mainly through its CTD, or the nascent transcript. The complexes perform specific RNA processing functions during specific stages of polymerase movement along the gene. Dashed arrows denote interactions that stabilize the complexes or perform the respective enzymatic function. (Left box) 5′-End capping. Capping occurs as soon as the 5′-end of the RNA transcript emerges from the RNA polymerase Pol II; the capping enzymes are recruited via Ser5 phosphorylation of the CTD. Once the cap structure is on, the cap-binding complex, CBC, binds to it and recruits the transcription export complex, TREX. Splicing factors, SFs, and some of the CPA or cleavage and polyadenylation components also join the complex at this stage. (Middle box) Spliceosome assembly. Assembly at the first intron is enhanced by protein factors that bind to both the CTD and the nascent RNA, thus bringing the first and second exons into close proximity. The exon-junction complex or EJC is recruited by the splicing machinery and is deposited just upstream of the exon-exon junction. The TREX complex is now stably associated with nascent RNA through interactions with the CBC, SFs, and/or the EJC. (Right box) Splicing of the 3′-terminal exon and formation of the 3′-end of mRNA. These two processes occur when transcription approaches the end of the gene, after the final intron and 3′-end exons have been transcribed. Recruitment of the CPA machinery occurs on the CPA signal. The schematic on the far right shows the proteins bound to the processed mRNA when exported to the cytoplasm. Note that many of the proteins still remain bound and might affect subsequent processes. (Adapted from Pawlicki JM, Steitz JA [2010] *Trends Cell Biol*, 20:52–61. With permission from Elsevier.)

nucleotide state of the Ran GTPase, which is bound to GTP in the nucleus and is maintained in a GDP-bound state in the cytoplasm. In mRNA export, it is believed that the transition from one compartment to the other is associated with extensive remodeling of the mRNP complex by two distinct DEAD-box helicases, nuclear Sub2 and cytoplasmic Dbp5, which are thus the compartment-recognition molecules. Nuclear Sub2 is needed for remodeling of the cargo-carrier complex so that one of the adaptor proteins used to recruit Mex67-Mtr2 to the mRNP is released from the export-competent mRNA. The cytoplasmic DEAD-box helicase Dbp5 remodels the mRNP on the cytoplasmic side of the NPC so that the Mex67-Mtr2 carrier is released for recycling (see Figure 10.22).

10.5.3 RNA sequence can be edited by enzymatic modification even after transcription

In some situations, the sequences of pre-mRNAs are edited, actually changed by insertion, deletion, or chemical modification of residues. The insertion and deletion of residues appear to be restricted to mitochondrial RNA of certain protozoa, such as trypanosomes. In these cases, insertion or deletion of short oligomers of U can occur at specific locations in the sequence. The sequences are dictated by guide RNAs, oligomers complementary to the target RNA but with a mismatched bulge. The guide RNA maintains connection while the mRNA is cleaved and oligo(U) is inserted or deleted. A ligase then reseals the modified message. A quite different kind of editing is observed in some higher organisms, including mammals. In some cases, adenosine can be converted to inosine, or cytidine to uridine (Figure 10.23). The enzymes that catalyze these reactions contain RNA-binding domains that recognize specific sequences a few nucleotides away from the site of modification. Although not common, the amino acid sequence changes effected by such editing may have significant effects. For example, there exists evidence that amyotrophic lateral sclerosis or ALS, also known as Lou Gehrig's disease, may involve a defect in editing of the mRNA encoding a protein involved in calcium conductance in neural membranes. Finally, tRNA

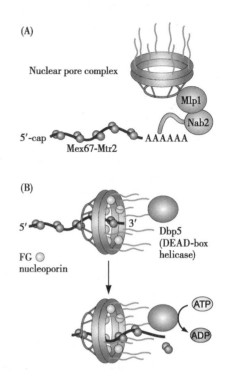

Figure 10.22 Nuclear export of mRNA in yeast. (A) Export-competent mRNA, with some export proteins bound to it. Additional proteins accompany mRNP to the cytoplasm but are not directly involved in the transport, such as the TREX complex that coordinates many of the steps during transcription and processing, exon-junction complexes, cap-binding complex, etc. (see Figure 10.21). The carrier complex Mex67-Mtr2 binds mRNA only weakly and is recruited to the message by proteins serving as adaptors, Yra1 and Nab2. Yra1 is dissociated from mRNA following recruitment of the Mex67-Mtr2 carrier and before export. The dissociation requires the action of a nuclear DEAD-box helicase, Sub2, not shown here, which somehow remodels the mRNP, probably by changing the conformation of mRNA. Nab2 positions the 3'-end of the export-competent mRNA at the Mlp1 component of the nuclear basket to help mRNA thread into the entrance of the NPC channel. (B) Brownian ratchet model for mRNA transport through the NPC. The Mex67-Mtr2 carriers bound to export-competent mRNP interact with the FG nucleoporins that line the NPC channel to facilitate movement of the mRNP back and forth by thermal or Brownian motion. When one of the Mex67-Mtr2 carriers reaches the cytoplasmic face of the NPC, it is removed by the cytoplasmic DEAD-box helicase Dbp5, whose ATPase activity is stimulated by other factors bound to the same face of the pore. Removal of the carrier functions as a molecular ratchet, as it does not allow the portion of the mRNP to which it was bound to go back into the channel. Hydrolysis of ATP to ADP makes the ratchet work in one direction. These steps are repeated several times until the entire mRNP enters the cytoplasm. The released carrier molecules are recycled back into the nucleus to participate in the export of another mRNP complex. (Adapted from Stewart M [2007] *Mol Cell*,25:327–330. With permission from Elsevier.)

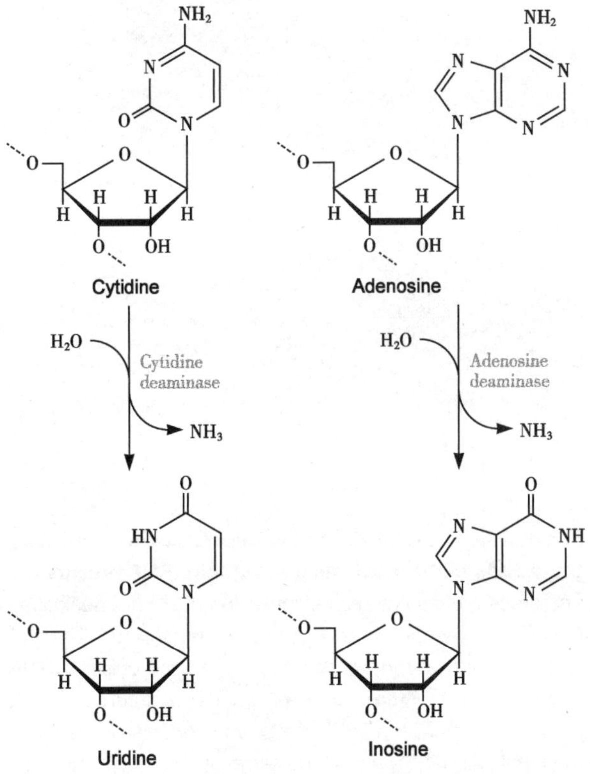

Figure 10.23 Editing by deamination in mammals.

molecules undergo extensive post-transcriptional modification involving a variety of base modifications at numerous sites.

10.6 Processing of constitutive noncoding RNA, tRNA and rRNA

10.6.1 tRNA processing is similar in all organisms

The processing of tRNAs always involves nuclease cleavage of pre-tRNAs from primary transcripts. In bacteria such as *Escherichia coli*, some tRNA genes are embedded in the gene cluster that encodes rRNAs, but the majority of tRNA genes are clustered together in groups of one to seven, surrounded by lengthy flanking sequences. The processing of the primary transcripts originating from these tRNA gene clusters is a multistep process that involves several endo- and exonucleases (Figure 10.24).

After the polynucleotide chain cleavage reactions that trim the chain to its final length, tRNA nucleotidyltransferase comes into play to add the universal triplet CCA to the 3′-end of the trimmed chain. The 2′- or 3′-OH groups of the terminal adenosine, A, serve as attachment sites for the amino acid during protein synthesis. The final step in tRNA processing involves chemical reactions that modify some of the bases; these modifications include methylation, thiolation, reduction of uracil to dihydrouracil, and pseudouridylation.

The enzyme that creates the mature 5′-end of all tRNAs, RNase P, was the first ribozyme ever to be described. Sidney Altman made the discovery and was honored with a portion of the 1989 Nobel Prize in Chemistry for his work. RNase P is a most unusual enzyme: it consists of a 377-nucleotide-long, highly structured RNA molecule and a small protein (see Figure 10.24). The big surprise came when it was proven, albeit under nonphysiological conditions, that the protein is not needed for the reaction: catalysis is carried out by the RNA component.

10.6.2 All three mature ribosomal RNA molecules are cleaved from a single long precursor RNA

All three mature ribosomal RNA molecules are cleaved from a single long precursor RNA The three rRNA molecules 23S, 16S, and 5S, which together with numerous ribosomal proteins form the structure of the bacterial ribosome, are initially transcribed as a very long pre-rRNA molecule (Figure 10.25). The ribosomal structure contains only one copy of each of the three molecules, so having them in a single transcript ensures that they are generated in the right stoichiometry. The long primary transcript is processed to give rise to the individual rRNA molecules. The first enzyme involved is RNase III, which introduces double-strand breaks at the bases of the stem-loop structures that contain the 16S, 23S, and 5S sequences. Further processing steps involve endonucleases that are specific for each of the three rRNA sequences and take place only after the first steps of ribosomal assembly, which occur co-transcriptionally. Processing of eukaryotic rRNAs follows a very similar path from a long precursor to yield the 28S, 18S, and 5.8S RNA molecules present in the eukaryotic ribosome. It relies extensively on the action of the exosome complex.

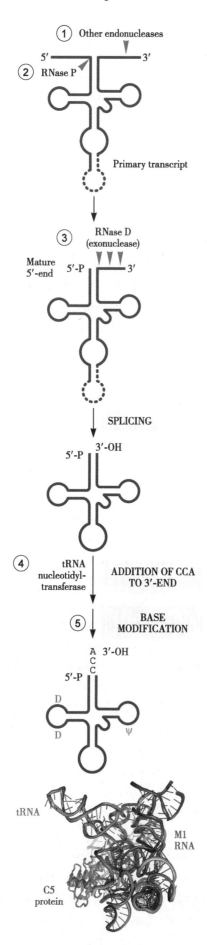

Figure 10.24 **Prokaryotic tRNA processing from primary transcripts that contain several tandem tRNA precursors.** The steps are catalyzed as follows: (Step 1) Endonucleases cleave the pre-tRNAs from primary transcripts. (Step 2) RNase P cleaves the transcript on the 5'-end of each tRNA sequence, releasing monomeric tRNA precursors with mature 5'-ends. (Step 3) RNase D, an exonuclease, reduces the length of the 3'-end. At the same time a loop, shown as a dashed line, is removed from the tRNA precursor and the cut ends are respliced. (Step 4) tRNA nucleotidyltransferase adds the universal CCA triplet to the 3'-end of the tRNA; this is an example of the addition of nucleotides without template DNA. (Step 5) The final step in the process, base modification, involves various specific enzymes that catalyze specific modification reactions. (Bottom) Crystal structure of *E. coli* RNase P bound to tRNA. The structure of the RNA component of the enzyme, M1 RNA, reveals a number of coaxially stacked helical domains; this structure is highly conserved in archaea, bacteria, and eukaryotes. The tRNA substrate is shown in green, and the protein component of the enzyme, C5, is shown in blue. *In vivo*, both RNA and protein components are required for activity. Under certain *in vitro* conditions, such as high Mg^{2+} concentrations, the protein is not needed for catalysis and the RNA acts as a true ribozyme. (Crystal structure from Wikimedia.)

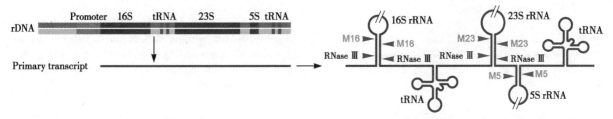

Figure 10.25 Processing of ribosomal RNA precursors in *E. coli*. The primary transcript contains a copy of each of the three ribosomal RNAs and may also contain several interspersed tRNA precursors. The 5'- and 3'-ends of mature rRNA are found in base paired regions; initially, these base-paired regions are processed by RNase III, followed by specific endonucleases: M16 endonuclease for 16S rRNA, endonuclease M23 for 23S rRNA, and M5 endonuclease for 5S rRNA. Processing is coupled to ribosome assembly.

10.7 Biogenesis and functions of small silencing RNAs

10.7.1 All ssRNAs are produced by processing from larger precursors

Small silencing RNAs, ssRNAs, are a heterogeneous group of short-length regulatory noncoding RNAs that perform distinct functions in silencing gene expression at the level of transcription, posttranscriptional regulation, or translation. The defining features of ssRNAs are short length, usually not exceeding 30 nucleotides, and association with members of the Argonaute (Ago) family of proteins, which serve the effector function in the silencing pathways. Ago proteins are ribonucleases characterized by two domains: Piwi, a ribonuclease domain, and PAZ, an ssRNA-binding domain. The function of the ssRNA is to guide the effector Ago proteins to their nucleic acid targets.

Table 10.3 lists some characteristics of the three major ssRNA classes. It should be noted that the mechanism of action of these small RNAs cannot be strictly defined as postsynthetic RNA processing as it does not represent a covalent change in RNA. However, their biosynthesis, in every case, involves processing of larger RNA precursors, which warrants the discussion of their biogenesis here.

As far as the regulation of target mRNA is concerned, all ssRNAs use a similar mechanism of interaction with their targets. As Figure 10.26 illustrates, the ssRNA interacts with its target by base pairing, which can be either very extensive or partial. Actually, the extent of base pairing determines the mechanical outcome of the silencing reaction: extensive base pairing leads to destruction of mRNA, whereas partial base pairing results in inhibition of translation. In either case, the biological outcome is the silencing of gene expression.

There are three major types of small silencing RNAs:

(1) MicroRNAs or miRNAs

The first miRNA to be discovered was the *lin-4* gene product in *Caenorhabditis elegans*, which is involved in regulating the expression of two important developmental genes, *lin-14* and *lin-28*. This discovery was

Table 10.3 Three major types of small silencing RNAs

	Type	Organisms	Length (nt)	Function
miRNA	microRNA	viruses, protists, algae, plants, animals	20–25	mRNA degradation; inhibition of translation
siRNA	Small interfering RNA	all eukaryotes, mainly plants	21–24	post-transcriptional silencing of transcripts and transposons; in some cases, silencing of transcription
piRNA	Piwi-interacting RNA	metazoa	21–30	transposon regulation and unknown functions; maintain germline stem cells and promote their division

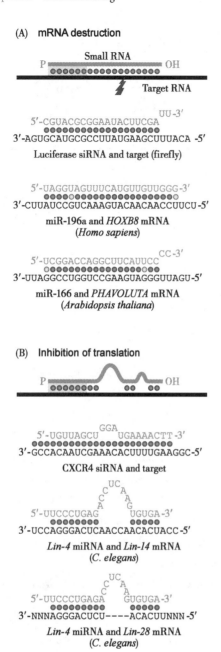

Figure 10.26 **Two modes of small RNA binding to their target mRNAs determine distinct mechanisms of action.** (A) Extensive base pairing of a small RNA to the 3′-UTR of a target mRNA guides catalytically active Argonaute proteins to specific mRNA molecules; the Argonaute proteins then cut a single phosphodiester bond in the mRNA, triggering its destruction. This mode of binding is customary for plants and some mammalian miRNA. The three examples given show the extensive base pairing that occurs between three ssRNAs and their respective targets. (B) Partial base pairing between a small RNA and the 3′-UTR of a target mRNA tethers an Argonaute protein to its mRNA target; the miRNA-Argonaute complex prevents translation. The first miRNA to be discovered, *Lin-4* miRNA in *Caenorhabditis elegans*, acts in this way on two closely related mRNAs, transcribed from the *Lin-14* and *Lin-28* genes. Note that the complexes are slightly different in the miRNA nucleotides that base-pair with the corresponding mRNAs. During larval development, *Lin-4* coordinates the downregulation of LIN-14 and LIN-28 protein concentrations, which in turn regulates the expression of stage-specific developmental events. In both parts of the figure, miRNA contains a short seed sequence at its 5′-end that contributes most of the energy for target binding; that is, it is the specificity determinant for target selection. The small size of the seed region allows a single miRNA to regulate many different genes. (Adapted from Zamore PD, Haley B [2005] *Science*,309:1519–1524. With permission from American Association for the Advancement of Science.)

quickly followed by descriptions of a whole range of miRNA molecules in a wide variety of organisms. To date, ~7000 miRNA genes have been identified in animals and their viruses; at least another ~1600 exist in plants.

1) Biogenesis of miRNAs. We illustrate miRNA biogenesis using the example of silencing of the *Hand2* gene in developing cardiac tissue in *Drosophila* (Figure 10.27). The process is very similar for all miRNAs, in terms of both the succession of steps and the proteins involved. In general, the process is characterized by the action of the cytoplasmic RNase III endonuclease Dicer.

The miRNAs are produced as large RNA primary transcripts called pri-miRNA that are self-complementary and can automatically fold into a double-strand hairpin structure, usually with some imperfect base pairing. The pri-miRNA is processed in a two-step reaction. The first step is catalyzed by Drosha, an RNase III superfamily member endonuclease, in the nucleus. Drosha reduces the pri-RNA to about a 70-bp, hairpin-shaped precursor fragment, pre-miRNA, which has a phosphate group at the 5′ end. This cleavage determines the 5′ and 3′ ends of the precursor. After export from the nucleus to the cytoplasm, the second step, pre-miRNA to miRNA, is catalyzed by a second RNase III family member, Dicer, by counting from the 3' end to produce a short, double-stranded ~22-bp segment. The miRNA now has a short, 22-nucleotide single-stranded 3' end, which is then usually modified by adding a 2'-0-methyl group for stability. Dicer has an N-terminal helicase activity, which enables it to unwind the double-stranded region, and two nuclease domains that are also related to the bacterial RNase III. Related enzymes are nearly universal in eukaryotes. In plants, the Dicer-like enzyme performs both the pri-miRNA and pre-miRNA processing steps in the nucleus.

These short, double-stranded RNA fragments are delivered to, or loaded onto, a complex called RISC (RNA-induced silencing complex). Proteins in the Argonaute (Ago) family are components of this complex and are required for the final processing to a single strand, by the elimination of the passenger strand, which is denoted as miRNA*. RISC then (usually) delivers the miRNA to the 3'UTR of its target mRNA. Humans have eight Ago family members, *Drosophila* has five, plants have 10, and *C.elegans* has 26. These proteins have an ancient origin and are found in bacteria, archaea, and eukaryotes.

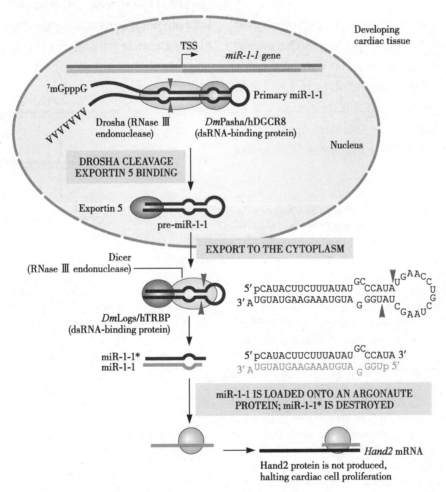

Figure 10.27 **Processing of a microRNA involved in gene regulation.** This microRNA example is from developing cardiac tissue in *Drosophila*. The consecutive stages of processing are as follows: (Step 1) Transcription factors SRF or serum response factor and MyoD stimulate Pol II transcription of the *miR-1-1* gene. (Step 2) The primary transcript is processed by the RNase III endonuclease Drosha with its dsRNA-binding partner Pasha, or partner DGCR8 in humans. The product, pre-miR-1-1, contains a 2-nt single-strand 3'-overhang, which is recognized by exportin 5 for transport into the cytoplasm. (Step 3) In the cytoplasm, a second RNase III endonuclease, Dicer, with its own dsRNA-binding partner, makes a second pair of cuts to liberate the miRNA-miRNA* duplex. The mature 21-nt-long miRNA is loaded, by a specialized loader, onto an Argonaute family member protein; the miRNA* chain is destroyed. (Step 4) The Argonaute protein, an effector of small-RNA-directed silencing, is now guided to the 3'-untranslated region of the mRNA, in this case Hand2 mRNA, by the miRNA, where the protein represses translation of the Hand2 protein. This halts cardiac cell proliferation. (Adapted from Zamore PD, Haley B [2005] *Science*, 309:1519–1524. With permission from American Association for the Advancement of Science.)

2) miRNAs target mRNA for degradation and translational inhibition. The degree of base pairing and the sequence of the ends (determined by Dicer cleavage) of the duplex dictate which of the multiple Ago family members pick up the RNA duplex and which strand is selected as the passenger strand to be degraded. The RISC complex is now in a position to use the mature miRNA to guide it to its target mRNA. Selection of the class of target by RISC lies with the specific Argonaute protein, while the specific RNA target itself is determined by the miRNA.

(2) siRNAs

Small interfering RNAs (siRNAs) are also called short interfering RNAs or silencing RNAs. A characteristic distinguishing feature of siRNAs is that they are derived from double-stranded RNA by the RNase activity of Dicer (Figure 10.28). Mammals and *C. elegans* have a single Dicer that participates in the biogenesis of both miRNAs and siRNAs; *Drosophila*, on the other hand, has two Dicer forms. Dicer-1 is involved in making miRNA, whereas Dicer-2 is the ribonuclease that cleaves the double-stranded RNA in the early stages of siRNA biogenesis.

(3) piRNA

These were discovered in 2001 in the *Drosophila* germline, where they repress transposons, thus stabilizing the germline genome. piRNA sequences are very diverse, with more than 1.5 million distinct piRNAs identified in *Drosophila*. These are clustered to a few hundred genomic regions. It remains to be established how many of these identified piRNAs have physiological significance. Their source seems to be extremely long ssRNA transcripts, 100 000–

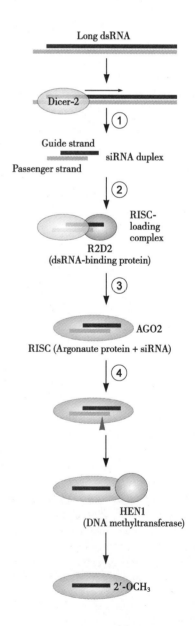

Figure 10.28 Processing of siRNAs in *Drosophila*. The consecutive stages of processing for small interfering RNAs are as follows: (Step 1) Double-stranded RNA precursors, dsRNA, are processed by Dicer-2 to generate siRNA duplexes containing a guide strand, which is of physiological relevance as it directs the silencing, and a passenger strand, which is subsequently degraded. The two strands, each ~21 nt, carry a 5′-phosphate and a 3′-OH group, and each has a 2-nt overhang at its 3′-end. (Step 2) Dicer 2 partners with a dsRNAbinding protein, R2D2, to form the RISC-loading complex. (Step 3) The active entity in gene silencing is RISC, an RNA-induced silencing complex of AGO2, an Argonaute protein, and siRNA. siRNA guides Argonaute proteins to their RNA targets. Argonaute proteins are the catalytic components of RISC: they possess endonucleolytic activity that degrades the mRNA that is recognized by the siRNA by complementary base pairing. The Argonaute proteins are also partially responsible for selection of the guide strand and destruction of the passenger strand of the siRNA. Argonaute proteins are characterized by two domains: an N-terminal PAZ domain of ~20 kDa and a C-terminal Piwi domain of ~40 kDa. The PAZ domain interacts with RNA, serving as the anchor for the 3′-end of siRNAs. (Step 4) Afterward, the passenger strand is destroyed, and the DNA methyltransferase HEN1 adds a methyl group to the 2′-OH group of the siRNA, which stabilizes the RNA. Finally, the siRNA carrying the catalytic Argonaute protein interacts with the target mRNA to cleave it; this step is not shown.

200 000 nucleotides, usually antisense. piRNAs are distinct from other small interfering RNAs in that they bind Piwi proteins, a clade of the Argonaute protein superfamily, and do not require Dicer for their biogenesis. The process of piRNA biogenesis is described in Figure 10.29.

10.7.2 LncRNAs are produced by processing from larger precursors

Long non-coding RNAs (lncRNA) are a type of non-coding RNAs (ncRNAs) that exceed 200 nucleotides in length and its transcripts is lacking a canonical open reading frame (ORF), without protein-coding potential. Most lncRNAs were first discovered and characterized by throughout sequencing, and identified the transcriptional signature from RNA Pol II binding and epigenetic modifications of chromatin. They are polyadenylated, and can be located within nuclear or cytosolic fractions.

According to their relative location to with respect to the genomic nearest protein-coding transcripts, lncRNAs are categorized as intergenic, overlapping, intronic, and exonic. The overlapping lncRNAs are subdivided into sense, antisense, and bidirectional according to their transcriptional loci relative to the overlapped transcripts (Figure 10.30).

LncRNAs are a relatively abundant component of the mammalian transcriptome. The number of lncRNAs is larger than the number of protein-coding RNAs. The GENCODE lncRNA catalog consists of 14 880 transcripts grouped into 9277 gene loci in the human genome. There are more and more evidence showed that lncRNAs are not by-products of gene transcription, neither gene trash nor transcriptional noise. They are independent and active participants in multiple important biological processes, such as cell proliferation, differentiation, migration, and apoptosis. Advanced studies implied that the working mechanisms of lncRNA mostly converge on their ability of regulating gene expression at almost every level, including chromosome remodeling, transcription and post-transcriptional processing. Moreover, the dysregulation of lncRNAs has increasingly been linked to many human diseases.

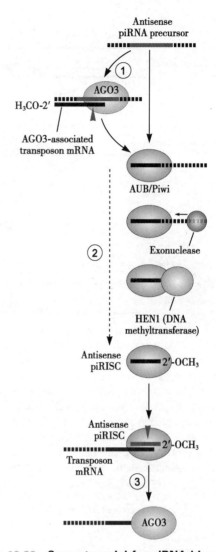

Figure 10.29 **Current model for piRNA biogenesis.** piRNAs are thought to derive from ssRNA precursors and do not go through a dicing step. The current model for their biogenesis is derived from the sequences of piRNAs bound to three Argonaute proteins belonging to the Piwi Argonaute clade: Piwi, Aubergine or AUB, and AGO3 in *Drosophila*. piRNA bound to Piwi and AUB is typically antisense to transposon mRNA, while piRNA bound to AGO3 corresponds to a portion of the transposon mRNA itself; that is, sense. The first 10 nucleotides of antisense piRNAs are frequently complementary to the sense piRNAs found in AGO3. This unexpected complementarity is part of a mechanism for piRNA amplification that is activated only after transcription of transposon mRNA. The consecutive stages of processing are as follows: (Step 1) The piRNA precursor binds AGO3-associated transposon mRNA and cleaves it across from position 10 of the antisense piRNA guide. The sense piRNA can, in turn, guide cleavage of the antisense piRNA precursor transcript. (Step 2) Several substeps lead to creation of the antisense piRISC, which can further interact with transposon mRNA. (Step 3) The 5′-end of the cleaved transposon mRNA product loads onto AGO3.

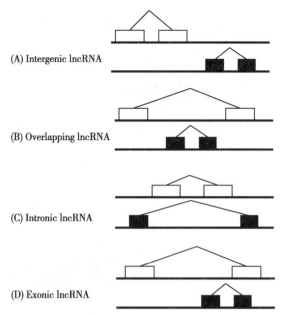

Figure 10.30 **LncRNAs were subclassified into four classes.** LncRNAs were subclassified into four classes on the grounds of their intersection with protein-coding genes: (A) intergenic lncRNAs: do not intersect with any protein-coding loci, (B) overlapping lncRNAs: contain a coding gene within an intron (sense or antisense), (C) intronic lncRNAs: contained within protein-coding introns completely (sense or antisense), and (D) exonic lncRNA: at least one of its exons intersects a protein-coding exon by at least 1 bp. Protein-coding genes (black); lncRNAs: long noncoding RNAs (white).

Similar to miRNA, the gain/loss of function strategies can be used to explore the functions of lncRNA. Phenotypes were observed after overexpression or silencing of lncRNA.

10.7.3 CircRNAs is formed by backsplicing

Circular RNAs (circRNAs) are a novel type of RNA that, unlike linear RNAs (with 5' and 3' ends), form a covalently closed continuous loop, with neither 5'-3' polarities nor polyadenylated tails. CircRNAs are highly represented in the eukaryotic transcriptome. CircRNAs's expression are not affected by RNA exonuclide, and are more stable and not easy to degrade. Because circRNAs are not sensitive to nucleases, they are more stable than linear RNA, which makes circRNAs have obvious advantages in the development and application of new clinical diagnostic markers.

CircRNA is formed by backsplicing in a non-classical way (Figure 10.31). Recent studies have discovered thousands of endogenous circRNAs in mammalian cells. CircRNAs are largely generated from

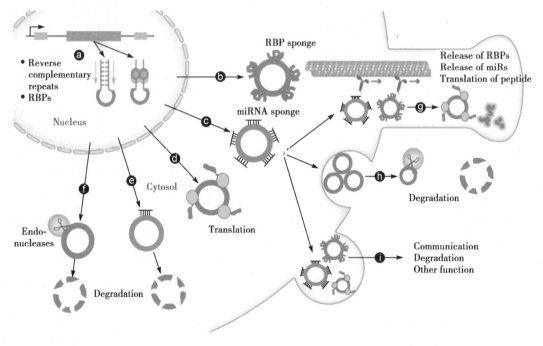

Figure 10.31 **Biogenesis and function of CircRNAs.** CircRNA can be produced either with the help of reverse complementary repeats or RNA-binding proteins and exported from the nucleus (a). The circRNA might be bound in the cytoplasm by multiple factors, such as RNA-binding proteins (b), miRNAs as sponge (c), ribosomes (d), or for direct degradation (e), or endonucleases that would cause circRNA degradation (f). The circRNAs or circRNA complexes have different fates. Of them, the non-degradative binding might diffuse in the cytoplasm or been actively transported in into particular regions of the cell (g) where it can release its bound cargo or starts to be translated. Those in vesicle will be released into the extracellular space and removed them from the cytoplasm (h). Other could reach other cells or tissues by vesicles/exosomes and act as messenger molecules (i).

exonic or intronic sequences, and reverse complementary sequences or RNA-binding proteins (RBPs) are necessary for circRNA biogenesis. The majority of circRNAs are conserved across species, are stable and resistant to RNase R, and often exhibit tissue/developmental-stage-specific expression.

Similar to miRNAs and long noncoding RNAs (lncRNAs), circRNAs are becoming a new research hotspot in the field of RNA and could be widely involved in the processes of life. Some functionally characterized circRNAs have critical roles in gene regulation through various actions, including sponging microRNAs and proteins as well as regulating transcription and splicing (Figure 10.31). The circRNA molecule is rich in microRNA (miRNA) binding sites, which acts as a miRNA sponge in the cell. In this way, the inhibition effect of miRNA on its target gene is removed and the expression level of target gene is increased. This mechanism of action is called competitive endogenous RNA (ceRNA) mechanism (Figure 10.31).

10.8 RNA quality control and degradation

10.8.1 Bacteria, archaea, and eukaryotes all have mechanisms for RNA quality control

The importance of fully functional RNA molecules has led to the evolution of numerous quality-control mechanisms in all three domains of life. Quality control acts at many steps before, during, and after translation. If, for example, mRNA is incompletely or incorrectly spliced or polyadenylated, it will be degraded in the nucleus instead of being exported to the cytoplasm. This applies to aberrant rRNA and tRNA molecules also. RNA molecules are also degraded in the cytoplasm at the end of their useful life. We first introduce the structure of the main protein complexes used for RNA degradation in both circumstances (Figure 10.32 and Figure 10.33). In archaea and eukaryotes, the complexes that accomplish this are known as exosome complexes.

Exosome complexes are related, both in sequence and structure, to the bacterial RNase PH, an enzyme that uses inorganic phosphate to mediate cleavage of RNA. All RNA degradation complexes of this class contain a ring of six protein subunits, with a hole in the middle through which the RNA to be degraded passes. The archaeal and eukaryotic exosome rings are not stable on their own and require additional cap proteins to stabilize their structures. The exosome complexes degrade RNA exonucleolytically, in a 3′→5′ direction.

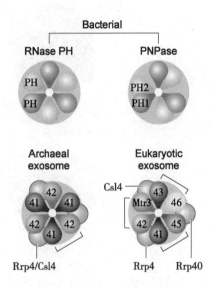

RNA_n + inorganic phosphate (P_i) → RNA_{n-1} + nucleoside 5′-diphosphate

Figure 10.32 Conserved architecture of RNA degradation complexes among bacteria, archaea, and eukaryotes. Bacteria contain two distinct complexes. Lighter and darker shades of green in the bacterial RNase PH complexes indicate the inverse orientation of neighboring subunits; the combined shapes in PNPase indicate PH domains from the same polypeptide chain. All schematics show the structures from the bottom; note the ring shape of the complexes with a hole in the middle. The RNA-binding caps of the archaeal and eukaryotic exosomes are depicted in green behind the subunits forming the rings. The archaeal structure is a homotrimer of heterodimer Rrp41–Rrp42; the three cap proteins that stabilize the ring structure can be either Rrp4, Csl4, or a combination of these. The eukaryotic exosome is a complex of three distinct heterodimers, Rrp41–Rrp45, Rrp46–Rrp43, and Mtr3–Rrp42; the dimers are held together by the cap proteins Rrp40, Csl4, and Rrp4. In addition to stabilizing the complex, the cap proteins contain RNA-binding domains that interact with the RNA to be degraded. The archaeal and eukaryotic ring subunits are related by sequence and structure to the bacterial RNase PH enzyme. The PH domains in these enzymes function in a phosphorolytic reaction, as shown below the schematics. (Adapted from Lykke-Andersen S, Brodersen DE, Jensen TH [2009] *J Cell Sci*, 122:1487–1494. With permission from The Company of Biologists.)

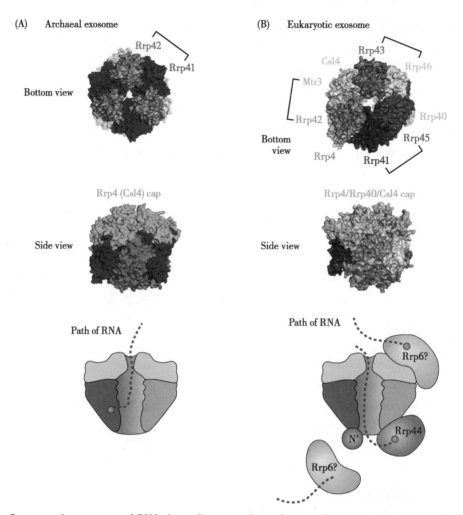

Figure 10.33 **Conserved structures of RNA-degrading complexes from archaea and eukaryotes.** The individual proteins are color-coded. (A) Structure of the *Archaeoglobus fulgidus* exosome containing Rrp41, Rrp42, and Rrp4. (B) Human exosome structure. The schematics below the structures represent sliced side views, showing the central cavity and the path of the RNA to the active site, marked by a red dot. The eukaryotic exosome does not possess enzymatic activities in the main ring structure; rather, the active center is located on Rrp44, and the N-terminal head domain of Rrp44 interacts with Rrp41. RNA may access the active site through the central channel or through a path that is independent of the exosome core. The position of the other subunit with enzymatic activity, Rrp6, is unclear. (Adapted from Schmid M, Jensen TH [2008] *Trends Biochem Sci*, 33:501–510. With permission from Elsevier.)

RNA degradation in bacteria occurs as a succession of steps (Figure 10.34). Four distinct enzymatic processes are involved, each catalyzed by one or more enzymes: endonucleolytic cleavage, oligoadenylation, exonucleolytic cleavage, and helicase action. The addition of an oligo(A) tail and the helicase activity are required for the removal of stable secondary structures that would otherwise preclude normal exonucleolytic degradation.

10.8.2 Archaea and eukaryotes utilize specific pathways to deal with different RNA defects

During evolution, the pathways involved in degrading RNA have become more complex, highly regulated, and divergent, with distinct pathways dedicated to the degradation of RNA carrying particular molecular defects. Degradation can occur either in the nucleus or in the cytoplasm. Some degradation occurs co-transcriptionally or immediately after transcription is terminated, as soon as processing and/or RNA packaging defects are detected. Nuclear degradation can occur in all three major RNA classes: rRNA, tRNA, and mRNA. A detailed description of nuclear degradation of faulty RNA molecules is presented in Figure 10.35.

In the cytoplasm, degradation occurs either at

the end of the useful lifetime of an RNA molecule or to destroy faulty RNA molecules that have escaped nuclear surveillance and should not be translated. The three major processes, nonsense-mediated decay or NMD, no-go decay or NGD, and non-stop decay or NSD, are all intimately involved with translation and are discussed.

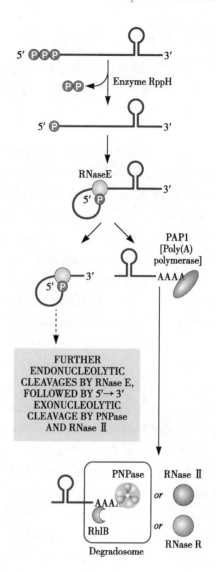

Figure 10.34 **RNA degradation pathways in bacteria.** RNA turnover is initiated by removal of a pyrophosphate, PP, from the 5′-end of the RNA and recognition of the resulting 5′-monophosphate by the endonuclease RNase E, which cleaves the mRNA into two pieces. The 5′-fragment is further degraded by the combined action of RNase E and the exonucleolytic activities of PNPase and RNase II. When the 3′-end happens to be protected by a stem-loop structure and proteins associated with it, it is first oligoadenylated by the bacterial poly(A) polymerase PAP1. The oligo(A) tail recruits the bacterial degradosome, PNPase plus the helicase RhlB, or is degraded by one of two other hydrolytic 3′→5′ exonucleases, RNase II or RNase R. Even with the wide range of phosphorolytic and hydrolytic enzymes available, degradation in bacteria seems to be predominantly hydrolytic. (Adapted from Lykke-Andersen S, Brodersen DE, Jensen TH [2009] *J Cell Sci*,122:1487–1494. With permission from The Company of Biologists.)

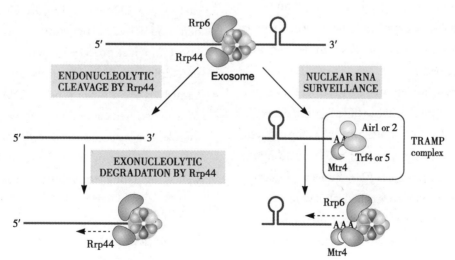

Figure 10.35 **Nuclear RNA degradation pathways in eukaryotes.** In eukaryotes, RNA degradation begins with endonucleolytic cleavage by exosome component Rrp44 and 3′-oligoadenylation by the nuclear TRAMP complex, which consists of a helicase, Mtr4; one of two poly(A) polymerases, Trf4 or Trf5; and one of two RNA-binding proteins, Air1 or Air2. The Mtr1 helicase can also associate directly with the exosome. TRAMP is also involved in 3′-end processing of stable RNAs that are not destined for degradation. Thus, it is likely that TRAMP and the exosome survey the entire RNA population but degrade only transcripts that lack protective secondary structures or certain RNA-binding proteins. (Adapted from Lykke-Andersen S, Brodersen DE, Jensen TH [2009] *J Cell Sci*,122:1487–1494. With permission from The Company of Biologists.)

Key Concepts

- Many kinds of RNA molecules must undergo one or more forms of post-transcriptional processing before they can play their appropriate roles in the cell.
- The pre-mRNAs synthesized in eukaryotic nuclei undergo a series of modifications before they are exported to the cytoplasm. These include 5′-capping, removal of introns and splicing of exons, 3′-polyadenylation, and in some cases editing: insertion or deletion of nucleotide residues and/or chemical modification of bases.
- Capping is the addition of GMP at the 5′-end of a message, in a 5′–5′ orientation. The cap recruits proteins that protect the message from exonucleases, aids in transport of the processed message to the cytoplasm, and serves as the site of ribosomal attachment during initiation of translation.
- Polyadenylation involves the addition of a poly(A) tail to the 3′-end of the message. The tail, with recruited proteins, protects this end from exonucleolytic attack.
- Removal of most introns and resplicing of exons require complex nuclear machines called spliceosomes. These recognize 3′- and 5′-splice sites, plus an internal intron site, catalyze cleavage, and then religate adjacent exons. The spliceosome is a huge complex of a number of RNA molecules and many proteins.
- In a few cases, introns are self-splicing; they excise themselves from the RNA while ligating exons. In such cases, the RNA in the intron has the catalytic power, even though proteins may help to maintain the structure. Such RNA molecules that act like protein enzymes are called ribozymes.
- In many cases, splicing can take alternate routes, adding or excluding exons or using alternative or cryptic splice sites. Such alternative splicing may occur within a given gene or may even involve exons from distant genes. Alternative splicing has the effect that one gene can often produce different protein products in different tissues or at different stages of development.
- Regulation of alternative splicing involves a number of factors, including splice-site strength, protein enhancers or silencers, RNA secondary structure, and chromatin structure.
- Processed eukaryotic mRNAs are exported to the cytoplasm through nuclear pores, via a ratcheted Brownian motion mechanism.
- After mRNA molecules have been processed in all of these ways, they can still be degraded if they are imperfect or if the cell no longer requires the protein product. Such degradation can occur in either the nucleus or cytoplasm.
- The m^6A methylation is the most abundant internal modification on mRNA, which can accelerate the processing time of mRNA precursors, and accelerate the transport speed and the speed of nuclear export of mRNA in cells.
- In bacteria, mRNA can be used directly for translation without processing, but functional tRNAs and rRNAs are generated by cleavage and trimming of tandem transcripts. tRNA molecules also undergo a variety of base modifications.
- Small silencing RNAs such as microRNA, small interfering RNA, and Piwi-interacting RNA, are generated from much larger gene products.
- LncRNA are a type of non-coding RNAs (ncRNAs) that exceed 200 nucleotides, which are independent and active participants in multiple important biological processes, such as cell proliferation, differentiation, migration, and apoptosis.
- CircRNAs are a novel type of RNA that form a covalently closed continuous loop, which have critical roles in gene regulation through various actions, including sponging microRNAs and proteins as well as regulating transcription and splicing.

Key Words

adenosine(腺苷)
catalysis(催化作用)
cytoplasmic surveillance systems(胞质监测系统)
cytoplasmic(胞浆的,细胞质的)
de-adenylation(脱腺苷化)
decay(降解)
eukaryotic(真核的)

gene cluster(基因簇)
gene regulation(基因调控)
gene silencing(基因沉默)
ligation(连接反应)
lncRNA(长链非编码 RNA)
methylation(甲基化)
miRNA/microRNA(微小 RNA)
nuclear exosomes(核外泌体)
nucleotide(核苷酸)
polyadenylation(聚腺苷酸化)
polynucleotide(多聚核苷酸)
post-transcription(转录后)
prokaryotic(原核的)
quality control(质量控制)
ribosome(核糖体)
ribozyme(核酶)
RNAi/RNA interfere(RNA 干涉)
RNPs(核糖核蛋白)
siRNA(小干扰 RNA)
splicing(剪接)
split gene(断裂基因)
termination codon(终止密码子)

Questions

1. What are the four general categories of processing?
2. What is splicing?
3. What is RNA interfere?
4. How do the Cytoplasmic Surveillance Systems control the quality of RNA?
5. What are the two decay pathways of General mRNA?

References

[1] Black DL (2003) Mechanisms of alternative pre-messenger RNA splicing. *Annu Rev Biochem*,72:291–336.

[2] Bentley DL (2005) Rules of engagement: Co-transcriptional recruitment of pre-mRNA processing factors. *Curr Opin Cell Biol*,17:251–256.

[3] Chen M, Manley JL (2009) Mechanisms of alternative splicing regulation: Insights from molecular and genomics approaches. *Nat Rev Mol Cell Biol*,10:741–754.

[4] Doma MK, Parker R (2007) RNA quality control in eukaryotes. *Cell*, 131:660–668.

[5] Donald DR, John SV, Neil S, et al. (2009) siRNA vs. shRNA: similarities and differences. *Adv Drug Deliv Rev*, 25;61(9):746–759.

[6] Ghildiyal M, Zamore PD (2009) Small silencing RNAs: An expanding universe. *Nat Rev Genet*,10:94–108.

[7] Mihye L, Boseon K, Narry K (2014) Emerging roles of RNA modification: m(6)A and U-tail. *Cell*,28;158(5):980–987.

[8] Nilsen TW, Graveley BR (2010) Expansion of the eukaryotic proteome by alternative splicing. *Nature*, 463:457–463.

[9] Patop IL, Wüst S, Kadener S(2019) Past, present, and future of circRNAs. *EMBO J*, 15;38(16):e100836.

[10] Roundtree IA, Evans ME, Pan T, et al.(2017) Dynamic RNA Modifications in Gene Expression Regulation. *Cell*, 169(7), 1187–1200.

[11] Shi XF, Sun M, Liu HB, et al.(2013)Long non-coding RNAs: A new frontier in the study of human diseases. *Cancer Letters*,339(2): 159–166.

Dongmin Li

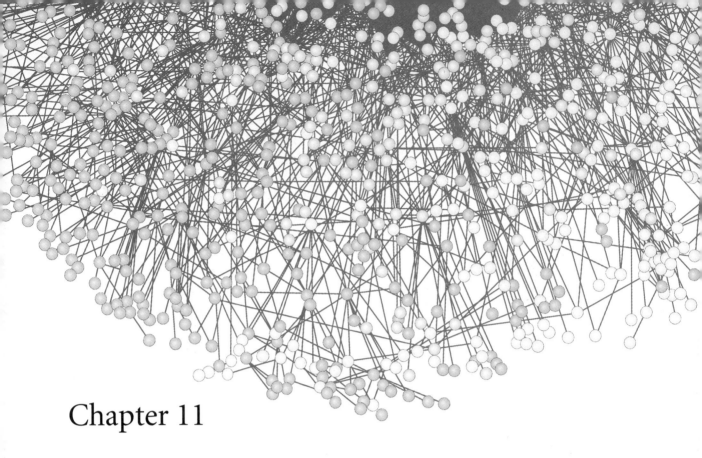

Chapter 11

Mechanism of Translation

11.1 Introduction

The transfer of information from RNA to proteins is named translation. This process is at base fundamentally simple which involves making a polynucleotide strand complementary to another, using base-pairing rules. And it also requires the coordinated use of the nucleic acid language during codon-anticodon recognition and also recognition between nucleic acids and proteins during the initial state of creating charged transfer RNAs, tRNAs. Three participants are needed for translation to occur: messenger RNA or mRNA, which provides the message to be translated; a set of tRNAs, each charged with an appropriate amino acid; and a ribosome, as a platform and catalyst for the process. Before reaching the key step where a peptide bond forms between two amino acids, many events must take place. Ribosomes must be assembled in the nucleolus of the cell from a diversity of different gene products. Transfer RNAs must be "charged" with their appropriate amino acid, and all the players must join together. The overall process of translation occurred on the ribosome, a complex RNA-protein machine that acts as both stage and director for translation. The overall process of translation is usually divided into three major phases: initiation, elongation, and termination (Figure 11.1). Translation is under strict regulation and control. Translational regulation can occur at several levels, from control of the availability of ribosomes to adjustments to the rate of translational initiation and elongation. The abundance, accessibility, and stability of mRNAs are also factors. Newly synthesized polypeptide chains are usually not biologically active. Post-translational processing of newly synthesized polypeptide chain is necessary to form proper spatial conformation to make proteins biologically active. And then the functional proteins are frequently targeted transport to appropriate subcellular sites to perform function.

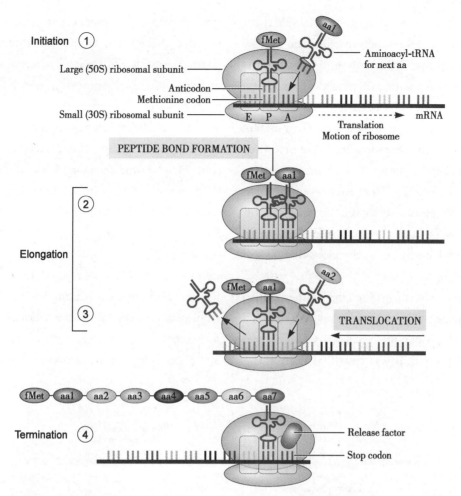

Figure 11.1 Overview of translation. Schematic of the ribosome with its three sites: exit or E site interacts with uncharged, deacylated tRNA; peptidyl or P site interacts with tRNA that carries the nascent polypeptide chain; and acceptor or A site interacts with tRNA that carries the incoming amino acid (aa). (Step 1) A special initiator tRNA, carrying the amino acid formylmethionine, fMet, in bacteria or methionine in eukaryotes, binds to the P site of the ribosome that accommodates the start codon. A new aminoacyl-tRNA or aa-tRNA then joins the complex, entering into the A site. The incoming amino acid must be cognate: that is, the anticodon on the carrier tRNA must correspond to the codon in the message, to ensure truthful transmission of the information encoded in the mRNA into the sequence of amino acids in the polypeptide chain. (Step 2) Peptide bond formation. (Step 3) Translocation of the ribosome, with its bound tRNAs, with respect to the mRNA, so that the next codon is now in the A site, ready to accept the next amino acid specified by the mRNA codon. (Step 4) When a stop codon enters the A site, translation is terminated with the help of special release factors.

11.2 Messenger RNA

11.2.1 Structure of mRNA

Messenger RNA, template for protein biosynthesis, is a major participant in translation. mRNAs in both bacteria and eukaryotes are of variable length and stability. Bacterial mRNAs are ready for use as soon as they are transcribed, but eukaryotic mRNAs have undergone extensive processing by the time they are accepted for translation in the cytoplasm. The length of an mRNA after splicing is, of course, determined mainly by the length of the protein product encoded.

There is typically a transcribed but untranslated region (UTR) of the mRNA, 5'-UTR and 3'-UTR. Many bacterial mRNAs share a common element at their 5'-ends that helps them to bind to the ribosome during initiation and positions the initiation codon in close proximity to the P site of the ribosome. This is known as the Shine-Dalgarno sequence or SD sequence and is located several nucleotides upstream of the start codon (Figure 11.2). It acts by base-pairing with a conserved region near the 3'-end of 16S rRNA in the small ribosomal subunit. And mRNAs may contain hairpin structures formed by self-complementary stretches of the polynucleotide chain to make mRNA small enough to entry the narrow channel of ribosome. The situation in eukaryotes is quite different. Eukaryotic mRNA undergoes several important modification steps during its transition from a primary transcript in the nucleus to a mature functional mRNA in the cytoplasm. These modifications include 5′-end capping, 3′-end polyadenylation, and frequently splicing. Here we present a much more detailed view of a typical eukaryotic mRNA, also describing functional regions and elements in both the 5′- and 3′-untranslated regions or UTRs of the molecule (Figure 11.3).

The coding regions (open reading frames, ORFs) in mRNA include the nucleotides sequences to act as genetic codes, which are translated to amino acid sequences of protein. It was clear that the translation of nucleotide sequences is carried out in units of three nucleotides. The unit of three nucleotides codes one amino acid or acts as a signal for Initiation or termination in the process of translation, which is known as codon or triplet code. The genetic code is a set of three-base codons that instruct the ribosome to incorporate specific amino acids into a polypeptide. A triplet code allows $4^3 = 64$ permutations, more than enough to code for all 20 amino acids. All 64 codons are used but three codons–UGA, UAA, and UAG–that do not represent amino acids are commonly used as stop codons to termination translation (see Table 5.2). One of these stop codons marks the end of every open reading frame. Under special circumstances, the three stop codons can also code for selenocysteine and pyrrolysine, the so-called 21st and 22nd amino acids, respectively. Sixty-one codons

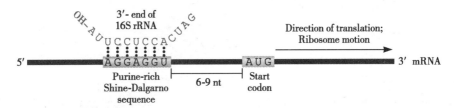

Figure 11.2 **Ribosome-binding sites at the 5′-end of prokaryotic mRNA.** The ribosome binding site, also known as the Shine-Dalgarno sequence, is located 6–9 nucleotides upstream of the start codon and helps to position the codon at the ribosome's P site, that is, to establish the correct reading frame. It does so by base-pairing with the 3′-end of 16S rRNA in the small subunit of the ribosome.

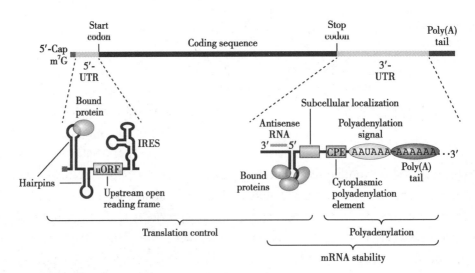

Figure 11.3 **Structure of a typical mature eukaryotic mRNA.** The coding region and two flanking untranslated regions are included. Note the existence of a cap structure at the 5′-end and a poly(A) tail at the 3′-end. The expanded regions below the mRNA depict structures that can be found in the untranslated regions and their functions. IRES, internal ribosome entry sites, are usually found in noncapped mRNAs. (Adapted from Mignone F, Gissi C, Liuni S, et al. [2002] *Genome Biol*, 3:reviews0004. With permission from BioMed Central.)

are recognized by tRNAs for the incorporation of the 20 common amino acids. AUG is an initiation codon, also called the start codon, which determines the point in the mRNA coding for the first amino acid methionine, and also the only codon that specifies methionine.

Several points of the code become obvious:

- The genetic codons in mRNA are read always in the direction from 5'-ending to 3'-ending until the stop codons when translation, corresponding to the amino acid sequence of a polypeptide read from N-terminus to C-terminus. The initiation codon AUG lies in the 5'-ending.

- The codons are read successively (Figure 11.4). The genetic code is non-overlapping. An individual nucleotide is part of only one codon. And the code is unpunctuated: all nucleotides need to be read in succession and no gaps are allowed between successive codons. Insertion or deletion of nucleotides with the length not an exact multiple of three nucleotides into the message would shift the phase in which the ribosome reads the triplet codons and, consequently, alters all of the amino acids downstream from the site of the mutation. Such mutations are called frameshift mutations which usually lead to total loss of function of the encoded protein. The deletion or insertion of three base pairs in the nucleotide sequence results in the loss or addition of one amino acid residue, but not shift of the reading frame in the mRNA.

- The code is highly degenerate: There are 61 meaningful codons that specify the 20 amino acids. Most amino acids are represented by more than one codon, with tryptophan and methionine being the only exceptions. Codons that have the same meaning are said to be synonymous. Degeneracy lies mainly in the third letter of the codon; in other words, the first two letters of codons alone often specify the amino acid. The reduced specificity at the last position is known as third-base degeneracy. It increases the probability that a single random base change will result in no amino acid substitution or in one involving amino acids of similar character. For example, a mutation of CUC to CUG does not change the resulting polypeptide because both codons represent leucine. But the frequencies with which different codons

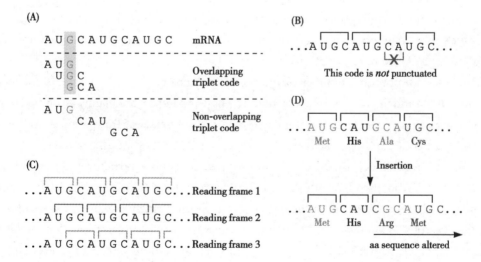

Figure 11.4 **Basic features of the genetic code.** (A) Schematic of the two possible ways a triplet code can be read: in the overlapping code, the same base can be part of three codons, whereas in the non-overlapping code, each base is part of only one codon. The genetic code is actually non-overlapping. (B) The code is unpunctuated: all nucleotides need to be read in succession and no gaps are allowed between successive codons. (C, D) The concepts of reading frame and frameshift mutations. (C) Each nucleotide sequence has to be read starting from a particular nucleotide. Specification of this first nucleotide defines the succession of codons and thus the primary structure of the polypeptide chain. If a different first nucleotide is specified, the same sequence can be read in a different frame, producing a different polypeptide. (D) Insertion of one nucleotide into the message would change the amino acid sequence in the downstream portion of the polypeptide chain. Such frameshift mutations usually lead to total loss of function of the encoded protein.

are used vary significantly between different organisms and between proteins within the same organism. This is referred to as codon bias.

- The wobble hypothesis: Some amino acids are encoded by more than one codon. This doesn't mean that there are 61 different types of tRNA in each cell. There are fewer tRNAs than sense codons, meaning that several anticodons, and thus several tRNAs, must be capable of recognizing the same codon during translation. It turns out, as proposed by Crick and later proven experimentally in 1966, that the 5′-base of the anticodon can wobble, making alternative, non-Watson-Crick hydrogen bonds with several different bases at the 3′-position of the codon. The pairing between codon and anticodon at the first two codon positions always follows the usual rules, but that exceptional "wobbles" occur at the third position. The third base of a codon is allowed to form a non-Watson-Crick base pair with the anticodon, such as the GU wobble pair or the modified base inosine with U, C and A (Figure 11.5).

- The code is almost, but not strictly, universal. Its near universality implies that all living organisms are descended from a single pool of primitive cells in which this occurred. Exceptions to the universal genetic code are rare. Exceptions to the universal genetic code occur in certain eukaryotic nuclei and mitochondria and in mycoplasma, codons that cause termination in the standard genetic code can code for amino acids such as tryptophan and glutamine, for example, the employment of UGA to encode tryptophan. In several mitochondrial genomes and in the nuclei of yeast, the sense of a codon is changed from one amino acid to another (see Table 5.3).

11.2.2 Overall translation efficiency depends on a number of factors

It is well established that gene expression is controlled at many different levels. This becomes obvious when the abundance of mRNAs and protein levels in a cell are compared. The abundance of some

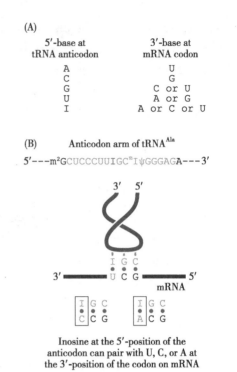

Figure 11.5 **Wobble rules.** (A) Wobble rules as proposed by Crick in 1966. These rules have been extended on the basis of experimental data to include differently modified bases in the anticodon region. (B) The partial sequence presented is that of the anticodon arm. The wobble base hypoxanthine, which is the base portion of the modified nucleoside inosine in the anticodon can pair with U, C, or A in the codon, according to the rules in part A.

proteins, including many secreted proteins, correlates with the abundance of their encoding mRNA. For other proteins, there can be huge differences in the abundance of the mRNA and its encoded protein. These differences are mainly attributed to differences in the frequency of translation initiation for different mRNAs; they are also influenced by mRNA and protein stability. Structural features at the 5′-UTR control mRNA translation efficiency, mainly through the presence of stable secondary structures, and sometimes proteins bound to them, that inhibit translation initiation. Additional control is exerted by elements known as internal ribosome entry sites or IRES. Under conditions in which normal cap-dependent translation initiation is impaired—for example, during stress, apoptosis, or mitosis—some cellular mRNAs can initiate translation at IRES. Use of IRES for initiation does not require the cap structure or any of the factors needed to remove secondary structures.

3′-Untranslated regions contain the conventional signal for polyadenylation. In addition, they may contain a cytoplasmic polyadenylation element or CPE, which is responsible for the lengthening of poly(A) tails on mRNAs that have already lost a portion of the tail and need to be further stabilized, a process that takes place in the cytoplasm. CPE action requires the binding of a specific protein factor, CPE-binding protein. 3′-Untranslated regions also contain sequences that determine the subcellular localization of the mRNA.

Finally, an interesting feature of some 3′-UTRs is that the cell transcribes small antisense RNAs that are complementary to portions of the UTR; these are known as microRNAs. These antisense RNAs are believed to participate in translational control, either by directly inhibiting translation or by decreasing the stability of the mRNA.

11.3 Ribosome structure and assembly

11.3.1 Structure of ribosomes

Ribosome, a complex RNA-protein machine, acts as a platform and catalyst for the translation process. In all types of cells, one small and one large subunit, which have different functionalities, combine to form the intact ribosome (Figure 11.6). The intact bacterial ribosome is referred to as the 70S ribosome, which composed of 50S and 30S subunits. The ribosome subunits are ribonucleoprotein or RNP particles containing one or more ribosomal RNAs, rRNAs, and many different proteins. In bacteria, the 50S subunit is assembled from a 23S rRNA and a 5S rRNA, whereas the 30S subunit contains a single 16S rRNA. The large and small subunits of mammalian ribosomes are known as 60S and 40S, respectively. The two main functions, however, are carried out by the ribosome itself: decoding the genetic code in the mRNA and catalyzing the formation of peptide bonds between amino acids, resulting in a polypeptide. The peptidyl transferase center, the catalytic site, is in the large subunit. The small subunit serves as the assembly guide for all the factors needed in protein synthesis. Decoding the mRNA occurs on the small subunit. The ribosome carries three sites called P or peptidyl, A or acceptor, and E or exit, each of which is occupied in succession by a particular tRNA during the protein synthesis cycle. Each site extends across the interface to involve both subunits.

11.3.2 Ribosome biogenesis

Ribosomes function in the cytoplasm, but their assembly occurs in the nucleus, within a special subcompartment called the nucleolus. The ultrastructure of the nucleolus shows three subcompartments: a fibrillar center, a dense fibrillar component, and a granular component. The fibrillar center contains the rRNA genes. The dense fibrillar component of the nucleolus is thought to be the location of active ribosomal RNA genes. It contains a high concentration of rRNA, and the first steps of rRNA processing take place here. The fibrillar center and dense fibrillar component are embedded in a mass of 15-nm granules comprising the granular component, which represents, at least in part, the preribosomal particles and is the site of later steps of pre-rRNA processing.

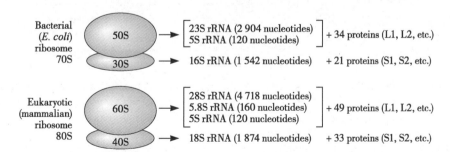

Figure 11.6 **The ribosome: an assembly of one large and one small RNP subunit.** Each subunit contains rRNA molecules and a set of bound ribosomal proteins. The two subunits and their RNAs are usually referred to by their sedimentation coefficients or S values. The sequence of the eukaryotic 5.8S rRNA corresponds to the 3′-end of the bacterial 23S rRNA.

Formation of the ribosomal particle involves a complex series of highly coordinated steps. For a ribosome to form, the cell needs to first synthesize numerous proteins and rRNA. And then the components need to be assembled into a functional particle. Ribosome assembly occurs in steps and that proteins bind to the rRNAs at different times and in a fixed order of succession, which is cooperative and hierarchical. There is an obligatory directionality in the assembly reaction, with the 5-end of the rRNA binding proteins first and the 3-end last.

In contrast to bacteria, where ribosome biogenesis occurs spontaneously, eukaryotes use a very complex process to form mature ribosome subunits, which requires numerous auxiliary factors. Eukaryotic large and small ribosomal subunits are assembled in the nucleolus of the cell from a diversity of different gene products before export to the cytoplasm in a highly coordinated, dynamic process that requires synthesis, processing, and modification of pre-rRNAs, assembly with ribosomal proteins, and transient interaction of more than 150 nonribosomal factors with the maturing pre-ribosomal particles. The earliest precursor particle in ribosome biogenesis is the 90S precursor which is split into the 40S and 60S preribosomal subunits. After export into the cytoplasm via the nuclear pore complexes, the remaining nonribosomal factors dissociate from the mature 60S and 40S ribosomal subunits (Figure 11.7). Despite this amazing diversity of participants, the assembly is highly efficient. In a human cell, as many as 14 000 ribosomal subunits are assembled and leave the nucleoli per minute.

11.4 Transfer RNA

11.4.1 tRNA molecules fold into four-arm cloverleaf structures

Each cell, whether bacterial, archaeal, or eukaryotic, contains a set of tRNA molecules, each of which mediates incorporation of one of the 20 canonical amino acids present in proteins into the polypeptide

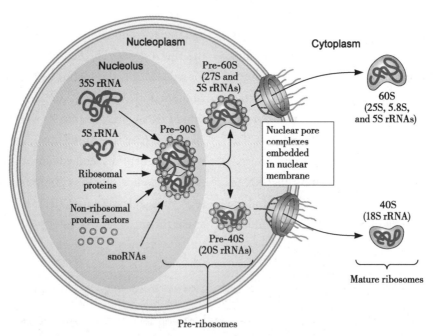

Figure 11.7 **Schematic for initial assembly, maturation, and export of the 40S and 60S ribosomal subunits in yeast.** After the pre-90S complex forms in the nucleolus, the constituent primary transcript 35S rRNA is cleaved in the spacer regions between the sequences of 18S and 5.8S rRNA, which leads to formation of the pre-ribosomal subunits pre-60S and pre-40S. The subsequent maturation of the two subunit precursors and their transport to the cytoplasm occur independently for the two subunits. The numerous auxiliary factors needed for various steps in the reaction leave the particles following export from the nucleus. Note that the mature large subunit contains 25S rRNA, in contrast to the 28S rRNA characteristic of mammals. (Adapted from Tschochner H, Hurt E [2003] *Trends Cell Biol*, 13:255–263. With permission from Elsevier.)

chain during translation. We know that there are 61 meaningful codons that specify the 20 amino acids. Because of the redundancy of the genetic code, some amino acids are encoded by more than one codon. Does this mean that there are 61 different types of tRNA in each cell? The answer is no, because some tRNAs can recognize more than one codon; the molecular basis for this is defined by the wobble hypothesis proposed by Crick in 1966. In *Escherichia coli*, for example, there are 40 tRNA types to accommodate the 20 amino acids and the 61 meaningful or sense codons. tRNA molecules containing different anticodons that specify the same amino acid are termed isoacceptor tRNA molecules. The specificity of a tRNA in terms of the amino acid it carries is designated by a superscript, for example, tRNA^{Thr}.

tRNAs are relatively short RNA molecules, usually 73–74 nucleotides long, whose sequence allows the formation of four different arms by intramolecular base pairing. The cloverleaf structure representations of two-dimensional (2D) tRNA structures and the three dimensional (3D) folding of tRNA are shown in Figure 11.8. Two of the arms are named to reflect their function: the anticodon and acceptor arms. The other two, the TψC arm and the D arm, are given their respective names because they contain a substantial fraction of invariant positions and modified bases in which tRNAs are especially rich. The anticodon arm carries the anticodon in the loop of the arm; the three bases that form the anticodon are exposed to the solution, which facilitates their interaction with the mRNA codon during protein synthesis. It is important to understand that when the anticodon pairs with the codon, a very short stretch of double-stranded A-form RNA helix is formed, following the rules of complementarity and anti-parallelism that govern double-helical nucleic acid structure.

11.4.2 tRNAs are aminoacylated by a set of specific enzymes, aminoacyl-tRNA synthetases

The overall fidelity of translation is dependent on the accuracy of two processes: codon-anticodon recognition and aminoacyl-tRNA synthesis. Aminoacyl-tRNAs are synthesized by the 3′-esterification of

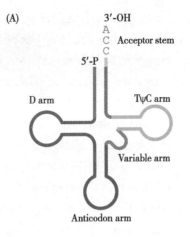

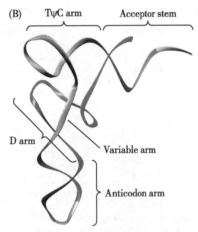

Figure 11.8 **Generalized cloverleaf secondary structure and three-dimensional (3D) folding of tRNA.** Distinct regions in (A) the cloverleaf model and (B) the 3D structure. Some arms derive their names from specifically modified residues that occur frequently in these parts of the polynucleotide: D, dihydrouridine; T, ribothymidine; ψ, pseudouridylate. The amino acid attaches through its COOH group to the 3′- or the 2′-OH groups of the ribose of the terminal adenine, A, residue in the universal CCA triplet at the acceptor stem's terminus. The anticodon arm contains the triplet anticodon, which specifies the amino acid that is to be incorporated into the growing peptide chain by interacting with the codon of the mRNA.

tRNAs with the appropriate amino acids. Aminoacylation of an "uncharged" tRNA to generate a "charged" tRNA is a high-fidelity reaction which is catalyzed by a family of enzymes known as the aminoacyl-tRNA synthetases that attach amino acid to tRNA, generating aminoacyl-tRNA in a two-step reaction that uses energy from ATP.

The aminoacyl-tRNA synthetases make an intricate series of contacts with both their amino acid and tRNA in the enzyme active site, which ensures for the most part that only the correct substrates are selected from the large cellular pool of similar candidates (Figure 11.9). Aminoacylation of tRNA is a two-step enzymatic process. (a) activation of the amino acid and (b) its subsequent addition to the CCA end of the tRNA. Both steps take place on the aminoacyl-tRNA synthetase and can occur on either the 2′- or 3′-hydroxyl of the A residue. The amino acid and a molecule of ATP enter the active site of the enzyme. In the first step, the amino acid reacts with ATP to become adenylated by the addition of AMP, and pyrophosphate is released. The amino acid is attached by a high-energy ester bond between the carbonyl group of the amino acid and the phosphoryl group of AMP. In the second step, AMP is released and the amino acid is transferred to the 3′ end of tRNA via the 2′-OH for class I enzymes and via the 3′-OH for class II enzymes. Class I synthetases contain a characteristic Rossmann fold catalytic domain, act as monomers, and couple the aminoacyl group to the 2′-OH of the tRNA. Class II synthetases share an anti-parallel β-sheet fold flanked by α-helices, are mostly dimeric or multimeric, and prefer the 3′-OH group of the tRNA. If the amino acid is attached initially to the 2′-OH group, it is then shifted to the 3′-position in an additional step, as only 3′-attached amino acids can serve as substrates in translation.

11.4.3 Proofreading activity of aminoacyl-tRNA synthetases

The aminoacyl-tRNA synthetases display an overall error rate of about 1 in 10 000. The existence and proper functioning of proofreading during charging of tRNAs with their cognate amino acids is of extreme importance to the fidelity of translation. The proofreading activities of aminoacyl-tRNA synthetases serve to hydrolyze the mismatched amino acid either before or after transfer to tRNA.

The proofreading mechanism involves water-mediated hydrolysis of the mischarged tRNA. Quality control that prevents tRNA charging with a noncognate amino acid acts at either of two different stages, pre- and post-transfer, that is, before and after attachment of the activated amino acid to the tRNA (Figure 11.10). First, the synthetic site excludes amino acids that are larger than the cognate one or that cannot establish proper specific contacts with the enzyme. Another pre-transfer mechanism may act by simply hydrolyzing the labile mischarged aminoacyl (aa) adenylate. The hydrolysis can occur at the editing site of aminoacyl-tRNA synthetase or spontaneously in solution after the noncognate aa-AMP is expelled from the enzyme. Second, if a noncognate amino acid becomes bound to the tRNA, then there is a movement of the amino acid from the synthetic site into the editing site, where the RNA-amino acid ester linkage is hydrolyzed.

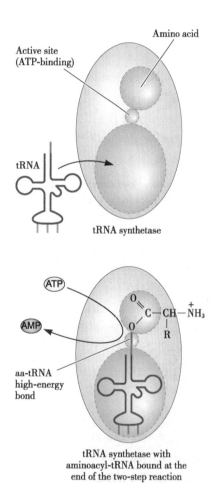

Figure 11.9 Schematic of the overall structure of tRNA synthetase. The catalytic ATP-binding site and binding sites for tRNA and amino acid are depicted in the upper part. ATP and the amino acid bind first, followed by tRNA; the blue line depicts the single-stranded universal CCA end. The lower part illustrates the structure at the end of catalysis, before release of the aminoacyl-tRNA. Often, a single aminoacyl-tRNA synthetase recognizes several isoacceptor tRNAs.

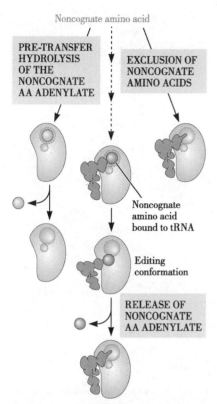

Figure 11.10 **Proofreading activity of aminoacyl-tRNA synthetases.** Quality control that prevents tRNA charging with a noncognate amino acid acts at either of two different stages, pre- and post-transfer, that is, before and after attachment of the activated amino acid to the tRNA. First, the synthetic site excludes amino acids that are larger than the cognate one or that cannot establish proper specific contacts with the enzyme. Another pre-transfer mechanism may act by simply hydrolyzing the labile mischarged aminoacyl (aa) adenylate. The hydrolysis can occur at the editing site of aminoacyl-tRNA synthetase or spontaneously in solution after the noncognate aa-AMP is expelled from the enzyme. Second, if a noncognate amino acid becomes bound to the tRNA, then there is a movement of the amino acid from the synthetic site into the editing site, where the RNA-amino acid ester linkage is hydrolyzed. This movement is possible because of the flexibility of the CCA end of the acceptor stem of the tRNA.

11.5 The process

The overall process of translation is usually divided into three major phases: initiation, elongation, and termination. Each stage of protein synthesis involves multiple accessory factors and energy from GTP hydrolysis. Here, we describe these steps in the order of occurrence during translation.

11.5.1 Initiation of translation

Initiation is the most complex and the most tightly controlled of the steps in protein synthesis. Initiation involves the reactions that precede formation of the peptide bond between the first two amino acids of the polypeptide. It requires the ribosome to bind to the mRNA, which forms an initiation complex that contains the first aminoacyl-tRNA. This is a relatively slow, rate-limiting step in translation and usually determines the rate at which an mRNA is translated.

(1) Initiation in Bacteria Needs 30S Subunits and Accessory Factors

The 30S subunit is involved in initiation. But it is not sufficient by itself to undertake the reactions of binding mRNA and tRNA. Initiation factors, which bind to 30S subunits, are also required. Bacteria use three initiation factors, numbered IF-I, IF-2, and IF-3. IF-1 binds to 30S subunits as a part of the complete initiation complex. It binds in the vicinity of the A site and prevents aminoacyl-tRNA from entering. Its location also may impede the 30S subunit from binding to the 50S subunit. IF-2 binds a special initiator tRNA and controls its entry into the ribosome and has a ribosome-dependent GTPase activity. IF-3 has multiple functions: It is needed to stabilize (free) 30S subunits and to inhibit the premature binding of the 50S subunit; it enables 30S subunits to bind to initiation sites in mRNA; and as part of the 30S-mRNA complex, it checks the accuracy of recognition of the first aminoacyl-tRNA (Figure 11.11).

1) Initiation of bacterial translation requires separate 30S and 50S ribosome subunits. IF-3 binds to free 30S subunits that are released from the pool of 70S ribosomes. The presence of IF-3 prevents the 30S subunit from re-associating with a 50S subunit.

2) A 30S subunit binds directly to a ribosome-binding site on mRNA. Initiation occurs at a special sequence on mRNA called the ribosome-binding site which consists of the Shine-Dalgarno sequence and the AUG initiation codon (in bacteria GUG or UUG may also be used less often). Approximately 10 bases upstream of the AUG is a sequence that corresponds

Figure 11.11 Formation of bacterial 70S initiation complex. Initiation requires the ribosome to position the initiator fMet-tRNAi^fMet over the start codon of mRNA, which is located in the P half-site of the ribosome. This precise positioning involves the action of three initiation factors, IF1, IF2, and IF3, the roles of which remain largely unclear. The process probably begins with binding of initiation factor IF3 to the small subunit. The intact ribosome will have split into its small and large subunits at the end of the previous round of translation, following translation termination. IF3 binding occurs following the dissociation of the 50S and 30S subunits, which stimulates the release of deacylated tRNA and mRNA from the small subunit at the end of translation. IF3 prevents the large subunit from re-associating, that is, it acts as an anti-association factor; in addition, it ejects any tRNA other than fMet-tRNAi^fMet from the P site. The role of IF1 is less clear. At the next step, IF2-GTP specifically recognizes the charged initiator tRNA; it binds to the 30S subunit and facilitates binding of fMet-tRNAi^fMet. The 30S complex interacts with mRNA by recognizing the Shine-Dalgarno sequence. The complex containing the small subunit, fMet-tRNAi^fMet, and mRNA is known as the 30S initiation complex; the three initiation factors are still bound. In the next step, the 50S subunit joins the complex to form the 70S initiation complex. IF3 and IF1 must leave before the 50S subunit joins. IF2 promotes subunit joining and most probably remains bound to the 70S complex until GTP bound to IF2 is hydrolyzed. The structure of fMet is shown at the bottom.

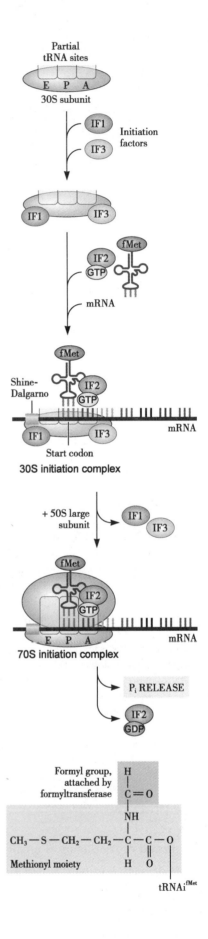

to part or all of the hexamer: 5'...AGGAGG...3'. This polypurine stretch is known as the Shine-Dalgarno sequence. The 3' end of 16S rRNA of the 30S bacterial ribosomal subunit has a highly conserved sequence, 3'...UCCUCC...5' (written in reverse direction), which is complementary to the Shine-Dalgarno sequence.

3) A Special Initiator tRNA—fMet-tRNA^fMet Starts the Polypeptide Chain. Translation starts with a methionine amino acid usually encoded by AUG. In bacteria, the initiator tRNA carries a methionine residue that has been formylated on its amino group, forming a molecule of N-formyl-methionyltRNA. This tRNA is used only for initiation. It recognizes the codons AUG or GUG (occasionally UUG). Formylation improves the efficiency of the recognition by the factor IF-2 that binds the initiator tRNA. IF-2, bound to GTP, binds the initiator fMet-tRNA, and allows it to enter the partial P site on the 30S subunit. IF-1 binds to the A site of 30S subunits to prevent aminoacyl-tRNA from entering.

4) Formation of initiation complex. Initiation factors are found only on 30S subunits and they must be released to allow 50S subunits to join the 30S-mRNA complex to generate 70S ribosomes. IF-2 has a ribosome-dependent GTPase activity: It sponsors the hydrolysis of GTP in the presence of ribosomes, releasing the energy stored in the high-energy bond to release three initiation factors. Thus complete initiation complex is formed. The initiator fMet-tRNA^{fMet} occupies the P site and the A site is available for entry of the aminoacyl-tRNA complementary to the second codon of the mRNA.

(2) Initiation in eukaryotes

Initiation of translation in eukaryotic cytoplasm resembles the process that occurs in bacteria, but the order of events is different and the number of accessory factors is greater (Figure 11.12). Eukaryotic cells have more initiation factors than bacteria: The current list includes I2 factors that are directly or indirectly required for initiation. They act at all stages of the process, including: forming an initiation complex with the 5′ end of mRNA; forming a complex with Met-tRNA; binding the mRNA-factor complex to the Met-tRNA; enabling the ribosome to scan mRNA from the 5′ end to the first AUG; detecting binding of initiator tRNA to AUG at the start site; mediating joining of the 60S subunit.

Some of the differences in initiation are related to a difference in the way that bacterial 30S and eukaryotic 40S subunits find their binding sites for initiating translation on mRNA. In eukaryotes, small subunits first recognize the 5′ end of the mRNA and then move to the initiation site, where they are joined by large subunits.

The initiation stage in eukaryotes can be further subdivided into four steps: ternary complex formation and loading onto the 40S subunit, loading of the mRNA, scanning and start codon recognition, and joining of the 40S and 60S subunits to form the functional 80S ribosome.

1) Ternary complex formation and loading onto the 40S ribosomal subunit. The first step in the initiation pathway is the assembly of a ternary complex of eukaryotic initiation factor 2 (eIF2), GTP, and the amino acid-charged initiator tRNA (Met-tRNA) that binds to the 40S subunit before it associates with mRNA. Eukaryotic initiator tRNA is a Met-tRNA, the methionine is not formylated. This complex binds to the 40S ribosomal subunit, in association with other initiation factors, to form a 43S pre-initiation complex.

2) Loading the mRNA on the 40S ribosomal subunit. A cap-binding complex binds to the 5′ end of mRNA prior to association of the mRNA with the 40S subunit. The factor eIF4F mediates the loading of mRNA on the 43S pre-initiation complex. eIF4F is a protein complex that contains three of the initiation factors, eIF4E eIF4A and eIF4G. eIF4A, eIF4B, eIF4E, and eIF4G bind to the 5′ end of the mRNA to form the cap binding complex. This complex associates with 3′ end of the mRNA via eiF4G, which interacts with poly(A)-binding protein (PABP). Together, the 5′ and 3′ initiation factor complexes work to load the mRNA onto the 43S complex.

3) Scanning and AUG recognition. The majority of eukaryotic initiation events involve scanning from the 5′ cap. Once the mRNA is loaded, the 43S complex then scans along the message from 5′ to 3′ looking for the AUG start codon. This process requires expenditure of energy in the form of ATP, and thus factors associated with ATP hydrolysis (efF4A, IF4B, and e1F4F) also play a role in this step. The small subunit stops when it reaches the initiation site, at which point the initiator tRNA base pairs with the AUG initiation codon, forming a stable 48S complex. The codon-anticodon interaction occurs by complementary base pairing, with an antiparallel orientation between the tRNA and mRNA. Usually the first AUG is embedded in a favorable sequence context called the Kozak consensus sequence (ACCAUGG, sequence context around AUG).

4) Joining of the 40S and 60S ribosomal subunits. Junction of the 60S subunits with the initiation complex cannot occur until eIF2 and eIF3 have been released from the initiation complex. This is mediated by eIF5 and causes eIF2 to hydrolyze its GTP. The reaction occurs on the small ribosome subunit and requires the base pairing of the initiator tRNA with the initiation codon. All of the remaining factors likely are released when the complete 80S ribosome is formed. Finally, the factor eIF5B enables the 60S subunit to join the 40S/Met-tRNA/mRNA complex, forming an intact ribosome that is ready to start elongation.

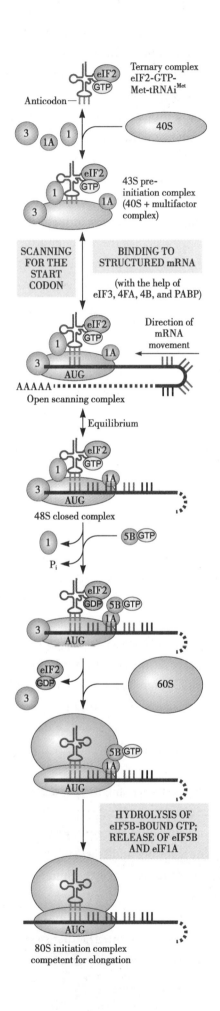

Figure 11.12 **Initiation of translation in eukaryotes.** In the first step of this multistep process, the ternary complex eIF2-GTP-Met-tRNAi^{Met} binds to the small 40S ribosomal subunit. This involves the active participation of three initiation factors, which bind cooperatively to the small subunit. eIF1A is the homolog of bacterial IF1 and eIF3 is a scaffolding factor for the assembly, made up of 13 subunits in mammals. eIF1 is a functional homolog of bacterial IF3: although the two proteins share no sequence homology, they bind at the same location on the small subunit and can function in heterologous initiation assays. This first complex has a sedimentation coefficient of 43S. If the mRNA contains no stable secondary structures, the 43S complex can bind directly to the message without the need for additional factors. Additional factors are required, however, to bind and scan structured mRNA; for simplicity, these factors are named but not depicted in the figure. In either case, this open complex scans along the mRNA from the 5'-end, until Met-tRNAi^{Met} base-pairs with the AUG codon in the partial P site of the small subunit. The open complex is in dynamic equilibrium with the closed complex, which is not capable of scanning but investigates the codon in the P site; when an AUG codon is identified, the equilibrium shifts toward the closed state and tRNA binds to AUG. Upon recognition of the start codon, a conformational change occurs in the ribosomal complex, and eIF1 dissociates. eIF1's departure from the complex, or at least its moving away from its rRNA binding site while still interacting with other factors, is instrumental in phosphate (P_i) release, making hydrolysis irreversible and thus committing the pre-initiation complex to initiation at the codon currently in the P site. In the next step, eIF5B-GTP binds to the complex, making important contacts with the C-terminal tail of eIF1A. The binding of eIF5B-GTP to the 48S complex catalyzes the association with the large subunit to form the 80S initiation complex; at this stage, the bound initiation factors are displaced by the large subunit. In the final step, GTP hydrolysis, catalyzed by the large ribosomal subunit, leads to eIF5B dissociation, leaving the complex ready for elongation.

11.5.2 Elongation and events in the ribosome tunnel

Once the complete ribosome is formed at the initiation codon, the stage is set for a cycle in which an aminoacyl-tRNA enters the A site of a ribosome whose P site is occupied by a peptidyl-tRNA. Elongation includes all the reactions from formation of the first peptide bond to addition of the last amino acid. Amino acids are added to the chain one at a time; the addition of an amino acid is the most rapid step in translation. Aminoacyl-tRNAs enter the acceptor (A) site, where decoding takes place. If they are the correct (cognate) tRNAs, the ribosome catalyzes the formation of a peptide bond between the incoming amino acid and the growing polypeptide chain. After the tRNAs and mRNA are translocated such that the next codon is moved to the A site, the process is repeated. This cycle is repeated until a stop codon is reached and the process of termination begins (Figure 11.13).

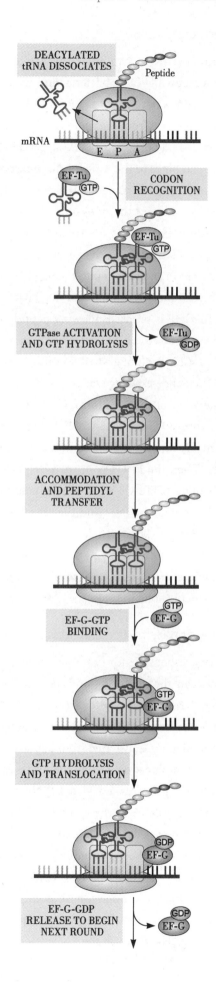

Figure 11.13 **Translation elongation in bacteria.** During initiation, the initiator tRNA carrying fMet is bound to the P site. Elongation is a cyclic process in which the addition of each incoming amino acid to the growing polypeptide chain occurs in three major steps: positioning of the correct aminoacyl-tRNA (aa-tRNA) into the A site of the ribosome, formation of the peptide bond, and translocation or shifting of the mRNA with respect to the ribosome over the length of a codon; note that during translocation the two tRNAs in the P and A sites also translocate. Other steps that occur alongside these three major steps are necessary to create favorable conditions for the major steps. An incoming acylated or charged tRNA is delivered to the A site by elongation factor EF-Tu bound to GTP. Correct base-pairing or recognition between the codon and the anticodon activates the GTPase center of the ribosome and GTP hydrolysis occurs, leading to a conformational change in EF-Tu and its release from the ribosome. This release in turn leads to conformational changes in the rRNA and the tRNA, which optimally orient the two tRNAs for the peptidyl transfer reaction to occur. During the peptidyl transfer reaction, the nascent polypeptide is moved from the tRNA in the P site to the aa-tRNA in the A site. Once the peptidyl transfer reaction occurs, there is a need to shift the ribosome in the 3'-direction on the mRNA, so that the next codon becomes available for decoding, that is, binding of a cognate aa-tRNA. GTP-bound EF-G is involved in this step: GTP hydrolysis leads to the movements of the two tRNAs as depicted. At this state, the ribosome is ready for a new round of elongation.

(1) Decoding the message

Any aminoacyl-tRNA can enter the A site by an elongation factor EF-Tu in bacteria, and its pairing with the codon is necessary for EF-Tu to hydrolyze GTP and be released from the ribosome. Aminoacyl-tRNA is loaded into the A site in two stages. First, the anticodon end binds to the A site of the 30S subunit. Then, codon-anticodon recognition triggers a change in the conformation of the ribosome. This stabilizes tRNA binding and causes EF-Tu to hydrolyze its GTP. Upon GTP hydrolysis, the aminoacyl-tRNA is released by EF-Tu and swings into the peptidyl transferase center of the 50S subunit and the binary complex EF-Tu-GDP is released, allowing peptide bond formation to proceed. This form of EF-Tu is inactive and does not bind aminoacyl-tRNA effectively. The guanine nucleotide exchange factor, EF-Ts, mediates the regeneration of the used form, EF-Tu-GDP, into the active form EF-TuGTP. First, EF-Ts displaces the GDP from EF-Tu, forming the combined factor EF-Tu-EF-Ts. Then the EF-Ts is in turn displaced by GTP, reforming EF-Tu-GTP. The active binary complex binds aminoacyl-tRNA, and the released EF-Ts can recycle (Figure 11.14). In eukaryotes, the factor eEF1α is responsible for bringing aminoacyl-tRNA to the ribosome, also in a reaction that involves cleavage of a high-energy bond in GTP.

(2) Peptide bond formation and translocation

The activity responsible for synthesis of the peptide bond between the incoming amino acid and the peptidyl-tRNA is called peptidyl transferase. It is a function of the large ribosomal subunit. The nascent polypeptide chain is transferred from peptidyl-tRNA in the P site to aminoacyl-tRNA in the A site. Peptide bond synthesis generates deacylated tRNA in the P site and peptidyl-tRNA in the A site (Figure 11.15). Translocation moves deacylated tRNA into the exit (E) site of the large ribosomal subunit and peptidyl-tRNA into the P site and empties the A site. And the mRNA is moved through the ribosome by three nucleotides to place the next codon of the mRNA into the A site. The tRNA in the E site is released as

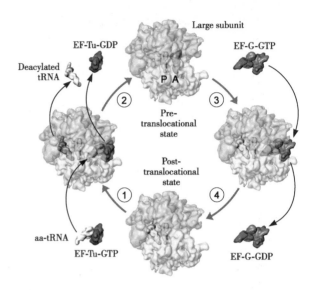

Figure 11.14 **Elongation cycle.** The cycle begins with the complex at the bottom of the figure in the post-translocational state. It is ready to accept a new aminoacyl-tRNA which is presented to the ribosome as a ternary complex that also includes the elongation factor EF-Tu and a GTP. This step produces the structure on the left: the ribosome with abound ternary complex. It has been suggested that when the ternary complex binds to the ribosome, the E-site tRNA moves further away from the P site to the E2 site. The snapshot shown here represents part of the decoding step when the aminoacyl-tRNA whose anticodon matches the next codon in the mRNA is selected to enter the A site, accompanied by GTP hydrolysis and conformational changes. If the codon and anticodon match, EF-Tu with the GDP formed by the hydrolysis of GTP leaves the ribosome, accompanied by a relaxation in the formerly strained conformation of the A-site RNA, termed accommodation in; the E-site tRNA also leaves the ribosome. The ribosome is now in the pre-translocational state, shown at the top, with tRNAs in the A and P sites, respectively. After peptidyl transfer, the nascent peptide, not shown, is covalently attached to the A-site tRNA. On the right, the elongation factor EF-G in complex with GTP has bound to the ribosome to facilitate the translocation of the peptidyl-tRNA to the P site and the deacylated tRNA to the E site. The translocation is induced by GTP hydrolysis accompanied by large transient conformational changes in the EF-G and the ribosome. The release of EF-G after GTP hydrolysis leaves the ribosome in the post-translocational state, shown at the bottom, to close the elongation cycle.

a new tRNA enters the A site. Translocation requires GTP and another elongation factor, EF-G. The structure of EF-G mimics the overall structure of EF-Tu bound to the amino acceptor stem of aminoacyl-tRNA (Figure 11.16). The eukaryotic homolog of EF-G is eEF2.

Figure 11.15 **Formation of a peptide bond.** Decoding, or matching of the codon with a cognate anticodon, occurs at the decoding center, close to the A site in the small subunit, whereas actual peptide bond formation occurs on the 23S rRNA of the large subunit. For simplicity, only the small subunit A and P half-sites are depicted here.

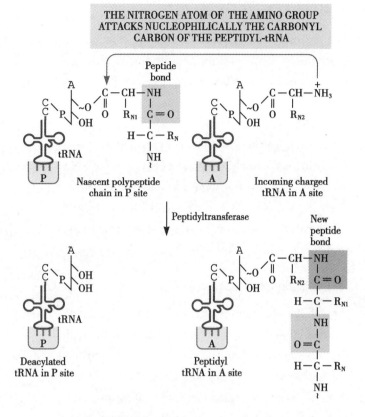

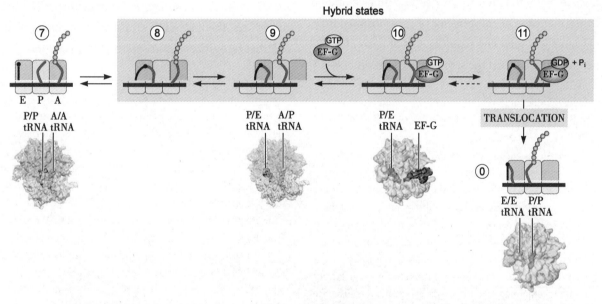

Figure 11.16 **Discernible distinct states and reversible versus irreversible steps in translocation.** State 7 shows pre-translocational complex. The nascent polypeptide is covalently linked to the A-site tRNA, the P-site tRNA is deacylated, and the E site is unoccupied. The tRNAs are in their classical A/A and P/P positions, and the L1 stalk is in an open conformation. (State 8) The pre-translocational complex is in an intermediate state of ratcheting: the ribosome is in a semirotated state and the tRNAs are in an intermediate state with classical A/A and hybrid P/E positions. The L1 stalk is in a closed position, forming a direct contact with the hybrid P/E tRNA. (State 9) The ribosome is in a rotated state, tRNAs are in hybrid A/P and P/E configurations, and the L1 stalk is in a closed conformation where it directly interacts with the hybrid P/E tRNA. State 9 can be reached directly from state 7, bypassing the intermediate state 8. (State 10) Elongation factor EF-G-GTP is bound, stabilizing the hybrid complex. (State 11) Following GTP hydrolysis, EF-G is bound in the GDP-P_i state. The transition to the 0 state occurs upon release of the GDP-bound EF-G and is irreversible. The ribosome returns to the nonrotated position, the newly formed peptidyl-tRNA and the newly deacylated tRNA move into the classical P/P and E/E configurations, and the L1 stalk moves back into the open position. (State 0) Post-translocational complex, ready to begin a new cycle. (Adapted from Frank J, Gonzalez RL Jr [2010] *Annu Rev Biochem*, 79:381–412. With permission from Annual Reviews.)

(3) Peptidyl transferase activity

The chemical reaction catalyzed by the ribosome is a simple one—the joining of amino acids through peptide bonds. The ribosome accelerates the rate of peptide bond formation by at least 10^7-fold. A central question for many years was whether the "peptidyl transferase activity" that catalyzes peptide bond formation is the result of a protein enzyme or an RNA ribozyme. Peptidyl transferase activity resides exclusively in the 23S rRNA in prokaryotes (Box 11.1), and in the 28SrRNA in eukaryotes, although proteins are probably needed to acquire the correct structure.

(4) Events in the ribosome tunnel

One of the key features of the ribosome is the long "tunnel" that conducts the nascent protein through the large subunit from the site of peptidyl transferase activity to the peptide exit hole. The tunnel is a dynamic structure that can expand to allow polypeptides to begin to fold into their rudimentary globular conformation. The components of the ribosomal tunnel somehow sense the requirements of nascent polypeptides and connect them to downstream processes.

11.5.3 Termination of translation

Termination takes place when a stop codon is encountered, and the completed polypeptide chain is released from the ribosome. The final steps in protein synthesis also involve expulsion of the tRNA from the E site, dissociation of the ribosome from the mRNA, and dissociation of the ribosome into its subunits, leaving them ready for another round of initiation.

The termination of protein synthesis by the ribosome requires release factor (RF). Termination codons are recognized by protein release factors, not by aminoacyl-tRNAs. In *E. coli*, two release factors are specific for different sequences. RF1 recognizes UAA and UAG; RF2 recognizes UGA and UAA, and activate the ribosome to hydrolyze the peptidyl-tRNA. RF3 are GTPases that stimulate the activity of RF1 and RF2 (Figure 11.17). The dissociation of the remaining components (tRNA, mRNA and the ribosome) requires ribosome releasing factor (RRF). In eukaryotes, there is only a single release factor called eRF that recognizes all three termination codons.

Box 11.1 Evidence that peptidyl transferase activity resides in rRNA

Biochemical evidence that 23S rRNA is a ribozyme

The involvement of rRNA was first indicated because a region of the 23S rRNA is the site of mutations that confer resistance to antibiotics that inhibit peptidyl transferase. Noller's group treated 50S ribosomes from the bacterium *Thermus aquaticus* with proteinase K and sodium dodecyl sulfate (SDS) to degrade proteins and to remove proteins from nucleic acids. They showed that after protein extraction, the remaining 23S rRNAs still had peptidyl transferase activity. Importantly, they also showed that this catalytic activity was inhibited by treatment with ribonucleases and by known peptidyl transferase inhibitors. Their findings strongly suggested that 23S rRNA alone can catalyze peptide bond formation.

Structural evidence that rRNA forms the active site of the ribosome

Atomic resolution views of the large ribosomal subunit revealed that about two-thirds of the ribosome's mass is composed of RNA. The structures showed that rRNA both creates a structural framework for the ribosome and, at the same time, forms the main features of its functional sites. The rRNA forms most of the inter subunit interface, the peptidyl transferase center, the decoding site, and the A and P sites. In striking contrast, the ribosomal proteins are abundant on the exterior of the ribosome, but not in the active site. The ribosomal proteins follow the contours of the RNA. No protein is located within 15 Å of the site of catalysis. The approximately 10^7-fold rate of enhancement of peptide bond formation by the ribosome appears to be mainly due to substrate positioning by the 23S rRNA within the active site, rather than to chemical catalysis. The 23S rRNA is organized in six major domains composed of connecting loops and helices. The peptidyl transferase center is located in domain V of the 23S rRNA. What are the functions of ribosomal proteins in the ribosomes? Ribosomal RNA folds correctly only by assembling with ribosomal proteins, which orchestrate the order of folding to avoid kinetic misfolding traps. In addition to protecting the rRNA from nuclease attack and helping the rRNA to fold into a more effective conformation, the ribosomal proteins appear to make contributions to efficient and accurate protein synthesis.

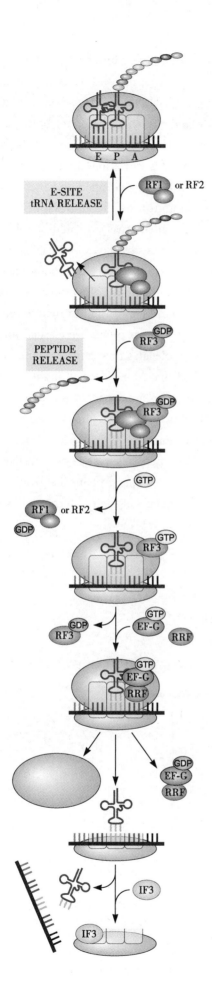

Figure 11.17 **Termination of translation in bacteria.** A stop codon, indicated by red nucleotides, recruits either release factor 1 or 2, RF1 or RF2, to mediate hydrolysis and release of the peptide from the tRNA in the P site. This functions as a signal to recruit RF3-GDP. Exchange of GDP for GTP on RF3, followed by GTP hydrolysis, is thought to release RF1 or RF2. The residual complex between the ribosome, mRNA, and a deacylated-tRNA in the P site is disassembled by binding of ribosome releasing factor RRF and elongation factor EF-G. GTP hydrolysis causes 50S subunit dissociation. Initiation factor 3, IF3, is required for tRNA dissociation from the P site and mRNA release, which occur at the same rate, presumably simultaneously.

11.6 Translational and post-translational control

Translational control is another important level of gene regulation. Translational regulation can occur at several levels, from control of the availability of ribosomes to adjustments to the rate of translational initiation and elongation. Although translation elongation is very similar in bacteria and eukaryotes, initiation is rather different in these domains of life, especially in terms of the way the mRNA is presented to the appropriate site of the ribosome. The abundance, accessibility, and stability of mRNAs are also factors. The mechanism centers on protein phosphorylation and thus also provides an excellent illustration of post-translational control by protein modification.

11.6.1 Regulation of translation by controlling ribosome number

Ribosome numbers in bacteria are responsive to the environment. Depending on the environmental conditions, which dictate appropriate rates of growth, bacteria can have as few as 2000 ribosomes or as many as 100 000 ribosomes per cell. This number will depend upon both the rate of synthesis of ribosomal components and the rate of degradation of ribosomes. Control of these processes relies on the transcription of rRNA genes and the translation of mRNAs that encode r-proteins.

11.6.2 Regulation of translation initiation

Controlling translation initiation is the fastest and most economical way to change the protein profile in a cell. This simply means that the 5′ UTR of the mRNA has a poor sequence context that does not allow rapid ribosome binding or movement onto the ORF. Bacteria and their viruses, or phages, make proficient use of secondary structures in the mRNA to control initiation. Bacteria also make use of riboswitches and of small noncoding regulatory RNAs, such as microRNAs and short interfering RNAs. In eukaryotic cells, regulation may depend on protein factors binding to the 5′- or 3′-ends of mRNA.

11.6.3 mRNA stability and decay in eukaryotes

One factor that determines protein levels in the cell is the stability of their respective mRNAs which is carefully controlled so that the encoded protein is made available only when it is needed and so that protein synthesis can respond to external and internal signals. These mRNAs are constantly being synthesized and degraded, and messages that encode different proteins may have very different half-lives, ranging from minutes to hours and, in exceptional cases, even months. The two major pathways of degrading mRNA, 5′→3′ and 3′→5′, are both pathways start with deadenylation, shortening of the poly(A) tail. Following deadenylation, the pathways diverge: the mRNA could be decapped and then degraded through the action of the abundant 5′→3′ exonuclease Xrn1; alternatively, it could be subjected to 3′→5′ exonucleolysis by the exosome (Figure 11.18). Unused mRNA is sequestered in P bodies and stress granules. Some of the mRNA molecules accumulated into P bodies, and then into stress granules, are degraded, whereas others are rescued for future use. mRNA molecules that contain premature stop codons which have long been thought to produce truncated proteins are degraded through nonsense-mediated decay or NMD. Non-stop decay or NSD functions when mRNA does not contain a stop codon-if these messages were to be translated, they would give rise to longer proteins that would contain additional sequences at their C-termini, now translated from the 3′-UTRs.

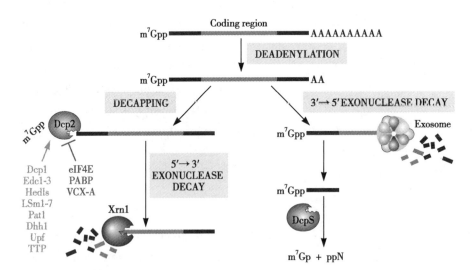

Figure 11.18 **Two major mRNA decay pathways in eukaryotes.** Both pathways initiate with shortening of the poly(A) tail, followed by either 5′→3′ or 3′→5′ exonucleolytic cleavage of the message. 5′→3′ decay, shown on the left side, is preceded by decapping of the mRNA by the Dcp2 decapping enzyme, which produces m⁷Gpp and 5′-monophosphate RNA. Dcp2 is a member of the NUDIX or nucleotide diphosphate linked to moiety X family of hydrolases; it will only hydrolyze cap structures that are linked to an RNA moiety of, in general, more than 25 nt in length. Thus, Dcp2 recognizes mRNA substrates by simultaneously interacting with their cap and the RNA body. Dcp2 is regulated by numerous proteins, either positively or negatively. Once mRNA is decapped, it is further degraded by the 5′→3′ exonuclease Xrn1. 3′→5′ decay, shown on the right side, involves the exonucleolytic action of the exosome. When the length of the RNA body becomes less than 10 nt, decapping occurs through the action of the DcpS or scavenger decapping enzyme.

11.6.4 Phosphorylation

Translational control by eIF2 was the first example of regulation of eukaryotic gene expression at the level of protein synthesis. The fundamental mechanism for translational control by eIF2 centers on protein phosphorylation and thus also provides an excellent illustration of post-translational control by protein modification (Figure 11.19).

(1) Phosphorylation of eIF2α blocks ternary complex formation

Many different types of stress, such as hypoxia, viral infection, amino acid starvation, and heat shock, repress translation by triggering the phosphorylation of the α-subunit of eIF2 at residue Ser51. This inhibits the exchange of GDP for GTP on the eIF2 complex, which is catalyzed by eIF2B, and thereby prevents formation of the ternary complex.

(2) eIF2α phosphorylation is mediated by four distinct protein kinases

The phosphorylation of eIF2α is mediated by four distinct protein kinases in mammals: heme-regulated inhibitor kinase (HRI), protein kinase RNA (PKR), PKR-like endoplasmic reticulum kinase (PERK), and the mammalian ortholog of yeast general control

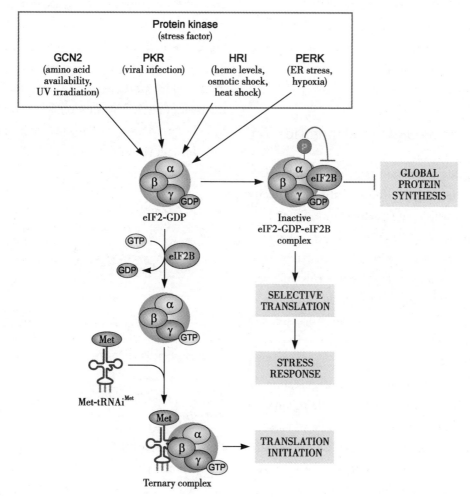

Figure 11.19 **Phosphorylation of eIF2α integrates the translational responses of eukaryotic cells to various stress factors.** Four distinct protein kinases—GCN2, general control non-derepressible 2; PKR, protein kinase RNA; HRI, heme-regulated inhibitor kinase; and PERK, protein kinase RNA-like endoplasmic reticulum (ER) kinase—sense different types of stress signals and phosphorylate eIF2α. eIF2α is one subunit of the three-subunit eIF2 complex. Together with GTP and the methionine-charged initiator tRNA, eIF2 forms the ternary complex that delivers the initiator tRNA to the ribosome during initiation. Phosphorylation of eIF2α inhibits GDP-GTP exchange by reducing the dissociation rate of nucleotide exchange factor eIF2B. As a result, global initiation is inhibited. However, selective translation of a subset of mRNAs continues; these messages code for proteins that mediate the stress response.

non-derepressible 2 (GCN2). These protein kinases share extensive homology in their kinase domain and integrate diverse stress signals into a common translational control pathway. The actions of PKR and HRI are highlighted to illustrate the diversity of signaling pathways involved in translational control.

From the above, a number of very different mechanisms are involved in translational control to regulate the level of protein production in cells. These include regulating the number of ribosomes available for synthesis, control of initiation of translation, or regulation by riboswitches or microRNAs. Finally, the possibility of translation depends on the stability of individual mRNAs. The multitude of transcriptional and translational regulators is amazing in its total complexity and subtlety. Perhaps we should not be amazed, as this is what it takes to allow any complex and adaptable organism to function in life: to survive, replicate, and evolve.

11.7 Protein processing and modification

Nascent peptide chains are not biologically active which must be properly folded into biologically active three-dimensional structures. And many protein structures are modulated after the gene sequence has been read by transcription and translation. This further modulation has the general title of post-translational processing, and it takes a number of forms. The consequences of protein processing are enormously varied: processing can dictate the cellular destination of a new protein molecule, modify its functional properties, or prescribe its degradation.

Most proteins require molecular chaperones to fold properly within the cell, either while still associated with the ribosome, or after release in the cytoplasm or in specific compartments such as the endoplasmic reticulum. Misfolded proteins are targeted for degradation by the ubiquitin-mediated protein degradation pathway. In this pathway, a chain of ubiquitin molecules is attached to the misfolded protein by means of a series of enzymatic reactions. Polyubiquitin targets the protein to the proteasome for proteolytic degradation.

Many of the modifications take place in specific, membrane-bound compartments such as the endoplasmic reticulum and the Golgi apparatus. The proteins that are processed must get into and out of these organelles. In fact, the passage of proteins through membranes in a directed way often determines where the protein will end up in the cell, or even outside the cell. Thus, we must consider that the first stage in protein post-translational processing frequently involves transport into or through membranes.

11.7.1 Structure of biological membranes

Biological membranes are protein-rich lipid bilayers. All biomembranes have a common structure: a bilayer of lipid molecules with hydrophobic tails buried in the interior of the bilayer and hydrophilic head groups on the two surfaces, with also attached or embedded proteins. While lateral motion within the membrane is easy, motion across the membrane by either head groups or any hydrophilic molecule is very difficult (Figure 11.20).

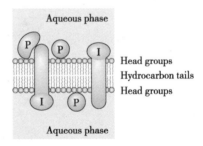

Figure 11.20 **The common bilayer structure in all biomembranes.** P denotes peripheral membrane proteins; I denotes integral membrane proteins.

11.7.2 Protein translocation through biological membranes

Protein translocation can occur during or after translation. Some proteins will be cytoplasmic, some are destined for various organelles or their membranes, and still others are secreted. The mechanisms of protein translocation across membranes can generally be classified as post-translational or co-translational, depending on whether the translocation occurs after synthesis is complete and the polypeptide is released from the ribosome or during polypeptide chain elongation in the ribosome (Figure 11.21). Proteins are targeted to their destined organelles the signal peptide in N-terminal (Figure 11.22). And translocation involves specialized channels in membranes, known as translocons. Translocons are evolutionarily conserved heterotrimeric protein complexes that undergo significant motions during translocation. Thus, the central pore through which the polypeptide passes is usually closed by a plug, a portion of one of the transmembrane domains; for transloca-

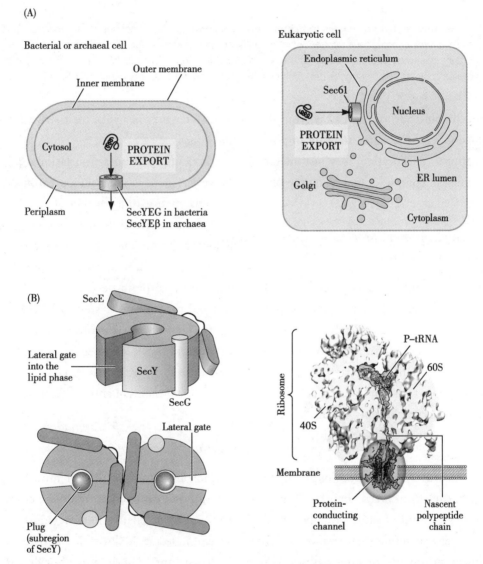

Figure 11.21 **Schematic of the conserved principle of protein exports through membranes and evolutionarily conserved protein translocation machinery.** (A) In bacteria and archaea, shown on the left side, proteins are transported from the cytosol into the periplasm or the extracellular environment. In eukaryotes, shown on the right side, proteins are transported from the cytoplasm, where synthesis is initiated, to the endoplasmic reticulum (ER) lumen, then to the Golgi complex, and finally, depending on the nature and site of action of the particular protein, to the cellular membrane, extracellular space, or other cellular compartments such as the lysosome. (B) Transport across membranes can occur post-translationally or co-translationally. Translocons are universal conduits through membranes for both co-translational and post-translational protein delivery. In bacteria, shown on the left side, the translocon is a heterotrimeric complex consisting of SecY, SecE, and SecG, located within the inner plasma membrane. When there is no protein in the translocon, its aqueous conduit or pore is plugged by a portion of the second transmembrane domain of SecY; the plug moves away during translocation. The arrangement of the transmembrane helices in SecY forms a lateral gate into the lipid phase of the membrane. This gate is used for insertion of integral membrane proteins into the membrane. In eukaryotes, shown on the right side, the complex that forms translocons in the endoplasmic reticulum membrane consists of Sec61α, Sec61β, and Sec61γ, which correspond to SecY, SecG, and SecE, respectively. The schematic represents a translating and translocating ribosome-Sec61 complex from yeast. The translocon, circled part, is at the end of the exit tunnel of the ribosome.

Figure 11.22 **N-terminal signal sequences in bacteria and eukaryotes.** The signal peptides are typically 15–30 amino acids long. There is no simple consensus sequence; in general, there are three distinct compositional zones: the N-terminal region often contains positively charged residues; the hydrophobic region contains at least six hydrophobic residues and the C-terminal region consists of polar uncharged residues. The lightning bolts denote cleavage sites.

tion to occur, the plug has to move away from the pore. Translocons serve the same function in post- and co-translational translocation; they just interact with several different complexes that bring the protein to the pore (Figure 11.21B).

11.7.3 Proteolytic protein processing: cutting, splicing, and degradation

Some proteins are synthesized as precursor molecules that need to undergo post-translational processing, sometimes involving proteolytic cleavage, to attain their functional mature form. Others are subject to protein splicing, a reaction very much analogous to the post-transcriptional splicing that occurs on RNA primary transcripts. Finally, there are situations where proteins need to be destroyed, either selectively or wholesale.

Proteolytic cleavage is sometimes used to produce mature proteins from precursors. Thus, signal peptides at the N-termini of some proteins, which help them to traverse membranes, are removed early in the translocation process. In some cases, this is the only proteolytic cleavage needed to produce the final form of the protein. Other protein precursors undergo a series of very specific cleavages by specific enzymes to reach their functional form. When translation is complete, the signal peptide is also cleaved and degraded by a special signal peptidase.

In protein splicing, different portions of a polypeptide chain are cut and then spliced together to form a new molecule, thus create circular proteins of extremely high stability and activity, which can withstand boiling, extremes of pH, and the action of proteolytic enzymes with their structure and function intact.

An example of post-translational processing is shown in Figure 11.23, from a preproinsulin precursor to the mature functional insulin.

When a particular protein is no longer needed or no longer appropriate, it will be degraded. An important example of a mechanism for highly regulated and specific protein destruction is the ubiquitin-initiated proteasome system, which uses covalent tagging of unneeded proteins to mark them for destruction. Much less specific proteolysis is a part of apoptosis, or programmed cell death, which can destroy the proteins of entire cells or tissues.

11.7.4 Post-translational chemical modification of side chains

A much more subtle way of modifying protein structure and/or function is through chemical modification of side chains on a previously synthesized protein molecule. Major post-translational chemical modifications include phosphorylation, acetylation, glycosylation, ubiquitylation, and sumoylation. Modification of side chains can affect protein structure and function.

11.7.5 The genomic origin of proteins

In every case in which a protein is post-translationally modified, the modification occurs at specific sites in the sequence, and these are gene-dictated. Furthermore, all of these modifications occur through catalysis by specific enzymes, which seek these sites and promote the given modification. These enzymes

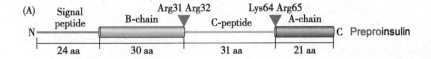
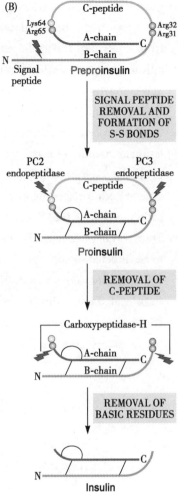

Figure 11.23 **Post-translational processing from a preproinsulin precursor to the mature functional insulin.** (A) Structure of preproinsulin and (B) processing pathway that creates mature insulin. Folding of preproinsulin probably requires prior removal of the signal peptide. Only then can the correct disulfide bonds be formed between SH groups.

are themselves gene-dictated. It seems that all of the information, for both synthesis and processing of proteins, still has its origin in the genome.

Key Concepts

- Translation occurs by Initiation, Elongation, and Termination Elongation is a repetitive cyclic process, with one cycle for the addition of each amino acid to the nascent polypeptide chain.
- There are three major participants in translation: the mRNA that carries the message, the tRNA or adaptor molecules that carry the amino acids to be incorporated in the chain, and the ribosome, which provides an active platform on which the process occurs.
- The code is three-letter or triplet, non-overlapping, and unpunctuated. All 64 possible three-letter codons are used, which means the code is degenerate, with different amino acids having from one to six possible codons. Most variation is in the third place of the codon.
- One codon, AUG, is used as a start signal for translation, and the three codons UGA, UAA, and UAG are commonly employed as stops.
- The code is almost universal from microbes to humans, with a few exceptions, mostly in mitochondria.
- tRNA molecules carry both an anticodon that matches the codon on the message and an amino acid to be incorporated in the nascent peptide chain. The accuracy of translation depends on the precision of two successive independent matchings: first, that of amino acids with tRNAs, and second, that of charged tRNAs with ribosome-linked messenger RNA.
- The molecular machine on which translation occurs is the ribosome.
- All ribosomes are ribonucleoprotein particles,

each containing several RNAs and many proteins.
- Translation can be regulated at several levels, from control of the availability of ribosomes to adjustments to the rate of translational initiation and elongation. The abundance, accessibility, and stability of mRNAs are also factors.
- Further processing and modification after translation are required for many proteins to be folded properly. Such processing may involve covalent cleavage and/or splicing of the chain, or modification of residue side chains.
- Transport of proteins across membranes is required to be directed to appropriate organelles or to fulfill their physiological function. Proteins can be translocated either directly from the ribosome, in co-translational translocation, or from the cytoplasm, in post-translational translocation.

Key Words

aminoacyl-tRNA synthetase（氨酰-tRNA 合成酶）
elongation（延长）
elongation factor（延长因子）
genetic code（遗传密码）
initiation（起始）
initiation factor（起始因子）
open reading frame，ORF（开放阅读框架）
peptidyl transferase（肽酰转移酶）
peptide bond formation（成肽）
post-translation processing and modification（翻译后加工修饰）
protein targeting（蛋白质靶向输送）
ribosome（核糖体）
release factor（释放因子）
Shine-Dalgarno sequence（S-D 序列）
termination（终止）
translation initiation complex（翻译起始复合物）
translocation（转位）
translation（翻译）
untranslated region，UTR（非翻译区）

Questions

1. What is genetic code? What are the features of genetic code?
2. What are the main steps of translation?
3. How to ensure the fidelity of translation?
4. What are the main mechanisms involved in translational control to regulate the level of protein production in cells?

References

[1] Aitken CE, Petrov A, Puglisi JD (2010) Single ribosome dynamics and the mechanism of translation. *Annu Rev Biophys*, 39:491–513.
[2] Beringer M, Rodnina MV (2007) The ribosomal peptidyl transferase. *Mol Cell*, 26:311–321.
[3] Caldarola S, De Stefano MC, Amaldi F, et al. (2009) Synthesis and function of ribosomal proteins: Fading models and new perspectives. *FEBS J*, 276:3199–3210.
[4] Crick FHC, Barnett L, Brenner S, et al. (1961) General nature of the genetic code for proteins. *Nature*, 192:1227–1232.
[5] Gromadski KB, Rodnina MV (2004) Kinetic determinants of highfidelity tRNA discrimination on the ribosome. *Mol Cell*, 13:191–200.
[6] Harigaya Y, Parker R (2010) No-go decay: A quality control mechanism for RNA in translation. *Wiley Interdiscip Rev RNA*, 1:132–141.
[7] Ling J, Reynolds N, Ibba M (2009) Aminoacyl-tRNA synthesis and translational quality control. *Annu Rev Microbiol*, 63:61–78.
[8] Rodnina MV, Wintermeyer W (2001) Fidelity of aminoacyl-tRNA selection on the ribosome: Kinetic and structural mechanisms. *Annu Rev Biochem*, 70:415–435.
[9] Rodnina MV, Beringer M, Wintermeyer W (2007) How ribosomes make peptide bonds. *Trends Biochem Sci*, 32:20–26.
[10] Simonetti A, Marzi S, Myasnikov AG, et al. (2008) Structure of the 30S translation initiation complex. *Nature*, 455:416–420.
[11] Steitz TA (2008) A structural understanding of the dynamic ribosome machine. *Nat Rev Mol Cell Biol*, 9:242–253.
[12] Zaher HS, Green R (2009) Fidelity at the molecular level: Lessons from protein synthesis. *Cell*, 136:746–762.

Sumei Zhang

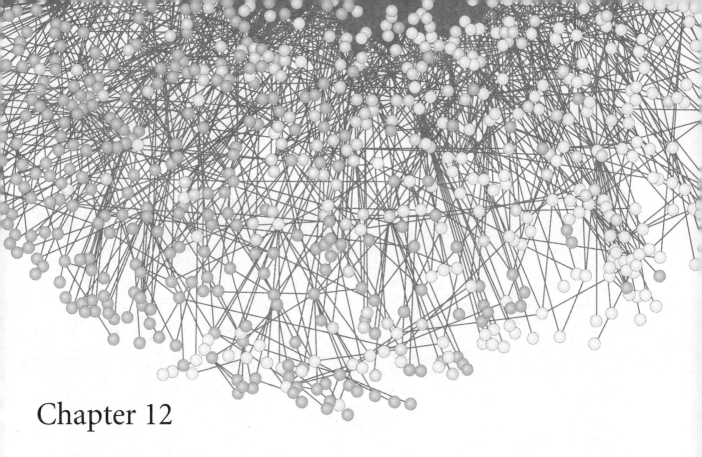

Chapter 12

DNA Replication

12.1 Introduction

When organisms reproduce, they must pass their genetic information accurately to their descendants. Within each organism, dividing cells need to replicate their DNA in a precise manner, so that the two daughter cells inherit exactly the same genetic information from the mother cell. Organisms, and viruses, have evolved complex and highly regulated replication machineries to do just that. In this chapter, we describe how DNA replication occurs in bacterial cells and viruses, and more complex processes involved in eukaryotic replication, in both mitosis and meiosis. The remaining chapters are devoted to the questions of how DNA is repaired.

Features of DNA replication shared by all organisms

As we know from Chapter 4, the double-helical model of DNA structure, put forth by Watson and Crick in 1953, immediately suggested semiconservative replication as one possible mode of DNA replica-tion. The Meselson-Stahl experiments demonstrated that DNA does indeed replicate in this way (Box 12.1). Because both strands of the parental DNA helix are replicated, the helix must be unwound and the two strands copied separately. Thus, replication occurs at a Y-shaped structure, termed the replication fork (Figure 12.1A); the fork moves steadily through a parental DNA helix, producing two daughter helices behind it, which form the two arms of the Y.

The chemistry of the reaction is such that each new nucleoside monophosphate is added to the free 3′-OH group of the ribose of the preceding nucleoside monophosphate (Figure 12.1B). Each new nucleoside monophosphate is added to the free 3′-OH group of the ribose of the preceding nucleoside monophosphate; a phosphodiester bond is created by nucleophilic attack of the 3′-OH group of the nascent DNA strand on the α-phosphate of the dNTP. The nucleotides to be added to the nascent DNA chain are selected according to the base-pairing rules that govern DNA structure. To prevent reversal of the reaction, the released pyrophosphate is quickly

Box 12.1 The Meselson-Stahl experiment

In 1958, Matthew Meselson and Franklin Stahl performed a very elegant experiment to establish the mode of semiconservative replication because the products are two double-stranded DNA molecules, each of which contains one of the parent DNA strands and one strand made of newly synthesized DNA. (Figure 1). This seems like a reasonable way to copy DNA but is not the only possibility, and, in 1958, at least two other modes were considered. First, replication might be conservative: that is, the whole parental DNA duplex might somehow be copied into a new duplex, composed entirely of new DNA. Second, the DNA might be copied in a patchwork fashion, by dispersive replication, so that all four resulting strands would be mixtures of old and new. To resolve the question of how DNA is replicated, Meselson and Stahl made clever use of an ultracentrifuge technique, in which DNA is centrifuged in a salt solution that forms a density gradient within the ultracentrifuge cell. This method can separate DNA molecules of only very small density difference, differences as slight as those produced by different isotopic compositions of the DNAs.

Meselson and Stahl grew the bacterium *E. coli* on a nitrogen source that contained only the heavy isotope ^{15}N. The DNA from these bacteria banded at a precisely defined point in the gradient, whereas bacteria grown on the common isotope ^{14}N banded at a different position (Figure 2). Next, the researchers took bacteria grown in ^{15}N for 14 generations and switched them, for one, two, three, or four generations, to ^{14}N. The result was DNA that all banded at an intermediate density after one generation; after two generations, half of the material had the intermediate density and the other half was light. The DNA distribution in the gradient changed with each generation as depicted in Figure 2. These results are consistent with semiconservative replication and not consistent with conservative replication, which would predict two bands of high and low density after the first generation. Finally, dispersive replication was ruled out by repeating the centrifugation experiment at high pH, where the DNA strands separate. In this case, only two bands, corresponding to ^{15}N and ^{14}N DNA, were found in any generation. Strands were copied whole. Rarely has science given such unequivocal answers.

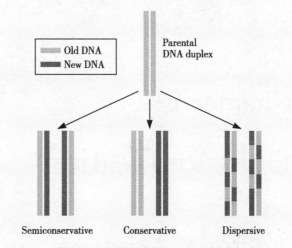

Figure 1 **Three modes of DNA replication that were considered as possibilities.**

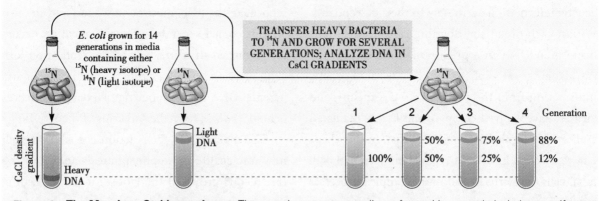

Figure 2 **The Meselson-Stahl experiment.** The experiment was actually performed in an analytical ultracentrifuge, but the principal is just as schematized here. (Adapted, courtesy of Mariana Ruiz Villarreal, Wikimedia.)

converted to inorganic phosphate by the activity of pyrophosphatase, a very abundant enzyme. Thus, replication of each strand is always in the $5' \rightarrow 3'$ direction. This, and the anti-parallel directions of the two strands in the double helix, raises a serious problem for replication; continuous DNA synthesis is possible on only one of the parental strands. DNA synthesis on the other strand must be by a discontinuous mechanism (Figure 12.2). The strand synthesized continuously is known as the leading strand, and its synthesis moves in the same direction as the replication fork. The other strand, the lagging strand, is synthesized in the opposite direction by a discontinuous mechanism: short Okazaki fragments are

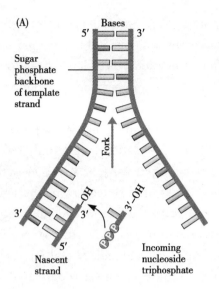

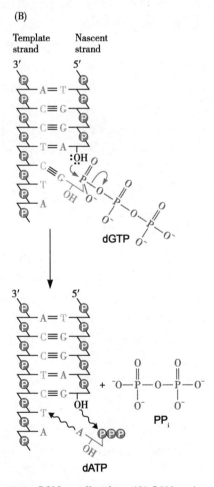

Figure 12.1 **DNA replication.** (A) DNA polymerase at the replication fork uses deoxynucleoside triphosphates (dNTPs) as substrates for the polymerization reaction. (B) Chemistry of the DNA polymerization reaction. (B, adapted from Mathews CK, van Holde KE, Ahern KG [1999] *Biochemistry*, 3rd ed. With permission from Pearson Prentice Hall.)

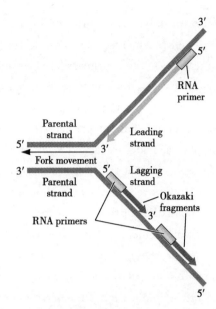

Figure 12.2 **Replication fork and priming of DNA synthesis.**

first synthesized and then ligated together to form an uninterrupted polynucleotide chain. Leading-strand synthesis requires only one primer, which is synthesized during replication initiation. Lagging-strand synthesis requires multiple primers, each synthesized when a new Okazaki fragment is initiated at the fork. The RNA primers in bacteria are ~10 nucleotides long and are synthesized at the fork by primase, a DNA-dependent RNA polymerase.

There are three basic requirements for DNA synthesis: (a) a template strand, which provides the nucleotide sequence information needed for the synthesis of a complementary strand; (b) a polymerase, which catalyzes the addition of residues to form the new strand; and (c) a primer, which is the source of a free 3′-OH group needed to start the action of the DNA polymerase.

In all organisms, two DNA polymerase complexes are active at the replication fork at any time (see Figure 12.2). One of them moves in a continuous fashion, following the direction of the replication fork movement, to produce the leading strand, a complementary copy of the leading-strand template. The leading strand is equivalent to the RNA formed during transcription. The other polymerase synthesizes the lagging strand as a series of short Okazaki fragments using the lagging-strand DNA template.

Note that both polymerases proceed in the 5'→3' direction, as all polymerases do. Okazaki fragment initiation requires multiple primers (see Figure 12.2). These fragments are later connected to each other, or matured, into the continuous lagging strand. This maturation requires that the primers are removed. After this, the fragments are ligated together by DNA ligase. The two copies of the elongating polymerase, one for each template strand, are anchored to their templates by auxiliary proteins that prevent them from falling off. Thus, DNA polymerase acts as a highly processive enzyme. This processivity accounts for the small number of enzymes needed and allows the rapid rate of replication. Processivity is due to sliding clamp subunits, which form a ring surrounding the DNA. Sliding clamps are evolutionarily conserved. Three different proteins provide the same-shaped clamp for DNA.

The machinery that advances the replication fork along the parental DNA duplex consists of a number of proteins in addition to the DNA polymerase complex. Each fulfills a specific function. Unwinding of the DNA helix is needed to expose the bases, which are otherwise hidden in the interior of the helix, so that the polymerases can copy the sequence. Unwinding is performed by a DNA helicase, an enzyme that moves quickly along one of the template strands to force open the parental helix using the energy of ATP hydrolysis for its translocation. It is not exactly clear whether the helicase acts by a passive or an active mechanism. Another protein that is important in replication is DNA topoisomerase. This enzyme relieves the positive superhelical stress that accumulates in front of the moving DNA helicase. Recall from Chapter 4 that unwinding of one turn of the double helix results in the production of one compensatory positive superhelical turn. The replication fork moves very rapidly, at about 1000 nucleotides per second. Without topoisomerase activity, the level of positive supercoiling would quickly rise to levels that would prohibit further unwinding and movement of the replication fork. Finally, single-strand DNA-binding proteins, or SSB proteins, cover the lagging-strand template while it is temporarily single-stranded, protecting it from degradation. In addition, SSB binding prevents the formation of unwanted secondary structures; it also holds the DNA in an open conformation with the bases exposed for copying.

The interactions among all of these proteins during replication elongation are not only extremely complex but also very dynamic. Accordingly, the use of simpler replication systems, which recapitulate the main features of bacterial and eukaryotic systems, has been instrumental in defining the minimal requirements for rapid and faithful replication of dsDNA molecules. The most widely used model systems are those derived from bacteriophages, the T7 system, which requires only four proteins, three virus-encoded and one host factor, to replicate DNA. The T4 bacteriophage system has provided important information about the sliding clamp and clamp loader. It is important to note, however, that replication in some viruses proceeds by mechanisms that are quite different from those in other organisms.

12.2 Bacterial DNA replication

12.2.1 DNA replication in bacteria

In most bacteria, the chromosome is a single circular DNA molecule, containing a single origin of replication. Replication initiated at this specific sequence is bidirectional: two replication forks form and move in opposite directions (Figure 12.3). Replication is initiated at a specific sequence, ori C, with two replication

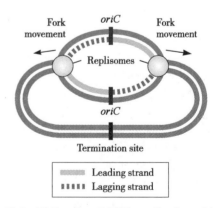

Figure 12.3 **Bidirectional DNA replication of bacterial chromosomes.** Schematic of the circular chromosome and the process of replication in *E. coli* are shown.

forks moving in opposite directions (Figure 12.4). The protein complex that functions at each fork is called the replisome, the protein complexes at the forks that contain the enzyme activities needed for replication to occur. The similarity of the replicating structure to the Greek letter θ has led to this process being called the θ mode of replication. The bidirectional manner of DNA replication has been convincingly demonstrated in cytological experiments.

As in all replication processes, many biochemical activities are in play here. We consider the most important players. DNA polymerase III, a multi-subunit complex, is the major replicative enzyme in *Escherichia coli* and other bacteria (Figure 12.5, Table 12.1). The asymmetric DNA Pol III holoenzyme, 900 kDa, contains (a) two copies of the catalytic core, which consists of α subunit with polymerase activity, ε subunit with 3'→5' proofreading exonuclease activity, and θ subunit that stimulates exonuclease; (b) two copies of the clamp, each a homodimer of β subunits; and (c) one copy of the five-subunit DnaX complex or clamp loader, which assists in assembly of the replisome and in the binding and dissociation cycle of the lagging-strand polymerase. The χ subunit is not essential for clamp loading; it links the clamp loader to SSB and primase. The ψ subunit is not essential for clamp loading either; it serves as a connector to χ and stabilizes the clamp loader. The E.

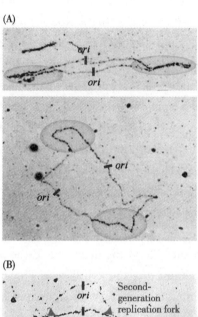

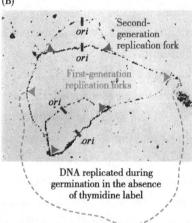

Figure 12.4 **Bidirectional replication and re-initiation in *Bacillus subtilis*.** (A) *B. subtilis* spores were germinated in the presence of low amounts of radioactive [*methyl*-³H]-thymine in order to label lightly the newly formed replication eyes; these are the portions of the DNA between two replication forks moving in opposite directions. The cells were then given a high-radioactivity pulse, to heavily label the DNA portions that are replicated during this pulse. Note that both forks have highly labeled portions, shown by yellow ovals, indicating that both replication forks were active during the high-radioactivity pulse. In other words, replication is bidirectional. (B) To visualize only that portion of the circular chromosome that is the product of replication, spores of *B. subtilis* thy⁻ trp⁻ strain were germinated in the absence of thymine for 150 min and then grown in medium containing [*methyl*-³H] thymine for 30 min. Under these conditions, three replication eyes can be seen: the larger eye results from label that is incorporated during the first round of replication following the shift to labeled medium, while the two smaller eyes indicate re-initiation of replication on the already partially replicated chromosome.

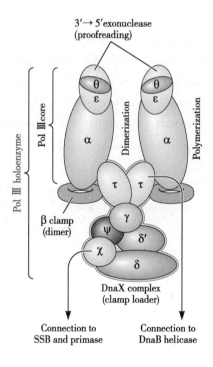

Figure 12.5 **Composition of the asymmetric DNA Pol III holoenzyme complex.**

coli replisome uses twin Pol III core enzymes to copy leading and lagging strands simultaneously.

The structure of the β clamp is presented in Figure 12.6. Sliding clamp β, or processivity factor, is highly conserved in structure: the ring shape that is necessary to confer processivity to the DNA Pol III core is ubiquitous. In some bacteria, such as E. coli, it is a dimer of a three-domain protein, whereas in some phages and in eukaryotes it is a trimer of a two-domain protein (Figure 12.6A). The clamp is a ring-shaped dimer containing two identical protomers, A and B, each containing three individual globular domains and together forming a six-domain ring. The protomers are arranged head-to-tail, which results in structurally distinct surfaces on the two faces of the clamp. The C-terminal face, from which the C-termini project, is implicated in many of the interactions of β clamp with other proteins. The clamp loader and the polymerase compete for binding to the C-terminal face of the β-ring. The DNA is sharply tilted within the central channel; the tilt allows DNA to make contacts with R24 and Q149 on the C-terminal face. Sliding clamps act by embracing the newly formed duplex, comprising the newly synthesized DNA chain and its single-stranded template, and then moving along the template strand together with the associated polymerase (Figure 12.7). The structure of the quadruple complex of polymerase, sliding clamp, single-stranded DNA template, and double-stranded helix of template and nascent DNA strands provides a clear picture of how the clamp ensures polymerase processivity (Figure 12.6B).

How are all of these elements structurally organized to allow synchronous replication of two oppositely oriented strands by a single fork? The key organizer is the clamp loader, also known as the DnaX complex in bacteria. The clamp loader is a five-subunit structure that is responsible for loading the clamp onto the DNA at the primer-template junctions. For quite a while, there was uncertainty about exactly which subunits in what stoichiometry combine to form the loader; see Figure 12.8 for more details. In an interesting twist, it was discovered that two of the five subunits are encoded by the same gene, dnaX; subunit τ is a full-length gene product, whereas subunit γ is a shortened version missing two of the C-terminal domains. The extra domains IV and V of τ endow the subunit with the ability to simultaneously bind the helicase and the polymerization subunit α of the Pol III core. These protein-protein interactions are absolutely required for the formation of a functional twin polymerase complex at the replication fork. Researchers in the field have finally agreed that the E. coli loader is composed of subunits τ, γ, δ, and δ', in the stoichiometry τ2γδδ'. This composition seems logical because two τ subunits are needed to keep the twin Pol III core polymerases moving on both leading- and lagging-strand templates simultaneously. It should be noted that this composition and stoichiometry were derived from in vitro experiments; the in vivo situation still remains to be elucidated.

The present understanding of how the clamp loader acts to load the β clamp onto DNA in an

Table 12.1 **Properties of the two replicative DNA polymerases in E. coli.**

Polymerase	Gene	Mol mass (kDa)	Family[a]	No. of molecules/cell	Max speed (nt/s)	Processivity	Biochemical activity	Biological function
Pol I	polA	103	A	400	16–20	100–200	polymerase; 3'→5' exonuclease; 5'→3' exonuclease	Okazaki fragment maturation with primer degradation
Pol III	polC	130	C	10	250–1000	500 000	polymerase; 3'→5' exonuclease	replicative chain elongation

[a]The polymerases are classified in several different protein families on the basis of similarities in primary structures.

Figure 12.6 **Sliding clamps are evolutionarily conserved.** (A) All sliding clamps are ring shaped structures comprising six domains that surround the double-stranded portion of the newly synthesized double helix. (B) Model of bacteriophage RB69 DNA polymerase bound to DNA and to the C-face of the gp45 sliding clamp. The 3′-end of the nascent DNA strand is positioned in the active center, with the single-stranded region of the template strand extending leftward. The movement of the polymerase with the bound sliding clamp is indicated by the arrow.

β clamp of *E. coli* Pol III
(a dimer of a three-domain protein)

Human PCNA
(Proliferating Cell Nuclear Antigen)
(a trimer of a two-domain protein)

Gp45 from phage T4
(a trimer of a two-domain protein)

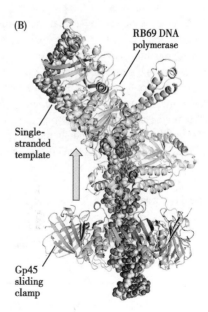

energy-consuming process is presented in Figure 12.9. Note that the loader and the polymerase compete for the same face of the clamp. In order for the polymerase to bind the β clamp, the clamp loader needs to leave the clamp. Thus, the clamp loader has two essential functions at the replication fork: it places the clamp onto the appropriate position at the primer-template junction of the DNA; through the τ subunits, it also cross-links between the leading and lagging-strand polymerases and binds to the replicative helicase, thus serving as a central organizer for the entire replisome.

The overall organization of the core proteins, together with the helicase and the primase at the replication fork, is presented in Figure 12.10. This complex constitutes the functional replisome. In addition to the key interactions between the β sliding clamps and Pol III α, other specific molecular interactions are important for stability of the holoenzyme at the replication fork. These include interactions between Pol III α and τ and, importantly, between τ and the DnaB helicase. The specific requirement for simultaneous synthesis of the two DNA strands imposes a very peculiar structure on the lagging-strand template: the template folds into a loop, known as the trombone. The need for loop formation was first recognized by Bruce Alberts as early as 1983 and was later directly visualized by electron microscopy. The loop allows movement of the twin Pol III core polymerases in the same physical direction, that of the replication fork, despite the fact that the two strands have opposite polarities. The requirement for a free 3′-OH group as the site of elongation, and the opposite polarities

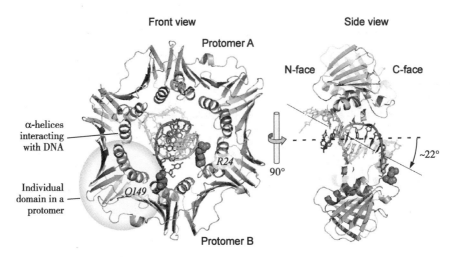

Figure 12.7 Structure of processivity factor, sliding clamp β. The structure shown is the *E. coli* factor complexed with primed DNA.

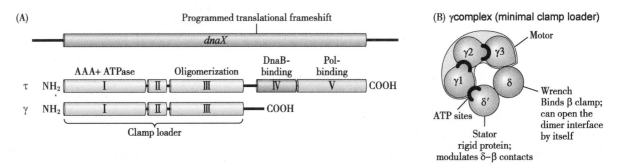

Figure 12.8 *E. coli* clamp loader. (A) Domain structure of the τ and γ subunits, both encoded by the *dnaX* gene. τ is the full-length protein product of 71 kDa, whereas γ is a truncated product of 47 kDa, resulting from a programmed translational frameshift that produces a stop codon. The two polypeptide chains share the first three clamp loader domains; these are ATPases whose activity is needed for the clamp-loading function. Domains IV and V in τ bind the helicase DnaB and the α polymerization subunit of the Pol III core. Thus, only τ has the capacity to bind helicase and core polymerase and thus to serve as a central organizer for the replisome. (B) Generalized structure of the minimal clamp loader or γ complex from *E. coli*. At one time, it was believed that the clamp loader complex contained just three γ subunits, hence the name γ complex. Indeed, the minimal γ₃δδ′ complex shown here, reconstituted from recombinant subunits, is capable of loading β clamps on appropriate DNA structures. Now a more general term, DnaX complex, is used to refer to the clamp loader since each loader is expected to contain at least two τ subunits to be able to bind two molecules of Pol III core for leading- and lagging-strand synthesis. δ and δ′ also contain domains IV and V. The roles of the subunits in DnaX complex are as follows: the three τ/γ subunits bind and hydrolyze ATP and constitute the motor of the complex, the δ subunit is the wrench that cracks open the β clamp at the dimer interface, and the δ′ subunit is the stator because of its rigidity; its domains assume the same orientation in the free protein and its complexes. (B, adapted from Pomerantz RT, O'Donnell M [2007] *Trends Microbiol*,15:156–164. With permission from Elsevier.)

of the template strands, would otherwise result in the two new strands being synthesized in opposite directions. The only way to circumvent this problem is to fold the lagging-strand template into the loop structure, so that both strands are now in the same orientation with respect to the polymerases (see Figure 12.10).

Replication in eukaryotes is tightly coupled to the cell cycle. We present a brief discussion of the cell cycle in Chapter 2, and a more detailed picture, including cycle regulation, is given in Box 12.2. Specific origins are activated during specific phases—early, middle, or late—of the S phase of the cell cycle, producing the changing pattern of appearance of replication foci as S phase progresses. Figure 12.11 illustrates the temporal activation of replication origins during S phase. Complexes that are assembled at the origins during the M phase of the preceding

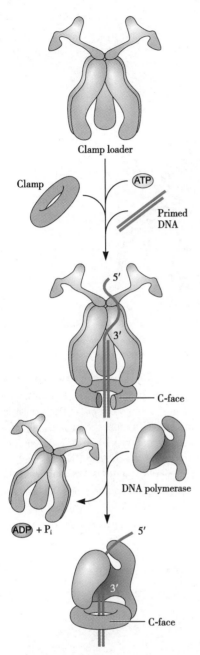

Figure 12.9 **Generalized mechanism of clamp loader action in E. coli.** The multiprotein clamp loader in the presence of ATP binds and opens the ring-shaped sliding clamp. In the ATP-bound state, the clamp loader has a high affinity for primer-template junctions. Binding of DNA stimulates ATP hydrolysis and the clamp loader is ejected, leaving the closed clamp on DNA, properly oriented for use by a replicative DNA polymerase. The loader and the polymerase compete for the same C-terminal face of the clamp. Because of this competition, it is necessary for the clamp loader to leave the clamp so that the replicative DNA polymerase can bind. Various protein partners involved in cell-cycle control, DNA replication, DNA repair, and apoptosis bind to the clamp. (Adapted from Indiani C, O'Donnell M [2006] *Nat Rev Mol Cell Biol*, 7:751–761. With permission from Macmillan Publishers, Ltd.)

mitosis and during G_1 phase can fire or become activated at different stages of the S phase. The pertinent question here is, what regulates the selection of the origins that fire during those different stages of S phase? We do not have a clear picture yet, but it is obvious that replication timing has multiple complex connections with static properties of genomic regions such as GC content, the subnuclear location of the regions, and their transcriptional activity.

Pre-replication complexes or pre-RC are assembled on origin sequences during M and G_1 phases. Activation or firing of these origins subsequently occurs throughout S phase; origins are classified as early-, mid-, and late-activating depending on whether they become active during the early, mid, or late stages of S phase (see Figure 12.11). As seen in Figure 12.12, foci were visualized by immunodetection of DNA that has been pulse-labeled for 30 min with bromodeoxyuridine, BrdU. Bulk DNA was stained with Hoechst dye. Early in S phase, hundreds of small foci are distributed throughout the nuclear volume; in mid S phase, foci are preferentially located around the nucleoli and the nuclear periphery; finally, in late S phase only several clusters of foci are seen in heterochromatic regions.

Pol I was the first DNA polymerase to be discovered and characterized in Arthur Kornberg's laboratory. It plays a crucial role in maturation of the Okazaki fragments that are synthesized discontinuously, from multiple primers, on the lagging-strand template (see Figure 12.12 and Figure 12.13). This process requires two of Pol I's three catalytic activities, all residing in the same polypeptide chain. The protein is a $5' \rightarrow 3'$ polymerase, which extends the Okazaki fragment synthesized by Pol III, and a $5' \rightarrow 3'$ exonuclease, which removes the RNA primer during the nick-translation phase of the maturation process. Pol III does not have such activity. By treating purified Pol I with the protease subtilisin, it is possible to remove the $5' \rightarrow 3'$ exonucleolytic activity responsible for primer removal during replication. The large, 605-amino-acid polypeptide fragment that is obtained as a result of subtilisin cleavage is known as the Klenow fragment after Hans Klenow. In addition,

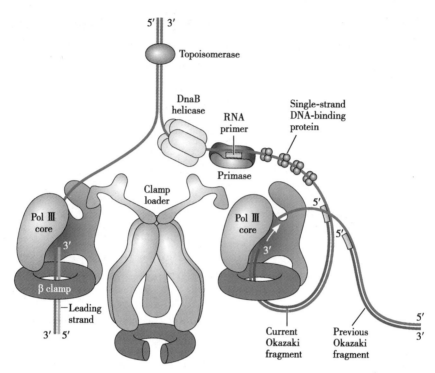

Figure 12.10 **Core proteins at the DNA replication fork.** Two DNA core polymerases are active at the fork at any one time, allowing synthesis of leading and lagging strands simultaneously. Note the loop or trombone structure formed by the lagging-strand template that allows the two core polymerases to move in the same direction. Both polymerases are anchored to their templates by auxiliary proteins, a sliding clamp and a clamp loader, so that they do not fall off. The other obligatory protein factors at the fork are DNA helicase, which unwinds the parental helix in an ATP dependent manner; DNA topoisomerase, which relieves the superhelical stress accumulating in front of the moving DNA helicase; primase or DNA-dependent RNA polymerase, which synthesizes the RNA primers; and single-strand DNA-binding or SSB proteins, which cover the single stranded lagging-strand template to protect it from degradation and hold the DNA in an open conformation with the bases exposed. (Adapted from Pomerantz RT, O'Donnell M [2007] *Trends Microbiol*, 15:156–164. With permission from Elsevier.)

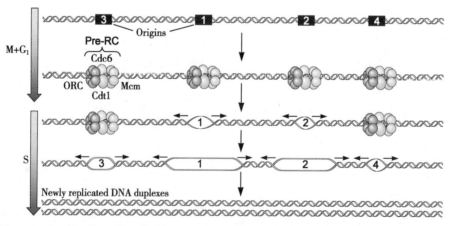

Figure 12.11 **Temporal activation of replication origins during S phase.** The schematic represents a portion of the genome where all potential origins are activated. In reality, only a small fraction of all potential origins is used at each cell cycle.

Pol I possesses a 3'→5' exonuclease activity. This serves in proofreading, an activity also present in subunit ε of Pol III. Proofreading is essential during replication, as it ensures the very low error frequency that is necessary during the reproduction of genetic information. During elongation, a wrong base is incorporated once every 105 polymerization steps, but the overall error frequency is four orders of mag-

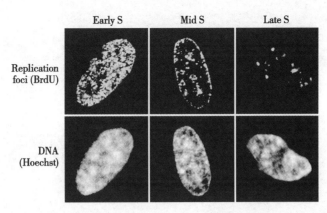

Figure 12.12 Patterns of replication foci as the cell progresses through S phase.

Box 12.2 Regulation of the cell cycle

Mitotic cell division was discovered by cytogeneticist Walther Flemming in 1879, and within a few years the general features of the cell cycle were recognized. It was not until the 1950s, however, that the role of DNA replication in this scenario could be fully appreciated. Even with this clarification, the mechanisms that regulate the cell cycle remained obscure for decades and in fact were little studied. The first major advances came from the work of Leland Hartwell in the 1970s. Hartwell was studying *Saccharomyces cerevisiae*, the budding yeast, using the appearance of buds and increase in cell size to measure progress through the cell cycle. A particularly interesting subset of temperature-sensitive mutants corresponded to modifications, such as blocks or premature transitions, in the cell cycle itself. These Hartwell termed cell division cycle mutants, abbreviated cdc. Over the years, a large collection of cdc mutants in various organisms was assembled, but their mode of function remained obscure. Important clues began to come from the work of Paul Nurse. Nurse used the fission yeast *Schizosaccharomyces pombe*, whose linear growth could be more easily measured than the expansion and budding of *S. cerevisiae*. An important discovery, in 1980, was a wee mutant that went into mitosis at a much smaller cell size, though in the same metabolic state, as the wild type. Mutants like this clearly indicated the existence of checkpoints in the normal cycle. Equally important was the discovery that many such mutations affected protein kinases, strongly suggesting that cell-cycle regulation involved phosphorylation and dephosphorylation events.

Just how this might be controlled was revealed by Timothy Hunt in one of the great serendipitous discoveries in molecular biology. Hunt was not even looking at cell-cycle regulation; he was examining the control of translation in marine invertebrates. As a model, he and colleagues studied the accumulation of proteins in sea urchin eggs following fertilization. The first cleavage divisions following fertilization in this organism are synchronous and Hunt was surprised to see a class of proteins that not only increased in amount at a certain point in the cell cycle but were also specifically degraded at a later point. These he called cyclins, and he soon found them in other organisms as well. It was soon revealed, in many labs, that cyclins were associated with cdc kinases. In summation, the work of Hartwell, Nurse, and Hunt has provided a mechanistic basis for understanding cell-cycle regulation. In 2001 they were awarded the Nobel Prize in Physiology or Medicine "for their discoveries of key regulators of the cell cycle."

Now we know that the cell-cycle regulation involves a complex interplay between kinases and phosphatases. Cyclin-dependent kinases, or CDKs, are serine or threonine protein kinases that are involved not only in cell-cycle regulation (Figure 1) but also in other processes such as transcription. The activity of CDK is regulated through phosphorylation by other upstream kinases and, significantly, by association with function-specific cyclins. In turn, the CDK-cyclin complexes are inhibited through reversible binding of CDK inhibitors and through cyclical degradation of cyclins during the cell cycle. Oscillations in the levels of the four cell-cycle-dependent cyclins are presented in Figure 2. The inhibitors act on CDK-cyclin complexes that spontaneously adopt an active conformation upon heterodimerization. Heterodimers that do not spontaneously remodel into an active conformation upon complex formation are regulated via co-factor or substrate binding. An example of the latter is the CDK4-cyclin D1 complex, which may be activated by nuclear translocation, substrate binding, or phosphorylation of specific residues. This complex is of special interest because of its connection to numerous human cancers. An unchecked or hyperactivated CDK4-cyclin D1 pathway may be responsible for enhanced cellular proliferation due to the role of this complex in the G_1 checkpoint.

The cyclin subunits of the CDK-cyclin complexes are degraded by highly specific ubiquitin-mediated proteolysis.

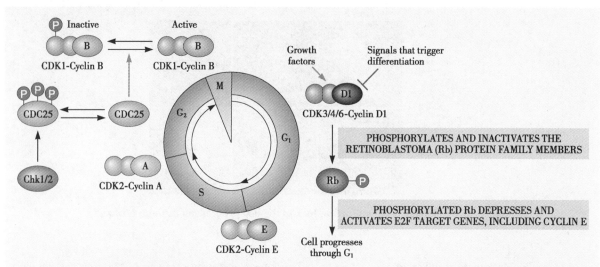

Figure 1 **Cell-cycle regulation: an overview.** Entry and progression through the four phases of the cell cycle is controlled by distinct CDK-cyclin heterodimeric complexes. D-type cyclins such as cyclin D1, acting in complexes with CDK3, CDK4, or CDK6, regulate events in early G_1 phase; CDK2-cyclin E triggers S phase; the CDK2-cyclin A complex regulates the completion of S phase; finally, CDK1-cyclin B is responsible for transition to mitosis. Some complexes can be inhibited by specific inhibitors, for example, cyclin D-associated kinases are inhibited by a group of proteins belonging to the INK4 or inhibitor of CDK4 family, whereas cyclin E and cyclin A kinases are inhibited by $p21^{waf1}$, $p27^{kip1}$, and $p57^{kip2}$. The decision of whether a cell should proliferate, in response to growth factors, or differentiate, in response to differentiation signals, is made during the G_1 phase of the cell cycle. The decision to initiate mitosis is regulated by the CDK1-cyclin B complex, which needs to be activated from its phosphorylated form by the phosphatase activity of CDC25, cell division cycle 25, which in turn is activated by another phosphatase, not shown here. CDC25 activity is inhibited by phosphorylation by Chk1/2 or mitotic checkpoint kinase 1/2, preventing premature entry into mitosis.

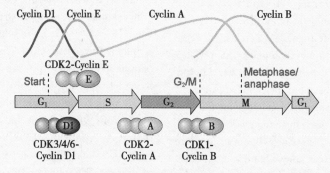

Figure 2 **Levels of cyclins during phases of the cell cycle.**

nitude lower, because of the combined proofreading activities of both Pol III and Pol I. Some errors are also corrected postreplicatively, in DNA repair processes. In perspective, during each round of replication of the human genome, some 3.2×10^9 bp, only about one error, on average, is transmitted to one of the two daughter cells. The proofreading activity of Pol I is well understood because high-resolution information is available to describe the structure of the crystallized Klenow fragment with suitable DNAs complexed with each of its two active sites (Figure 12.14). Nevertheless, understanding the Pol I proofreading process required finding answers to a puzzling question: how can the polymerase active site and the 3′→5′ exonuclease active site work together to assure that mismatched base pairs incorporated at the polymerase active site are edited out at the 3′→5′ exonuclease active site, some 25–30 Å away? Research from Thomas Steitz's laboratory suggested that the two active sites might communicate by virtue of the DNA sliding between them. The path that the 3′-terminus must follow to proceed from the polymerase active site to the exonuclease active site involves 4 bp of duplex DNA plus 4 bases of single-stranded frayed end. A hint is given by the fact that the 3'→5' exonuclease activity of Pol I excises ~10% of all correctly incorporated nucleotides; thus, the polymerase and exonuclease activities of Pol I are in what Steitz called a "delicately posed competition" for the newly formed 3'-terminus. How does the enzyme discriminate between proper and mismatched bases at the 3'-terminus, so that it knows whether to continue polymerization or go for excision? The fact that melting of 4 bp at the 3'-end is required for this end to reach the exonuclease site means that the structural basis for discrimination could be the increased

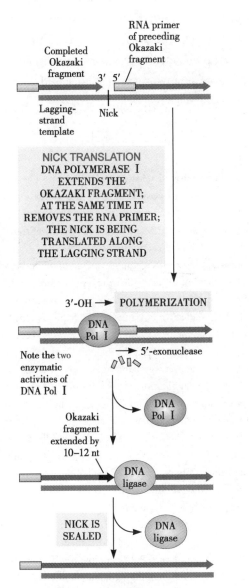

Figure 12.13 **Maturation of Okazaki fragments: combined action of DNA Pol I and DNA ligase.**

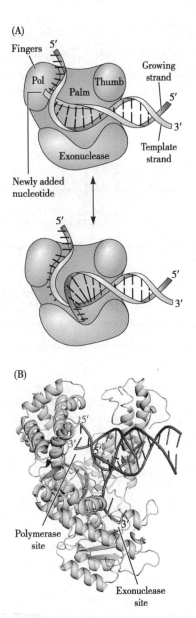

Figure 12.14 **Proofreading: removal of a mismatched nucleotide during DNA synthesis.** (A) Schematic model of DNA polymerase I–DNA structure during polymerization (top) and during proofreading (bottom). The model shows the fingers, palm, and thumb domains common to all DNA polymerases. When a mismatched nucleotide is added to the growing 3'-end, the polymerase is retarded, leaving the mismatched nucleotide at the end of the duplex between the template strand and the nascent strand. The slowing down of the polymerase movement allows spontaneous melting of the end of the DNA duplex and releases the 3'-end to contact the exonuclease site, where the excision of the wrong nucleotide occurs. (B) Space-filling model of Klenow fragment with two separate active sites, polymerase and exonuclease. The 3'-end of the primer strand in polymerizing mode, shown in red, is double-stranded and lies near three catalytically important carboxylates in the polymerase active site; the 3'-end of the primer strand in editing mode, at the exonuclease site, shown in blue, is single-stranded. (A, adapted from Baker TA, Bell SP [1998] *Cell*, 92:295–305. With permission from Elsevier.)

propensity for melting of a duplex containing a mismatch. Finally, intact DNA Pol I also has laboratory applications. For example, it is used for internal labeling of DNA strands (Figure 12.15). While DNA polymerase III, a multisubunit complex, is the major replicative enzyme in *Escherichia coli* and other bacteria (Figure 12.5, Table 12.1). Here, we present a brief history of the discovery of bacterial DNA polymerases in Box 12.3.

12.2.2 The processes of Bacterial DNA replication

The need for priming of DNA synthesis stems from the fact that DNA polymerase is not capable in itself of starting a new chain; it can only extend an already existing DNA or RNA chain that carries the 3'-OH group to which the incoming deoxyribonucleoside triphosphate can be attached (see Figure 12.1B). The free 3'-OH groups that are necessary for replication to begin are usually provided by RNA primers. The primers are synthesized by a primase, a DNA-dependent RNA polymerase. In bacteria the primase is encoded by the *dnaG* gene. In eukaryotes, the primers are two-part oligonucleotides composed of RNA followed by DNA.

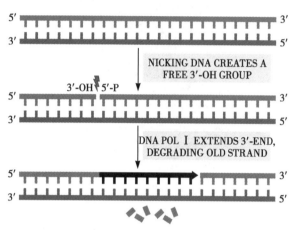

Figure 12.15 Use of DNA Pol I for nick translation in laboratory manipulations. The method is primarily used for internal labeling of DNA strands with labels such as radioactivity or fluorescence.

Box 12.3 The discovery of bacterial DNA polymerases

The final sentence in Watson and Crick's classic 1953 paper on DNA structure hints at the kind of mechanism that might be involved in replication. Indeed, by 1958, Meselson and Stahl had demonstrated that bacterial replication was semiconservative, with each new strand being copied from an old strand that is used as template. But even earlier, others were seeking the mechanism of replication by searching for enzymes that might catalyze the process. Foremost among these was Arthur Kornberg, a young biochemist working at the Washington University School of Medicine in St. Louis.

Starting in 1955, Kornberg used the newly available ^{14}C-labeled thymidine as a marker. He found that in the presence of DNA and an *E. coli* cell extract, a very small fraction of the label was incorporated into acid-insoluble material. That this was DNA was established by the fact that treatment with DNase released the label into the acid-soluble fraction. Using this assay, Kornberg and co-workers were able to purify the relevant enzyme, presently called DNA polymerase I or Pol I. In an impressive series of papers, whose publication began in 1958, they demonstrated that the enzyme needed both a template and primer and that it added deoxyribonucleoside triphosphates in the 5'→3' direction with release of diphosphate. The enzyme was also shown to possess a 3'→5' exonuclease activity. Arthur Kornberg was awarded a share of the 1959 Nobel Prize in Physiology or Medicine for "discovery of the mechanisms in the biological synthesis of DNA."

Ten years later, the whole challenge of understanding DNA replication suddenly became more complicated. John Cairns and Paula de Lucia isolated an *E. coli* mutant that contained a defective Pol I gene. Nevertheless, this bacterial strain was viable; it reproduced but exhibited deficiency in DNA repair. This indicated that Pol I could not be the enzyme primarily responsible for bacterial chromosome replication. Clarification came within a couple of years from the work of Malcolm Gefter and Thomas Kornberg (son of Arthur), who demonstrated the existence of two other bacterial polymerases, Pol II and Pol III. They then created a number of temperature-sensitive mutants for each of these polymerases. Only the Pol III mutants showed restricted growth at high temperatures. Thus, Pol III is the enzyme mostly responsible for replication of the bacterial genome. The major function of Pol I is now known to be in the maturation of Okazaki fragments on the lagging strand. Both Pol I and Pol II are involved in DNA repair. Although Arthur Kornberg's enzyme is not the vital one, his work opened the whole field.

There are well-defined mechanisms for starting DNA replication. For example, the bacterial cell possesses a single, well-regulated replication origin, known as *oriC* in *E. coli*. As we have seen, replication forks proceed bidirectionally from this position, and in the circular bacterial chromosome (Figure 12.3). In considering the process of initiation, the factors that participate in it, and how it actually occurs, we shall focus on the process as defined in *E. coli*, where it is best understood. The origin of replication is a sequence of ~250 bp, composed of multiple 9-bp repeat elements. These elements are called DnaA boxes, as they provide the binding sites for the sequence-specific initiator protein DnaA. Initiation absolutely requires DnaA and two additional proteins: the DNA hexameric helicase DnaB and the helicase loader DnaC. DnaA and DnaC form multisubunit right-handed protein filaments. Oligomerization of DnaA is instrumental in melting the AT-rich DNA unwinding elements in the DUE region at *oriC* (Figure 12.16) by forming a positive toroidal DNA superhelix on the outside of the DnaA filament. Why DnaC would form a very similar structure is less clear. Both DnaA and DnaC are ATPases of the AAA+ (ATPases Associated with various cellular Activities) superfamily. ATP hydrolysis by both DnaA and DnaC is a prerequisite for function. The similarities in the structure and enzymatic activities of DnaA and DnaC suggested that the cell may use DnaC as a molecular adaptor that employs ATP-activated DnaA as a docking site for recruitment and correct orientation of the helicase DnaB at replication origins. Two molecules of the helicase are bound, on opposite strands, and oriented in opposite directions. These will provide the docking sites for the two replisome complexes to be assembled at *oriC*. A model for possible cross-talk between DnaA and DnaC filaments, as suggested by James Berger and collaborators, is presented in Figure 12.17.

It is becoming increasingly clear that abundant bacterial nucleoid proteins, known primarily for their ability to compact the bacterial chromosome by bending and bridging the DNA, also play significant roles in the initiation of replication. Thus, the non-sequence-specific heat-unstable or HU protein has been shown to dramatically enhance DnaA-mediated

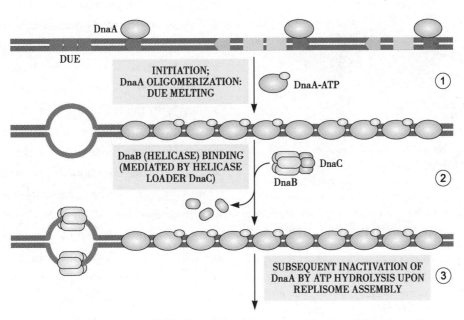

Figure 12.16 Initiation of replication at *oriC* in *E. coli*. DnaA is bound to the three high affinity DnaA boxes throughout the cell cycle; it interacts with the weaker sites only at the start of initiation, when ATP binds to DnaA. Initiation of replication involves a number of steps: (Step 1) DnaA-ATP molecules bind to the origin, where they oligomerize into the large nucleoprotein complex and facilitate the melting of the adjacent DNA unwinding elements or DUE. (Step 2) Helicase DnaB binds to the unwound single strands with the help of the helicase loader, DnaC. (Step 3) DnaA is inactivated by ATP hydrolysis, which is stimulated by additional regulatory mechanisms, upon replisome assembly.

oriC unwinding *in vitro*, probably through its ability to bend and destabilize double-stranded DNA. The participation of two other DNA-bending proteins, IHF or integration host factor and Fis or factor for inversion stimulation, seems to be more DNA-sequence specific, because binding sites for these two proteins are interspersed among DnaA-binding sites at *oriC*.

Having considered the components of the replisome and how they individually function during elongation, we now move to how they work together in a dynamic process. We have noted that the two twin Pol III core polymerases are in constant contact through the τ subunits of the holoenzyme. The situation is actually more complex, as there is constant recycling of the lagging-strand polymerase, which must move from a completed Okazaki fragment to a new clamp assembled at the 3′-end of the primer that initiates the next Okazaki fragment. The clamp loader plays a central role by maintaining the overall integrity of the replisome, despite its complex internal dynamics.

Another problem arises at the site of primer synthesis. We know that the primase must be associated with the helicase in order to function, but the primers are synthesized in the same direction as synthesis of the lagging strand, the direction opposite to the movement of the helicase. Moreover, primer synthesis is slow in comparison to the movement of the helicase. There are three possible ways in which this problem could be solved (Figure 12.18), and it seems that different organisms use one or another of these mechanisms. First, the replisome may pause to wait for the primer to be synthesized. Second, there could be a temporary release of the primase from its interactions with the helicase. Finally, as recently demonstrated by single-molecule experiments, a small priming loop can form, nested in the larger lagging-strand template's trombone loop. The priming loop eventually collapses into the trombone loop. There is still another level of complexity to the dynamic replisome. It turns out that a second mechanism, in addition to the sliding clamp's movement together with the polymerase, contributes to the high processivity of the replisome complex (see Figure 12.6). The replisome might actually carry three, not two, core polymerases and there could be free exchange, or switching, between the lagging-strand polymerase and a spare polymerase that is bound to the helicase. The spare core polymerase might take over

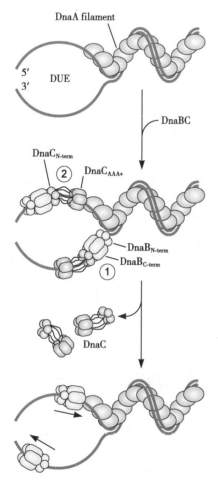

Figure 12.17 Model for DnaC-DnaA cross-talk and deposition of DnaB helicase. The DnaA nucleoprotein forms to one side of a DNA unwinding element, DUE; the question then arises as to how an asymmetric initiation assembly facilitates the symmetric loading of two DnaB helicases. The model suggests that this is done through the ability of DnaC to directly engage an available end of the DnaA oligomer. (Top) DnaA has assembled at *oriC* and has melted the DUE. (Middle) (Step 1) Helicase-loading on the bottom DUE strand is facilitated by direct DnaA-DnaB interactions. (Step 2) The helicase destined for the top strand is recruited through a specific interaction between DnaC and ATP charged DnaA. (Bottom) ATP hydrolysis leads to loss of DnaC; both DnaB helicases are now free to migrate to their proper positions at the fork. (Adapted from Mott ML, Erzberger JP, Coons MM, et al. [2008] *Cell*, 135:623–634. With permission from Elsevier.)

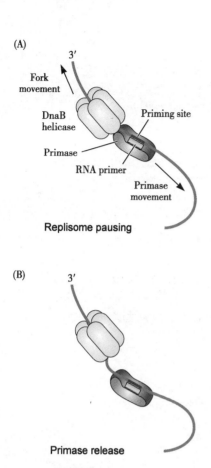

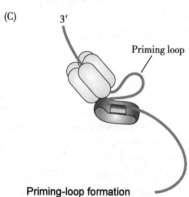

Figure 12.18 Models for DNA priming. All three models for solving the directionality problem have experimental data to support them. (A) The pausing scenario envisions that the replisome pauses for primer synthesis to occur; this has been shown to occur in the T7 phage replisome (B) In the second model, the primase, once clamped onto the lagging-strand template by the helicase, can be temporarily released from its interaction with the helicase to synthesize the primer; this release is known to occur in the *E. coli* replisome. (C) The third model envisions the formation of a temporary loop, which allows the replisome to move forward while the normal primase-helicase interaction persists. The priming loop eventually collapses into the lagging-strand trombone loop, probably when the primer is transferred from the primase to the lagging-strand polymerase. This model has been supported by single-molecule experiments in T7 and T4 replication systems. (Adapted from Dixon NE [2009] *Nature*, 462:854–855. With permission from Macmillan Publishers, Ltd.)

lagging-strand synthesis when the active polymerase becomes temporarily arrested, for whatever reason. This mechanism is especially important when the low concentration of Pol III in the cell is considered; with just ~10 copies per cell, it would be very hard to continue synthesis if the lagging-strand polymerase is lost into solution. A final interesting twist in the dynamics of the replisome has to do with situations in which the fast-moving replisome catches up with, or collides with, the ~20-timesslower RNA polymerase when the two polymerases are using the same DNA strand as template for DNA replication and transcription. This situation, described as codirectional or rear-end collision, contrasts with the head-on collision that occurs when the two polymerases move in opposite directions on each of the two strands. While head-on collisions lead to replication fork arrest and induce DNA recombination, co-directional collisions do not block fork progression. RNA complexes are simply bypassed in an interesting mechanism that involves permanent release of RNA polymerase, temporary dissociation of Pol III from the DNA, and subsequent use of the nascent mRNA molecule as a primer for continued DNA replication (Figure 12.19). It is worth pointing out that the existence of this bypass mechanism could explain why essential genes and most transcription units in bacteria have been observed to be encoded by the leading-strand template. It also suggests natural selection against head-on collisions. Co-directional collisions are also selected for in human cells.

Despite the importance of termination of replication and its role in chromosome segregation and cell division, the location of the sites where replication terminates is ill-defined, and both the nucleotide sequences and the proteins that interact with them differ from bacterium to bacterium; that is, there is hardly any evolutionary conservation. In fact, researchers today talk about termination zones rather than termination sites, and these zones can occupy a significant region of the entire chromosome, at least 5%. Termination, in general, is not merely the coincidental arrival of two replication forks traveling in opposite directions, which then merge. Details of the

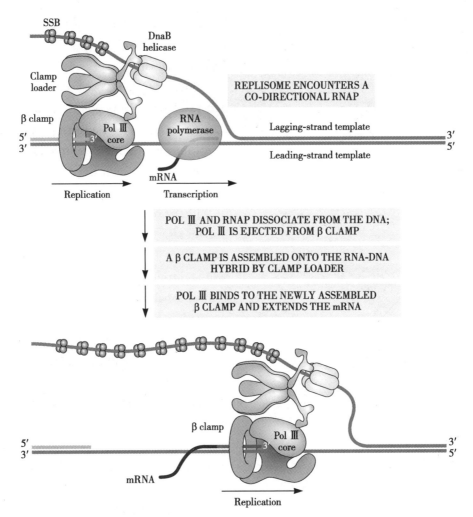

Figure 12.19 **Model for bypass of co-directional RNA polymerase by the replisome.** Co-directional collision of the replisome with RNA polymerase (RNAP) results in mRNA takeover. RNA polymerase transcribes the leading-strand template in the same direction as movement of the DNA replication fork. Collision of Pol III and the RNA polymerase causes displacement of both polymerases from the DNA template, but the newly formed mRNA transcript remains in position and is then used as a primer for DNA extension by the newly assembled leading-strand polymerase. The model has precedent in normal synthesis of the lagging strand, in which Pol III rapidly hops from a clamp on the completed Okazaki fragment to a newly assembled clamp on an RNA primer-DNA hybrid. (Adapted from Pomerantz RT, O'Donnell M [2008] *Nature*, 456:762–766. With permission from Macmillan Publishers, Ltd.)

organization of the termination zone in *E. coli* are presented in Figure 12.20. Interestingly, the DNA in the bacterial chromosome has strand compositional skew: the base composition of the two strands is not the same, which divides the chromosome into two segments bisected by *oriC* and the *dif* loci. The *dif* site is the chromosomal nexus for the recombination and decatenation reactions that complete chromosome separation. When replication forks meet, a considerable amount of supercoiling will have accumulated between the fronts of the forks, despite the action of topoisomerase. Movement of RNA polymerase along the double helix leads to accumulation of supercoiling stress, positive in front and negative in its wake; an analogous principle is in operation here. The consequence is that the two daughter helices will end up catenated or linked to each other and must be separated or unlinked before they can be apportioned, one chromosome each, to the daughter cells.

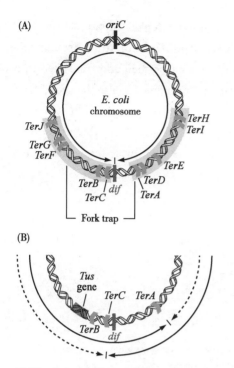

Figure 12.20 Termination of DNA replication in *E. coli*. (A) Ten *Ter* sites have been identified in *E. coli*, known as *TerA-TerJ*. The two opposed groups of polar *Ter* sites form the structure known as a fork trap: *TerC*, *TerB*, *TerF*, *TerG*, and *TerJ* are oriented so as to block clockwise-moving forks, whereas *TerA*, *TerD*, *TerE*, *TerI*, and *TerH* are oriented to block counterclockwise fork movement. (B) Expanded view of the inner region of the fork trap. The continuous arrows indicate the movement of the first fork to be halted, at *TerA* for a counterclockwise-moving fork or at *TerC* for a clockwise-moving fork, depending on which fork arrives first at the respective site. Dashed arrows indicate the movement of the second fork to arrive in each case. Thus, termination would occur at different sites: at *TerA* if the clockwise replication fork was delayed for some reason or at *TerC* if the counterclockwise fork were delayed. Most frequently, replication will terminate in the region between *TerA* and *TerC*. (Adapted from Duggin IG, Wake RG, Bell SD, et al. [2008] *Mol Microbiol*, 70:1323–1333. With permission from John Wiley & Sons, Inc.)

12.2.3 Bacteriophage and plasmid replication

The mechanisms described above apply in general to bacteria and eukaryotes, but the specialized genomes and lifestyles of some viruses and plasmids require special mechanisms. Unlike bacteria or eukaryotic cells, bacteriophage must rapidly produce multiple copies of their genomes within the host cell. Sometimes, this has the consequence that the viral replication machinery is much simpler than that found in bacteria or eukaryotes, which has been used to facilitate fundamental studies of replication. A case in point is bacteriophage T7, which utilizes a replication complex consisting of only a few proteins. This is very useful for *in vitro* studies. We describe two examples of DNA replication mechanisms in phages and plasmids here.

(1) Rolling-circle replication is an alternative mechanism

The genomes of certain small phages are organized as circles of single-stranded DNA or ssDNA. These phages use rolling-circle replication to replicate their genomes. Two systems have been extensively studied: the spherical phage φX174 (Figure 12.21) and the filamentous phage M13. The process in φX174 involves multiple events that are usually classified into three steps: (a) conversion of the ssDNA genome to a double-stranded form, known as replicative form I or RFI; (b) rolling-circle replication of RFI; and (c) generation of an ssDNA genome for packaging into phage particles. Note that the entire replication process depends heavily on the use of host proteins; the only essential phage-encoded protein is gpA, the initiator endonuclease that introduces a site-specific nick into the double-stranded RFI to provide bacterial Pol III with the free 3′-OH group needed for elongation.

Many small plasmids of Gram-positive bacteria, which do not have ssDNA phages, are multiplied by the rolling-circle mechanism. These plasmids encode an initiator protein that introduces a site-specific nick. The initiator protein shares sequence similarity with gpA and recognizes a similar nucleotide sequence for cutting. These similarities are interpreted as an indication of evolution from a common ancestor.

(2) λ phage replication can involve both bidirectional and rolling-circle mechanisms

In certain cases, replication is a combination of the bidirectional replication typical of circular bacterial chromosomes and rolling-circle replication. One well-studied example is the phage λ genome (Figure 12.22). The genome in the virion is linear. As we see in the next chapter, the replication of linear genomes encounters what is called the end problem. If an RNA primer is added at each end to begin replication, the removal of these primers leaves each daughter strand incomplete. The phage λ genome evades this problem in the following way: first it undergoes circularization upon entry into the host cells. The now-circular genome is initially replicated bidirectionally in order to quickly produce numerous circular genomes, which can be transcribed and translated to provide essential viral proteins. Later in the cycle, the genome switches to rolling-circle replication, which results in long concatemeric structures that are subsequently cut into genome-size linear fragments, suitable for packaging into new phage particles.

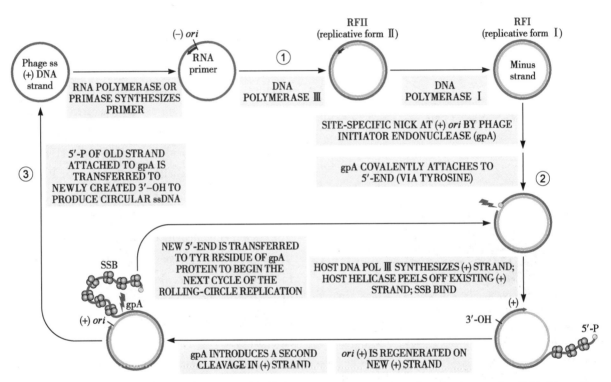

Figure 12.21 **Rolling-circle replication of ssDNA phage φX174.** The process of replication can be divided into three steps: (Step 1) conversion of the ssDNA phage genome into a double-stranded form, known as replicative form I or RFI; (Step 2) multiplication of RFI by a rolling-circle mechanism; and (Step 3) generation of an ssDNA genome for packaging into new phage particles. RFI is used as a template for transcription, which in turn initiates synthesis of the viral proteins. The DNA genome is synthesized in ~10s, with more than 20 circles released from a single rolling-circle intermediate. For simplicity, the schematic shows the immediate production of a single-length genome; however, usually this mechanism gives rise to long concatemers, which are then cleaved to fragments of single-genome size. The process relies mainly on the use of host proteins, with the important exception of the phage gpA protein, an initiator endonuclease involved in two processes, as indicated in the schematic. The first cleavage reaction requires superhelical RFI, produced by host gyrase, whereas the second occurs on a relaxed template.

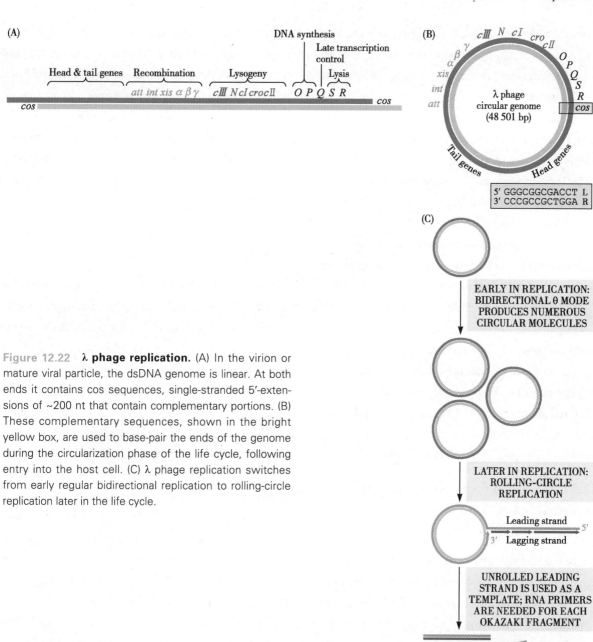

Figure 12.22 λ **phage replication.** (A) In the virion or mature viral particle, the dsDNA genome is linear. At both ends it contains cos sequences, single-stranded 5′-extensions of ~200 nt that contain complementary portions. (B) These complementary sequences, shown in the bright yellow box, are used to base-pair the ends of the genome during the circularization phase of the life cycle, following entry into the host cell. (C) λ phage replication switches from early regular bidirectional replication to rolling-circle replication later in the life cycle.

12.3 DNA replication initiation in eukaryotes

12.3.1 Replication initiation in eukaryotes proceeds from multiple origins

The multiplicity of replication origins in eukaryotes ranges from hundreds in yeast to tens of thousands in metazoa, between 30 000 and 50 000 according to recent estimates. There are numerous ways to visualize active origins, such as use of radioactive nucleoside triphosphate or NTP precursors and autoradiography, precursors carrying fluorescent dyes, or derivative precursors such as bromodeoxyuridine, BrdU, that can be visualized by fluorescently labeled antibodies. More recent methods make use of recombinant constructs expressing protein components of the replisome tagged with green fluorescent protein, GFP. At the molecular level, origins can be located by a number of techniques; at the genome wide level, they can be located by chromatin immunoprecipitation, ChIP, against protein components of the initiation complexes.

The large number of eukaryotic origins can be classified in three general categories: constitutive, which are active in all cells under all conditions as set by transcription and/or chromatin structure constraints; inactive or dormant, which are practically always inactive under normal conditions but can be woken or activated in stress conditions or during cell differentiation; and flexible, which are activated randomly in individual cells of the same cell population. These origins show flexibility in another aspect, too: if some origins are mutated or permanently inactivated through another mechanism, nearby origins can be activated or become more efficient. Flexible origins form the largest category and are usually clustered along the DNA. The molecular mechanisms involved in the random activation of a particular flexible origin are still under investigation.

12.3.2 Eukaryotic origins of replication have diverse DNA and chromatin structure depending on the biological species

There is a vast variety of eukaryotic origins, depending on the species. In budding yeast, *Saccharomyces cerevisiae*, origins correspond to defined DNA sequences termed autonomous replicating sequences or ARSs. Sequences fused to an ARS gain the ability to replicate in yeast. This can be used to advantage when large eukaryotic genes must be cloned under conditions in which they will be properly expressed, replicated, and processed. The routinely used bacterial cloning vectors cannot accomplish the cloning of large genes, since they only allow the insertion and propagation of relatively small DNA fragments. To clone such large genes, it is necessary to construct yeast artificial chromosomes or YACs that contain, in addition to the usual selective markers, centromeric and telomeric sequences that create a small functioning chromosome in the recipient cell. The replication of these small chromosomes is ensured by introducing ARS into the constructs. In *Schizosaccharomyces pombe*, ARS elements do not share a specific consensus sequence as they do in *S. cerevisiae*. Origins are located in AT-rich islands, as established by both BrdU labeling and ChIP assays. These AT-rich islands are targeted by Orc4, a specific subunit of the origin recognition complex or ORC; this subunit contains an AT-hook domain, a region that preferentially interacts with AT-rich sequences, and is absent from other species.

12.3.3 There is a defined scenario for formation of initiation complexes

A defined sequence of events prepares origins for activation. Pre-replication complexes or pre-replicative complexes, pre-RCs, are assembled on origin sequences during M and G_1 phases, and these are then activated at various times during S phase (Figure 12.23). First, the origins are recognized by the ORC: while still in M phase, the ORC recruits two proteins, Cdc6 and Cdt1, that are needed for recruiting of the hexameric Mcm2–7 complexes. One Mcm2–7 complex is bound on each side of the bound ORC, in opposite orientations. These complexes are the helicases that serve to unwind DNA in the two divergent replication forks. The site is now said to be licensed for initiation. In practice, this means that it can now bind other essential factors to produce first a pre-RC and subsequently, at the junction between G_1 and S phase, the pre-initiation complex or pre-IC. The more detailed structural views presented in Figure 20.8 should be helpful in understanding the process at a more mechanistic level. These figures also provide information on the conformational changes experienced by Mcm2–7, as well as the role of the recently discovered GINS complex, in establishing the final clamping of the helicase about the DNA. GINS stands for go, ichi, nii, and san: five, one, two, and three in Japanese, after the four related subunits of the complex, Sld5, Psf1, Psf2, and Psf3. The pre-initiation complex is now ready to accept the polymerase and to begin replication.

12.3.4 Re-replication must be prevented

There is one additional aspect of initiation that

needs to be mentioned. With many origins of replication simultaneously active, it is essential that none of them fires twice in a single round of replication. Recall that any origin sites already reproduced on the daughter duplexes are potentially capable of accepting ORCs. If one of these were to initiate another round prematurely, the DNA structure would be disastrously tangled. There are mechanisms to prevent just this possibility. The major set of mechanisms that prevent re-replication involves members of the cyclin-dependent kinase or CDK family, which phosphorylate and inactivate, or destabilize, a number of protein factors in the pre-RC: Cdc6, Cdt1, Mcm2–7, and/or ORC (Figure 12.23). Note that Mcm2–7 can also be phosphorylated by another kinase, Cdc7. CDK activity peaks in G_2 and M phases, thus preventing the formation of new pre-RCs in G_2 or destabilizing old pre-RCs in M. Another pathway involves Cdt1 degradation in S phase; such degradation occurs in a replication-dependent way. A more recently recognized mechanism for preventing re-replication uses the protein geminin as an inhibitor of Cdt1 function at origins. The mechanism depends on cycling of the amounts of geminin available throughout the cell cycle: its levels are high in S, G_2, and M phase and low in G_1 phase. The degradation of geminin occurs during the metaphase-anaphase transition and is proteasome/ubiquitylation-mediated. Low levels of geminin in G_1 allow binding of Mcm2–7 to its recruiter, Cdt1.

12.3.5 Histone methylation regulates onset of licensing

Recent reports have identified the methylation of histone H4 at lysine 20, H4K20me1, and the enzyme that puts this modification mark in place, PR-Set7, as key positive regulators of the onset of licensing in mammalian cells. Levels of both the enzyme and the modification are cell-cycle-regulated, being high during M and G_1 phases and dropping when S phase begins. Proteolytic degradation of PR-Set7 is needed to prevent re-replication. Even more interesting is

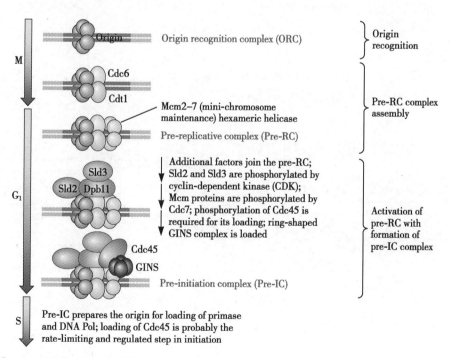

Figure 12.23 **Sequence of events leading to initiation of replication in eukaryotes.** In eukaryotes, the origins of replication are set by a three-step process: (Step 1) recognition of the origin by the origin recognition complex, ORC; (Step 2) assembly of the pre-replication complex, pre-RC; and (Step 3) activation of the pre-RC, leading to formation of the pre-initiation complex, pre-IC. The pre-IC is ready to accept primase and DNA polymerase. The pre-RC contains two Mcm2–7 helicases, one on each side of the bound ORC; only one Mcm2–7 complex is shown here for simplicity. The schematic also depicts the phases of the cell cycle when the respective events occur. This figure does not attempt to show the structural aspects of the complexes. (Adapted from Boye E, Grallert B [2009] *Cell*, 136:812–814. With permission from Elsevier.)

the fact that targeting of the enzyme to non-origin sites on chromatin fibers is sufficient to induce the H4K20 modification and assembly of the pre-RC.

12.4 Histone removal at the origins of replication

Chaperones may be envisioned as being actively engaged in removing histones, playing a facilitative or storage role, or both. There are two chaperones that accept H2A-H2B dimers: FACT and NAP1 (Table 12.2). The former was shown to facilitate transcription. Because FACT also interacts with the helicase, it could provide a mechanism for the reconstitution of nucleosomes by transmitting H2A-H2B dimers across the fork and/or by recruiting newly synthesized dimers to the replication machinery.

Another chaperone, ASF1, has an affinity for H3-H4 and certainly can escort newly synthesized H3-H4 dimers to the fork. Its role with parental (H3-H4)$_2$ tetramers is controversial. During nucleosome dissolution, the linker histone H1 must also be lost, but it is apparently picked up by another chaperone, NASP. At present, we visualize the replication fork with a clear space ahead, formed by the dissociation of a parental nucleosome, and the histones from that nucleosome sequestered by appropriate chaperones (Figure 12.24).

12.5 Replication of chromatin

It is evident from just a glance at chromatin structure that the replication of DNA in eukaryotes faces severe complications. Not only must the parental

Table 12.2 **Major roles of histone chaperones.**

Histone cargo	Chaperone	Recognized functions in DNA replication	Additional interactions with
H3-H4	ASF1	chromatin assembly and disassembly; promotes H3K56 acetylation	CAF1, RFC, Mcm
	CAF1	chromatin assembly; heterochromatin silencing	ASF1, PCNA, Rtt106
	Rtt106	chromatin assembly; heterochromatin silencing	CAF1
H2A-H2B	FACT	chromatin assembly and disassembly	Mcm, RPA, DNA Pol I
	NAP1	chromatin assembly and disassembly	
H1	NASP	chromatin assembly	

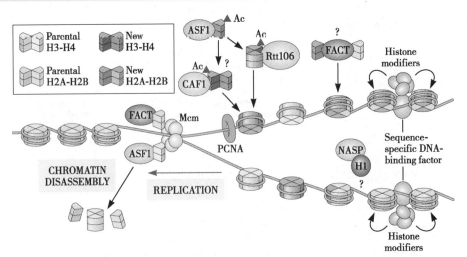

Figure 12.24 **Model for replication of chromatin during DNA replication.** Much of the model is speculative, as indicated by the question marks. For example, it does not deal with the uncertainty regarding the transfer of intact (H3-H4)$_2$ tetramers from parental chromatin. The model postulates that nucleosome modification in new chromatin is determined by sequence specific DNA-binding factors, which then recruit specific histone modifiers, as well as chromatin remodelers, not shown here. After being so marked and modified, both daughter fibers are considered to be old chromatin. (Adapted from Ransom M, Dennehey BK, Tyler JK [2010] *Cell*, 140:183–195. With permission from Elsevier.)

nucleosomal structure be traversed, it also must be reproduced with fidelity in both daughter duplexes, even maintaining epigenetic information. It is perhaps no wonder that the process has been, until recently, characterized by uncertainty and controversy on fundamental issues. Now a consensus is beginning to emerge, largely on the basis of the discovery of histone chaperones that play a major role in proper chromatin replication (Table 12.2).

First, it should be clear that the bulky DNA replication apparatus cannot pass through an intact nucleosome during elongation. The nucleosome must be displaced or dissociated. Electron microscopy of the well-defined simian virus 40 or SV40 replication system reveals a nucleosome-free region just ahead of the fork, as well as evidence of the perturbation of the next nucleosome. These effects could arise from any or all of the following: (a) unwinding of DNA by the helicase; (b) destabilization of the nucleosome by the accumulating positive stress in front of the moving DNA polymerase, in a scenario reminiscent of the situation in front of an elongating RNA polymerase; or (c) acceptance of histones by chaperones.

For a long time, it has been clear that parental histones are recycled randomly onto the leading and lagging strands. However, these histones represent at best only half the requisite amount needed for both daughter duplexes. New histones must be synthesized in the cell and delivered to the replication fork to complete chromatin reconstruction. There are two kinds of histone synthesis in most eukaryotes: replication-dependent synthesis and replacement synthesis. It is the former that we are concerned with at this point. Replication-dependent histone synthesis occurs only in S phase and produces the canonical histone variants that make up most nucleosomes: H3.1, H4, H2A, H2B, and the family of linker histone variants H1. The other set of histones are called replacement variants and include H3.3, CENP-A, H2A.Z, and H2A.X. These are synthesized throughout the cell cycle and may be inserted into preexisting chromatin.

From the structure of the nucleosome, it seems evident that histones H3 and H4 must be the first to be assembled on the daughter DNA duplexes, forming the (H3-H4)$_2$ tetramer, the core of the nucleosome. The tetrasome is a well-defined particle with DNA wrapped about (H3-H4)$_2$, but it seems that new H3 and H4 are first presented to the DNA as heterodimers bound to the chaperone ASF1. The X-ray structure of this complex reveals that the interaction between ASF1 and H3-H4 blocks the histone surface that would otherwise interact with another H3-H4 dimer to form the tetramer. Thus, ASF1 cannot deliver an intact tetramer to start a new nucleosome. It has been suggested that another essential histone chaperone, CAF1, can carry the H3-H4 dimer partner and accept a dimer from ASF1, thereby producing a tetramer that can be transferred to newly replicated DNA (see Figure 12.24).

Such a mechanism seems reasonable for the insertion of all-new tetramers, but it does not answer the question of how parental tetramers are transferred to daughter duplexes behind the fork. Are they broken down by ASF1 and then reassembled as described above? Or are they transferred intact, perhaps via CAF1? The first mode might result in some mixed old-new tetramers, and there is evidence against this. Nevertheless, intact transfer might lead to difficulties in conserving epigenetic information. It must be emphasized that the cell has clear ways to tell old H3 and H4 from newly synthesized H3 and H4. Post-translational marks are attached to the new histones by acetylation at lysines 9, 14, and 56 on H3 and lysines 5 and 12 on H4. These modifications are removed as chromatin matures, but they are essential for proper chromatin replication.

Transfer of the linker histone, H1, has not been as thoroughly studied. There is evidence that H1 binding to the chaperone NASP (see Figure 12.24) facilitates loading of the histone onto H1-depleted chromatin.

Much of the information governing gene expression in transcription appears to be coded into modifications of chromatin structure. This information, which is specific to cell and tissue types in *metazoans*, must be preserved or sometimes specifically modified through somatic cell divisions. Such epigenetic information can be stored in chromatin in a variety of ways, from the placement of nucleosomes along the DNA and higher-order chromatin folding, to the placing of histone variants, to post-translational modifications of histones or methylation of DNA. How is such a variety of often very specific changes preserved through the wholesale process of DNA replication?

The positioning of nucleosomes on DNA is dictated, at least in part, by certain motifs in the DNA sequence itself, yet these do not appear to be strong determinants. It seems likely that the most favorable final positioning can be achieved only with the aid of remodeling factors, which are known to play a role in chromatin reconstitution *in vivo*. The placement of linker histones and non-histone proteins may also participate in nucleosome placement. A particular example is the protein HP1, which is abundant in heterochromatic regions.

Preservation of the multitude of very specific covalent marks, such as acetylation, methylation, phosphorylation, *etc.*, is difficult to explain. Note that most of these marks are found on H3 and H4. When chromatin is being replicated, only the parental H3 and H4 will carry such modifications. These are to be distinguished from the special acetylation marking that is given to newly synthesized H3 and H4 to mark them as new. These marks are not a part of the epigenetic marking pattern and are removed in chromatin maturation. The question is, how are all of the other modifications that exist on the old histones reproduced on the new ones? Possible answers depend on the model assumed for the transmission of old H3-H4. If old tetramers are split into dimers before being transferred intact to daughter duplexes and each pairs with a new dimer, then all nucleosomes will have at least one set of markers. These could presumably recruit enzymes to the proper places to reconstruct the original pattern. There is, however, good evidence against mixed tetramers, so we must consider the other alternative. If H3-H4 tetramers are transferred intact from the parental strands, then half of the nucleosomes on the daughters will be properly marked, but half will be naïve. Marking these naïve tetramers would seem practicable only in the domain sense; a region could be acetylated in a certain way, for example, depending on the preponderance of such acetylation on old H3 and/or H4 in that domain. It is hard to see how very specific, nucleosome-to-nucleosome patterns could be regenerated, but we know little about the patterns at this level.

There is evidence that the processes by which chromatin receives its various epigenetic marks are interrelated and in some sense cooperative. For example, the formation of heterochromatic regions is favored by a very specific H3 methylation, which helps recruit the protein HP1. HP1, in turn, recruits the specific methylase to catalyze this modification. Thus, the formation of heterochromatin domains with their characteristic condensed structure can spread over broad regions.

12.6 The DNA end-replication problem and its resolution

Each chromosome in a eukaryotic nucleus contains a single, linear DNA duplex, which must have two ends. The existence of chromosome ends causes problems in replication, because they should suffer gradual shortening with each cycle of replication. The DNA end-replication problem was first recognized in the early 1970s, independently by James Watson and by Alexey Olovnikov, when it was realized that the requirement of all cellular DNA polymerases for a primer meant that DNA replicated by a lagging-strand mechanism would shorten, when the terminal RNA primer is degraded.

During S phase, linear chromosomal DNA is copied by replication forks that move from an interior position on the chromosome toward the ends. Leading-strand synthesis can theoretically copy the parental strand all the way to its last nucleotide. Discontinuous lagging-strand synthesis by polymerase/primase copies the respective parental strand (Figure 12.25), primed by RNA primers. The RNA primers are removed from each Okazaki fragment, and the inter-

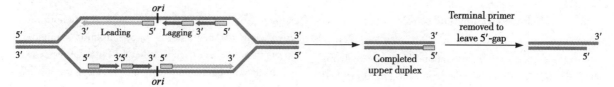

Figure 12.25 The DNA end-replication problem. The requirement of all cellular DNA polymerases for a primer should lead to shortening of DNA that is replicated by a lagging-strand mechanism, once the terminal RNA primer is degraded. This DNA end-replication problem is illustrated to the right for the upper of the two daughter duplexes. Following subsequent rounds of DNA replication, if only the semiconservative DNA replication machinery operates, as shown here, this gap will result in progressively shorter daughter strands.

nal gaps are filled in by extension of the discontinuous DNA and subsequent ligation. Removal of the most distal RNA primer, however, leaves a gap at the 5′-terminus, in the telomeric region of the chromosomes. Following subsequent rounds of DNA replication, this growing gap will result in progressively shorter daughter strands. Eventually, this erosion could extend into essential, coding regions of the genome. Additional erosion of telomeric DNA results from post-replicative processing of chromosomes ends. How does the cell deal with this problem?

The key to understanding why chromosome ends are not progressively shortened came with the discovery that telomeres consist of multiple repeats of simple sequences. The subsequent discovery of telomerase, the RNA-protein enzyme complex that has the ability to add those sequences so as to elongate chromosome ends, provided the complete solution; see Box 12.4 for a historical account of this Nobel Prize-winning discovery. Each telomerase complex contains a small RNA molecule characterized by a sequence complementary to that of the telomere repeat sequence. This portion of the RNA always remains single-stranded, despite the relatively complex secondary and tertiary structures of the rest of the RNA molecule.

In addition, the protein part of the telomerase has the thumb-palm-fingers structure typically present in all DNA and RNA polymerases. Its reverse transcriptase activity synthesizes stretches of DNA using the telomerase RNA as a template. Figure 12.26 depicts the proposed mechanism by which multiple copies of the telomere tandem repeats are synthesized by the telomerase. Multiple repeats are added by a slippage mechanism in which one repeat is synthesized and then the enzyme slips along and repositions itself at the new end of the chromosome and repeats the process. This continued re-extension

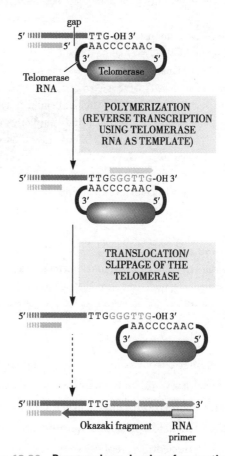

Figure 12.26 Proposed mechanism for synthesis of telomeric DNA. Each repeat can be added by a slippage mechanism, in which one repeat of telomeric ssDNA is synthesized and then the enzyme slips along, repositions itself at the new end of the chromosome, and repeats the process. The extended 3′-end can then act as template for new Okazaki fragment synthesis. Note that the telomere is extended but still has a 3′-overhang when the end RNA primer is removed. (Adapted from Greider CW, Blackburn EH [1989] *Nature*,337:331–337. With permission from Macmillan Publishers, Ltd.)

Box 12.4 Telomeres, aging, and cancer

The DNA end-replication problem resisted explanation for over a decade, until the groundbreaking work by Elizabeth Blackburn, in the lab of Joseph Gall. Blackburn and Gall cleverly chose the protozoan *Tetrahymena*, which has two nuclei, for their studies. Each *Tetrahymena* cell has a micronucleus, with five normal chromosomes, and a macronucleus, where the five chromosomes are chopped into hundreds of bits. This means lots of chromosome ends and thus telomeres. By 1978, Blackburn had shown that the telomeric DNA in *Tetrahymena* consisted of multitudes of repeats of the simple sequence TTGGGG. About this time, collaboration was established with Jack Szostak, who was able to demonstrate a similar but more complex situation in yeast. With her own laboratory, and with a strong collaborator in Carol Greider, then a graduate student, Blackburn was ready to attack the question of how these repeated sequences were added to the DNA ends. Progress was rapid: in 1985 they demonstrated the existence of the enzyme telomerase; in 1987, they showed that it contained RNA; and two years later, they had sequenced the RNA and demonstrated that it could serve as a template to add the repeats successively. A few years later, Greider demonstrated that the enzyme was processive. But it took until 1996 before Joachim Lingner and Thomas Cech were able to purify the enzyme so that its structure could be determined. Blackburn, Greider, and Szostak were awarded the Nobel Prize in Physiology or Medicine in 2009 "for the discovery of how chromosomes are protected by telomeres and the enzyme telomerase."

Telomeres and aging

Most cell types are deficient in telomerase, except briefly in S phase; exceptions are germ cells, stem cells, and cancer cells. This means that most somatic cells are slowly losing telomere length as they repeatedly divide throughout life. When telomeres shrink beyond a certain limit, about 100 repeats, processes are triggered that lead to cell senescence and death. An obvious implication of this is that many of the degenerative processes we associate with aging may have their cause in wilting telomeres. Could a dose of telomerase provide extended life?

Many researchers have become fascinated with this idea. Initial studies were not encouraging. Mice in which the telomerase gene had been knocked out seemed to do very well without it. But then it was found that the mouse lines from which the knockout mice had been derived had unusually long telomeres.

When the experiment was repeated on mice with human-length telomeres, degenerative disease and earlier death was marked. Most impressive are experiments in which telomerase has been switched on in aging mice. These mice lived about 40% longer than controls and had improved cognition and fertility. It must be emphasized that we have no evidence that these results extrapolate to humans.

Telomerase and cancer

Cancer is characterized by unlimited cell division, to the point at which certain cancer cell lines are virtually immortal. Apparently, the usual progression to senescence and death does not apply here. Furthermore, the great majority of cancer cells have high levels of telomerase. They express the telomerase gene constitutively, not just in S phase. All of this suggests that if we could find a nontoxic telomerase inhibitor, we might have a useful cancer drug. This perception has not been lost on drug companies, which are engaged in a massive competition for such a find. There are, in fact, promising candidates now undergoing clinical trials.

---

of the telomeric DNA compensates for the chromosome-end losses and prevents erosion into coding regions of telomere-proximal genes. If chromosomes are degraded below a critical telomere length of around 100 nucleotides, replicative potential is lost.

Most normal somatic cells do not need infinite replicative potential and hence repress telomerase activity to limit cell replication. By contrast, telomerase is up-regulated in many cancer cells (see Box 12.4), which enables their unlimited proliferation. Many types of cancer cells, however, are telomerase-minus. In such cells homologous recombination or HR is used to increase telomere length through the alternative lengthening of telomeres or ALT pathway. ALT produces highly heterogeneous telomere lengths.

ALT functions in the context of the shelterin complex that is present at all telomere ends. The shelterin complex includes TRF1 and TRF2 dimer proteins that bind to the double-stranded portion of the telomere and POT1 protein that binds to the single-stranded G-rich overhang, as well as TIN2 and TPP1 proteins that serve to bridge these proteins. POT1 suppresses normal DNA repair processes that would recognize the single-stranded overhang as DNA damage. Shelterin also protects the ends from degradation.

The mechanism(s) involved in ALT are very poorly understood. It is believed that ALT occurs through the formation of an intertelomeric D-loop in which the 3′-overhang of one telomere invades the telomeric duplex of another chromosome; lengthening occurs by use of the sequence information in the second chromosome. The resulting intertelomeric D-loop requires resolution, which involves helicases such as Werner (WRN) and Bloom (BLM) helicases. WRN and BLM interact physically and functionally with the critical telomere-binding and maintenance protein TRF2, a subunit of shelterin. Mutations in WRN and BLM lead to the premature-aging Werner syndrome and to Bloom syndrome, respectively.

It is known for certain that ALT requires dimers of two coiled-coil proteins SMC5 and SMC6, as well as a set of proteins that interact with them (Figure 12.27). Two other proteins belonging to the structural maintenance of chromosomes or SMC family of proteins, SMC1 and SMC3, form the backbone of cohesin. Cohesin is the complex that forms in S phase around the two sister chromatids in a replicated chromosome to keep these chromatids together until mitosis. Two other SMC proteins, SMC2 and SMC4, are the major structural component of condensin I, the complex that is instrumental in the formation and maintenance of the compacted structure of mitotic chromosomes.

The SMC5-SMC6 complex contains a protein subunit, MMS21/NSE1, which is an E3 SUMO ligase. It is believed that this subunit sumoylates numerous components of the promyelocytic leukemia bodies or PML, the cytologically recognizable structures where ALT occurs (see Figure 12.27). PML bodies facilitate the homologous recombination process by bringing together shelterin-decorated telomeres, the SMC5-SMC6 complex, and the numerous proteins that perform the homologous recombination reaction.

12.7 Alternative modes of DNA replication

The mechanisms described above apply in general to bacteria and eukaryotes, but the specialized genomes and lifestyles of some viruses and plasmids require special mechanisms. Unlike bacteria or eukaryotic cells, bacteriophage must rapidly produce multiple copies of their genomes within the host cell. Sometimes, this has the consequence that the viral repli-

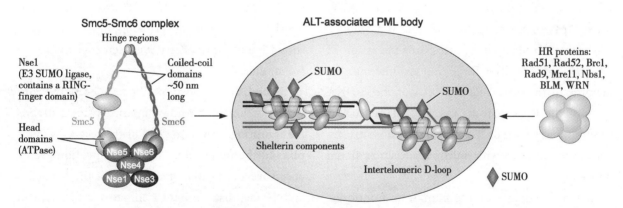

Figure 12.27 **Smc5-Smc6 and promyelocytic leukemia or PML bodies in ALT.** Smc5-Smc6 plays roles in double-strand break repair, restart of collapsed replication forks, and maintenance of rDNA integrity. The structure presented here is hypothetical and is based on the Smc1-Smc3 cohesin complex. NSE stands for non-SMC element proteins; Nse5 and Nse6 are not found in humans. Loading of Smc5-Smc6 to chromatin is likely coupled to replication. In cells undergoing ALT, telomeres associate with PML bodies known as ALT-associated PML bodies or APBs. PML bodies are dynamic nuclear structures that are involved in numerous cellular processes. They facilitate post-translational modifications and can localize proteins to their sites of action. Many components of PML bodies are sumoylated. Sumoylation affects protein stability, protein-protein interactions, and subcellular localization. PML bodies facilitate homologous recombination, HR, by bringing together telomeres, the Smc5-Smc6 complex, and HR proteins. In ALT cells, the Smc5-Smc6 complex and HR proteins associate with PML bodies in the G_2 phase, when sister chromatids are available for HR. Sumoylation of shelterin components RAP1, TIN2, TRF1, and TRF2 by MMS21 recruits or maintains telomeres at APBs and promotes telomere HR. (Adapted from Murray JM, Carr AM [2008] *Nat Rev Mol Cell Biol*, 9:177–182. With permission from Macmillan Publishers, Ltd.)

cation machinery is much simpler than that found in bacteria or eukaryotes, which has been used to facilitate fundamental studies of replication. A case in point is bacteriophage T7, which utilizes a replication complex consisting of only a few proteins. This is very useful for *in vitro* studies.

12.7.1 Replication in viruses that infect eukaryotes

The large number of viruses that infect eukaryotic organisms maintain their genomes in a number of different ways. Some, like SV40, have circular double-stranded DNA genomes, so they can replicate by the θ mechanism in the manner we have described. RNA viruses, however, have only RNA for a genome. These include many of the common human pathogens, causing illnesses from the common cold to AIDS to cancer. These viruses may have single-stranded or double-stranded RNA genomes, and there are a remarkable variety of mechanisms to replicate them. We discuss here an important class of RNA viruses, the retroviruses, because they use a DNA intermediate in replication.

A very unusual replication mechanism has evolved for replication of the RNA genomes of retroviruses. These viruses have an ssRNA molecule as a genome. The retroviral life cycle involves the integration of a dsDNA copy of this ssRNA viral genome into the genome of the host. The synthesis of the dsDNA copy of the RNA genome requires the action of reverse transcriptase or RT, an enzyme whose discovery was initially met with much skepticism and even derision.

Retroviruses include the human immunodeficiency virus type 1, HIV-1, as well as a number of oncogenic viruses. The viral genome carries three classes of genes: gag genes, which code for internal capsid proteins; pol, which provides functional enzymes; and env, coding for envelope proteins. There are usually two copies of the RNA genome within the capsid, accompanied by the reverse transcriptase protein.

The process of creating the dsDNA copy of the retroviral genome is extremely complex (Figure 12.28).

It involves the synthesis of two distinct single strands of DNA, a (−)-strand and a (+)-strand. The (−)-strand is synthesized first by using the viral RNA genome as a template; the (+)-strand is a copy of the (−)-strand and is initiated later in the process, while the synthesis of the (−)-strand is still ongoing. Complicated as it is, synthesis of the dsDNA copy depends on mechanisms and polymerase activities with which we are already familiar: the incoming NTP is attached to the free 3′-OH group of the nascent polynucleotide chain, primers are needed for chain initiation, and the RT enzyme possesses the thumb-palm-fingers structure typical of polymerase enzymes. The DNA copy that is ultimately synthesized possesses two long terminal repeats or LTRs that are not present in the viral genome. These LTRs are required for integration of the dsDNA into the host genome.

There are, however, several features that are unique to RT activities that deserve special mention. First, two different kinds of primers are used for synthesis of the two strands. The (−)-strand uses a host tRNA molecule as a primer, whereas the (+)-strand is initiated on a specific sequence of the viral RNA, the polypurine tract or PPT. This sequence is not susceptible to cleavage by the RNase H activity of the RT, which is responsible for degrading the original RNA strand immediately after it has been copied into the (−)-strand.

Second, there are two specific translocation events, when portions of already synthesized single strands are transferred from one end of the respective template to the other end. These translocation events occur because of the linear nature of the templates and the fact that synthesis is initiated at sites located close to the end of the templates; thus, the newly synthesized stretch of ssDNA must jump to the other end of the template to complete the synthesis. How can this occur? The answer lies in the existence of sequences that are repeated in the template and can serve as sites where base-pairing between the ssDNA and its respective template can occur, presumably upon transient circularization of the molecule. For example, the first translocation event involves the R sequences located at the 5′- and 3′-ends of the RNA;

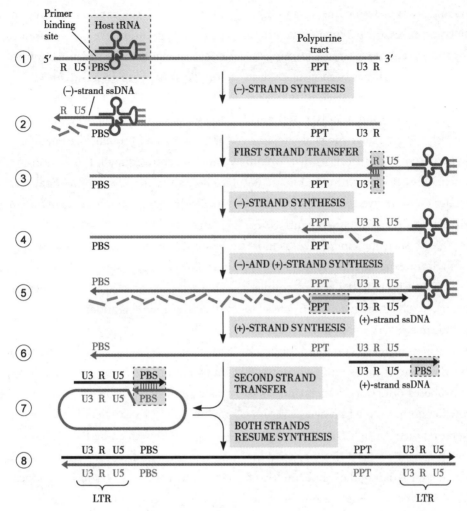

Figure 12.28 **Expansion of HIV genomic RNA into DNA to create LTRs.** The unusual features of the process are boxed in yellow. The steps are as follows: (Step 1) Formation, by reverse transcription, of a complementary ssDNA strand, the (−)-strand, is initiated on the 3′-end of a primer of host tRNA. (Step 2) Extension of the primer continues to the 5′-end of the RNA genome to generate (−)-strand ssDNA; the 5′-end of the RNA is degraded by RNase H. (Step 3) (−)-Strand DNA is translocated to the 3′-end of genomic RNA to complete (−)-strand DNA synthesis; base-pairing between R at the 3′-end of the RNA genome and R at the 3′-end of the (−)-strand ssDNA mediates the translocation reaction. (Step 4) (−)-Strand DNA synthesis continues; the RNA strand of the hybrid formed between the (−)-strand DNA and the 3′-end of the genomic RNA is susceptible to the action of RNase H, which degrades the RNA strand. (Step 5) (−)-Strand DNA synthesis continues; (+)-strand synthesis is initiated from the PPT or polypurine tract fragment of genomic RNA. PPT is resistant to RNase H action and serves as an efficient primer for the initiation of (+)-strand synthesis. (Step 6) (+)-Strand synthesis copies a portion of the primer tRNA to create a DNA copy of the primer binding site, PBS, at the 3′-end of (+)-strand ssDNA; the primer tRNA is removed by RNase H. (Step 7) (+)-Strand ssDNA must transfer to the 3′-end of (−)-strand DNA to complete viral replication; the transfer is mediated by base-pairing between the complementary copies of PBS on (+)-and (−)-strand DNA. (Step 8) Following the second strand transfer, both strands resume DNA synthesis, with each strand using the other as a template until the dsDNA, with the LTR ends needed for integration into the cellular genome, is fully synthesized. (Adapted from Basu VP, Song M, Gao L, et al. [2008] *Virus Res*,134:19–38. With permission from Elsevier.)

copying of the 5′-R sequence into the (−)-strand DNA creates the complementarity needed for interaction between the (−)-strand and the R sequence at the 3′-end of the RNA template. Similar complementarity and base-pairing is utilized during the second translocation event (see step 7 in Figure 12.28).

12.7.2 Models for organelle DNA replication

Replication is continuous, unidirectional from both origins of replication, and begins in the D-loop region. The two DNA strands of the mitochondrial genome are distinguishable by their density during centrifugation and are thus termed heavy and light. As replication commences, the parental heavy strand is displaced by the nascent heavy strand, forming the displacement or D-loop that has been visualized by electron microscopy. When replication of the heavy strand has progressed around a substantial part, nearly two-thirds, of the circular molecule, replication of the light strand begins from its own origin. The initiation requires RNA priming: remnants of inefficient primer excision give rise to the few ribonucleotides frequently found with mature mtDNA.

Recent years have seen the emergence of more models resulting from experiments on various biological systems. These include the strand-coupled replication model, in which the two strands are synthesized simultaneously, very much like replication of the nuclear genome; the RITOLS model, RNA incorporation throughout the lagging strand; and a model that proposes initiation from various sites along mtDNA. In another model known as recombination-driven replication or RDR, homologous recombination is considered as a likely process for mtDNA replication initiation in the mitochondria of yeast and plants, and more recently in human heart cells.

Mitochondria and chloroplasts contain their own DNA. We discuss the more thoroughly studied mitochondrial genome. Although the size of the mitochondrial genome accounts for only 0.0005% of the genome in humans, it is densely packed with ribosomal and tRNA genes, as well as essential genes that code for components of the respiratory chain. Together, these genes make up almost 0.1% of the total number of human genes. Because mitochondria are so universally important and because they are presumably derived from bacterial symbionts, their DNA and its replication are of considerable interest. Like bacteria, mitochondria do not have histones or the chromatin structure of eukaryotic nuclear DNA. Instead, mitochondrial DNA, mtDNA, is condensed with HU-like proteins similar to those of bacteria.

For years, the established view on the physical structure of mitochondrial genomes was that they were circular, like those of bacteria. For the purpose of discussing the mechanisms of mitochondrial genome replication, we have chosen to present this conventional view of the mitochondrial genome in human cells, with its resident RNA- and protein-coding genes. The D-loop region is believed to contain promoters and replication origins and to interact with certain proteins that maintain the structure. More recently, though, this notion has been shaken to its core by the advent of new data acquired by new methodology. In the yeast *Candida albicans*, some plants, and human heart cells, the mitochondrial genomes are complex networks of highly branched, variably sized subfragments. Genome-size molecules are very rare and circular molecules are not detected at all in such preparations.

There are very few examples of processes that have been so controversial as the replication of the mitochondrial genome, and, to this day, the mechanism remains unresolved. The only point of agreement among researchers is that the mitochondrial genome is replicated by Pol γ, a homotetramer with polymerase and 3′→5′ exonuclease activities and high processivity. In terms of its biochemical activities, Pol γ is similar to the Pol III core in bacteria and to Pol δ, the lagging-strand polymerase in eukaryotes. Early replication models proposed that replication occurred via the formation of a θ structure, similar to that found in the bidirectional replication of λ phage DNA. Further research led to the most widely accepted transcription-initiation strand-displacement model or SD, in which replication initiates at two unidirectional origins, each specific for one strand (Figure 12.29).

Recent years have seen the emergence of more models resulting from experiments on various biological systems. These include the strand-coupled replication model, in which the two strands are synthesized simultaneously, very much like replication of the nuclear genome; the RITOLS model, RNA incorporation throughout the lagging strand; and a model that proposes initiation from various sites along mtDNA. In another model known as recombination-driven replication or RDR, homologous recombination is considered as a likely process for

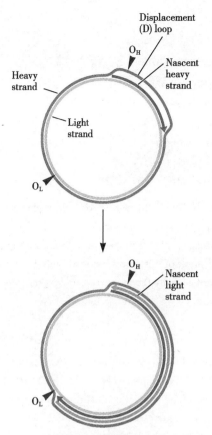

Figure 12.29 Strand-displacement model for circular mitochondrial DNA replication. Replication is continuous, unidirectional from both origins of replication, and begins in the D-loop region. The two DNA strands of the mitochondrial genome are distinguishable by their density during centrifugation and are thus termed heavy and light. As replication commences, the parental heavy strand is displaced by the nascent heavy strand, forming the displacement or D-loop that has been visualized by electron microscopy. When replication of the heavy strand has progressed around a substantial part, nearly two-thirds, of the circular molecule, replication of the light strand begins from its own origin. The initiation requires RNA priming: remnants of inefficient primer excision give rise to the few ribonucleotides frequently found with mature mtDNA. (Adapted from Brown TA, Cecconi C, Tkachuk AN, et al. [2005] *Genes Dev*, 19:2466–2476. With permission from Cold Spring Harbor Laboratory Press.)

mtDNA replication initiation in the mitochondria of yeast and plants, and more recently in human heart cells. Complex networks of linear molecules like those found in the mitochondria of human heart tissue seem to be replicated by both RDR and SD mechanisms, with SD occurring at the branch points, exposing single-stranded regions that persist in mitochondrial and chloroplast genomes. Interestingly, RDR and SD synthesis may account for the abnormal process of telomere lengthening in some cancer cells.

Key Concepts

- Replication in bacteria, eukaryotes, and most viruses proceeds in very similar ways; it is semiconservative, copying both complementary strands.
- Because polymerases copy only in the $5' \rightarrow 3'$ direction, one strand, the leading strand, can be synthesized continuously whereas the other, lagging strand must be synthesized discontinuously.
- The major enzyme for bacterial DNA replication is DNA polymerase III. Two copies of this enzyme are present in the protein complex at the replication fork, the replisome. One synthesizes the leading strand, and the other, the lagging strand.
- Maturation of the lagging strand requires the enzyme DNA polymerase I, which has an exonuclease activity to remove the primer and can then fill in the gaps with DNA chains. The fragments are then connected by a ligase.
- The structure of the replisome, and the necessity to allow for coordinated synthesis of both strands, require the formation of a trombone-loop conformation in the lagging-strand template.
- Initiation of bacterial replication proceeds bidirectionally from, usually, a single initiation region. This contains binding sites for the initiation proteins DnaA and DnaC. These recruit the helicase DnaB to begin unwinding the duplex.
- Termination of replication in bacteria occurs in zones containing *Ter* sites, each capable of binding the protein Tus, which can halt a replication fork traveling in a specific direction.
- While the above description applies, in the broadest sense, to DNA replication in bacteria, eukaryotes, and many viruses, some viruses employ a quite different strategy called rolling-circle replication or utilize a combination of bidirectional and rolling-circle replication.
- Elongation in eukaryotes has many similarities to elongation in bacteria but uses three polymerases: one for priming, one for the leading strand, and one for the lagging strand.

- Chromatin structure introduces many complications into eukaryotic replication: nucleosomes must be disassembled ahead of the fork and reassembled on both daughter duplexes. It is also necessary that nucleosome arrangement and epigenetic marking be reestablished on the new chromatin. How this occurs is not yet fully understood.
- An unusual form of replication is found in retroviruses. These have an ssRNA genome that is replicated into dsDNA, in a process catalyzed by an enzyme called reverse transcriptase. The dsDNA copy of the RNA genome can then be integrated into the cellular genome, where it can replicate as part of the cellular DNA.

Key Words

DNA dependent DNA polymerase（依赖 DNA 的 DNA 聚合酶）
DNA ligase（DNA 连接酶）
DNA topoisomerase（DNA 拓扑异构酶）
exonuclease（核酸外切酶）
helicase（解旋酶）
integration（整合）
Klenow fragment（Klenow 片段）
leading strand（领头链）
lagging strand（随后链）
Okazaki fragment（冈崎片段）
primer（引物）
primosome（引发体）
primase（引物酶）
replicon（复制子）
reverse transcription（逆转录）
retrovirus（逆转录病毒）
semiconservative replication（半保留复制）
telomerase（端粒酶）
template（模板）

Questions

1. Describe briefly the features of DNA replication and the process of DNA replication.
2. Describe briefly the process of reverse transcription. Give an example of this process in organisms.
3. Describe briefly the telomerase and its relationship with diseases.

References

[1] Burgers PM (2009) Polymerase dynamics at the eukaryotic DNA replication fork. *J Biol Chem*,284:4041–4045.
[2] Burgess RJ, Zhang Z (2010) Histones, histone chaperones and nucleosome assembly. *Protein Cell*,1:607–612.
[3] Lovett ST (2007) Polymerase switching in DNA replication. *Mol Cell*,27:523–526.
[4] Okazaki R, Okazaki T, Sakabe K, et al. (1968) Mechanism of DNA chain growth. I. Possible discontinuity and unusual secondary structure of newly synthesized chains. *Proc Natl Acad Sci USA*,59:598–605.
[5] Pomerantz RT, O'Donnell M (2007) Replisome mechanics: Insights into a twin DNA polymerase machine. *Trends Microbiol*,15:156–164.

She Chen

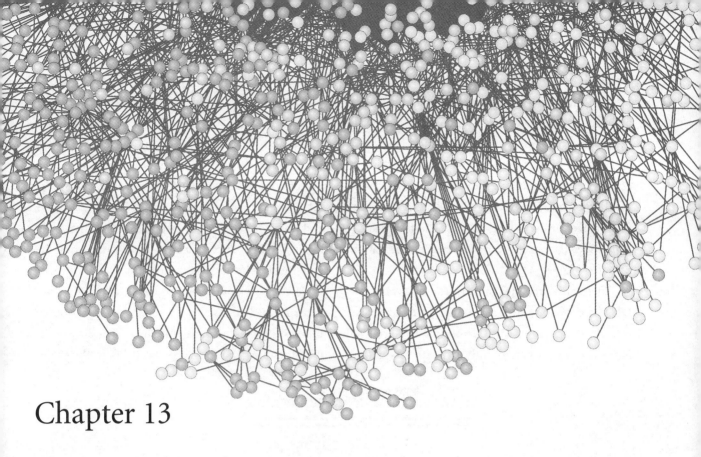

Chapter 13

DNA Recombination and DNA Repair

13.1 Introduction

It has been known for some time that genes and other genetic elements can change their locations within the genome, and such changes may bring about alterations in gene expression. Any change in location must, by necessity, involve cutting of both strands of the DNA, recombining the pieces, and then reestablishing the continuity of the phosphodiester back bones by the action of ligases. Exchanges of stretches of nucleotide sequences are broadly defined as DNA recombination. The process is widespread; it occurs in all known organisms and takes several distinct forms. Genomes are subject to spontaneous and induced DNA lesions throughout the life cycle of the host organism. It has been estimated that each cell of the human body receives tens of thousands of DNA lesions per day. Lesions in DNA can have deleterious effects on the cell, blocking replication and transcription. DNA lesions can be repaired to the original DNA sequence without a change or altered through mutagenesis and DNA recombination. This chapter discusses the biological roles and mechanisms involved in DNA recombination and DNA repair.

13.2 Homologous recombination

We discriminate between homologous recombination or HR, site-specific recombination, and non-homologous recombination depending on whether or not regions of homology between the exchanging partner DNA duplexes are required and, to some extent, on the length of these homologous regions. Homologous recombination involves the exchange of DNA sequences within large regions of homology between DNA molecules. It can provide a means of introducing genetic variability into the genome and can be a mode of DNA repair.

13.2.1 Homologous recombination in bacteria

Homologous recombination is the basis for major DNA repair processes in bacteria; it occurs only following DNA replication, when an intact copy of a sequence is available to serve as a source of information for the repair process. Homologous recombination is also used in horizontal gene transfer to exchange genetic material between different strains and species of bacteria and viruses. The ability of cells in natural populations to take up and incorporate homologous DNA from outside the cell has been maintained during evolution; it is beneficial because it provides the genetic variation needed for natural selection.

There are three mechanisms for acquiring new DNA: conjugation, transformation, and transduction. In conjugation, DNA from one bacterial cell is transferred, concomitantly with its replication, to another cell, through a process akin to the sexual process in eukaryotes. it is beneficial because it provides the genetic variation needed for natural selection. In transduction, pieces of DNA from one bacterium are transferred to another bacterium via a bacteriophage intermediate. Viruses that infect bacteria can accidently pick up pieces of bacterial DNA and transfer them into a newly infected bacterial cell; the newly introduced piece of DNA can recombine with the host bacterial chromosome.

HR is best understood in bacteria, mainly from experiments with *Escherichia coli*. An overview of the process is presented schematically in Figure 13.1. HR is initiated by the formation of DSBs in the DNA;

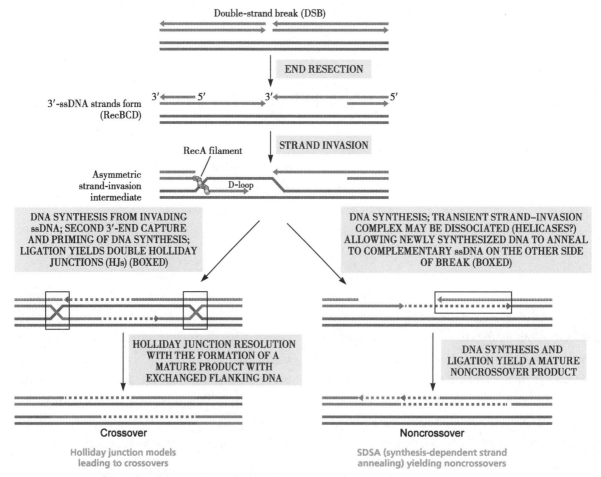

Figure 13.1 **Overview of homologous recombination.** HR is initiated by double-strand breaks or DSBs in the DNA. Once DSBs are introduced, the protein complex RecBCD resects the 5′-ends of each strand in the resulting double-stranded fragments, forming single-stranded 3′-overhangs. The 3′-ssDNA strand is covered by a helical filament of RecA, which allows the strand to invade into an intact homologous dsDNA, from a sister chromatid or a homologous chromosome. The resulting D-loop or displacement loop can then be processed by two different pathways. The formation of Holliday junctions and their resolution lead to the creation of crossovers. The alternative pathway, known as synthesis-dependent strand annealing or SDSA, yields noncrossover products. Note, however, that even in such noncrossover products there is a short stretch of hybrid DNA. (Adapted from San Filippo J, Sung P, Klein H [2008] *Annu Rev Biochem*, 77:229–257. With permission from Annual Reviews.)

these breaks result from either DNA-damaging agents or, during meiotic recombination, from a programmed process. Long single-stranded regions known as 3′-overhangs are then created in each of the resulting DNA fragments in a process known as end resection, which is catalyzed by the protein complex RecBCD. The single-stranded 3′-overhang is covered by a helical protein filament formed by multimerization of individual subunits of the protein RecA. The RecA-covered single strand then invades the intact homologous dsDNA—this second copy is present in the cell only following DNA replication—and performs a search for the region of homology. Once the homologous regions are properly aligned, the process can take one of two alternative pathways. The first pathway involves the formation and resolution of four-way junctions, or 4WJ, also known as Holliday junctions, or HJ, between the two duplexes with the formation of crossovers. Resolution of the junctions by cutting and religating the appropriate strands will yield duplexes that have exchanged segments. The second pathway has come to be known as synthesis dependent-strand annealing or SDSA and leads to noncrossover products.

13.2.2 Holliday junctions are the essential intermediary structures in HR

Holliday junctions, HJ, also known as four-way junctions, 4WJ, are formed as intermediates in the pathway that leads to the formation of crossover products (see Figure 13.1). They were first recognized on the basis of purely genetic studies performed by Robin Holliday in the early 1960s. The results revealed that the 4WJ is highly dynamic, switching between closed and open forms, depending on the environmental conditions and on the presence of protein ligands.

Once formed, the joint in the 4WJ can move along the duplex in a process called branch migration (Figure 13.2). The extent of branch migration determines the location and length of strand exchanges. Branch migration is an extremely fast, ATP hydrolysis-dependent process, ~5000 bp/s. It can occur in either direction and is mediated by a complex of two proteins, RuvA-RuvB complex. RuvA recognizes and binds the open form of the 4WJ. This binding is followed by RuvB binding on both sides of the junction. RuvB is the molecular motor that rotates two of the four DNA arms of the junction in opposite directions. This action results in rotational movement of the other two strands that are thus being drawn into the junction. Rotatory molecular motors, while not common, are not unknown in biology; consider the motor for the bacterial flagellum, for example. The process that ends the branch migration is the resolution of the junction by a resolvase, RuvC, which cleaves the DNA at symmetrical positions across the junction (see Figure 13.2).

13.2.3 Homologous recombination in eukaryotes

(1) Proteins involved in eukaryotic recombination resemble their bacterial counterparts

The eukaryotic recombinase, the counterpart of the bacterial RecA, is Rad51. Rad51 belongs to a group of proteins encoded by the *RAD52* epistasis group. Epistatically grouped genes participate in the same biological pathway and are most frequently defined by analysis of double mutants. Mutations in the *RAD52* group of genes lead to hypersensitivity to DNA-damaging agents, more specifically to those causing the formation of DSBs. This presumably reflects the importance of HR in eukaryotic DNA repair. In 1970, the term rad, referring to their X-ray irradiation sensitivity, was introduced for these mutants, with the numbers 50 and above reserved for the individual genes that, when mutated, exhibit the irradiation-sensitivity phenotype (Table 13.1).

The eukaryotic recombinase Rad51 possesses all the features characteristic of RecA. Rad51, like its bacterial counterpart RecA, needs to bind to the recessed ssDNA ends to form the helical filament for strand invasion. However, these ends are already covered, to protect them from degradation, by replication protein A, RPA, or its bacterial counterpart SSB. Recombination mediator proteins come to the rescue. These proteins share common features: they physically interact with their respective recombi-

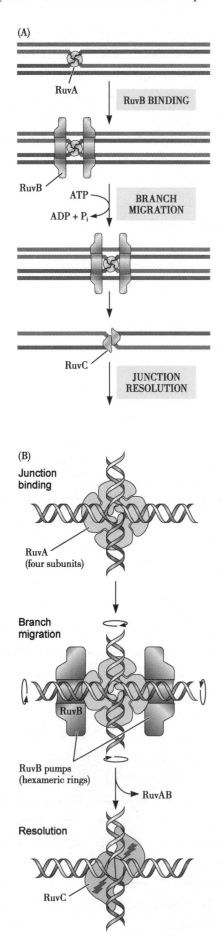

Figure 13.2 **Branch migration during the Holliday junction pathway of HR.** (A) Branch migration starts with the junction being recognized by RuvA. RuvA binding is followed by binding of the two RuvB pumps or motors. RuvAB promotes branch migration in a reaction requiring hydrolysis of ATP, and RuvC cleaves Holliday junctions. (B) Arrangement of protein and DNA components during the three stages of recombination outlined in part A. The proteins are shown schematically, based on their known structures. The two RuvB hexameric rings are presented in cross-section to visualize the DNA passing through their centers. RuvC active sites are marked as zigzags. (B, adapted from Rafferty JB, Sedelnikova SE, Hargreaves D, et al. [1996] *Science*, 274:415–421. With permission from American Association for the Advancement of Science.)

nase, Rad51 or RecA, and they preferentially bind to ssDNA over dsDNA. Only a very small amount of mediator is needed to overcome the inhibitory effects of the ssDNA-binding proteins, presumably because the mediators are only necessary for the removal of one or two of these molecules, the rest being removed by the growth of the Rad51 or RecA helical filament. Several such proteins have been described (see Table 13.1), but we focus on the two best-studied mediators, Rad52 and BRCA2.

Rad52 is a well-understood example of a recombination mediator. Detailed biochemical and structural studies have revealed that Rad52 has two distinct DNA-binding sites (Figure 13.3): the first site binds to ssDNA, and the second site binds to either ds- or ss- DNA. In addition, Rad52 binds to Rad51, forming a stable complex that is now capable of displacing RPA from the ssDNA; remember that Rad51 cannot, on its own, bind to RPA-coated ssDNA. Studies in which amino acid residues in either of Rad52's two DNA-binding sites were mutated showed that both sites are necessary for D-loop formation. The present understanding is that Rad52 serves to catalyze the replacement of RPA by Rad51.

(2) HR malfunction is connected with many human diseases

Because HR is known to participate in a wide variety of processes, it comes as no surprise that improper functioning of HR mechanisms leads to numerous

Table 13.1 **Homologous recombination factors that function with the recombinases Rad51 and/or Dmc1 in yeast and humans.**

Saccharomyces cerevisiae	Homo sapiens	Biochemical function and characteristic features
MRX complex (Mre11-Rad50-Xrs2)	MRN complex (Mre11-Rad50-NBS1)	DNA binding and nuclease activities; associated with DSB and resection
	BRCA2	Recombination mediator; binds to ssDNA and interacts with RPA, Rad51, Dmc1
Rad52[a]	RAD52	Recombination mediator; binds to ssDNA and interacts with RPA, Rad51
Rad54	RAD54	ATP-dependent dsDNA translocase
Rdh54	RAD54B	Induces superhelical stress in dsDNA; stimulates D-loop formations; interacts with Rad51
Rad55-Rad57	RAD51B-RAD51C RAD51D-XRCC2 RAD51C-XRCC3	Rad55-Rad57 and RAD51B-RAD51C are recombination mediators

[a]Recombination-mediator activity has been found in the yeast protein only.

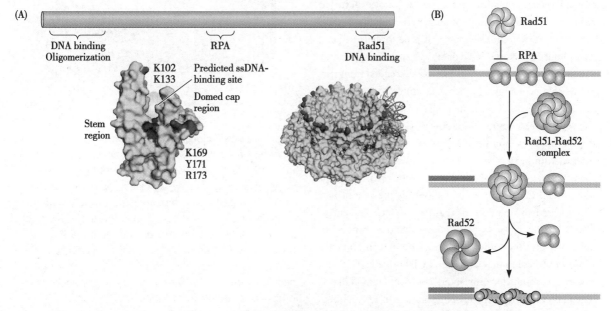

Figure 13.3 **Recombination mediator Rad52: structure and role in the delivery of Rad51 to ssDNA substrate during HR.** (A) Top, domain structure of the Rad52 monomer. Bottom, surface views of human Rad52$^{1-212}$ oligomerization domain protomer and of the 11-subunit ring of Rad52. The basic residues, shown in dark blue, clustered at the bottom of the groove formed between the stem and domed cap region constitute the ssDNA binding site, shown in yellow in the ring structure. The basic residues shown in magenta participate in binding of dsDNA. dsDNA runs along the rim of the stem region, with sites for ssDNA and dsDNA binding closely aligned with each other. (B) To perform its recombination mediator role, Rad52 forms a complex with Rad51 that delivers Rad51 to RPA-coated ssDNA; this seeds assembly of the presynaptic complex. Polymerization of additional Rad51 molecules results in further displacement of RPA from the DNA. (B, adapted from San Filippo J, Sung P, Klein H [2008] *Annu Rev Biochem*,77:229–257. With permission from Annual Reviews.)

human disorders. We illustrate this point by discussing the involvement of HR in hemophilia A. Two defective HR mechanisms lead to the formation of a nonfunctional product of a gene that encodes protein factor 8 in the blood-clotting biochemical cascade. Box 13.1 further discusses blood clotting and the role of defective HR in hemophilia.

Box 13.1 Hemophilia A and genetic recombination

Hemophilias are genetic diseases that interfere with blood clotting. Clotting can be initiated along two pathways. One, known as the intrinsic pathway, is started by tissue damage; the other, initiated by damage to blood vessels, is known as the extrinsic pathway. In each case a successive series of proteolytic activation steps leads to the polymerization of fibrin to form a clot (Figure 1). A critical point in these pathways is that at which factor X, denoted F10 in current nomenclature, is activated by either F7 or F9. Here the pathways merge, and either one now leads to clot formation. For F9 to function, it needs the participation of the anti-hemophilic factor F8. F8 must itself be activated by cleavage by thrombin (Figure 2); since active thrombin is a later product in the series, there is strong positive feedback at this point. F8 is the determinant of many varieties of hemophilia A, the classic and most dangerous form. This is because the gene for F8 is carried on the X chromosome: thus, females have two copies but males have only one. A woman heterozygous for aberrant F8 will not experience severe hemophilia but she will be a carrier for the disease, since she may transmit her aberrant gene to male descendants. Queen Victoria of England was such a carrier, and many of the crowned heads in Europe descended from her. Some were hemophiliac; others transmitted the condition to their male heirs. Thus hemophilia A is sometimes called the royal disease. Hemophilia A can result from any of a large number of point mutations in F8 or proteins that interact with it. However, the most disastrous form results from the presence of sequences in F8 introns that have almost identical sequences located outside the gene. There are at least two large introns that harbor such sequences. One is intron 1 (Figure 3A). Looping and recombining splits the F8 gene into two pieces, oriented in opposite directions; these cannot code for any F8 protein. This site, however, accounts for only about 1% of all hemophilia A. The real culprit is in intron 22 (Figure 3B). Recombination of this sequence produces the same kind of result but accounts for around 40% of all hemophilia A.

Hemophilia has long been a target of therapy. Unfortunately, concentrates of F8 from blood donors turned out too frequently to be contaminated with HIV. When recombinant DNA technology was developed, it became possible to produce cloned F8; this is used today. At the same time, efforts are underway to introduce somatic gene therapy.

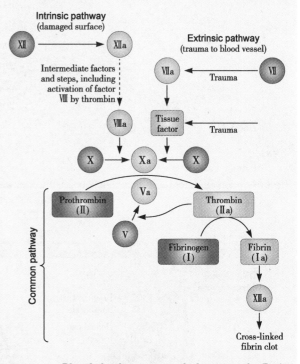

Figure 1 Blood clotting or coagulation cascade. Each factor in the pathway is a serine protease that exists in an inactive form: activation of the protease occurs by proteolysis by an already activated upstream protease. Of the two pathways depicted, the primary pathway for initiation of blood coagulation is the tissue factor or extrinsic pathway. The coagulation factors are usually indicated by Roman numerals, with a lowercase a suffix designating the active form. In our schematic, the inactive proteases are boxed in red and the active ones in green. (Adapted from Mathews CK, van Holde KE, Appling DR, et al. [2012] *Biochemistry*, 4th ed. With permission from Pearson Prentice Hall.)

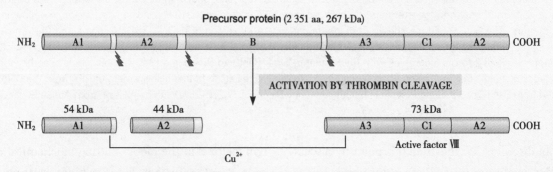

Figure 2 Activation of coagulation factor VIII by stepwise proteolytic cleavage by thrombin. The final three peptides remain associated, stabilized by a copper ligand.

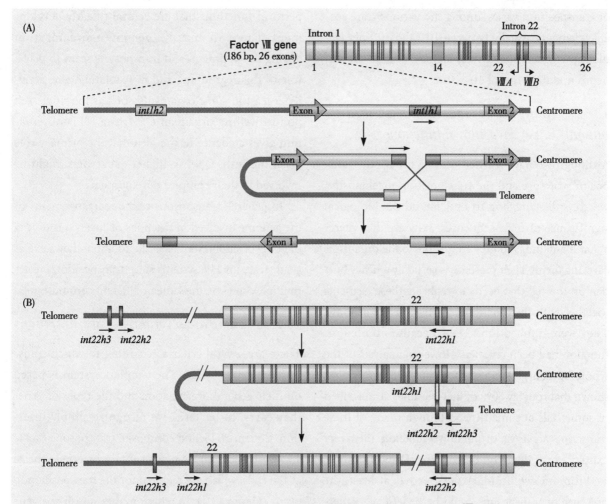

Figure 3 **Homologous recombination events lead to pathological inversions of factor VIII gene.** The gene contains 26 exons, shown as green boxes, and 25 introns; the two critical introns, 1 and 22, are marked. Intron 22 contains two further genes, VIIIA and VIIIB; arrows indicate the direction of their transcription. Horizontal arrows in the figure indicate the orientation of the repeated sequences. (A) Intron 1 contains a sequence, *int1h1*, that is repeated outside of the gene, *int1h2*. Homologous recombination between these two repeats explains the origin of pathological gene inversion. (B) Intron 22 contains a sequence, *int22h1*, with similarities to two sequences, *int22h3* and *int22h2*, that are distal to the gene. Intra chromosomal HR leads to the formation of crossover structures, with inversion of exons 1–22 with respect to exons 23–26. For simplicity, only the consequences of recombination with *int22h2* are shown, but *int22h3* can also recombine, with a similar outcome. (Adapted from Graw J, Brackmann H-H, Oldenburg J, et al. [2005] *Nat Rev Genet*, 6:488–501. With permission from Macmillan Publishers, Ltd.)

13.3 Nonhomologous recombination

13.3.1 Transposable elements or transposons are mobile DNA sequences that change positions in the genome

Nonhomologous recombination is the major mechanism that moves DNA sequences around the genome. It permits a DNA sequence to be lifted or copied from one site and placed into another that exhibits no homology to the original location. The existence of discrete, independent, mobile sequences, known as transposable elements or transposons, was discovered in maize, Zea mays, in 1948 by Barbara McClintock. Transposons are abundant, scattered throughout the genomes of many plants and animals, and can constitute a considerable portion of the DNA. Thus, in mammals, transposons occupy ~40%–45% of the genome; Transposons are also present in bacteria and unicellular eukaryotes but comprise a much smaller portion of the genome: transposons in *E. coli* constitute ~0.3% of the genome, and in *Saccharomyces cerevisiae*, 3%–5%. The frequency of transposition varies among the various transposable elements, usually between 10^{-3} and 10^{-4} per element per generation. This is higher than the spontaneous mutation rate of 10^{-7}–10^{-5}.

13.3.2 Many transposons are transcribed but only a few have known functions

Although we have learned quite a lot about transposon structure and the transposition mechanisms, we are still struggling to understand the biological significance of these sequences. Why are they there? What if anything, do they contribute to the organisms carrying them? Their presence is certainly a heavy burden on the cell that needs to replicate these elements together with the functional portion of the genome. They were dubbed junk DNA because no obvious function had been revealed. However, analysis of the whole human genome in the ENCODE project has shown that roughly 80% of that genome is transcribed in some cells at some time. Although many of these transcripts are as yet of unknown function, their very existence casts doubt on the concept of junk DNA.

From an evolutionary perspective, at least, the presence of transposons could be beneficial, conferring a selective advantage by DNA rearrangements. Obviously, mobile DNA is a slow but potent source of mutations, so it may have provided a method, over millions of years, to shuffle and rearrange the genome, giving rise to the diversity that drives evolution. Thus, transposons may form the DNA pool needed for natural selection to occur.

In addition, newer research has provided that, junk DNA may not be junk after all but may actually perform functions that are central to early development. There are massive genome reorganization during development, and transposons seem to play a role in this regrouping and in regulating gene activity at least in some organisms, such as the single-cell pond-dwelling organism *Oxytricha*. This observation gives credence to the idea that the primary role of all the extra DNA in higher eukaryotes might be involved in their complex development.

In general, transposons create rearrangements of the genome involved in deletions or inversions during the transposition event itself. In addition, they serve as a substrate for HR systems since transposition creates multiple copies on the same or different chromosomes.

13.3.3 There are several types of transposons

There are several criteria according to which transposons are classified. The simplest system is based on their general organization and the types of genes they carry. In bacteria, we recognize simple insertion sequences, IS, and composite transposons (Table 13.2). All bacterial elements contain inverted repeats at the ends, which are required for the transposition to occur (Figure 13.4). In addition, they all carry a gene encoding a transposase, the enzyme that performs the cleavage and ligation necessary for transposition. The composite transposons can also carry other genes, which usually confer antibiotic resistance. These extra genes are located in the central region, which is flanked by an IS element on each end. The IS elements can be oriented in the same direction or inverted; some IS elements may be functional, that is, capable of mediating transposition, whereas others are nonfunctional.

Table 13.2 **Characteristic features of some insertion sequences, IS, or composite transposons in *Escherichia coli*.**

Transposon	Size(bp)	Target(bp)	Inverted repeat(bp)	Resistance conferred
IS1	768	9	23	
IS2	1327	5	41	
IS4	1428	11–13	18	
IS 10R	1329	9	22	
Tn5	~5700	9		Kanamycin
Tn10	~9300	9		tetracycline
Tn2571	~23 000	9		chloramphenicol, streptomycin, sulfonamides, mercury

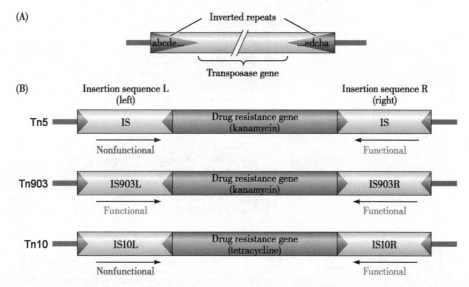

Figure 13.4 **Schematics of two types of transposable elements in bacteria.** (A) Insertion sequences have inverted repeats at their ends and a transposase gene at their center. (B) Composite transposons: three well-studied examples, Tn5, Tn903, and Tn10, are shown.

The number and types of transposons in eukaryotic cells are staggering, and the most useful classification is based on their mechanism of transposition (Figure 13.5 and Table 13.3). Some elements use a cut-and-paste route: in other words, a DNA element in the genome is excised from its original position and then inserted into a different position in the genome. Others copy the elements without excising the mother copy and then insert the daughter copy into the new position. These transposons are known as DNA transposons or class II transposons because they involve only DNA transactions. No currently active DNA transposons have been identified in mammals.

The second major type of eukaryotic transposons, known as retrotransposons or class I transposons, are those that pass through an RNA stage, also multiplying by a copy-and-paste mechanism but now involving an RNA intermediate. In the first stage of the process, the DNA element is transcribed into RNA; in the second stage, the RNA is reverse-transcribed into DNA, which is then integrated into the genome at a new position. Reverse transcription is catalyzed by a reverse transcriptase, which is encoded by the transposon itself. Retrotransposons behave very similarly to retroviruses, such as HIV.

Retrotransposons fall into two major categories, autonomous retrotransposons and nonautonomous retrotransposons, depending on whether they con-

Table 13.3 **Human transposons.**

Element	No. of copies (x1000)	Total length (Mb)	% of genome	Activty
LTR retrotransposons	443	227	8.3	
LINEs	868	558	20.4	
LINE-1	516	462	16.9	active
LINE-2	315	88	3.2	
LINE-3	37	8	0.3	
SINEs	1558	360	13.3	
Alu	1090	290	10.6	active with LINE-1
MIR and MIR3	468	69	2.5	
SVA	2.76	4	0.15	active with LINE-1
DNA transposons	294	78	2.8	

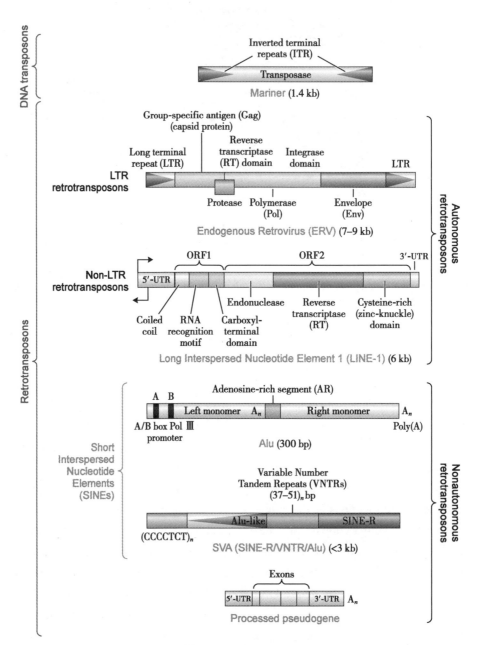

Figure 13.5 **Classes of transposons and specific examples in the human genome.** Examples are classified according to their mechanisms of transposition. In addition to the specific information for each class of transposons, provided by the figure itself, several features of the long interspersed nucleotide element 1 or LINE-1 are worth mentioning. LINE-1 possesses two open reading frames or ORFs, both of which are required for transposition. ORF1 encompasses a coiled-coil domain, an RNA recognition motif, and a basic carboxyl-terminal domain: the coiled-coil domain participates in ORF1p protein trimer formation, while the other two domains bind to nucleic acids. The endonuclease and reverse transcription activities of ORF2 are also essential for transposition, but the function of the C-terminus is unclear. ORF1p and ORF2p preferentially associate with their respective mRNAs to form ribonucleoprotein or RNP particles that participate in transposition. The LINE-1 5′-untranslated region or UTR contains two promoters. There is an internal Pol II promoter that directs sense transcription of the element. In addition, it contains a potent antisense promoter whose transcript contains a portion of the 5′-UTR and genomic sequences flanking the 5′-end. The products of the sense and antisense transcription presumably form dsRNAs that regulate LINE-1 transposition by RNA-interference-based mechanisms. (Adapted from Beck CR, Garcia-Perez JL, Badge RM, et al. [2011] *Annu Rev Genomics Hum Genet*, 12:187–215. With permission from Annual Reviews.)

tain all the information needed for transposition or require other functional transposable elements to help them transpose. Each of these two categories is subdivided according to the presence of specific sequence elements. Autonomous retrotransposons may or may not carry long terminal repeats, hence the LTR and non-LTR distinction; both encode reverse transcriptase. LTR retrotransposons include endogenous retroviruses, relics of past infection of the germline that have lost their ability to re-infect due to a nonfunctional envelope gene.

Most of the members of the non-LTR retrotransposon group, whose number is estimated at half a million, ~17% of the entire human genome, have lost their ability to act in transposition. A well-understood non-LTR transposon is the long interspersed nucleotide element 1, also known as LINE-1 or L1, which is the only currently active transposon that can mobilize or transpose itself and all other elements. LINEs are transcribed by Pol II.

The nonautonomous category contains short interspersed nucleotide elements or SINEs, inactive transposons that rely exclusively on LINE-1 for transposition. These elements fall into two categories: Alu elements and SVA elements. Alu elements account for ~10% of the human genome; they contain two monomeric sequences derived from the signal recognition particle 7SL RNA and an internal A/B box Pol III promoter. The monomeric sequences bind SRP9–14 proteins, like they do in the signal recognition particle itself. SVA elements, short for SINE-R/VNTR/Alu, have a composite structure and are probably transcribed by Pol II.

The nonautonomous category also includes processed pseudogenes. Processed pseudogenes arise through occasional usage of L1-encoded proteins to mobilize mature mRNAs to new genomic locations. Interestingly, most pseudogenes, ~10 000 copies in the human genome, are derived from genes highly expressed in the germline, such as housekeeping and ribosomal protein genes. Because pseudogenes lack functional promoters, most of them are not transcriptionally active, although some pseudogenes are expressed, probably by co-opting nearby promoters.

Representatives of the nonautonomous category do not code for reverse transcriptase; they are transcribed by Pol III.

It is curious to note that the relative amounts of DNA transposons and retrotransposons vary enormously among species (Figure 13.6). In humans, retrotransposons prevail, and the autonomous category is dominant over the nonautonomous one.

13.3.4 DNA class II transposons can use either of two mechanisms to transpose themselves

There are two mechanisms by which DNA transposons can move: nonreplicative and replicative (Figure 13.7). In the nonreplicative mechanism, which is a cut-and-paste-type mechanism, the transposase

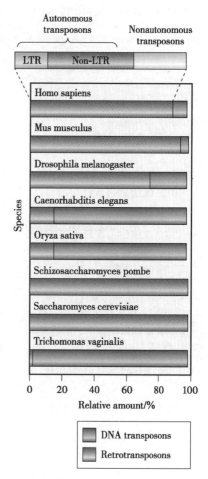

Figure 13.6 **Relative amount of DNA transposons and retrotransposons in some eukaryotic genomes.** The expansion of the human retrotransposon portion at the top depicts relative amounts of the three major retrotransposon types. (Adapted from Feschotte C, Pritham EJ [2007] *Annu Rev Genet*, 41:331–368. With permission from Annual Reviews.)

makes a staggered cut at the target site, which can be sequence-specific or random; this cut produces sticky ends. Then the transposase cuts out the DNA element from a distant genome location and ligates it into the protruding DNA ends of the target site. The resulting gaps are filled by DNA polymerase, and the integrity of the DNA backbone is restored by the action of a DNA ligase. This results in target-site duplication; actually, the existence of short direct repeats serves to identify insertion sites in specific genome regions or genome wide.

The second mechanism, replicative transposition, involves a replication step, in which the sequence at the donor site is copied in S phase and inserted at a new target site that has not yet been replicated. When the target site with the insert is replicated at a later stage, the number of transposons actually doubles.

Thus, with time, replicative transposition leads to ever-increasing transposon numbers. This mechanism involves two types of enzymes: the usual transposase, which acts at the ends of the original transposon, and a resolvase, which acts on the duplicated copy.

Transposase is an enzyme that performs both the excision and reinsertion steps in transposition. The process can be described in three steps: (a) Two copies of the enzyme bind to the transposon DNA at the inverted repeats located at the ends of the transposon; they form a dimer following binding. (b) The two ends are brought together, closing the transposon into a big loop, and the transposase cuts the transposon DNA at both ends. (c) The enzyme finds a new location on the DNA and reinserts the transposon. To do so, it must cut the DNA at the new site. A diagram and structure of a sample bacterial transposase bound to the ends of the excised element are presented in Figure 13.8. The enzyme acts as a dimer,

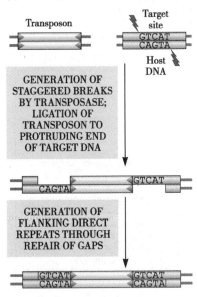

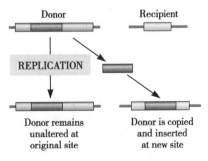

Figure 13.7 **Mechanisms of DNA transposition.** (A) Nonreplicative or cut-and-paste mechanism; (B) replicative or copy-and-paste mechanism.

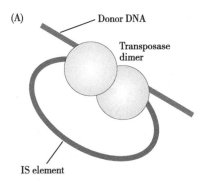

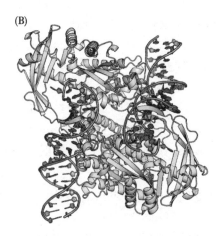

Figure 13.8 **Cut-and-paste mechanism of transposase action.** (A) The transposase cuts out the transposon DNA and moves it to a different place. (B) The structure shows two copies of the enzyme holding the two severed ends of the DNA; the actual loop of DNA is quite large, ~5700 bp long. The enzyme shown is a bacterial transposase that moves Tn5.

with each monomer bound to one of the inverted repeats at the ends of the insertion sequence. The dimerization is responsible for bringing the ends to be cut in close proximity, looping out the intervening IS sequences; the DNA is then cut at both ends, releasing the element for insertion into the target site.

13.3.5 Retrotransposons, or class I transposons, require an RNA intermediate

As might be expected, the mechanism for transposing retrotransposons is a more complicated process, as it involves both forward RNA transcription and reverse transcription of the RNA into a dsDNA copy, which can then be inserted into the genome. In addition, the process requires the participation of proteins encoded by the element, and these have to be translated from the mRNA after it has been exported to the cytoplasm. Transcription and reverse transcription subsequently occur in the nucleus. The mere spatial separation of these processes adds a level of complexity to an already complex process. We illustrate the process for the best-studied LINE-1 element in humans (Figure 13.9).

The integration process itself occurs by target-primed reverse transcription or TPRT (see Figure 13.9). Integration of the element into the genome is initiated by sequence specific nicking of the target site by the ORF2-encoded endonuclease. The newly generated 3′-OH group serves as a primer; thus, polymerization occurs directly onto the host DNA. The L1 mRNA serves as a template. For this to occur, the message needs to be in contact with the target

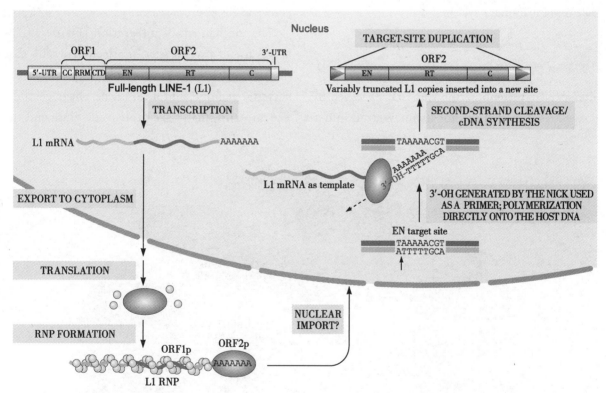

Figure 13.9 **Target-primed reverse transcription, TPRT, integrates L1 into the genome.** The process begins with transcription of L1 elements and export of the mRNA into the cytoplasm, where the two proteins encoded by ORF1 and ORF2 are translated. These proteins then interact with the mRNA to produce a ribonucleoprotein or RNP particle. Once this particle is imported into the nucleus by an unknown mechanism, nicking of the host chromosome at a specific sequence by the ORF2-encoded endonuclease, EN, initiates the integration of the element into the host genome by TPRT. The process uses the newly created 3′-OH group as a primer for the RT activity of ORF2; the mRNA is held in place by base-pairing interactions between the target site and the poly(A) tail at the 3′-end of the mRNA. The rest of the steps are poorly elucidated. The final result of integration is often a 5′-truncated L1 copy flanked by target-site duplications. The two L1-encoded proteins, ORF1p and ORF2p, are presumably hijacked by nonautonomous elements like Alu, SVA, and occasionally mature mRNAs to mediate their integration in trans. (Adapted from Beck CR, Garcia-Perez JL, Badge RM, et al. [2011] *Annu Rev Genomics Hum Genet*, 12:187–215. With permission from Annual Reviews.)

site; this steady contact is ensured by base pairing between the poly(A) tail of the message and the single-stranded T-rich portion of the target that was created by ORF2p.

13.4 Site-specific recombination

Site-specific recombination involves only limited sequence homology between recombining partners; thus, it occupies an intermediate position between homologous and nonhomologous recombination. The process does not require RecA or Rad51 recombinases. Two relatively well-understood processes that depend on site specific recombination are the integration of phage λ into the bacterial genome and the rearrangement of immunoglobulin genes that occurs during differentiation of antibody-producing B-cells in the immune system.

13.4.1 Bacteriophage λ integrates into the bacterial genome by site-specific recombination

Site-specific recombination was first observed in phage λ as a mechanism for its integration into a specific site of the host chromosome, known as lysogeny. Phage λ is the best-studied representative of temperate phages (Figure 13.10). These viruses can exist in one of two modes: they either replicate in the host bacterium immediately after infection and cause cell lysis, which is called lytic mode, or they can integrate their genome into the bacterial chromosome, remaining dormant for many generations while their DNA is replicated as part of the bacterial chromosome, which is called lysogenic mode. The integrated phage is known as a prophage, and the bacterium containing the integrated viral genome is known as the lysogen. Prophages can be induced to enter the lytic cycle by DNA-damaging agents, for example. Excision of the circular phage chromosome during induction is a reversal of the integration process; however, excision requires an additional protein known as Xis or excisionase.

The mechanism of integration (Figure 13.11) makes use of two *att* sites or attachment sites that share a short 15-bp region of homology: *attB*, in the bacterial genome, and *attP*, in the phage genome. Two proteins, the phage protein integrase and the

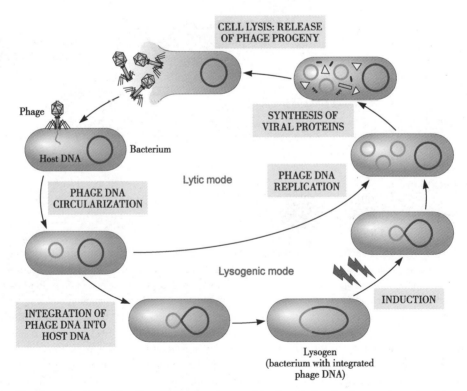

Figure 13.10 **Site-specific recombination.** This type of recombination during the lysogenic cycle of temperate phages is exemplified by phage λ.

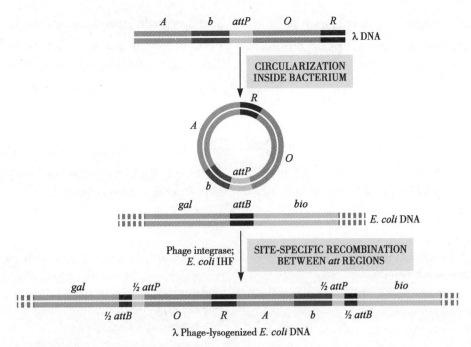

Figure 13.11 Site-specific recombination leads to establishment of lysogeny in bacteriophage λ. The λ phage linear chromosome circularizes upon entry into the bacterial cell by using the *cos* sites at the end of the linear chromosome. Recombination takes place between the *attP* site in the phage genome and the *attB* site in the bacterial genome; these sites share a limited 15-bp region of homology. *attB* is located between genes involved in galactose utilization and biotin synthesis. Two proteins, phage integrase and bacterial integration host factor or IHF, are needed for the reaction to occur. Detailed maps of where these proteins bind in both *attP* and *attB* are shown in Figure 13.12. (Adapted from Mathews CK, van Holde KE, Appling DR, et al. [2012] *Biochemistry*, 4th ed. With permission from Pearson Prentice Hall.)

bacterial protein integration host factor or IHF, are absolutely required. Additional bacterially encoded proteins bind to *attP* (Figure 13.12). The two homologous regions are recognized by the integrase, which creates a 4WJ or Holliday junction as an intermediate in the integration process. IHF participates in the integration reaction by bending DNA at the site of binding by 180°.

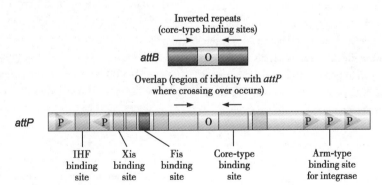

Figure 13.12 Binding of the site-specific recombinase λ integrase to *attP* and *attB* attachment sites. The integrase recognizes both *attB* and *attP*; these sites are quite different apart from the short stretch of identity, O, where crossing over occurs. *attB* is a simple site encompassing two short inverted repeats, called core-type binding sites, flanking the overlap region. *attP* is more complex, containing several adjacent binding sites called arm-type sites; it also contains binding sites for other factors involved in integration and excision. (Adapted from Groth AC, Calos MP [2004] *J Mol Biol*,335:667–678. With permission from Elsevier.)

13.4.2 Immunoglobulin gene rearrangements also occur through site-specific recombination

Vertebrates have developed a highly sophisticated immune system to fight foreign substances invading the organism, including viruses and bacteria. As part of the immune response, specialized B-cells produce immunoglobulins or antibodies; these are protein molecules highly specific to the invading molecule. A brief background review on how the immune system works is given in Box 13.2, The current estimate of the capacity of the human immune system is ~10 million distinct antibody molecules, each specific for a given antigenic determinant. This huge number of immunoglobulins cannot be encoded by individual genes; recall from Chapter 5 that the total number of protein-coding human genes, based on the complete sequence of the genome, is only ~20 500. Then

> **Box 13.2 A closer look: immunoglobulins and polyclonal and monoclonal antibodies**
>
> The production of antibodies, or immunoglobulin molecules, constitutes the major portion of the immune response of an animal to foreign substances. Sometimes, a molecule that normally occurs within the body is recognized as foreign or, in the terminology of immunology, as non-self; then antibodies are produced against it, leading to autoimmune diseases such as lupus erythematosus. Immunoglobulins, abbreviated Ig, are produced by specialized cells of the immune system, called B-cells, with one B-cell and its progeny producing only one type of antibody specific to only one antigenic determinant. An antigenic determinant or epitope is the entity that is recognized by the immune system as non-self and induces antibody production. There are two types of epitopes in proteins, those that consist of an uninterrupted portion of the primary polypeptide sequence, termed sequential, and 3D epitopes formed by the 3D proximity of amino acid residues that are not sequential in the chain. Immunoglobulins are made of two heavy chains and two light chains, all held together by disulfide bonds (Figure 1). Each chain contains a constant domain, C, and a variable domain, V. Constant domains are the same in all antibody molecules of a given class, whereas variable domains confer specificity to a given antigen. Carbohydrates attach to the heavy chain and help to determine the destinations of antibodies in tissues; they also stimulate secondary responses such as phagocytosis. Several different kinds of heavy chains exist, yielding the following classes of antibodies with different localization and functionalities:
>
> **IgG**, with γ heavy chains, can readily pass the blood vessel walls and the placenta to protect the fetus. One IgG variant is attached to B-cell surfaces. IgGs trigger secondary immune responses, known as the complement system, that destroy foreign cells. IgGs have the highest serum concentration, ~1g/dL, and the longest half-life, 21 days.
>
> **IgA**, with α heavy chains, is found in body secretions such as saliva, sweat, and tears and along the gastrointestinal and respiratory tracts, where the antibodies are arranged along the surface of cells to interact with antigens, preventing the antigens from directly attaching to cells. The invading substance is then swept out of the body together with IgA. They can trigger the complement system. IgAs are the main antibodies of colostrum and milk.
>
> **IgD**, with δ heavy chains, is found on the surface of B-cells, where the antibodies serve as antigen receptors. IgDs participate in class switching, in which a B-cell changes the class of antibodies it produces. During this process, the constant region of the heavy chain is changed, but the variable region stays the same. Since the variable region does not change, class switching does not affect antigen specificity. What changes is the interaction with different effector molecules that use different pathways to destroy the antigen.
>
> **IgE**, with ε heavy chains, is associated with allergic responses, known as immediate hypersensitivity. They bind to allergens and trigger histamine release from mast cells in epithelium and connective tissue. IgEs have the lowest serum concentrations, around 5 μg/dL, and the shortest half-life, 2 days.
>
> **IgM** is involved in the early response to invading microorganisms. They are the largest antibodies, pentamers, whose monomers are held together by disulfide bridges and a joining chain. The large size restricts IgMs to the bloodstream. IgMs trigger the complement system.
>
> Two types of antibodies can be produced; both are widely used in research and the clinic. Polyclonal antibodies or PAbs represent a mixed population of antibodies that recognize numerous epitopes. They are produced by an organism in the course of a normal immune response. By contrast, monoclonal antibodies or MAbs are produced by a single B-cell or its identical progeny. They are specific only for a given antigenic determinant and can be produced by a specific protocol in the laboratory (Figure 2). The 1984 Nobel Prize in Physiology or Medicine was awarded to Niels Jerne, Georges Köhler, and César Milstein "for theories concerning the specificity in development and control of the immune system and the discovery of the principle for production of monoclonal antibodies."

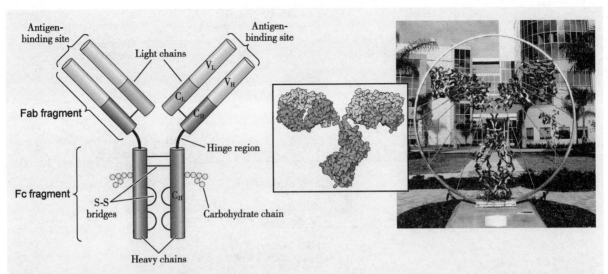

Figure 1 **Immunoglobulin structure.** Proteolytic cleavage at the hinge regions in the laboratory produces monovalent Fab fragments, which are widely used as research reagents. The Fc fragment in the intact Ig molecule functions as an effector to signal macrophages to attack. The boxed inset shows a space filling model of an antibody molecule, with the two heavy chains and the two light chains. The picture at the right is an artistic rendering of the molecule by quantum physicist-turned-sculptor Julian Voss-Andreae. The stainless-steel sculpture named Angel-of-the-West was erected in 2008 in front of the Florida campus of the Scripps Research Institute and symbolizes the protective or angel function of immunoglobulins. (Middle, courtesy of David Goodsell, The Scripps Research Institute. Right, courtesy of Julian Voss-Andreae, Wikimedia.)

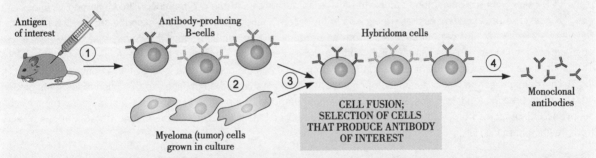

Figure 2 **Production of monoclonal antibodies by hybridoma technology.** The procedure involves several steps. (Step 1) A mouse is immunized with the antigen of interest; the mouse responds by producing B-cells in the spleen that secrete antibodies against the antigen. (Step 2) The spleen is removed and the B-cells are isolated. These are highly heterogeneous cell populations that contain individual B-cells, each producing an antibody specific to a single epitope on the antigen. These cells are, however, short-lived in culture. (Step 3) Isolated B-cells are fused to myeloma cells, cancerous B-cells that can multiply indefinitely in culture and are preselected to not produce antibodies. The resulting hybrid cells, called hybridomas, grow at the rate of myeloma cells but also produce large amounts of the desired antibody. Unfused plasma cells and myeloma cells die out because of the use of selective growing medium. (Step 4) Hybridomas that produce antibodies specific to a given epitope are selected and grown in bulk. Hybridomas can be frozen and shipped to other research and clinical laboratories, where they can be further propagated, serving as an unlimited source for the monoclonal antibody of interest.

how can such a daunting number of antibodies be encoded in the genome? The answer lies in the use of unique, albeit random, rearrangements of distinct portions of a gene cluster that exists in the precursors to the B-cells. This cluster contains multiple variants of the variable portion of the immunoglobulin gene. The rearrangements occur during differentiation such that each individual differentiated B-cell carries a final rearranged gene that encodes the synthesis of only one specific antibody molecule. The immune response then involves fast and robust proliferation of this specific B-cell type, by clonal

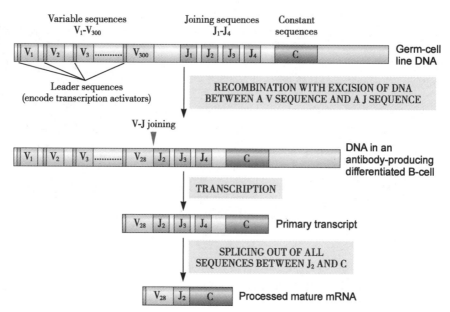

Figure 13.13 Gene rearrangements for production of κ light chains. Gene organization in a germ-cell line is shown at the top. Germ cells are not differentiated, so they do not produce antibodies. In antibodies, each light chain is encoded by noncontiguous sequences on the same chromosome: V, variable; J, joining; and C, constant. In humans there are ~300 different V sequences, each encoding the first 95 amino acid residues of the variable region. Each V sequence is preceded by its leader sequence, which contains transcription activator sequences that are not expressed in the germline; the V sequences cluster on the chromosome. Each of the four J sequences encodes the last 12 amino acid residues of the variable region; they form a separate cluster. Finally, there is one C region. During differentiation of one antibody-producing clone of B-cells, gene rearrangement occurs: the final mature mRNA and the polypeptide will contain one V, one J, and one C region. The DNA sequences that are excised in the recombination process—in the example shown, the sequences between V28 and J2—are permanently lost from all progeny of this cell line. Note, however, that the V sequences upstream of the junction, in this case V1–V28, and the downstream J sequences, in this case J_2–J_4, remain in the DNA. The other steps that lead to production of a functional antibody molecule occur at the level of transcription and RNA splicing. Transcription uses the leader sequence of only the V sequence that had been joined to the J sequence, thus producing an mRNA precursor that contains only one V sequence. Removal of the extra J sequences occurs during primary transcript splicing. (Adapted from Mathews CK, van Holde KE, Appling DR, et al. [2012] *Biochemistry*, 4th ed. With permission from Pearson Prentice Hall.)

expansion, to produce large amounts of antibodies to neutralize that specific antigen. Immunoglobulins consist of two heavy and two light chains connected by S-S bridges (see Box 13.2). Each chain comprises two different domains, variable and constant, connected by a short segment called the joining sequence. We illustrate the process of gene rearrangement on one class of light chains, κ or kappa; however, similar rearrangements are involved in the formation of the genes encoding the other classes of chains. The sequence of events involved in the production of mature mRNA for a specific κ chain is outlined in Figure 13.13. Note that in addition to rearrangements at the gene level, such as V-J joining, the primary transcript needs to undergo splicing of additional sequences, between the J sequence that happened to be joined to V and the C sequence. The final mature mRNA contains only one V, one J, and one C sequence.

Curiously, nature has come up with an additional mechanism that enlarges the antibody pool even further. This mechanism depends on the exact way that V and J sequences recombine. The process is akin to λ phage integration and depends on the existence of limited sequence homology on the 3′-side of each V and the 5′-side of each J (Figure 13.14); these sequences are used for site-specific recombination. The process begins by catalyzing DSBs between the homologous sequences by two proteins, RAG1 and RAG2. During the repair of these DSBs, additional diversity is created.

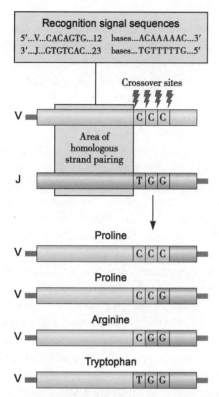

Figure 13.14 Generation of additional diversity by recombination of V and J sequences. Site-specific recombination between recognition signal sequences found at the 3′-side of each V and the 5′-side of each J produces additional diversity. The identity of the final product depends on the exact point of cutting and splicing, that is, location of the crossover sites, within the terminal trinucleotide sequences of the V and J that are being joined. The mechanism enlarges the antibody pool by a factor of 2.5, the average number of different amino acids encoded by four random triplets. (Adapted from Mathews CK, van Holde KE, Appling DR, et al. [2012] *Biochemistry*, 4th ed. With permission from Pearson Prentice Hall.)

13.5 Types of lesions and repair in DNA

In this section, we describe the kinds of lesions that are known to occur in DNA and the best-understood DNA repair pathways. A brief history of the development of this fascinating field of study is presented in Box 13.3. We begin with a brief, and by no means comprehensive, description of the most common types of lesions in DNA.

13.5.1 Natural agents, from both within and outside a cell, can change the information content of DNA

There are numerous agents that can affect the chemical structure of DNA and thus its informational content. In addition to the well-known external agents, physical and chemical, coming from the environment, numerous attacks on DNA may come from within the cell itself. The intracellular culprits include, for example, both oxidative stress and the by-products of errors that occur during scheduled biological processes, such as DNA replication, V(D)J recombination in immunoglobulin genes, and meiotic recombination.

Lesions can come in a variety of shapes and forms (Figure 13.15). The DNA backbone may be broken, giving rise to single-strand or double-strand breaks: the latter usually appear if two single-strand breaks occur in close proximity, within about one helical turn

---

Box 13.3 A brief history of the early years in the DNA repair field

The remarkable thing about genetic or DNA repair is that the phenomenon was recognized long before it was realized that it was actually DNA that was being repaired, or even that DNA had much biological significance. The story can be traced back to the classical genetic studies of fruit flies by Thomas Hunt Morgan in the early twentieth century (see Chapter 2). This work pointed out the importance of spontaneous gene mutations. A major advance came in 1927, when Hermann Muller found that exposure to X-rays greatly increased the mutation rate in fruit flies. Within a few years, a number of other laboratories found that short-wavelength ultraviolet light could also induce mutations in a wide variety of organisms, including bacteria and fungi.

This instigated a new field, called radiation biology. The slow realization that organisms could repair radiation damage generated a new field of study, ultimately termed DNA repair. The early history of this field is rich in incidents of unexpected discoveries, and sometimes more than one group claimed to have made a discovery first (Figure 1). In 1935, Alexander Hollaender, then at the University of Wisconsin, made a most peculiar and puzzling observation. If *E. coli* bacteria were exposed to UV radiation and then plated onto nutrient agar, some colonies appeared only after a considerable delay. One explanation was that the damage to the genes caused by the UV irradiation was somehow being spontaneously repaired. However, these holding recovery experiments were hard to reproduce, and the phenomenon remained an obscure puzzle for many years.

In 1948, Albert Kelner was a young man just beginning his scientific career in the Cold Spring Harbor Laboratory on Long Island, NY. He decided to reinvestigate holding recovery on UV-irradiated *E. coli*. The results from one experiment to another seemed maddeningly inconsistent. Looking at the phenomenon in the fungus Actinomyces rather than *E. coli* did not help, nor did careful control of the temperature of incubation. But one day in September of 1948, Kelner noted a most curious correlation: whenever the UV-irradiated samples were subsequently held in full daylight, recovery was remarkably stronger. A series of light-controlled experiments quickly confirmed that recovery from UV damage was somehow stimulated by light in the visible region of the spectrum. Remarkably, at about the same time, Renato Dulbecco, a researcher in the laboratory of Salvatore Luria at the University of Indiana, stumbled upon the same discovery. In Dulbecco's case, it was because he forgot, one night, to turn out the lights in the laboratory. The discovery of photoreactivation, as it came to be called, illustrates two important points about scientific progress: first, that important discoveries often come from the most unexpected, even weird-seeming results; and second, that when the time has come for a discovery, because all of the preliminary data are in, it can often happen almost simultaneously in different laboratories. Note that photoreactivation was discovered before the Watson-Crick structure and the Hershey-Chase experiment convinced most scientists that DNA was the genetic material. Neither the target nor the mechanism of photoreactivation could be understood in 1948. It was not until 1961 that work in a number of laboratories demonstrated that the formation of thymine dimers and similar molecular species is the DNA lesion repaired by photoreactivation. The photoactivatable enzyme responsible was not isolated until 1983. Although photoreactivation dominated early interest in the repair of radiation damage, it soon became obvious that other mechanisms were at work. Experiments using chemical mutagens, as well as radiation studies in which light was excluded after treatment, also showed the steady accumulation of viable cell colonies with time of storage. Such liquid storage experiments were conducted in a number of laboratories in the 1950s. The mechanisms for nonradiative repair remained obscure until a crucial experiment was performed in 1963 by Richard Setlow, a researcher at the Oak Ridge National Laboratory in Tennessee. By this time, thymine dimers were recognized as a major product of UV damage, and their cleavage by the photosensitive reactivating enzyme was understood. Setlow used an *E. coli* mutant that was resistant to photoreactivation but would still reactivate in the dark. The presumption was that some other process was cleaving the dimers. Yet analysis of the reactivated cells indicated that the dimers were still there.

1927	H. Muller: X-rays induce mutations
1928	F. Gates: UV at DNA absorption maximum most lethal for bacteria
1935	A. Hollaender and J. Curtis: Recovery after UV irradiation
1941	A. Hollaender and J. Emmons: Action spectrum = UV absorption spectrum
1944	O. Avery, C. MacLeod, and M. McCarty: DNA is the genetic material
1949	A. Kelner: Photoreactivation R. Dulbecco: Photoreactivation
1953	J. Watson and F. Crick: Structure of B-DNA J. Weigle: Reactivation of phage λ: basis for Salt Overly Sensitive (SOS) model
1956	S. Goodgal: Evidence for role of enzymes in photoreactivation
1961	R. Setlow and J. Setlow: Thymine dimers shown to be DNA lesions
1964	R. Setlow and W. Carrier: Excision repair R. Boyce and P. Howard-Flanders: Excision repair
1974	M. Radman: Full description of SOS pathway T. Lindahl: Base excision repair
1975	J. Wildenberg and M. Meselson: Mismatch repair

Figure 1 **Major events in the first 50 years of DNA repair research.**

of the DNA molecule, on opposite DNA strands. The bases themselves can undergo chemical changes, such as deamination, oxidation, and alkylation; if unrepaired, these lesions lead to mutations that become fixed during DNA replication. Intra- and interstrand cross-links are also common forms of DNA lesion. Figure 13.15 also illustrates the intrastrand cross-links caused by UV light. Interstrand cross-linking also occurs frequently, sometimes as a result of a genetic disorder, and if unrepaired leads to severe disease conditions such as Fanconi anemia (see Box 13.4). Another major lesion is depurination, when a purine

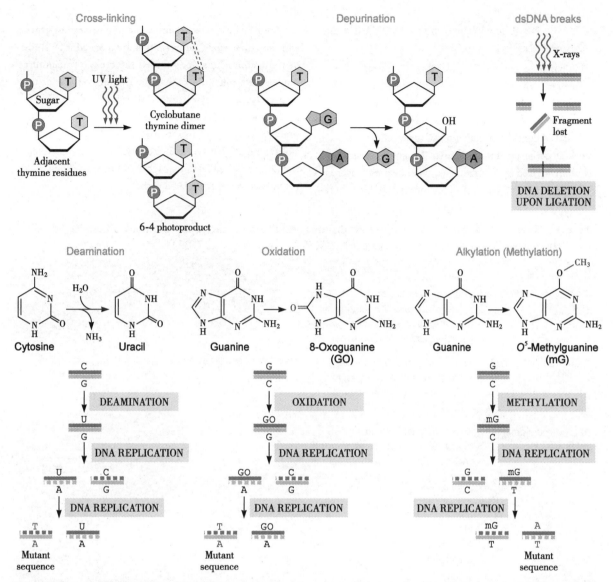

Figure 13.15 Changes that elicit DNA repair responses. These include changes in the chemical structure of DNA and the introduction of double-strand DNA breaks. If not repaired, most of these changes can affect the genetic information stored in DNA, after a couple of replication cycles.

base is cleaved off the backbone. This type of damage can also give rise to mutations, unless the repair machinery finds guidance as to what base, in the context of a dNTP, should be incorporated across from the missing base.

13.5.2 Introduction to pathways and mechanisms of DNA repair

DNA repair is a collective term that encompasses all biological processes during which alterations in the chemistry of DNA are removed and the integrity of the genome is restored. Many different, but still highly integrated, processes are used by cells to repair DNA damage. They are usually damage-specific (Figure 13.16) and some may be biological species-specific. Here, as an overview, we mention only the main characteristic features of each pathway.

- Direct repair: damaged DNA base, O^6-alkylguanine or a cyclobutane pyrimidine dimer, undergoes a chemical reaction to restore the original structure.
- Nucleotide excision repair, NER: repairs helix-distorting base lesions. The mechanism involves excision of a 22–30-nt fragment that contains the damage and use of the resulting ssDNA strand as a template for DNA polymerase action, followed

Box 13.4 **Do defects in the response to DNA damage contribute to aging?**

Aging has been thought for years that organisms age as a consequence of stochastic deterioration of biomolecules, caused by oxygen radicals and other endogenous and exogenous harmful compounds or exposure to physical agents such as UV light and X-radiation. A major effort has been aimed at understanding the causes of human diseases that are characterized by premature aging, known as progeroid syndromes. Table 1 lists some of the most prevalent of these conditions with their symptoms, the mutated repair genes, and the DNA repair pathways that are affected. The study of these conditions necessitated the creation of mouse models with phenotypes that closely mimic those of the respective human syndrome. Collectively, the studies on the human conditions and on the mouse models reveal that premature-aging syndromes are caused by defects in the ability to maintain genome integrity. Thus, the accumulation of unrepaired DNA damage might be the driving force behind aging and age-related pathology.

Table 1 **Most prevalent premature-aging syndromes and the gene and repair pathway affected.** (Adapted from Schumacher B, Garinis GA, Hoeijmakers JHJ [2008] *Trends Genet*, 24:77–85.)

Syndrome	Clinical features	Mutated genes	Repair processes affected
Cockayne syndrome (CS)	neuronal degeneration, loss of retinal cells, cachexia or wasting syndrome: loss of appetite and weight, muscle atrophy, fatigue, weakness	CSA, CSB	transcription-coupled NER[a]
trichothiodystrophy	neurological and skeletal degeneration, cachexia, ichthyosis or dry, rough, scaly skin, characteristic brittle hair and nails	XPB, XPD, TTDA	transcription-coupled NER
xeroderma pigmentosum (XP)	hypersensitivity to sun exposure, pigmentary alterations and premalignant lesions in sun-exposed skin areas, extremely high incidence of skin cancer	XPA-D, XPF, XPG	NER
Fanconi anemia (FA)	pancytopenia or low number of blood cells, bone marrow failure and renal dysfunction, abnormal pigmentation, short stature, cancer	FANC BRCA2	DNA-cross-link repair
Nijmegen breakage syndrome (NBS)	immunodeficiency, increased cancer risk, growth retardation	NBS1	DBS repair; telomere instability
Bloom syndrome	immunodeficiency, growth retardation, genomic instability, cancer	BLM helicase	Mitotic recombination
Werner syndrome (WS)	atrophic skin, thin gray hair, osteoporosis, type II diabetes, cataracts, arteriosclerosis, cancer	WRN helicase	DNA recombination, telomere maintenance
Rothmund-Thomson syndrome (RTS)	growth deficiency, graying hair, juvenile cataracts, skin and skeletal abnormalities, osteosarcoma, skin cancers	RECQL4 helicase	repair of oxidative DNA damage
ataxia telangiectasia (AT)	progressive cerebellar degeneration leading to severe ataxia or lack of voluntary coordination of muscle movements, telangiectasia or dilated blood vessels, immunologic defects, cancer	ATM	DSB signaling response

[a]NER, nucleotide excision repair.

by ligation. There are several subpathways, which will be described in detail. The process uses different mechanisms in bacteria and in eukaryotes.

- Base excision repair, BER: removes abnormal bases that result from chemical alterations in DNA bases. The first step involves cleavage of the glycosidic bond connecting a damaged base to the DNA sugar-phosphate backbone, with the removal of the base; then nucleases, polymerases, and ligases come into play.
- Mismatch repair, MMR: detection of mismatches and insertion or deletion loops triggers an incision of one of the DNA strands, which is further processed by nucleases, polymerases, and ligases. The proteins involved differ between bacteria and eukaryotes. In *Escherichia coli*, the process is directed by DNA methylation; the mechanism in eukaryotes is unknown.
- Homologous recombination repair, HR repair, and nonhomologous end-joining, NHEJ: error-free and error-prone mechanisms, respectively, to repair double-strand breaks, DSBs, and other lesions. HR can take place only in the late S and G_2 phases of the cell cycle, whereas NHEJ is active throughout the cell cycle. This distinction is due to the fact that HR needs a homologous intact DNA duplex, from a sister chromatid, as a template to fix the damage in an error-free way; such homologous duplexes exist only following DNA replication. On the other hand, NHEJ uses no template for the repair process and is thus error-prone and cell-cycle-independent. Numerous and different proteins are involved in each pathway. HR is most frequently initiated by generation of a DSB, followed by numerous steps that lead to the formation of crossover or noncrossover products. In NHEJ, existing DSBs are detected and bound by a damage sensor, which in turn leads to recruitment of other proteins. There is still another, relatively recently recognized mechanism that allows cells to continue functioning in the presence of lesions that would normally block DNA replication. This mechanism is appropriately termed translesion DNA synthesis.

13.5.3 Nucleotide excision repair is active on helix-distorting lesions

Nucleotide excision repair, NER, is the most versatile DNA repair system. It takes care of a wide array of DNA damage, from UV-induced lesions, to intrastrand cross-links and bulky chemical adducts, to lesions produced by reactive oxygen species, such as 8-oxoguanine. The unifying feature of these lesions is that they all disrupt the double helical DNA struc-

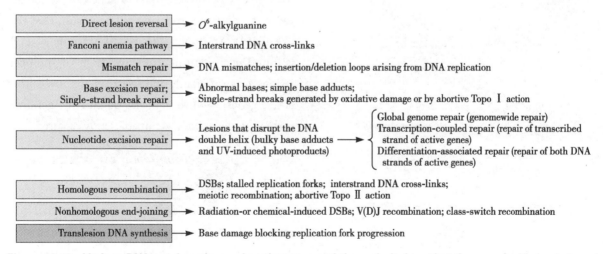

Figure 13.16 Various DNA repair pathways in eukaryotes and the main lesions that they repair. All classical repair pathways that lead to the repair of a lesion, with or without errors, are marked in yellow boxes. There is another, less desirable mechanism for dealing with lesions, known as translesion DNA synthesis, shown in the red box. Rather than repairing the lesion, this mechanism allows it to be bypassed during DNA replication. In addition to repairing harmful lesions in the DNA molecule, some of the pathways are responsible for normal physiological processes, such as meiotic recombination and rearrangement of immunoglobulin genes to create the repertoire of immunoglobulin molecules.

ture. The main NER pathway is the same in both bacteria and eukaryotes in that it involves four steps following lesion recognition (Figure 13.17): (Step 1) incision; (Step 2) excision of a short single-stranded segment of DNA spanning the lesion; (Step 3) DNA synthesis using the intact complementary strand as a template; (Step 4) DNA ligation.

In bacteria, a complex of UvrA and UvrB proteins tracks along DNA until it reaches a thymine dimer or other helix-distorting lesion, where it halts and forces the DNA to bend (Figure 13.18A). UvrA then dissociates, allowing UvrC, a nuclease, to bind. The UvrBC complex cuts on both sides of the dimer, in a somewhat asymmetric way. After the UvrBC complex leaves, helicase D unwinds the DNA to release the strand containing the lesion; then DNA polymerase and ligase act to seal the nick. The structure of the damage sensor complex UvrA$_2$B is presented in Figure 13.18B.

The process in eukaryotes is more complex and involves numerous proteins. We distinguish between global genome repair or GGR and transcription-coupled repair or TCR, two pathways that target the same lesions, either in the entire genome or in transcribed genes only. TCR occurs also in bacteria. Global repair is essential in proliferating cells because it ensures the integrity of the genome, which must be faithfully replicated and transcribed. The situation in terminally differentiated cells is different because the DNA in these cells is never replicated; these cells could arguably accumulate numerous lesions in the genome, as long as they are able to maintain the integrity of the genes needed for viable cell function. These genes are transcriptionally active; thus, they use a specialized pathway, TCR, which targets several repair systems to transcribed genes. Moreover, the transcribed strand is the one that is preferentially repaired. The other strand is also regularly checked, as its integrity is needed for its role as a template strand in DNA repair. As Figure 13.19 depicts, GGR and TCR use practically the same pathway, the difference being in the initial recognition of the damage. The pathway is well understood in humans because of extensive analyses of gene mutations that cause the disease xeroderma pigmentosum (see Figure 13.19). Another well-studied example of TCR is found in human neurons, in which UV-induced lesions are proficiently and selectively removed from active genes, even though the bulk of the genome, including silent genes, is not efficiently repaired.

13.5.4 Base excision repair corrects damaged bases

Base excision repair, BER, is the pathway that corrects for the presence of uracil, resulting from hydrolytic deamination of cytosine; methylated bases; and oxidized bases such as 8-oxoguanine. It also repairs some single-strand breaks generated by oxidative damage or by abortive topo I action. The active-site tyrosine of topo I forms a transient covalent bond

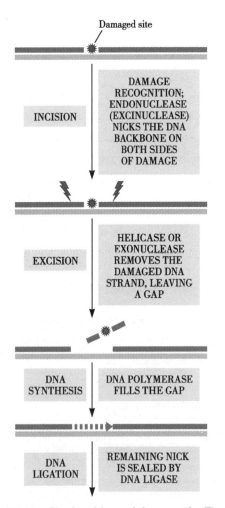

Figure 13.17 Nucleotide excision repair. The main pathway in both bacteria and eukaryotes involves four consecutive steps—incision, excision, DNA synthesis, and DNA ligation—which are performed by specific enzymes.

Figure 13.18 **A more detailed look at nucleotide excision repair in bacteria.** (A) NER pathway, with participating proteins. (B) Model of the damage sensor UvrA$_2$B, which has a flat and open structure. Note that the DNA-binding path on UvrA, shown in black, as proposed on the basis of site-directed mutagenesis studies, neatly aligns with the crystallographic position of DNA on UvrB. The model is based on the experimentally determined high resolution structure of the interaction domains of UvrA, aa 131–245, and UvrB, aa 157–250, boxed in the central part. Understanding the exact mechanism of damage recognition awaits further studies.

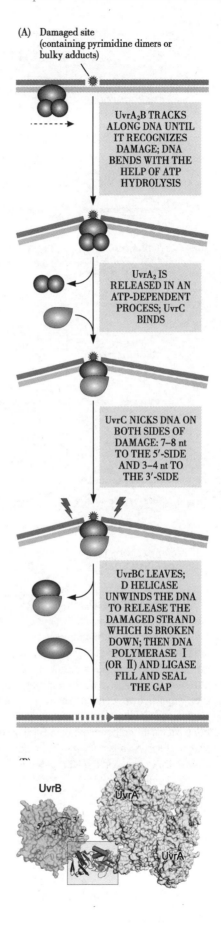

with the 5′-phosphate at the single-strand break it initially creates.

Base excision repair begins with cleavage of the glycosidic bond between the damaged base and deoxyribose by a specialized enzyme, N-glycosylase (Figure 13.20). The apurinic/apyrimidinic or AP site so formed is recognized by an AP endonuclease, which cleaves on the 5′-side of the AP site. This intermediate is then processed by two different pathways that involve different enzymes. Depending on the length of the stretch being replaced in the process, we distinguish between short patch repair, which occurs in 99% of cases, where only one nucleotide is replaced, and long patch repair, used in 1% of cases, where two or more nucleotides are replaced.

13.5.5 Mismatch repair corrects errors in base pairing

Despite careful proofreading, mismatched base pairs and insertion/deletion loops do arise during DNA replication. These errors are repaired promptly by a process that involves a single-strand incision. Mismatch repair, MMR, increases the overall fidelity of replication by 2–3 orders of magnitude. In both bacteria and eukaryotes, MMR requires Mut proteins; the genes encoding these proteins were identified by mutations in them that have a mutator phenotype, that is, increased frequency of spontaneous mutations.

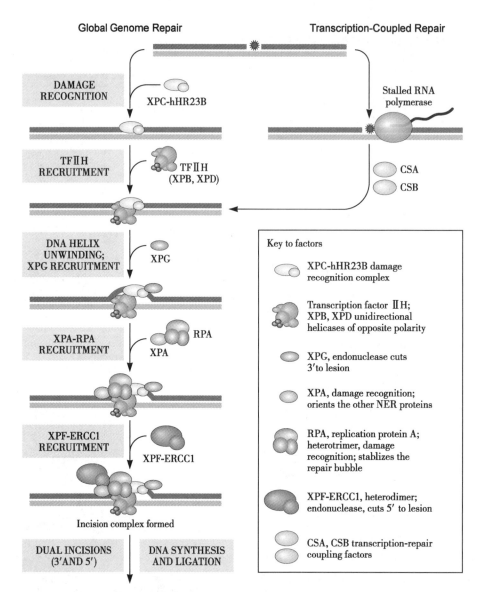

Figure 13.19 **Human NER excision complex assembly.** We distinguish between global genome repair and transcription-coupled repair, two pathways that target the same lesions, either in the entire genome or in transcribed genes only. They follow essentially the same pathway, the difference being in the initial recognition of the damage. Roles of the individual protein factors have been identified by studying xeroderma pigmentosum (see Box 13.4) patients and cells derived from them, identifying the genes affected by mutations, and then performing genetic complementation tests to see whether the mutated recessive genes can complement each other, with restoration of wild-type phenotypes. Seven complementation groups were identified, corresponding to seven genes participating in the pathway. For example, XP group or type A contains mutations in the gene coding for XPA, which participates in damage recognition and proper orientation of the other proteins in the complex; type B mutations affect the XPB gene, which encodes a helicase subunit of the transcription factor TFIIH. A variant form of the disease, XP-V, is associated with mutations in the gene for polymerase η, eta; this polymerase is not active in the NER pathway but rather performs translesion DNA synthesis.

13.5.6 Mismatch repair pathways in eukaryotes may be directed by strand breaks during DNA replication

The choice of which strand is repaired is clearly methyl-directed in bacteria, but the situation in eukaryotes is much less well understood. MutS and MutL are highly conserved but MutH is found only in Gram-negative bacteria; no functional homolog has been identified in other organisms. This fact led to the suggestion that mismatch processing in eukaryotes is directed by strand breaks that occur during replication, such as the 3′-end of the leading strand or the ends of Okazaki fragments (Figure 13.21). Repair is

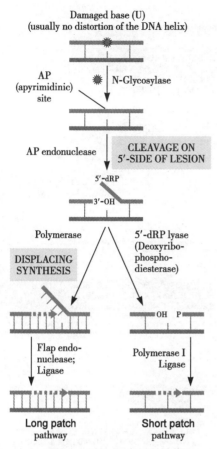

Figure 13.20 **Base excision repair.** The process begins with cleavage of the glycosidic bond between the damaged base, U, and deoxyribose; the apyrimidinic or AP site created is recognized by an AP endonuclease, which cleaves on the 5′-side of the site. Some enzymes have both glycosylase and AP endonuclease activities in a single polypeptide chain. Furthermore, two pathways—long patch and short patch, depending on the enzymes involved—can then complete the repair.

are also required. DNA resynthesis is catalyzed by an aphidicolin-sensitive polymerase, probably DNA polymerase δ.

13.5.7 Repair of double-strand breaks can be error-free or error-prone

Double-strand breaks, DSBs, are probably the most harmful of DNA lesions. The severe consequences of DSB lesions can be explained by two major factors: first, DSBs are difficult to repair without introducing errors or mutations in the DNA sequence, and second, the disruption of continuity in the molecule leads to numerous chromosomal translocations

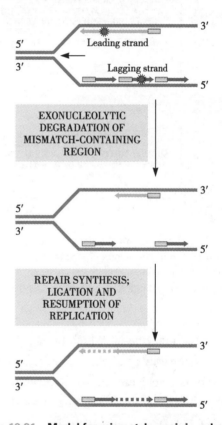

Figure 13.21 **Model for mismatch repair in eukaryotes.** Mismatch processing in eukaryotes is probably directed by the presence of ssDNA at strand breaks that occur during replication, at the 3′-end of the leading strand or at the ends of Okazaki fragments. The schematic shows two separate lesions, on the leading and lagging strands. Repair in the leading strand might begin at the 3′-terminus, with the polymerase resynthesizing the region; in the lagging strand, an entire Okazaki fragment may be removed, with degradation commencing at either end; extension of the fragment closest to the replication fork would replace the degraded one. (Adapted from Jiricny J [2006] *Nat Rev Mol Cell Biol*,7:335–346.With permission from Macmillan Publishers, Ltd.)

initiated when complexes of MutS homologs, MSHs, either MSH2–MSH6, known as MutSα, or MSH2–MSH3, known as MutSβ, bind to a mismatch. The process presented in Figure 13.22 has been proposed on the basis of studies in which the entire mechanism was reconstituted from recombinant proteins: either MutSα or MutSβ, plus MutLα; replication protein A, RPA; exonuclease 1, EXO1; proliferating cell nuclear antigen, PCNA; replication factor C, RFC; DNA polymerase δ, Pol δ; and DNA ligase. To date, no eukaryotic DNA helicase has been shown to participate in the repair of replication errors. As in *E. coli*, more than one eukaryotic exonuclease has been implicated in MMR, and several other proteins

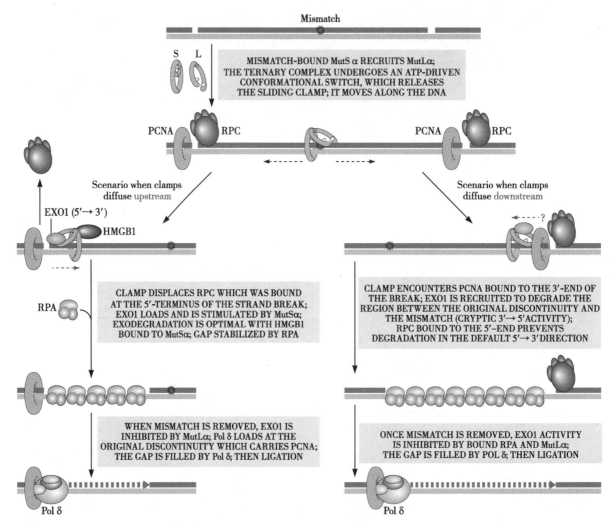

Figure 13.22 **Proposed mechanism of mismatch repair during DNA replication in eukaryotes.** Note that a cryptic activity of EXO1 in the downstream scenario does not need to be invoked; maybe Mre11 or MutSα, RFC, and PCNA activate a latent MutLα endonuclease activity in an ATP- and mismatch-dependent manner, as suggested by Paul Modrich. (Adapted from Jiricny J [2006] *Nat Rev Mol Cell Biol*, 7:335–346. With permission from Macmillan Publishers, Ltd.)

and other rearrangements, which pose a threat to the genomic integrity of a cell. Cells have developed highly sophisticated mechanisms to repair such damage, collectively known as the DNA damage response, DDR. DDR is also elicited in response to long stretches of ssDNA.

DDR is a major factor that controls the functioning of cell-cycle checkpoints. Genetic and biochemical studies have defined points in the cell cycle at which the cell stops temporarily to check whether conditions are ripe for the cell to transit to the next phase of the cell cycle. These conditions include the overall metabolic status of the cell; the presence of the correct molecules, such as the deoxyribonucleoside triphosphates needed as precursors for DNA replication, in the needed amounts; and last but not least, the integrity of the DNA molecules. When DNA lesions, especially DSBs, are detected, the cell cycle is halted until the damage is repaired.

Two major pathways are used to repair DSBs: homologous recombination, HR, and nonhomologous end-joining, NHEJ. The key proteins involved in each of these pathways are listed in Table 13.4.

13.5.8 Homologous recombination repairs double-strand breaks faithfully

Homologous recombination, HR, depends on the availability of intact DNA sequences in sister chromatids that can be used as templates for faithful repair. Thus, it can only occur in S and G_2 phases. In

Table 13.4 **Major proteins that participate in the two major pathways for DNA damage repair.** (Adapted from Mladenov E, Iliakis G [2011] *Mutat Res*, 711:61–72.)

Function	Homologous Recombination, HR	Nonhomologous End-Joining, NHEJ
Single-strand break sensor molecules	MRN[a]	Ku70-Ku80
DNA end-processing enzymes	MRN, CtIP Exo1, Dna2	Artemis, TdT[a], PNK[a]
recombinases	Rad51	
DNA repair mediators	Rad52, BRCA2, Rad51 paralogs	DNA-PKcs
polymerases	Pol δ, Pol ε	Pol μ, Pol λ
ligases	ligase I	ligase IV
ligase-promoting factors	PCNA?	XRCC4, XLF-Cernunnos

[a]MRN Mre11-Rad50-Nbs1 complex; TdT, terminal deoxynucleotidyltransferase; PNK, polynucleotide kinase; DNA-PKcs, DNA-dependent protein kinase catalytic subunit.

addition to DSBs, this pathway takes care of stalled replication forks and interstrand DNA cross-links. HR is also involved in physiologically relevant processes such as meiotic recombination and the formation of abortive topo II intermediates. We presented the general features of HR and its roles in such processes in. Here, we introduce another important player that was not discussed previously, the Mre11-Rad50-NBS1 complex, abbreviated MRN.

The importance of the MRN complex in the cellular response to DSBs was initially recognized in the study of two human conditions: ataxia telangiectasia-like disorder, caused by mutations in *MRE11*, the meiotic recombination 11 gene, and Nijmegen breakage syndrome, caused by mutations in the NBS gene; the yeast homolog of this gene is *Xrs2*. The MRN complex is used in both HR and NHEJ, despite their different mechanisms. This is the complex that initially senses the damage and interacts with the broken DNA ends in such a way as to keep the ends in close proximity with each other to allow repair to occur.

In addition to this purely structural role, the two major subunits of the MRN complex, Mre11 and Rad50, possess enzymatic activities that are essential to its function; the role of NBS1 is much less clear. Mre11 is a multifunctional protein that exhibits numerous activities *in vitro*: it is a nuclease that also has strand-dissociation and strand-annealing activities. The nuclease activity is highly regulated by the other two protein subunits, by ATP, and by sequence homology in the DNA substrate. Exactly how these activities are modulated and coordinated *in vivo* remains to be elucidated. It is clear that Mre11 participates in the initial processing of DNA ends, which contain adducts that could interfere with further processing, and in resolving possible secondary structures of DNA ends.

Rad50 binds and hydrolyzes ATP. The ATP-binding motifs at both ends of the Rad50 polypeptide chain are crucial to its function, as mutations in these motifs result in a null phenotype in yeast and in partial loss of nuclease activity in the human complex *in vitro*. The central role of MRN in both DSB repair pathways, as well as in the early steps of meiotic recombination and in maintenance of telomeres, warrants further studies of the complex. Understanding the exact mechanism of MRN action might also lead to important clues for the treatment and eventual cure of the genetic diseases associated with its malfunctions.

13.5.9 Nonhomologous end-joining restores the continuity of the DNA double helix in an error-prone process

Nonhomologous end-joining, NHEJ, is the major pathway that repairs DSBs. It simply restores the integrity of the DNA molecule by joining the two ends in an error-prone process. For NHEJ to occur, ligase-compatible ends must first be produced. The positive aspect of NHEJ that compensates for these sequence errors is that it quickly restores the structural integrity of the DNA molecule, and thus the chromosome, at sites where the breakage would otherwise result in loss of large segments of the genome.

Like other repair processes, the NHEJ pathway begins with recognition of the damage. NHEJ uses the highly abundant Ku70-Ku80 dimer, ~300 000 molecules per cell in humans, for this task. Ku can only load and unload onto DNA at DNA ends, and two Ku complexes form during repair, one at each of the two DNA ends to be joined. Ku is believed to be able to recruit the other players that have a role in NHEJ—nuclease, polymerase, and ligase—in any order.

The other important players in NHEJ are:
- The catalytic subunit of DNA-dependent protein kinase, DNA-PKcs, whose kinase activity is activated upon binding to the ends of the severed DNA molecule. DNA-PKcs phosphorylates a number of proteins involved in this pathway—RPA, DNA ligase IV or LigIV, and its partners XRCC4 and XLF/Cernunnos—as well as itself. It also binds to and regulates the activity of Artemis.
- Artemis, an endonuclease believed to be involved in the preligation processing of DNA ends. Other enzymes such as TdT, terminal deoxynucleotidyl-transferase, and PNK, polynucleotide kinase, are also involved in DNA end processing.
- Translesion DNA polymerases, Pol μ and Pol λ.
- LigIV-XRCC4-XLF complex, which is specifically involved in this pathway.

The steps in the classical NHEJ process—the alternative term canonical is also used—and the proteins involved are depicted in detail in Figure 13.23.

13.5.10 Translesion synthesis

Both endogenous and environmentally induced damage to the genome are primarily removed by DNA repair mechanisms, but the damage that remains blocks the progression of the replication fork. The activity of specialized, low-fidelity, low-processivity DNA polymerases enables cells to tolerate damage by replicating lesion-containing DNA without removing the lesion. The process has been termed translesion synthesis, TLS, and the many polymerases capable of TLS are called translesion synthesis or bypass polymerases.

Mammalian genomes encode 15 different DNA polymerases, most of which are involved in TLS. Normal replication is very fast—in bacteria ~1000 nucleotides are added to the nascent chain per second—and very accurate, with ~1 error per 10^6 incorporation steps. On the leading strand at least, it is highly processive. TLS enzymes, on the other hand, incorporate only a few nucleotides before they dissociate from the template. Interestingly, the presence of accessory proteins increases the *in vivo* processivity of high-fidelity enzymes thousands of times but increases the processivity of lesion-bypass polymerases only very slightly. Moreover, the low fidelity of lesion bypass polymerases leads to the creation of numerous mutations in the DNA across the lesion they bypass. Even when they copy normal, undamaged DNA, their intrinsic error frequency is in the range of 10^{-3}–1. How polymerases process lesions and whether they induce mutations are dictated by the type of lesions, as the bypass polymerases are lesion-specific; the sequence context; and the specific polymerase involved.

DNA polymerases are categorized into several families based on sequence similarity: A–D, X, Y, and RT. High-fidelity polymerases are members of the A and B families, whereas lesion-bypass polymerases belong primarily, but not exclusively, to the Y family. The TLS process itself involves switching from the classical polymerase, which will have stalled at a lesion, to a lesion-specific bypass enzyme, which extends the new chain only by a couple of nucleotides, and a further switch to a second bypass polymerase, usually Pol ζ, which belongs to the high-fidelity B family (Figure 13.24). This second switch is needed to extend the DNA molecule to a sufficient length beyond the lesion, so that the lesion-induced distortions of the DNA structure are no longer in the way of a classical polymerase. At this point, a third polymerase switch takes place, with the re-recruitment of the classical enzyme. This highly orchestrated process must involve many control steps, especially in determining the choice of the polymerase

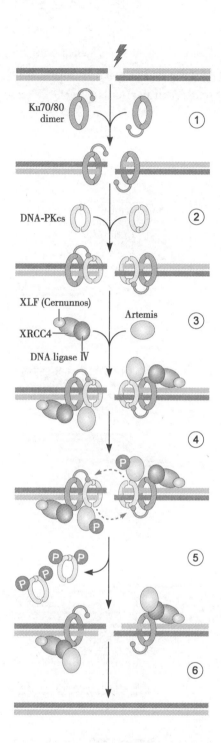

Figure 13.23 **Events leading to NHEJ of an ionizing radiation-induced DSB.** Ionizing radiation, IR, can induce single-strand breaks on opposite DNA strands, resulting in the formation of a DSB with short overhanging ends. The steps in the repair process are as follows: (Steps 1 and 2) The DSB is detected by Ku70-Ku80 heterodimer, which binds to the DNA ends and recruits DNA-PKcs through its flexible C-terminal region. Binding of DNA-PKcs induces inward translocation of the Ku dimer by about one helical turn, presumably to facilitate access of further repair proteins to the business end of the break, and mediates synapsis of the ends. (Step 3) DNA ends are processed by one or more enzymes that may include the following: (a) Artemis, a versatile nuclease capable of cleaving a wide variety of DNA structures. (b) DNA polymerases, not shown here: translesion polymerases Pol λ and Pol μ; and TdT, terminal deoxynucleotidyl transferase. Each of these polymerases can be used, depending on the structure of the ends to be repaired: 3′- or 5′-overhangs, blunt ends, or small single-stranded gaps. TdT can add untemplated nucleotides to DNA ends, whereas the translesion polymerases can fill gaps. (c) XRCC4-XLF (Cernunnos)-DNA ligase IV complex. DNA ligase IV is specific for this pathway and acts in a complex with two other proteins: XRCC4, which stabilizes the ligase and binds to DNA, and XLF, or XRCC4-like, whose function is still not well-defined but is important, because cells carrying mutations in XLF are radiosensitive and deficient in DSB repair. (Steps 4 and 5) Threading of ssDNA ends into cavities in the DNA-PKcs molecule activates the kinase. DNA-PKcs undergoes autophosphorylation, which opens the central DNA-holding cavity, releasing the protein from the DNA ends. This may provide preferential access to XRCC4-XLF-DNA ligase IV. DNA-PKcs phosphorylates Artemis, and possibly other proteins including Ku, activating its endonuclease activity. (Step 6) If the DNA ends are compatible, ligation occurs immediately. If the ends are not compatible, XRCC4-DNA ligase IV remains in the synaptic complex, while its polymerase and nuclease activities process the ends. As soon as the ends are processed into a compatible substrate, XRCC4-DNA ligase IV completes the joining reaction. (Adapted from Dobbs TA, Tainer JA, Lees-Miller SP [2010] *DNA Repair*, 9:1307–1314. With permission from Elsevier.)

capable of bypassing the particular lesion at hand. For example, of the three bypass polymerases that belong to the X family, Pol β bypasses 1-bp gaps with 3′-OH and 5′-PO$_4$, Pol λ bypasses the same lesion or a discontinuous template with paired termini, and Pol μ bypasses recessed DNA ends or ends where no complementarity of bases exists whatsoever. It can also perform template-independent DNA synthesis. Note from Table 13.4 that Pol λ and Pol μ are also the polymerases involved in NHEJ, reinforcing the point that DNA repair processes form a complex network in the cell.

13.5.11 Many repair pathways utilize RecQ helicases

The chicken-foot model has been postulated to describe replication restart, a process involving the action of RecQ helicases, a subfamily of DNA helicases that are highly evolutionarily conserved (Figure 13.25). The family was named after the recQ gene in *E. coli*, and RecQ is the sole member of the family in *E. coli*. Lower eukaryotes also have only one RecQ-type helicase: Sgs1 in *Saccharomyces cerevisiae* and Rqh1 in *S.pombe*. Humans have five members of the RecQ family, and genetic defects in three of them lead to severe diseases. Loss of RecQ helicase function leads to loss of genome integrity, resulting from hyper recombination in particular. RecQ helicases par-

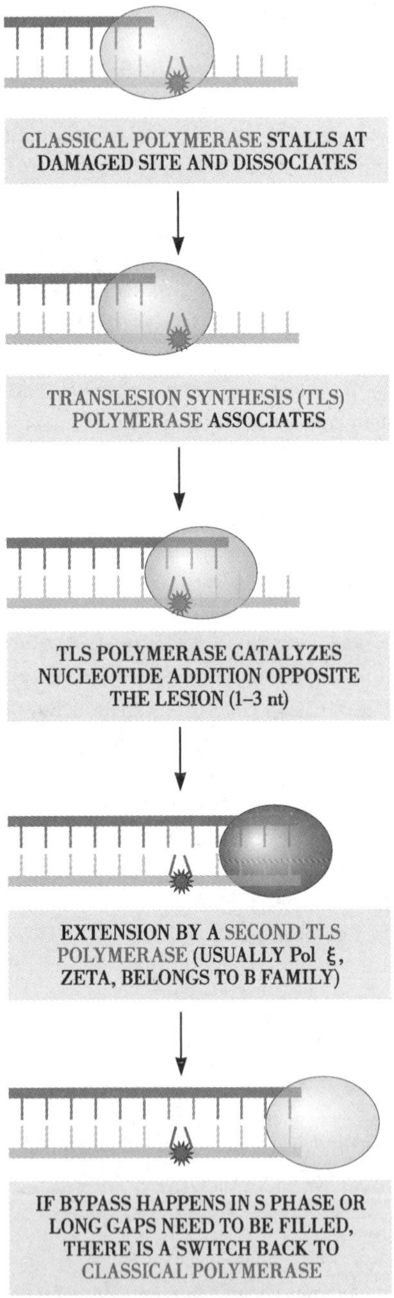

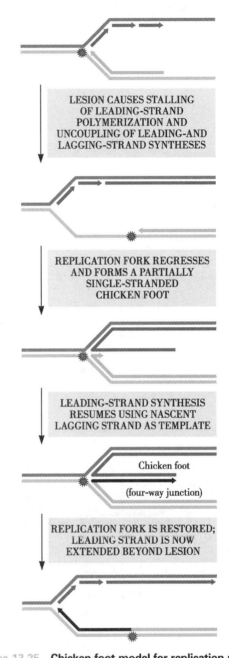

Figure 13.24 Polymerase switching in translesion synthesis. Switching back to replicative DNA polymerase may involve deubiquitylation of PCNA. (Adapted from Washington MT, Carlson KD, Freudenthal BD, et al. [2010] *Biochim Biophys Acta*,1804:1113–1123. With permission from Elsevier.)

Figure 13.25 Chicken-foot model for replication restart. A lesion in the leading-strand template, shown as a red starburst, blocks leading-strand synthesis. Lagging-strand synthesis continues for a short period. The fork regresses, allowing the nascent strands to anneal and creating a chicken-foot structure or four-way junction. The leading strand is then extended by use of the longer lagging strand as a template. The fork is reset, presumably by a RecQ helicase, via branch migration of the four-way junction, and now the leading strand has been extended beyond the lesion. Replication can recommence. (Adapted from Bachrati CZ, Hickson ID [2003] *Biochem J*,374:577–606. With permission from Biochemical Society.)

ticipate in many repair pathways, including double-strand repair through HR. We mentioned Werner and Bloom helicases, members of the RecQ helicase subfamily, in connection with ALT, alternative lengthening of telomeres; they are discussed further in Box 13.5.

13.5.12 Histone variants and their post-translational modifications are specifically involved in DNA repair

DNA repair in eukaryotic cells takes place in chromatin, so the repair machinery must deal with the presence of nucleosomes and higher-order chromatin structure (see Chapter 5). That chromatin plays a role in determining the distribution of DNA damage and the extent and rate of DNA repair has been known for some time. *In vitro* experiments in human cell extracts, using individual nucleosomes or nucleosomal arrays reconstituted on plasmid DNA as substrates for repair, have clearly indicated that repair of naked DNA is more efficient than repair of chromatin DNA, and linker DNA in chromatin is more accessible to repair than the DNA on nucleosomal core particles. It became clear that chromatin rearrangement is often required to allow access of repair proteins to DNA damage sites. This led to the formulation, in the early 1990s, of the access, repair, and restore model or ARR of NER.

Today, we know that chromatin not only is a passive barrier to repair but should be considered as an integral player in DDR. To accommodate this new information about the active participation of chromatin in the repair pathways, the ARR model is being replaced by the prime, repair, and restore model or PRR. Three major components of the chromatin toolbox are utilized in repair: (a) histone variants and the enzymes that place or remove post synthetic modifications on them (Table 13.5); (b) chromatin remodelers; and (c) chromatin chaperones. The interplay among these participants and repair depends on the context and specific repair pathway and is extremely complex. It is also very poorly understood.

Box 13.5 Human diseases linked to RecQ helicase mutations

Diseases caused by mutations in the RecQ helicase genes are rare autosomal recessive genetic disorders, occurring mostly through consanguineous marriages. All of these disorders show genomic instability associated with cancer predisposition. Expression of the respective helicase is highly up-regulated in rapidly growing or immortal cells. The clinical features of three major disorders and some facts about the protein defects are presented here.

Bloom syndrome
The disease was first described by dermatologist David Bloom in 1954. It is characterized by proportional dwarfism, narrow face and prominent nose and ears, male infertility and female subfertility, frequent infections, sun-induced erythema, type II diabetes, and predisposition to a wide range of cancers such as non-Hodgkin's lymphoma, leukemias, and carcinomas of the breast, gut, and skin, with mean age of onset of 24. The mutated protein is RecQ2 or BLM.

Werner syndrome
Werner syndrome was named after ophthalmologist Otto Werner, who recognized the disease in 1904. The disorder manifests itself in numerous features of premature aging and sequential appearance of age-related diseases, retarded growth, hypogonadism, immunodeficiency, type II diabetes, cancers of mesenchymal origin, and soft-tissue and osteogenic sarcomas; in the general population carcinomas to sarcomas are 10:1, while in Werner syndrome it is 1:1.

The gene affected in Werner syndrome patients encodes protein RecQ3 or WRN. Some WRN mutants are truncated versions of the protein that lack the C-terminal NLS, nuclear localization signal, and cannot migrate to the nucleus. Unfortunately, there is no good mouse model of the disease, since the mouse protein lacks the NoLS, nucleolar localization signal, adjacent to the NLS, and is not localized in the nucleolus, while the human WRN can traffic between the nucleoplasm and nucleolus.

Rothmund-Thomson syndrome
The syndrome was first described by ophthalmologist August von Rothmund in 1868. Patients affected exhibit short stature, skeletal abnormalities, skin pigmentation changes, skin atrophy, congenital cataracts and bone defects, premature aging, and cancers, primarily osteogenic sarcoma. The genetic defect has been identified as mutations in a single gene, the *RecQL4* gene.

Table 13.5 **Histone modifications affecting the DNA damage response.** Modifications are classified according to their participation in the consecutive steps in DDR: signaling, chromatin opening at the beginning of repair, and chromatin restoring following repair. (Adapted from Rossetto D, Truman AW, Kron SJ, et al. [2010] Clin Cancer Res, 16:4543–4552.)

Type of modification	Residue modified (human)	Enzyme responsible	Step in DDR affected
acetylation	H4/H2A.X	Tip60/yNuA4	opening
	H3 K9	Gcn5, CBP/p300	opening
	H3 K56	yRtt109, CBP/p300, Gcn5	restoring
	H3 K14, K23	Gcn5	restoring
	H4 K5, K12	Hat1	restoring
	H4 K91	Hat1	restoring
deacetylation	H3/H4 K	Sin3/Rpd3, Sir2, Hst1/3/4	restoring
methylation	H4 K20	Set8/Suv4–20	signaling
	H3 K79	Dot1	signaling
phosphorylation	H2A.X S139	ATM/ATR, DNA-PK	signaling
	H4 S1	casein kinase 2	restoring
	H2B S14	Ste20	restoring
dephosphorylation	H2A.X Y142	EYA1	signaling
	H2A.X S139	yPph3/hPP4, PP2A, PP6, Wip1	restoring
ubiquitylation	H2A/H2A.X	RNF8/RNF168	signaling
	H4 K91	BBAP	signaling
	H2A K119	Ring1b/Ring2	restoring
sumoylation	H2A.Z K126/133	?	signaling

Key Concepts

- The genome is not static but subject to many kinds of rearrangements. These are grouped under the general title of DNA recombination. Depending upon the homology or lack thereof between the moving DNA and its target site, we distinguish between homologous, site-specific, and nonhomologous recombination.
- Homologous recombination is involved in DNA repair and lengthening of telomeres, but perhaps its most important function is in meiotic cells, where it facilitates exchange of alleles and the proper alignment of homologous chromosomes.
- HR is generally initiated from double-strand breaks or DSBs, followed by resection of the 5′-ends. A single-strand invasion of the intact DNA duplex then occurs, in which the single-strand-binding protein RecA plays a major role.
- A fundamental intermediate structure in recombination is the Holliday junction, which allows branch migration and, ultimately, strand exchange in recombination.
- Although some nonhomologous recombination is site-specific, acting through the mutual recognition of small complementary sites, in most cases it occurs through the agency of mobile genetic elements called transposons. These require a special enzyme called transposase to move the transposon to new sites in the genome.
- In bacteria, two types of transposons are found: insertion sequences carry only the transposase gene, whereas composite transposons may also carry other genes, often conveying antibiotic resistance.
- In eukaryotes there are two very different classes of transposons. DNA or class II transposons cut and paste DNA segments, or copies thereof, into

new locations. Class I transposons, also called retrotransposons, act through a transcribed RNA that is then reverse-transcribed into dsDNA and inserted into the target site.

- Site-specific recombination, on the other hand, not only plays the roles mentioned above but also provides great diversity in protein products from a limited genome. Examples are the immune system in vertebrates and the defense against that system used by some parasites.

- Lesions occur in many different forms, including single-strand breaks, double-strand breaks, chemical damage to bases, and intra- or interstrand cross-linking.

- Repair can be direct, reversing the lesion reaction, or may involve excision of one or a number of nucleotides surrounding the lesion, removal of a base, correction of mismatches, and either homologous or nonhomologous recombination. There are several pathways of DNA repair such as Direct repair, Nucleotide excision repair, Base excision repair, Mismatch repair.

- Double-strand breaks, DSBs, are very harmful and difficult for the cell to deal with. There are two general mechanisms. Homologous recombination, HR, depends on the existence of a sister chromatid, which can be used to align the ends to join and is considered essentially error-free. This is possible only in the S or G_2 phases of the cell cycle. Nonhomologous end-joining, NHEJ, restores the integrity of the chromosome by more or less random rejoining. Clearly this can lead to the loss or gain of genetic information and is thus error-prone.

- Sometimes the cell simply ignores a lesion and replicates the DNA that includes it in a mechanism called translesion synthesis, TLS. This is accomplished by a gallery of minor polymerases, 15 in mammals, that can carry out error-prone replication.

Key Words

autonomous retrotransposons(自主逆转录转座子)
base excision repair, BER(碱基切除修复)
composite transposons(复合转座子)
conjugation(接合)
direct repair(直接修复)
DNA damage response, DDR(DNA 损伤反应)
double-strand breaks, DSBs(双链结合蛋白)
global genome repair or GGR(全基因组修复)
homologous recombination or HR(同源重组)
Holliday junctions, or HJ(Holliday 交叉)
homologous recombination repair, HR repair(同源重组修复)
insertion sequences, IS,(插入序列)
long terminal repeats, LTR(长末端重复)
long interspersed nucleotide element 1, also known as LINE-1 or L1(长散布核苷酸元件)
lytic mode(溶菌模式)
lysogenic mode(溶原模式)
lysogen(溶原性细菌)
mismatch repair, MMR(错配修复)
nucleotide excision repair, NER(核苷酸切除修复)
nonhomologous end-joining, NHEJ(非同源重组修复)
nonautonomous retrotransposons(非自主逆转录转座子)
nonhomologous recombination(非同源重组)
recombinase(重组酶)
replication protein A, RPA(复制蛋白 A)
retrotransposons(逆转录转座子)
short interspersed nucleotide elements or SINEs(短散布核苷酸元件)
site-specific recombination(位点特异性重组)
synthesis-dependent strand annealing or SDSA(合成依赖性退火)
transduction(转导)
transformation(转化)
transposon(转座子)
transcription-coupled repair or TCR(转录偶联修复)
target-primed reverse transcription or TPRT(靶向反转录)
translesion synthesis, TLS(跨损伤修复)

Questions

1. Diagram the recombinational events leading to the formation of a heteroduplex DNA region within a bacteriophage chromosome.

2. Diagram a homologous recombination process in bacteria and describe the possible two results.
3. Describe briefly the pathways of DNA repair in bacteria.
4. Explain why are transposons mutagenic agents?
5. List the classification of transposons in human genome.

References

[1] Babushok DV, Kazazian HH Jr (2007) Progress in understanding the biology of the human mutagen LINE-1. *Hum Mutat*,28:527–539.

[2] Barzel A, Kupiec M (2008) Finding a match: How do homologous sequences get together for recombination? *Nat Rev Genet*,9:27–37.

[3] Beck CR, Garcia-Perez JL, Badge RM, et al. (2011) LINE-1 elements in structural variation and disease. *Annu Rev Genomics Hum Genet*,12:187–215.

[4] Biémont C, Vieira C (2006) Genetics: Junk DNA as an evolutionary force. Nature,443:521–524.

[5] Cox MM (2007) Motoring along with the bacterial RecA protein. *Nat Rev Mol Cell Biol*,8:127–138.

[6] Bonner WM, Redon CE, Dickey JS, et al. (2008) γH2AX and cancer. *Nat Rev Cancer*,8:957–967.

[7] Chapman JR, Taylor MRG, Boulton SJ (2012) Playing the end game: DNA double-strand break repair pathway choice. *Mol Cell*,47:497–510.

[8] Cromie GA, Smith GR (2007) Branching out: Meiotic recombination and its regulation. *Trends Cell Biol*, 17:448–455.

[9] Deem AK, Li X, Tyler JK (2012) Epigenetic regulation of genomic integrity. *Chromosoma*,121:131–151.

[10] Lammens K, Bemeleit DJ, Möckel C, et al. (2011) The Mre11:Rad50 structure shows an ATP-dependent molecular clamp in DNA doublestrand break repair. *Cell*,145:54–66.

[11] Mladenov E, Iliakis G (2011) Induction and repair of DNA double strand breaks: The increasing spectrum of non-homologous end joining pathways. *Mutat Res*,711:61–72.

[12] Moreno-Herrero F, de Jager M, Dekker NH, et al. (2005) Mesoscale conformational changes in the DNA-repair complex Rad50/Mre11/Nbs1 upon binding DNA. *Nature*, 437:440–443.

[13] Moses RE, O'Malley BW (2012) DNA transcription and repair: A confluence. *J Biol Chem*,287:23266–23270.

Guanwu Li

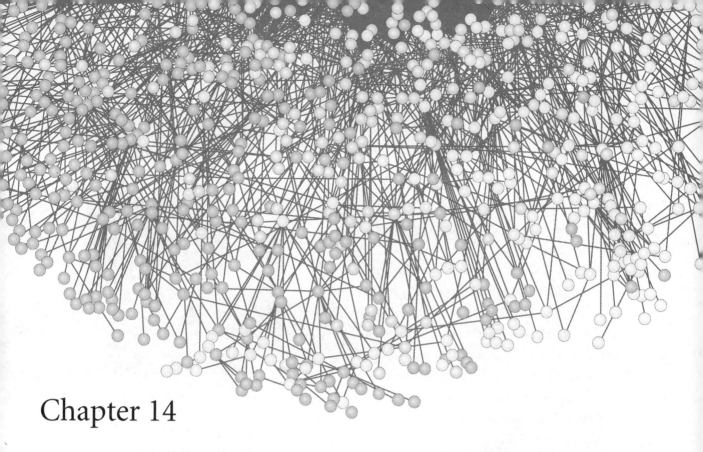

Chapter 14

Genetically Modified Organisms: Use in Basic and Applied Research

14.1 Introduction

Throughout this book, we describe molecular structure and dynamics at molecular or atomic level and show the mechanisms and cellular processes of life. These molecular mechanisms were first worked out in prokaryotes (bacteriophage lambda (λ) and the bacterium *E. coli*). These early studies provided the foundation to understand the more complex processes in eukaryotes. Many species could claim to be model organisms for eukaryotes.

Here, we will introduce the most widely used in the study of prokaryotic and eukaryotic molecular biology, include: Nematodes (*Caenorhabditis elegans*), *Drosophila*, Zebrafish, and Mouse. These model organisms share common attributes as well as unique characteristics. In general, they are easy to manipulate in the lab, have short gestation periods, produce large numbers of offspring, and are relatively cheap, inexpensive to house. Moreover, we have identified the genomes and understood the basic developmental and metabolic processes of almost all of these model organisms.

We will also concentrate on genetically modified (or genetically engineered) organisms, especially mice, constructed through transgenic techniques, gene targeting techniques and gene editing techniques. Genetically modified mice have been widely and deeply used in decades of basic researches and clinical applications in the medical field, such as the research platform of gene function *in vivo*, the animal models for human diseases and discovery platform of full human therapeutic antibodies.

In this chapter, the applications of other transgenic animals (primates and livestock) are also addressed.

14.2 Model organisms and genetically modified organisms

You will find throughout the book that certain organisms have been used again and again as convenient models for whole categories of organisms (Figure 14.1). Because of their well-characterized genetics, these species became model organisms, defined as organisms used for the study for basic biological processes. We describe some of these model organisms briefly below, with notes as to why they have been so often chosen. The genomic sequences of all of these organisms are now available.

Today bacteriophage λ has been employed largely as a cloning vector, but it played an important part in the early development of genetics, especially because it has two alternative life cycles, lytic and lysogenic. In other words, the phage can either destroy or lyse the host bacterium or become integrated into its genome, existing in a dormant state that is propagated from cell generation to cell generation without any signs of the viral DNA's presence, known as the lysogenic pathway.

The bacterium *Escherichia coli* has been labeled the workhorse of molecular biology. There is practically no fundamental biochemical process, from DNA replication to protein synthesis, which was not first elucidated in *E. coli*. It is extremely easy to grow, either in liquid culture or on solid agar plates, and metabolically very versatile, which has made it useful for studies of metabolic regulation.

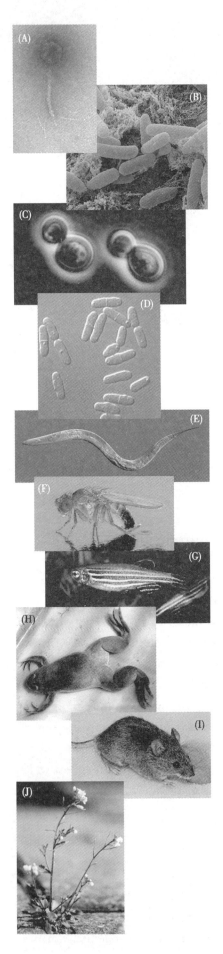

Figure 14.1 **Picture gallery of some of the model organisms used most frequently in genetics research.** (A) λ phage. (B) *Escherichia coli*. (C) *Saccharomyces cerevisiae*. (D) *Schizosaccharomyces pombe*. (E) *Caenorhabditis elegans*. (F) *Drosophila melanogaster*. (G) *Danio rerio*. (H) *Xenopus laevis*. (I) *Mus musculus*. (J) *Arabidopsis thaliana*. (A, courtesy of Bob Duda, University of Pittsburgh. B, courtesy of Peter Cooke and Stephen Ausmus, United States Department of Agriculture. C, courtesy of Maxim Zakhartsev and Doris Petroi, International University Bremen. D, from Gutterman JU, Lai HT, Yang P, et al. [2005] *Proc Natl Acad Sci USA*, 102:12771–12776. With permission from National Academy of Sciences. E, courtesy of Judith Kimble, University of Wisconsin. F, courtesy of André Karwath, Wikimedia. G, from Wikimedia. H, courtesy of Michael Linnenbach, Wikimedia. I, courtesy of George Shuklin, Wikimedia. J, courtesy of Brona Brejova, Wikimedia.)

The common budding or bakers' yeast *Saccharomyces cerevisiae* is almost the simplest eukaryotes. Unicellular and easy to grow in large quantities, it provides a bridge between bacteria and the more complex eukaryotes. Its genetics has been very thoroughly studied, with many knockout strains available. In a knockout strain, a particular gene is inactivated by recombinant DNA techniques. Studying such knockouts helps in elucidating the biological functions of genes. One difficulty in working with *S. cerevisiae* is the tough outer cell wall, which makes it difficult to insert substances. The fission yeast *Schizosaccharomyces pombe* is genetically similar to *S. cerevisiae* but lacks the tough outer layer. It divides, rather than buds, which is an advantage for some studies.

The free-living, primitive, unsegmented, and bilaterally symmetrical worm *Caenorhabditis elegans*, which was introduced to the field by Sydney Brenner, is a remarkably simple creature. The adult worm has only 1090 cells, and the lineage of each is precisely known. This makes it an outstanding candidate for developmental studies. Sydney Brenner, Robert Horvitz, and John Sulston were awarded the 2002 Nobel Prize in Physiology or Medicine "for their discoveries concerning genetic regulation of organ development and programmed cell death." Andrew Fire and Craig Mello investigated the regulation of gene expression in *C. elegans* and identified RNA interference (RNAi), a widely known and used mechanism of gene silencing by double-stranded RNA. This work was awarded the 2006 Nobel Prize in Physiology or Medicine.

Drosophila melanogaster was the organism that provided the seminal studies in modern genetics. Morgan's fruit fly is easy to grow, in enormous numbers, in a very short time. This, plus the availability of a great many mutant strains, including many with mutations that affect general developmental patterns, make it still a useful model. The embryos are also used, especially for biochemical studies.

The small zebrafish *Danio rerio* is easy to grow and is very fecund, so it provides a convenient vertebrate model. Its special attraction lies in the fact that the embryos are transparent so that development of internal organs can be followed in live embryonic fish.

The African frog *Xenopus laevis* is useful because of its large and abundant eggs, which can easily be manipulated for injection studies and the like. *X. laevis* can rapidly produce thousands of embryos. A disadvantage of this model is that it is tetraploid and it takes years for the frog to reach sexual maturity. Another frog species, *Xenopus tropicalis*, is diploid and matures in less than 3 months; these two properties make it very attractive for genetic research.

Arabidopsis thaliana is a weed commonly known as *thale cress*. The plant is easy to grow and sexually matures in less than 6 weeks, producing ~5000 seeds per plant. This is the most commonly used plant model. First, it has a small genome, five pairs of chromosomes, whose entire sequence has been reported. Second, a large number of mutant lines and genomic resources or databases are available. Third, it is easy to transform by use of recombinant DNA technology techniques, including *Agrobacterium tumefaciens*.

The house mouse *Mus musculus*, a common rodent, is the easiest mammal to study and has been used by generations of researchers in medical field, although it has a longer generation time compared with other simple model organisms. The gestation of mouse is 19–21 days, with an average of 1–10 pups per litter. It needs 6 weeks for reaching sexual maturity and have a lifespan of 1.5–3 years. Despite a large evolutionary separation between mice and humans, ~85% of the mouse genome is very similar to that of humans, and only less than 1% of mouse genes have no detectable homolog in humans. Mouse, as the only mammalian model organism, remains the preferred model of human disease because their physiology is at least similar to that of human. By now, many purebred strains including those with specific genetic modifications are readily available.

Over time, researchers created a large catalog of mutant or genetically modified strains for these species mentioned above, and these strains were carefully analyzed. Genetically modified organisms (GMOs) are living organisms whose genome has been artificially manipulated through genetic engineering technology. These genetically modification consist of virus, bacterium, plant, and animal genes that do not occur in nature or through traditional crossbreeding methods. Remember that each time you meet these genetically modified organisms, they not only have a rich history in basic life science and medical research but are also at the forefront in the study of human diseases and clinical applications.

14.3 Transgenics, gene targeting, and genome editing

The development of recombinant DNA techniques has given us the power to change, at will and in specific ways, the genomes of plants and animals. These are in fact new organisms that will breed true if they are homozygous for the genetic change. Undoubtedly, *Mus musculus* is the most widely used among these model organisms. Transgenics, gene targeting and genome editing techniques allow us to produce different genetically modified mice.

14.3.1 Transgenic mouse

A mouse that gains new genetic information from the addition of foreign DNA is described as transgenic mouse. It is easy to generate transgenics by transformation with DNA constructs containing sequences of interest for simple organisms, such as bacteria or yeast. However, transgenesis in multicellular organisms can be much more challenging.

In the early 1980s, a number of laboratories developed a technique for introducing new genetic material into eukaryotic genomes in a way that would be transmitted from generation to generation. The first visible phenotypic change in transgenic mice was described in 1982 by Richard Palmiter and Ralph Brinster. They created mice that integrated and expressed the rat growth hormone gene coding sequence successfully. The result was that some mice grew to be twice the size of normal siblings. Images of such transgenic mice are presented in Figure 14.2. Since then, transgenics has been a rapidly growing field, with many technological advances.

(1) Procedure for making a transgenic mouse

The standard procedure for making transgenic mice by microinjection of foreign DNA into fertilized eggs is shown in Figure 14.3. Microinjection results in the introduction of the transgene into the genomic DNA of a fertilized mouse egg. There are three main steps in the process: microinjection of foreign DNA into the pronucleus of a fertilized mouse egg; implantation of the microinjected embryo into a foster mother; and analysis of mouse pups and subsequent generations for the stable integration and expression of the transgene.

- Pronuclear microinjection. A transgenic mouse, represents a "gain of function" mouse model, is usually designed to overexpress a gene of interest. The first step in making a transgenic mouse is to obtain fertilized eggs from a superovulated female mouse surgically or by *in vitro* fertilization (IVF). The second step is microinjection of the foreign DNA, the fragments from linearized vectors (usually plasmids), into the fertilized egg (Figure 14.3). In general, the transgenic construct contains a promoter element, complementary DNA (cDNA) for the gene of interest, and a polyadenylation signal. The promoter sequence is required as a binding site for RNA polymerase and transcription factors. The expression of the foreign gene can be regulated spatially and temporally by choosing either a tissue-specific or inducible promoter.

To avoid the production of a mosaic mouse in which many cells do not possess the new DNA information, the foreign DNA must integrate into the mouse genome prior to the doubling of the genetic material that occurs before the first cleavage. Therefore, the timing of microinjection is critical. For this reason, the transgene should be introduced into the fertilized egg at the earliest possible stage.

For several hours following the entry of the sperm into the egg, the sperm nucleus and the egg nucleus called the male and female pronuclei which are microscopically visible as individual structures. Injections must be done before the haploid sperm and egg pro-

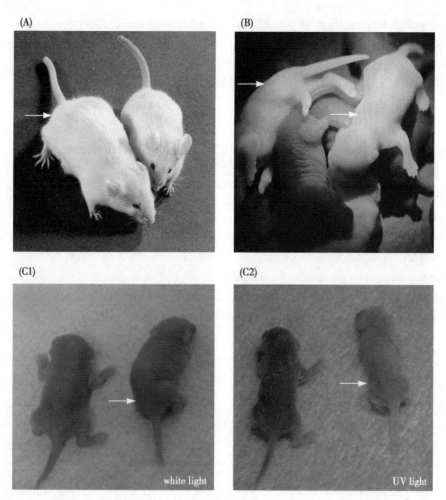

Figure 14.2 **Transgenic mice.** (A) Super mouse. Richard Palmiter from the University of Washington and Ralph Brinster from the University of Pennsylvania microinjected the human growth hormone gene into freshly fertilized mouse eggs. Expression was ensured by fusion of a mouse promoter, that of the *metallothionein I* gene, to the structural gene coding for the human growth hormone. A significant percentage of mice stably incorporated the fusion gene into their genome and exhibited high concentrations of the hormone in the blood. The transgenic mice grew to twice the size of normal mice. (B) Green mice. Masaru Okabe and colleagues from Osaka University in Japan engineered mice that exhibit green fluorescence by injecting into mouse embryos a jellyfish gene that codes for a protein called GFP, green fluorescent protein, that can fluoresce under UV illumination. The construct that was injected contained the GFP cDNA, an uninterrupted coding sequence from jellyfish, under the control of elements which ensure that the protein is expressed at high, easily detectable levels. The transgenic mice expressed GFP in every cell type tested—brain, liver, adrenal gland, *etc.* —except sperm, red blood cells, and hair. The newborn mice were all green under UV illumination until they grew fur, at which point only the naked body parts were fluorescent. The mice are not just a scientific curiosity; their green cells can be used in basic research to follow, for example, the fate of green cancer cells transplanted into normal mice. In addition, one can monitor the response of these cells to treatment. (C1) and (C2) Wild-type mice and *Alas2* transgenic mice under white and UV illumination. Under UV illumination, the transgenic mice exhibited a significant red fluorescence due to protoporphyrin IX accumulation. White arrows indicate transgenic mice. (A, courtesy of Ralph Brinster, University of Pennsylvania. B, from Okabe M, Ikawa M, Kominami K, et al. [1997] *FEBS Lett,*407:313–319. With permission from Elsevier. C1 and C2, courtesy of Xu Gao, Harbin Medical University.)

nuclei have fused to form a diploid zygotic nucleus. The transgene solution (1–2 ng/μL for standard DNA construct, about 5 kb) are loaded into a glass microneedle, and the DNA is microinjected directly into one of the pronuclei. The male pronucleus then fuses with the pronucleus of the egg to form the diploid nucleus of the embryo. The egg is only 100 μm in diameter (including zona pellucida), so this operation requires great skill and patience. The injections are very tedious and even the experienced technician may only be able to complete about 100 successful injections in a day.

- Implantation into foster mother. To be able to become live-born transgenic mice, the successful injected embryos must be transferred into the reproductive tract of a female mouse (Figure 14.3). Female mouse recipients for embryo transfer are prepared by mating with vasectomized males. Fifteen successfully injected embryos are surgically transferred to the oviduct ampulla of the recipient "pseudopregnant" mouse. Pregnancy is visible about 10 days after transfer and the offspring are delivered about another 10 days later. Mouse pups are weaned and analyzed at 3–4 weeks of age.
- Analysis of mouse pups. The main aim of analysis of mouse pups is to solve two important questions: (a) Does the transgene integrate into the mouse chromosome stably? (b) Does the transgene express appropriately? Here, "appropriate" means whether the transgene is expressed at the correct time during development, in the correct tissue(s), and in the correct amount.

Screening of transgenic mouse Tail top biopsies (3–5mm) are usually taken from about 4 weeks old mouse pups to obtain genomic DNA for analysis. The DNA is tested for the integration of the transgene by Southern blotting hybridization or polymerase chain reaction (PCR) (Figure 14.3). The mouse which has successful integration of the transgene is referred to as the founder (F0) mouse of a new transgenic lineage. The transgene will be heritable and passed on from generation to generation as part of the mouse genome. The success rate of stable integration is around 2.5%~6% in mice which develop from fertilized eggs after microinjection.

The integration of foreign DNA into the embryonic genome usually is random by non-homologous recombination. The mechanism is not clear thoroughly, but non-homologous recombination may be provoked by the action of DNA repair enzymes. These enzymes are believed to be induced by free ends of the transgene and create double-stranded breaks in the chromosomal DNA. Therefore, it is impossible to regulate the exact site of the integration of transgene and the number of copies of transgene which both affect the expression degree of transgene in mouse in traditional transgenic technique. The copy number is generally from one to 150 copies.

The first generation (F0) of transgenic mice is sometimes "mosaic" or "chimeric" which have foreign DNA in some, but not all, germline and somatic cells, although we microinject the transgene in the pronuclear period. The transgenic founder mice need to be inbred to produce a second generation, called the F1 generation. The F1 mice need also analysis for stable integration of the transgene. If the germ cells of the founder transmit the transgene stably, then all descendants of this mouse are "positive" for this unique transgene, called transgenic lineage. It is worth noting that those mice derive from the same F0 mouse should be classified as a transgenic lineage. In general, the transgenic mice from a lineage can be used as the same group in research for their same integration site and copy number.

In addition, the genotype of the F0 generation is described as hemizygous for the transgene rather

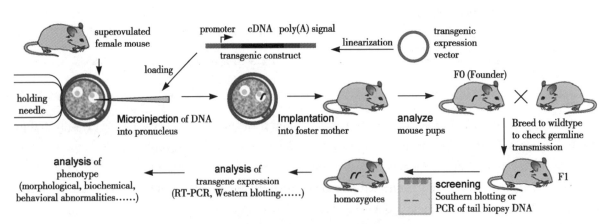

Figure 14.3 **Transgenic mouse procedure.**

than heterozygous, since the new transgenic locus is present in only one of a chromosome pair. Because the site of integration may be different, we should mate the mice with identical transgene in the same lineage. If two such hemizygous mice breed together, some of the progeny, 25%, will be homozygous for the new trait. In other words, a homozygous genotype, transgene alleles are present on each chromosome in a pair, may be produced by the mating of hemizygous F1 siblings.

- Analysis of transgene expression. If the transgene is identified to be present in the mouse genome, we must answer whether the transgene is regulated well enough to function in its new loci. To answer this question, gene expression at transcriptional and translational levels are analyzed. Northern blotting and RNA *in situ* hybridization can be used to assess the transcription of the transgene, and reverse transcription-PCR (RT-PCR) is more efficient and convenient. Western blotting and immunohistochemistry are often used for analysis of transgene expression at the level of translation.

Many studies have found dramatic different phenotype in a specific transgene between different lineage simply due to different sites of random integration. The researchers believe that additional gene-specific regulatory regions are most likely needed to achieve appropriately regulated expression. In addition, there are problems with expression of the transgene technique itself: frequently the transgene randomly inserts into an endogenous mouse gene and interferes with its expression. Such mutations may result in embryonic lethality, but they can also result in a viable mouse with a distinct mutant phenotype. The researchers must distinguish these mutations from the true transgenic phenotype because only a single transgenic lineage will exhibit the phenotype.

(2) Inducible transgenic mouse

The introduction of a transgene sometimes results in death or such reduced viability so that it is difficult, even impossible to maintain the transgenic mouse line by breeding. It is helpful to design a transgene that can be activated after establishment of the transgenic line. Inducible expression systems have allowed researchers to overcome these problems associated with transgenic studies. For example, "Tet-off" is one commonly used expression system. Transgene expression is dependent on the activity of an inducible transcriptional activator in this system. Expression of the transcriptional activator can be regulated both reversibly and quantitatively by exposing the transgenic animals to varying concentrations of the antibiotic tetracycline, or tetracycline derivatives such as doxycycline. Another example is the "super" mice described above, fertilized eggs were microinjected with a foreign DNA construct containing the rat growth hormone cDNA under control of the mouse metallothionein (MT) gene promoter. The MT promoter was chosen because it is inducible (turned on) by the metals zinc and cadmium.

Site-directed mutagenesis makes it possible, in principle, to insert a protein with a new or modified sequence into an organism and to study its effect on the whole organism. However, it must be remembered that this is the addition of a function; the original variant of the gene is still present and, in most cases, functional. If we want to take away a gene or replace it with an altered version, the challenge is somewhat more difficult.

14.3.2 Gene-targeted mouse models

In the last section, the production of transgenic mice by pronuclear microinjection was described. Transgenic mice engineered in this way are generally "gain of function" mutants because the transgene is designed to express a novel gene product. Another strategy for gene function research *in vivo* is by genetic "knockout" which a particular gene is deliberately disrupted. Genetically modified mice engineered by this approach are called "loss of function" mutants.

In 1980, Capecchi proposed the concept of gene targeting which can alter endogenous genes in mouse embryonic stem (ES) cells by homologous recombination with a foreign DNA sequence. Generation of the first "knockout" mouse using this gene targeting method was reported by Capecchi in 1990. Mice homozygous for the loss of the *Int-1* proto-oncogene showed severe abnormalities in the development of the midbrain and cerebellum. Since then, many variations of this technique have been developed. For example, "knock-in" and "knock-down" mice have been created where expression of an endogenous gene is altered, but not necessarily inactivated. The targeted changes are introduced into the sequence of the coding region or upstream regulatory elements of the gene. In addition, inducible and conditional expression systems have been designed for gene-targeted mouse models.

(1) Knockout mouse

The technique developed by Capecchi will be illustrated by the widely used procedure to produce knockout mice, in which a specific gene has been inactivated by interrupting it with an extraneous sequence. The knockout method involves five main stages: construction of the targeting vector based on the DNA sequence of the gene of interest; gene targeting in embryonic stem (ES) cells; selection of gene-targeted cells; introduction of targeted ES cells into mouse embryos and implantation into a foster mouse; and analysis of chimeric mice and inbreeding (Figure 14.4).

- Construction of the targeting vector. The procedure begins with the construction of a vector, called the "targeting vector", containing the interrupted gene, together with enough native flanking sequences to favor homologous recombination in the recipient cell. The cloned gene of interest is typically disrupted by insertion of the neo^r gene (serves as marker), which encodes the enzyme neomycin phosphotransferase. The existence and expression of neo^r gene can indicate that the vector DNA was integrated in a mouse chromosome. The neo^r insert is flanked by DNA sequences from the two ends of the gene of interest. These DNA regions provide the region of homology for recombination with the corresponding interested mouse gene. The targeting vector usually also carries the herpesvirus thymidine kinase (tk) gene at one end as the second marker. It is standard, but others could be used instead.

- Gene targeting in ES cells. The next step for creating a knockout mouse is to obtain ES cells from early mouse embryos. ES cells are pluripotent cells that have the potential to develop into any kind of cell of the organism if provided the right molecular cues from the environment. Furthermore, they can be cultured in the lab indefinitely.

The vectors carrying the altered gene are inserted, by chemical means or by electroporation, into embryonic stem cells taken from a mouse embryo. Three events can take place between targeting vector and EC cells chromosome: no recombination, homolo-

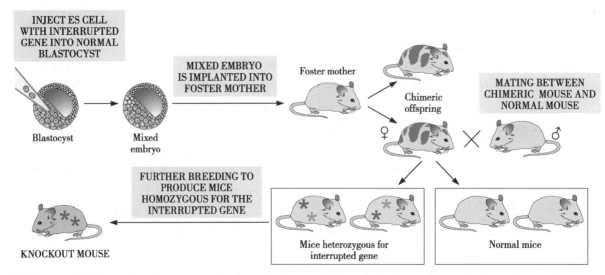

Figure 14.4 **Making a knockout mouse by introducing an *in vitro* altered gene into mice (gene-targeted mouse).** (Adapted from the Nobel Foundation.)

gous recombination, or non-homologous recombination. Homologous recombination is a very rare event in mammalian cells, occurring in only 1/1000 cells. Non-homologous recombination is a more common result and involves random integration.

- Selection of gene-targeted ES cells. Then, the cells are selected for recombination events during growth in a culture medium. This is a very time-consuming operation, lots of cells and trials are required to obtain a pure culture of targeted ES cells.
- Introduction of ES cells into a mouse embryo and implantation into a foster mother. Recombinant cells are then inserted into another blastocyst mouse embryo. The resulting mixed embryo that contains both normal cells and cells with the interrupted gene is implanted into the uterus of a foster mouse. This procedure is similar to that of transgenic technology.
- Analysis of chimeric mice and inbreeding. The newborn mice will be chimeras, with a mosaic of altered heterozygous cells and wild-type cells. If such females are allowed to grow to maturity and bred to wild-type males, some of the progeny will be wild-type while the rest will be heterozygous for the interrupted gene. However, homozygous mice can be obtained by interbreeding the heterozygous mice. These mice will survive and breed further, if the lack of the gene is not lethal.

The focus of this section is on genetically modified mouse, but knockout technology is not limited to this model organism. Genetic knockout zebra fish, fruit flies, and other eukaryotic systems have been created using similar methods. Recently, gene-targeted human embryonic stem cells were engineered, although in this case the experiments must end with the stem cells according to ethical principles.

Modifications of the above procedure can be used to generate knock-in mouse, in which a modified gene is specifically substituted for the wild-type gene, or knock-downs, in which the regulation of a particular gene is modified. All of these techniques are very time-consuming and relatively inefficient, but they can provide the most definitive evidence for the *in vivo* function of a particular gene. Capecchi shared the 2007 Nobel Prize in Physiology or Medicine with Sir Martin Evans and Oliver Smithies for their "discoveries of principles for introducing specific gene modifications in mice by the use of embryonic stem cells."

(2) Knock-in mouse

According to the research question, a researcher may be interested in expressing transgenes from known regulatory elements, or modifying rather than merely inactivating the targeted gene. Such expression of introduced DNA sequences can be achieved by generating a knock-in mouse. The knock-in approach is often used for *in vivo* site-directed mutagenesis. It is essentially the same as the procedure of generating knockout mice: a targeting vector with a foreign sequence insert is introduced into a specific locus by homologous recombination in ES cells. To study the correlation of structure and function of the gene product in the whole animal, single base substitutions or deletions can be created by knock-in. The knock-in allele replaces the coding region of the endogenous allele without disturbing the endogenous upstream regulatory elements. The knock-in allele is generally expressed precisely as the endogenous gene would have been expressed.

(3) Knock-down mouse

Making a "knock-down" mouse is analogous to "knockout" the endogenous gene expression in the whole animal. By modifying the promoter, such as the *cis*-regulatory elements, gene targeting strategies can also be applied to inactivate or modify expression levels of a gene. For example, single base modification, substitutions or deletions, can be introduced to study the regulatory elements required to drive tissue-specific or developmental stage-specifc gene expression. The knock-down targeting sequence disrupts endogenous regulatory elements upstream coding sequences, while keeping the endogenous coding region intact. Then, the effect of the targeted mutation on gene transcription is analyzed in the genetically modified mouse.

In addition, we can also overexpress short hairpin RNA (shRNA) to create knock-down mouse by transgenic which mentioned in the previous section. A shRNA can trigger RNA interference (RNAi) which suppress the expression of interested gene.

(4) Conditional knockout mouse

Conventional gene knockout technology by homologous recombination can provide useful information in elucidating the function of some genes. However, the deletions of many genes often result in embryonic lethality. It is useful to design a conditional mutation of the target gene to make it possible to study the roles of these genes in development or in the adult mouse. Genetic switches can be engineered to target expression or disruption of any gene to any tissue at any defined time. These genetic switches contain the Cre/LoxP system and FLP/FRT system (Figure 1 in Box 14.1). These two systems are based on the use of site-specific DNA recombination.

> **Box 14.1 Conditional knockout mouse: Cre/LoxP and FLP/FRT system**
>
> The Cre/LoxP system Cre/LoxP technology was introduced in the 1980s. It has been successfully applied in yeasts, plants, mammalian cell cultures and mice. This system is based on the ability of the P1 bacteriophage Cre (cyclization recombination) recombinase which can affect recombination between pairs of LoxP sites (a 34 bp DNA sequence). Such recombination in a "Cre/LoxP" mouse can either activate or inactivate a gene of interest, or produce tissue-specific expression.
>
> To use Cre/LoxP technology, an investigator has to produce a Cre/Lox mouse, typically by breeding a Cre mouse to a LoxP mouse. A Cre mouse contains a Cre recombinase transgene under the direction of a tissue-specific promoter, and a LoxP mouse contains two LoxP sites that flank a genomic segment of interest, the "floxed" locus, by gene targeting. After crossbreeding these two types of transgenic mice, depending on the location and orientation of the LoxP sites in a Cre-Lox mouse, Cre recombinase can result in deletions (excision), inversions, and translocations of the sequence between the floxed loci. Cre-mediated excision results in tissue-specific gene knockout.
>
> The FLP/FRT system is similar to the Cre-Lox system and is becoming more frequently used in mouse-based research. It involves using flippase (FLP), one of recombinases which derived from the yeast *Saccharomyces cerevisiae*. FLP recognizes a pair of FLP recombinase target (FRT) sequences that flank a genomic region of interest.

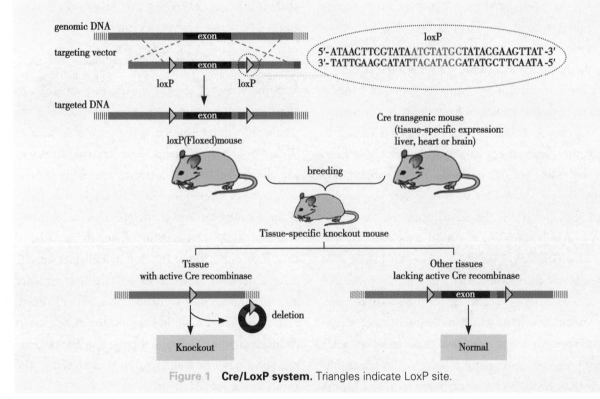

Figure 1 **Cre/LoxP system.** Triangles indicate LoxP site.

14.3.3 Genome editing by engineering nucleases

Recently, several technologies which are more convenient and efficient have emerged that allow direct editing of target sequences in the genome *in vivo*. These technologies are all based on endonucleases that can be targeted very specifically to genomic sites. These nucleases can create DNA double-strand breaks (DSB), then the cell's own repair machinery, homologous recombination or non-homologous end-joining (see Chapter 13), generate sequence alterations. These changes can include deletion, insertion, gene mutation, or even precise gene editing or correction based on a provided donor template.

Unlike the random integration of foreign DNA in the transgenic technique and the low efficiency of homologous recombination in gene targeting technique, the specificity and outcomes of these novel techniques depend on the specific targeting of endonucleases to only the site(s) of interest. Four general classes of nucleases include: zinc finger nucleases (ZFNs), transcription activator-like effector nucleases (TALENs), meganucleases, and most recently, the CRISPR/Cas9 system. The basic characteristics of these systems for genetically modification are summarized in Table 14.1.

(1) zinc finger nuclease (ZFN)

ZFNs are engineered proteins used for gene targeting. They consist of a DNA-cutting endonuclease domain fused to zinc finger (ZF) DNA binding domains. The ZF domains are modular domains that each recognize a 3 bp sequence and can be linked together into multifinger domains to recognize and bind longer DNA sequences (interested gene). The endonuclease domain derives from the *Fok I* restriction enzyme then dimerizes to make a DSB at the desired site guided by multifinger domains.

(2) transcription activator-like effector nucleases (TALENs)

TALENs operate on the similar principle as ZFNs. TALENs are fusions of transcription activator-like effector (TALE) proteins and a *Fok I* endonuclease. TALE proteins are composed of 34 amino acid repeating motifs with two variable positions (the 12th and 13th amino acid residues) that have a strong recognition for specific base. For example, the motif contains histidine (H) and aspartate (D) at 12th and 13th position can recognize "C (cytosine)" base. TALE motifs/repeats can be strung together to recognize virtually any DNA sequence.

By assembling arrays of these TALEs and fusing

Table 14.1 Multiple gene genetically modified techniques and their basic features

Tool	Principle	Characteristics	Effect
Transgenesis (Tg)	foreign DNA random integration	Vector (plasmid, BAC, YAC) carry transgene / Random	Overexpression
Gene targeting	DNA homologous recombination	ES cells Low efficiency Site specificity	Knockout Knock-in
Gene/genome editing	Endonucleases break DNA, DNA repair system (ZFN, TALEN, Meganuclease, CRISPR/Cas)	Site specificity Efficient and Flexible	Genome editing, etc.
Others			
Drug mutagenesis	DNA damage	Random mutation	Functional inactivation
Gene trap	random integration trap vector	High-throughput Random insertional mutation	Functional inactivation
Transchromosome	Chromosome serves vector	DNA super-large fragment	Overexpression

them to a *Fok I* nuclease, specific cutting of the genome can be achieved. Similarly, TALENs also needs dimerization to work. When two TALENs are close to each other and bind, the *Fok I* domains induce a DNA double-strand break which can inactivate a gene, or can be used to insert DNA of interest. TALENs are more specific than ZFNs; This simplicity of "assembling" as building blocks has facilitated automation of TALENs construction. However, each base in the target site needs the recognition of a 35 aa motif, TALENs are larger and somewhat harder to deliver into target cells or tissues.

(3) Meganucleases

The meganucleases, despite their name, are the smallest member (~40 kDa) in these editing nucleases and thus it is the easiest to transfer into cells or tissues, even several meganucleases with different target sites could be delivered simultaneously for multi-sites gene editing. These nucleases are derived from homing endonucleases, a family of nucleases encoded within introns. They have a large DNA recognition site (up to 40 bp), which occurs rarely, even in entire genomes. The large target sites are the origin of the name. Many meganucleases can be engineered or selected to recognize novel sequences, but because they lack the modular DNA-binding domain architecture found in ZF or TALE proteins, readily available customizable reagents, and clear rules for designing a construct that recognizes a specific DNA sequence, they have not achieved widespread adoption as tools for genome editing than the other strategies discussed here.

(4) CRISPR/Cas9

Just entering the second decade of the 21st century, a novel gene editing technique based on CRISPR/Cas (clustered regularly interspaced short palindromic repeats/CRISPR-associated proteins) systems has been widely and rapidly used, including life sciences, medicine, agriculture, and almost all fields related to molecular biology. CRISPR/Cas systems are genetic hallmarks of adaptive immunity in bacteria and archaea that have evolved to target and eliminate invading genetic elements, which enables them against viruses and plasmids.

Briefly, the CRISPR-Cas system can integrate invading nucleic acids into CRISPR loci of bacteria, where they are transcribed into CRISPR RNAs (crRNAs). Then, crRNAs form a complex with a *trans*-activating crRNA (tracrRNA) and Cas (CRISPR-associated) proteins as a surveillance system. The crRNA then targets cleavage of complementary DNA sequences which may be the genetic elements of the re-invading viruses (Figure 14.5C).

To develop this system for gene editing, firstly, the crRNA and tracrRNA are fused into a single guide RNA (sgRNA), and changes to a portion of this RNA sequence can be used to target interested DNA sequence. It is more convenient to synthesize a segment of sgRNA than to construct novel proteins (repeats of domain or motif) for every desired target sequence in previous other technologies. Then, the Cas9 protein from the bacterial type II CRISPR/Cas system can simply be delivered with a sgRNA or several sgRNAs designed against the interested site(s). In addition, Cas9 protein requires a short protospacer-adjacent motif (PAM) in 3′ of the target site, which is usually NGG ("N" represents A, G, C, or T) (Figure 14.5A).

Gene editing using CRISPR/Cas system requires only routine pronuclear or cytoplasm microinjection of sgRNA and Cas9 protein (or Cas9 mRNA) complex (sometimes it also includes donor DNA) in fertilized egg stage, rather than the complex operations of embryonic stem cells. Therefore, the production period of genetically modified animals using this technique is greatly shortened.

Simplicity, flexibility, efficiency and extensibility of CRISPR/Cas system have led to rapid adoption by many laboratories around the world for the editing of virtually any desired genome in mammalian cells and model animals. In addition, the system is also used for turning on and off gene expression and for imaging in live cells (Figure 14.5B). CRISPR system have almost recently swooped down and snatched the genome editing monopoly from TALENs and ZFNs. The developed techniques based on CRISPR/Cas

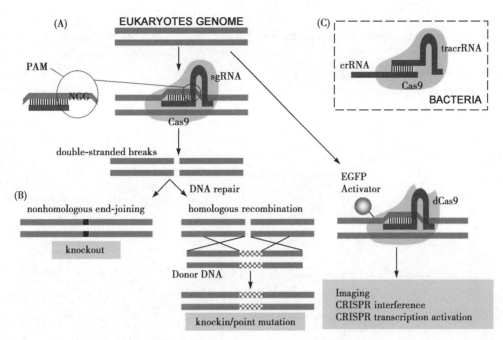

Figure 14.5 **CRISPR system in bacteria and eukaryotes genome editing.** (A) Cas9 was shown to use a sgRNA, fusion of the tracrRNA and crRNA, to direct DNA cleavage. Target recognition requires both base pairing to the sgRNA sequence and the presence of a PAM that is adjacent to the targeted sequence. DSBs were introduced into genome. (B) DSBs generated by Cas9 in genomic DNA are subsequently repaired by NHEJ to introduce gene disruptions or by HDR through the insertion of donor DNA sequences. We can also carry out gene expression regulation and cell imaging using dead Cas9 (dCas9). (C) crRNA-directed cleavage of foreign nucleic acid by Cas proteins (e.g. Cas9) at sites complementary to the crRNA spacer sequence.

systems are also reported to serve as rapid molecular diagnosis tool (named SHERLOCK) and RNA editing tool (CRISPR/Cas13). These techniques have already begun to show promise as a novel gene therapy tool to treat human genetic disorders and other diseases and are being utilized in preclinical studies(see Chapter 16). At the time of writing this book, Emmanuelle Charpentier and Jennifer Doudna are awarded the Nobel Prize in Chemistry 2020 for discovering CRISPR/Cas9 genetic scissors, a genome editing tool mentioned in this chapter.

In fact, almost all of the endonucleases and gene editing systems mentioned above come from prokaryotes. We continue to learn from nature, learn the molecular mechanism from prokaryotes, and use human's own DNA damage and repair mechanism to create new gene editing technologies. Furthermore, we can continuously improve our ability to understand and benefit ourselves through these technologies.

There are also gene trapping technique and chemicals, e.g., ethylnitrosourea (ENU), induced gene mutagenesis technique besides transgenesis, gene targeting and gene editing techniques for specific gene mentioned in this section. These methods can produce large-scale gene point mutations in animals and become the mutant mouse library. ENU mutagenesis involves exposing male mice to ENU and then mating the treated males to wild type (WT) females. Many progenies carry point mutations and are screened for phenotypes of interest. However, the mutation sites are random and require a complicated identification and screening process.

Generally, a simple plasmid vector can transfer foreign DNA smaller than 20 kb. Vectors derived from bacterial artificial chromosome (BAC) and yeast artificial chromosome (YAC) can carry DNA fragments up to ~200 kb and ~1 Mb, respectively. Recently, Chromosome engineering technology based on microcell-mediated chromosome transfer (MMCT) has also been developed. Oshimura M and his colleagues have created genetically modified mouse carrying the full length of human antibody genes clusters (~5 Mb) (Figure 1 in Box 14.2).

Box 14.2 **Fully human antibody and transgenic mouse**

Nowadays, most fully human antibody drugs in clinical testing stage and on market were discovered on the immunoglobulin genes humanized transgenic mouse platform, others were found by phage display technique. Some pharmaceutical companies (BMS, Amgen, REGENERON and Kirin, etc.) used Humab, XenoMouse, VelocImmune mouse and TC-mAb mouse to produce fully human therapeutic antibody. Some of these mice were replaced part of the mouse immunoglobulin genes into humanized DNA sequence by gene targeting, and some were introduced fully human immunoglobulin genes into inactivated endogenous genes mice by transchromosome technology (Figure 1).

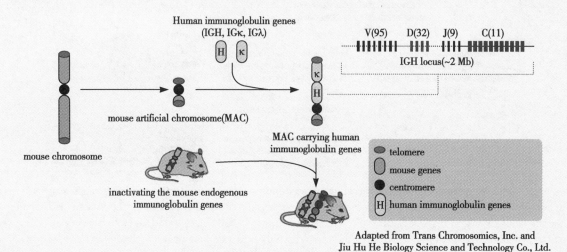

Figure 1 **immunoglobulin genes transgenic mouse based on transchromosome technique.** (Courtesy of Xu Gao, Harbin Medical University).

14.4 Applications of genetically modified mouse in medical field

Genetically modified mice, as an important part of GM model organisms, have developed worldwide in the past four decades, and have been successively applied in life science and basic medical research, humanized antibody drug evaluation, and fully human antibody drug discovery (Figure 14.6).

14.4.1 Gene function research

One of the most important tasks in the post-genomic era is to clarify the functions of genes. The GM mice provide experimental systems for understanding gene functions. Firstly, the creation of genetically modified mouse (of course, so do the other GM organisms) provides two platforms, gain of function (transgenic overexpression or knock-in) and loss of function (knockout), for researchers. Next, GM mouse experimental system is *in vivo* and can help us understand the gene function at the level of whole animal. *In vivo* studies include nervous control and endocrine control system, and intercellular interactions, which are more consistent with human physiology and pathology.

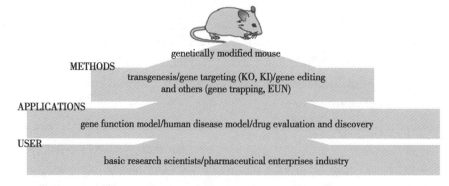

Figure 14.6 **The application of GM mouse in basic and applied research.**

The GM mice are examined carefully for morphological, biochemical, and/or behavioral abnormalities, also known as analysis of phenotype. The phenotype of the GM mouse displays the impact of the particular gene on mouse development and physiology. In this way, we can associate different phenotypes with the functions of genes and understand the biological functions of interested genes.

14.4.2 Animal models of human diseases

Genetically modified mice with specific genes can mimic human diseases very well for these genes are crucial in the generation and development of human diseases. In other words, the pathological features in these GM mice are similar to that of human. Researchers can use these GM mice as animal models of human diseases, and then explore the molecular mechanism of related diseases, new therapeutic methods and perform drug evaluation. For example, the amyloid precursor protein/presenilin 1 (APP/PS1) double transgenic mouse model provided robust neuropathological hallmarks of familial Alzheimer's disease (AD) at early ages. Over the years, this double transgenic mouse has been used as a mouse model of AD disease. Another example is apolipoprotein E (ApoE), a glycoprotein for a structural component of very low density lipoprotein (vLDL) synthesized by the liver and intestinally synthesized chylomicrons, knock-out mice. ApoE$^{-/-}$ mice are frequently used for the study of hyperlipidemia and atherosclerosis.

14.4.3 Gene knockout-based immunodeficient mice for PDX model

The genetically modified mouse can not only elucidate the pathological and physiological mechanisms underlying diseases but also offer the platform in preclinical research, such as the testing of new drugs for personalized therapies.

Many combined immunodeficient mice strains whose *Il2rg* gene is deleted, such as NOG mice [NOD/Shi-scid IL2rgamma(null)], NSG mice [NOD/LtSz-scid IL2r gamma(null)] and BRG mice [BALB/cA-Rag2(null) IL2rgamma(null)], lack T cells, B cells and NK cells and have functionally impaired macrophages and dendritic cells.

The multiple dysfunctions of innate and acquired immunity in these mice create a suitable environment for engraftment of human/patient cells or tissues. These humanized mice carrying human/patient cells or tissues serve as patient-derived xenograft (PDX) Models for personalized medicine.

14.4.4 Fully human antibody discovery and pharmacodynamic evaluation

Nowadays, the humanized mice have become a powerful tool of the human being for research about the problems of human development and disease. Another group of humanized mice generated by using gene knockout and transgenic (sometimes transchromosome) techniques carry only human functional genes. These genetic humanized mice can be used in fully human antibody discovery and pharmacodynamic evaluation.

(1) Fully human antibody discovery

Blocking membrane receptor protein by a monoclonal antibody (mAb) is a possible therapeutic strategy in targeted therapy and immunotherapy. In order to reduce the immunogenicity of antibody drugs, it is best to use fully human monoclonal antibodies. One of the common methods is to create the humanized mice with human immunoglobulin gene clusters (heavy chain and light chain).

Firstly, inactivating the mouse endogenous immunoglobulin genes by gene targeting technique. Then, human immunoglobulin gene clusters are introduced by transgenics or transchromosome technique. For example, TC-mAb mouse developed by Oshimura M with transchromosome technique, carrying full length of the human immunoglobulin genes, can produce fully human antibody and increase antibody diversity significantly for therapeutic antibody discovery.

(2) Pharmacodynamic evaluation

In order to solve the problem of animal species differences found in basic research and pre-clinical testing, some human genes including immune checkpoint related genes can be introduced into mice with gene targeting or gene editing techniques, such as CRISPR system. These humanized mice retained the transmembrane region of mouse protein and replaced the extracellular region, recognized by therapeutic antibodies, with that of human. These mice are useful for evaluating human drug pharmacokinetics, pharmacodynamic, and analysis of drug-drug interactions.

14.5 Applications of other transgenic animals

So far the focus of this chapter has been primarily on basic and applied research using transgenic or gene-targeted mouse. Transgenic technology has been applied to many other animals, including farm animals, primates, *Caenorhabditis elegans* and zebrafish. These transgenic animals have been explored successfully as tools for applied purposes, ranging from basic research to genetically engineered pharmaceuticals.

14.5.1 Transgenic nonhuman primates

Mouse does not always serve as an accurate model of human physiology or disease pathology. For example, Lesch-Nyhan syndrome is a rare X-linked disease in humans due to loss of activity of HPRT (hypoxanthine-guanine phosphoribosyltransferase). HPRT plays an important role in purine metabolism. This deficiency results in motor, behavioral, and intellectual abnormalities in children. However, the *Hprt* gene knockout mice only show minor neurochemical or functional changes because mice have alternative pathways for purine metabolism that are not present in humans.

Given such problems in developing mouse models, there has been many efforts in extending transgenic and gene-targeting studies to nonhuman primates. The nonhuman primate may offer platform for better understanding and therapy of human diseases, such as brain disorders for they are much closer to human in development and genetics than mice. However, the ethical and financial costs have to be weighed against the potential benefits from the use of primates.

In 2001, the first transgenic primate, a rhesus monkey, was engineered. Researchers first used a genetically modified retrovirus vector to introduce the GFP gene and the foreign gene and results in random integration of the transgenes into the chromosomes of unfertilized rhesus monkey eggs. Then, the eggs were fertilized by intracytoplasmic injection of sperm a few hours later. Fertilized embryos were then implanted into a female rhesus monkey. However, the first transgenic monkey failed to glow green in any accessible tissue for unknown reasons. In recent years, it is easier for us to acquire transgenic and gene-targeting nonhuman primates model with the development of new technologies, such CRISPR system. We believe that the application of transgenic nonhuman primates model will be more convenient for disease research in the future.

14.5.2 Gene pharming

Although bacteria have been widely used to produce therapeutic protein, there are many problems in synthesize eukaryotic proteins. To overcome these difficulties, many mammals like sheep and cows are popular for gene farming, as their milk can be used to generate pharmaceuticals. For example, a flock of transgenic sheep was engineered to produce an enzyme named α1-antitrypsin. This enzyme is used in the treatment of cystic fibrosis and hereditary emphysema. A small herd of lactating transgenic sheep can provide an abundant supply of this protein.

14.5.3 Transgenic livestock

Transgenic technology has been used widely in agriculture for increasing the performance of commercially important animals by adding new traits (e.g., faster growth rate, disease resistance and decreased body fat) or improving on existing ones (e.g., yield or quality of meat and milk). There is only limited success (less than 1%) to use pronuclear microinjection to produce transgenic livestock such as pigs, goats, sheep, and cattle. Some developed technique, such as linker-based sperm-mediated gene transfer (LB-SMGT), can improve the production efficiency of large transgenic animals. In this method, a monoclonal antibody (mAb) binds a special DNA sequence (transgene) through an ionic interaction which allow exogenous DNA to be linked specifically to sperm. mAb can bind to a surface antigen on sperm of the tested species, including mouse, chicken, cow, goat, sheep, pig, even human. In one example using pigs, the transgene was successfully transferred and integrated into the genomic DNA with germline transfer to the F1 generation at a highly efficient rate of 37.5%.

Model organisms are widely and deeply used for basic research and clinical application. Here, we mainly introduce the most widely used in the study of eukaryotic molecular biology: mouse. Researchers can create genetically modified mouse by using transgenesis, gene targeting, and gene editing technique. These animals serve as gain or loss of function platform in the research of gene function. Recently, these genetically modified animals play important role in pharmaceuticals. These combinations of different genetically modification technique will open up new avenues for drug development and gene therapy as well as for basic research.

Key Concepts

- Model organisms are easy to manipulate in the lab, have short gestation periods that produce large numbers of offspring, and are relatively cheap, inexpensive to house.
- Almost all fundamental biochemical processes, from DNA replication to protein synthesis, were first elucidated in *E. coli*.
- We can create genetically modified organisms by using DNA homologous recombination and DNA damage/repair mechanisms.
- Three main strategies for gene editing: transgenesis, gene targeting, and gene editing.
- CRISPR system is developing rapidly and has been widely used.
- Genetically modified mice can be applied for the evaluation and development of biopharmaceuticals as well as for basic research in medical field.

Key Words

conditional knockout（条件性敲除）
CRISPR（成簇的规律间隔的短回文重复序列）
fully humanized antibody（全人源抗体）
gain of function（功能获得）
gene humanized mouse（基因人源化小鼠）
gene pharming（基因制药）
gene targeting（基因打靶）
genetically modified organism（遗传修饰生物）
in vivo（体内）
knock-down（敲减）
knock-in（敲入）
knockout（敲除）
loss of function（功能丧失）
meganuclease（巨核酸酶）
model organism（模式生物）
PDX（病人来源肿瘤异种动物模型）
random integration（随机整合）
TALENs（转录激活因子样效应物核酸酶）
transgenesis（转基因）
ZFN（锌指核酸酶）

Questions

1. What are the characteristics of model organisms?
2. Describe briefly the application of genetically modified animals in medical research?
3. Why can the genetically modified animals be created? (DNA can be broken and repaired)
4. Describe briefly the advantages of CRISPR system in eukaryotic genome editing.

References

[1] Fields S, Johnston M(2005) Cell biology. Whither model organism research? *Science*,307(5717):1885–1886.
[2] Friedel RH, Wurst W, Wefers B, et al.(2011) Generating conditional knockout mice. *Methods Mol Biol*,693:205–231.
[3] Fujiwara S(2018)Humanized mice: A brief overview on their diverse applications in biomedical research. *J Cell Physiol*,233(4): 2889–2901.
[4] Li D, Qiu Z, Shao Y, et al.(2013) Heritable gene targeting in the mouse and rat using a CRISPR-Cas system. *Nat Biotechnol*,31(8):681–683.
[5] Ma N, Li Z, Shi J, et al.(2019) development and change of research and market in the field of genetically modified mouse in China in recent 20 years. *Science & Technology Review*,37(14):51–58.
[6] Palmiter RD, Brinster RL, Hammer RE, et al.(1982) Dramatic growth of mice that develop from eggs microinjected with metallothionein-growth hormone fusion genes. *Nature*,300(5893):611–615.
[7] Uno N, Abe S, Oshimura M, et al.(2018) Combinations of chromosome transfer and genome editing for the development of cell/animal models of human disease and humanized animal models. *J Hum Genet*,63(2):145–156.

Ning Ma

Chapter 15
Disease Genes

15.1 Introduction

The genetic information of all kinds of biological cells should keep relatively stable. Under the influence of internal and external environmental factors in the long-term evolution process, the structure of genetic material and its expressing products can change, which is a common phenomenon in the biological community. The variation of these genes and genomes may be harmful, beneficial or neutral. For animals, the variation may change the original balance obtained through long-term evolution. From a medical point of view, the attention is to pay more harm than a beneficial variation of genes and genomes for human individuals. The structural or functional abnormalities of genes and genomes are likely to lead to the occurrence of diseases and then endanger human life and health. In fact, almost all human diseases are directly or indirectly related to gene and genome abnormalities, such as malignant tumors, cardiovascular and cerebrovascular diseases, neurological diseases, diabetes and chronic immune diseases. The gene mutation directly causes the disease called genetic disease, and the mutated genes are called causative genes. We should keep in mind that most genes influence the susceptibility of the complicated disease also called common diseases, such as cancers, cardiovascular diseases and degenerative disease as well as autoimmune diseases, and the genes are called susceptibility genes. We do not confuse by the fact that called asthma genes do not mean the gene existing only in asthma patients, actually, it exists in all the people with correct function and only the difference is various alleles in a population. In this chapter, we firstly introduce the types and causes of gene and genome abnormalities, then discuss the resulting molecular and clinical outcomes, and finally discuss the strategies and approaches how to localize and clone the disease genes in or from the genome.

15.2 Genome and gene abnormalities

Gene and genomic abnormalities can be divided

into 2 types: structural abnormalities and expression abnormalities of functional genes. Changed genes then can lead to the occurrence of diseases. To identify the causes of disease occurrence and development and provide efficient treatment and prevention, we should first understand the pathogenesis of the gene and genome abnormalities.

15.2.1 Abnormal gene and genome structures show various types

There are many types of structural abnormalities of gene and genome, which may occur in germline cells or somatic cells, in DNA of nuclei or mitochondria, and also in coding or non-coding regions of genes. These differences in location and type of abnormalities can lead to distinct phenotypic and biological effects.

(1) The increase or decrease of chromosome number can have an irreversible effect on individuals

Humans have a karyotype of 46 chromosomes in the nuclei of normal people as a typical diploid organism consisting of 22 pairs of auto-chromosomes and one pair of sex chromosomes, X and Y chromosomes for men, and X and X chromosomes for women. The number of chromosomes can change in both auto-chromosomes and sex chromosomes, which show different clinical manifestations Aneuploidy is used to describe the number change of chromosomes including more than the normal and less than the normal. Since each chromosome contains a number of genes, aneuploidy could lead to the group of diseases named syndrome, e.g, Down syndrome patients have one more chromosome 21.

(2) Point mutation refers to the variation of a single base in a DNA strand

Among point mutations, purine nucleotide substituting for purine nucleotide (the exchange between A and G) and pyrimidine nucleotide substituting for pyrimidine nucleotide (the exchange between C and T) are called transition, and the exchange between purine nucleotide and pyrimidine nucleotide is called transversion.

Single nucleotide polymorphisms (SNPs) are common in human genome and occur in 1 out of 300 nucleotides in human genome. SNPs just indicate DNA sequence variance among the population and differ from the point mutations, since the frequency of the least allele of SNPs is more than 1% in human population without the detrimental effect to human health. The SNPs are often used as a DNA marker for genome-wide association study (GWAS) and forensic medicine.

The single nucleotide variance could show a different effect on the gene function. According to effects on the protein, point mutations can be divided into 3 types: the missense mutations show different amino acid codons; the same-sense mutations show different code without the change of amino acid, and non-sense mutations change to stop code. Missense mutations can lead to amino acid residue change that shows different effects to the protein function. The same-sense mutations do not change the amino acid sequence but change the RNA sequence, which may impact the RNA stereo-structure and the stability of RNA, since the codons show degeneracy and the usages of codons that means the different usage to the codon form for the same amino acid. Non-sense mutations can translate truncated peptides which probably do not play the function.

(3) InDel includes nucleotide insertion and deletion variation

InDel represents the Insertion and the Deletion of one or a segment of nucleotide in DNA chain. The size of DNA segment can be varied from one to several hundred base pairs. The effects depend on 3/3 multiple or other number InDel bases. In the former scenario, one or several more or less amino acid peptides can be expressed compared with wild-type genes; but no 3/3 multiple InDels lead to the peptide with wrong amino acid sequence and often with truncated peptides. Th InDels occur in the regulatory region which is more often happen than in the coding region and definitely influence gene expression throng effecting binding with a transcription factor, for example.

(4) Gene rearrangement is that segments of DNA chain changed from original sequence

Since different segments are arranged and combined in different directions and patterns to form a new transcription unit. Gene rearrangement is an important way of gene variation. The gene rearrangement can lead to a 180° change of gene sequence, namely inversion. Cross-over of homologous chromosomes is a frequent phenomenon during meiosis, and transfer of gene sequence between non-homologous chromosomes, is named translocation. Homologous recombination also occurs when pathologic microorganism genes integrate into human genome. All the recombination may form a novel transcriptional unit of interrupting other genes, resulting in different molecular outcomes.

(5) VNTR and CNV can be used as DNA markers

A variable number of tandem repeats (VNTR) is a special sequence that is composed of the same core nucleotide repeats and arranged in series according to the way of head-tail connection. The repeat sequence with 1–6 bp nucleotide as the core is also called short tandem repeat (STR). The increase of these tandem repeat copies can be expanded with the passage of generations, so it is called a dynamic mutation, which is an important mechanism to explain the severity of genetic early occurrence and clinical manifestations, such as Huntington disease. Sometimes this kind of disease is called genomic diseases.

Copy number variations (CNV) exist in human genome, which is a large DNA fragment with a size ranging from 3.1 kb to over 1 Mb. It has known that copy number difference among the human population. Some CNVs are believed in linking with mental disorders.

Importantly, both CNVs and VNTR, especially STR as well as SNP have commonly used as DNA markers for genetic study and forensic medicine, since they are highly polymorphic in the human population, dominant expression, known location on the chromosome, and easy detection.

(6) Abnormal expression of genes indicates low-expression or high-expression of genes

Gene expression changes are mainly reflected in the abnormal deletion or overexpression of functional gene expression level, which leads to the imbalance of biological activity of expression products. Abnormal gene expression often has a structural basis, such as the change of gene promoter sequence or the change of structure or expression of regulatory factors. There are also many diseases whose gene and genome structures are normal. Intervening gene transcription levels can lead to the occurrence and development of diseases. All kinds of cells in the same organism contain the same genetic information, but these identical structural genes are expressed selectively, programmatically, and moderately according to the needs of growth, development, and reproduction of the organism and also expressed with the fluctuation of the environment, so as to adapt to the environment and play physiological functions. The expression of most genes in the multicellular genome has strict time and space specificity, and the time and space specificity of gene expression determines the functional characteristics of cells and the health status of individuals.

15.2.2 Many factors can cause gene and genome abnormalities

There are many reasons for gene and genome abnormalities, including random mutation under natural conditions, pathogenic mutation induced by physical, chemical, and biological factors, natural selection, genetic drift, homologous recombination, and integration of foreign invaded genes.

(1) The random mutation occurs under natural conditions

In the process of DNA replication, the parental DNA is used as the template to complete the replication event according to the principle of complementary base pairing. Occasional base mismatch, tautomerism, deamination, and various base modifications may occur spontaneously. Although the $3' \rightarrow 5'$ exonuclease activity of DNA polymerases can correct the mismatched nucleotides to ensure the accuracy of replication, the overall mismatching rate of DNA molecules can be reduced to about 10^{-10}–10^{-9} for one generation. However, due to the large genomic content of human beings, it is difficult to avoid the base mismatch in individuals.

(2) Gene mutations can be induced by various physical, chemical and biological factors

A variety of physical, chemical and biological mutagenic factors can act on DNA molecules to cause changes in the primary structure, resulting in structural abnormalities of proteins, and consequently, the genetic change leads to the occurrence of diseases. Many factors caused mutations, including physical factors, such as ultraviolet radiation, ionizing radiation in which X-ray may be seen in the most common, chemical factors, such as hydroxylamine, acridine dye, an alkylating agent, nitrite and base analog, and biological factors, such as measles, rubella and other viruses, fungi and bacterial toxins like Aspergillus flavus. These factors can cause different changes of DNA structure called DNA damage by the various molecular mechanisms.

(3) Natural selection and genetic drift make genetic abnormality exist in special population

Natural selection is the key factor to promote evolution, and can also promote a certain genetic character to have different continuity in a specific region. In a specific environment, if a genetic abnormality can lead to an increase in the survival rate of offspring, the abnormality will show a selection advantage in population reproduction, which is called positive selection. Sickle cell anemia is an example of malignant genetic diseases. Heterozygous individuals with abnormal gene (allele) are generally unimpeded. Fatal malignant malaria is rampant in central and western Africa. Heterozygous individuals with an abnormal gene are less likely to suffer from malaria. Therefore, they have survival advantages over individuals with homozygotes, making the abnormal gene frequency quite high in central and western Africa. However, in other malaria-free areas, the abnormal gene no longer shows the advantage of selection, so the gene frequency is low.

In a specific population, especially in a small population, some gene abnormalities may fluctuate in generation transmission, leading to the disappearance of some alleles and the fixation of specific gene abnormalities, thus changing the genetic structure of the population. In a large population, all kinds of alleles are transmitted evenly in the process of individual free mating; but in a small population, due to the limited number of offspring, if the mating mode is controlled, some gene abnormalities can be accumulated. The most extreme example is inbreeding, which may lead to the disappearance of a closed small population.

(4) Homologous recombination and gene integration are the foundation of many genetic abnormalities

Homologous recombination is the recombination of DNA molecules with homologous sequences in sister chromatids or the same chromosome. Therefore, inversion, exchange, and translocation of gene segments can occur, which is the basis of many gene abnormalities.

The invasion of foreign genes is also an important factor leading to genetic abnormalities. The nucleic acid fragments of HIV and hepatitis B virus, for example, can be integrated into the host genome through homologous recombination, which can interfere with the transcription and translation of the host endogenous genes, resulting in abnormal expression of the endogenous genes, or directly change the genome structure and affect the stability of the host genome, and systematically affect the normal metabolism of the body.

15.3 Molecular outcome of genetic abnormalities

The effect of genetic abnormality on gene function can be divided into two molecular outcomes: gain of function (GOF) or loss of function of gene (LOF).

The abnormal loss or gain of gene function is unfavorable for maintaining the normal metabolism and life activity of cells, which can lead to the occurrence of disease when it is serious. The various mechanisms by which genetic abnormalities result in loss or gain of gene function are described as following.

15.3.1 Genetic abnormalities lead to loss of gene function

Various mechanisms are involved in the loss of a function of genes, which include dose effect of the gene and down-regulation of gene expression. If the genes responsible for suppressing tumors lose function, tumors could occur.

(1) Haplotype insufficiency shows gene dose effects on disease phenotype

Haplotype deficiency refers to the need for mutation or deletion of one of the two copies of a given gene and the expression product of the other copy (that is, 50% of the protein product of the gene) could not maintain the normal function of cells. In autosomal dominant genetic diseases such as familial hypercholesterolaemia, heterozygous mutations can reduce the number of LDL receptors by 50%. Compared with normal homozygous individuals, the cholesterol level of heterozygous individuals is almost twice as high as that of normal homozygous individuals, so the risk of cardiovascular disease is greatly increased. In mutant homozygotes, the disease is more serious.

(2) Dominant negative effect occurs in genes responsible for multiple-peptide proteins

The protein complex is an important functional unit of cell activity. The dominant-negative effect refers to the loss of normal functions of the proteins produced by a mutation in a subunit gene. So, dominant-negative effects generally occur in genes that encode protein complexes (complete proteins consist of two or more protein subunits). For example, the abnormal structure and dysfunction take place in spiral subunits caused by a site mutation in the type I collagen gene. The protein is composed of three chains; and an abnormal single chain can bind to other two normal helix chains, resulting in various distortions and then serious damage of the protein.

(3) Loss of heterozygosity can explain the dysfunction of some tumor suppression genes

The loss of heterozygotes refers to the genetic damage of heterozygous individuals with a certain allele mutation from the fertilized eggs inherited by their parents, which leads to the mutation or deletion of wild type dominant genes to form mutant homozygotes and the loss of the original heterozygous traits (Figure 15.1). The carcinogenesis of retinoblastoma is caused by the mutation of tumor inhibitor Rb to rb or deletion, both of which are deleted or mutated before carcinogenesis. In 1971, Knudson put forward the "secondary strike theory" according to the occurrence of retinoblastoma. In 1978, Yunis was based on the 14 zone of long arm of chromosome 13. Rb gene was located at 13q14 in Retinoblastoma. The researchers believe that a rb gene or germ cell inherited by a parent has been hit for the first time (Rb becomes rb), the Rb/rb hybrid is damaged again, resulting in the loss of the original normal Rb, which leads to carcinogenesis.

(4) Transcriptional RNA position effects and epigenetic modifications affect gene expression

The antisense RNA transcription has been found in the study of a special phenotype of the Mediterranean anemia family. There are two functional globin genes (HBA2 and HBA1) on one chromosome of the normal person, one of which is identified as the missing one HBA1 in the family case member, and the HBA2 gene point mutation is excluded, but it shows a similar hematological characteristic to a typical-ground-poor heterozygote case that is both deleted. The absence of a 23kp DNA fragment in downstream of the HBA2 gene in the patient resulted in the expression of the LUC7L gene in downstream of the gene of the HBA2 gene in the reverse direction of the HBA2 gene in the vicinity of the HBA2 gene, the transcription of the LUC7L gene, The antisense RNA and the transcription of the HBA2 form part of double-stranded RNA, so that the RNA of the HBA2 is degraded, and the CpG island of the upstream site of the HBA2 is mediated to be methylated so that the expression level of the structure intact HBA2 gene is reduced. The lack of globin leads to the occurrence of the disease.

(5) Gene variation of transcription factors regulates gene expression

The abnormality of the transcription factor gene can affect the normal transcription of the downstream gene in combination with it, and the biological effect of the reduced gene function is generated. The important regulatory protein GATA-1 in hematopoiesis is a transcription factor with a zinc finger structure, and the zinc finger motif of the amino-terminal of the protein is responsible for binding to the DNA molecule. The binding stability of the transcription factor to the DNA molecule is reduced when the 216th amino acid of the N-terminal zinc finger region of the GATA-1 protein is mutated by arginine to glutamine in the thalassemia. The mutation of the gene leads to a reduction of the protein binding with DNA, so the transcription level of the human-globin gene is decreased. An insufficient amount of hemoglobin is synthesized and shows anemia.

(6) The stability of mRNA Affects gene expression

The 3′-terminal non-translation region (3′-UTR) of the eukaryote-encoding protein gene is involved in mRNA processing and stability. The abnormality of the 3′-UTR can lead to the occurrence of the disease. In the thalassemia, the mutation of the conservative sequence of globulin 3′-UTR, AAUAAA, results in the inability to add a tail to the normal splice site of polyadenylate, producing a large amount of long mRNA processing products, which are considered to be "non-self" and rapidly removed by the intracellular mRNA degradation mechanism.

15.3.2 Genetic abnormalities lead to gain of function of genes

A number of genetic abnormalities through various molecular mechanisms lead to the enhancement of gene function or acquirement of the new function of the gene.

(1) The transcription regulation sequence of a specific gene can be changed to the transcriptional enhancement

When the transcriptional regulatory sequence of a specific gene is abnormal, the gene transcription can be enhanced, so that an abnormal phenotype is generated owing to overexpression of the gene. In the case of human genetic fetal hemoglobinemia, several point mutations that promote the transcription of the gene have been identified, which can increase the level of expression of the gene. The gene of the fetal globin g chain which has been closed in the adult stage is re-opened to cause the high-expression state of the chain in the adult red blood cells to produce the disease phenotype.

(2) Position effect of enhancers regulates gene overexpression

Enhancer is a DNA sequence that specifically enhances transcription with the distance of 1–30 kp from regulated target genes in human genome. This sequence can promote gene transcription through the binding between specific chromatin with advanced structure and specific protein factors to regulate target genes. A class of δ β-thalassemia found in the population is characterized by the continuous increase of fetal hemoglobin in the adult. The molecular basis of δ β-thalassemia is gene fragment deletion. The specific enhancer sequence was identified at the distal end of the 3′ end deletion site. The enhancer sequence was taken to the site adjacent to the G γ-gene because of the loss of large segments. G γ-gene can be activated by the "distance effect". The result is that G γ-gene can be opened up and the expression level of a gamma-globin chain can be upregulated.

(3) Gene copy number in human genome sometimes influences gene expression

In dose effect, gene copy number variation is one of the main mechanisms to enhance gene function. In the process of tumorigenesis, tumor inhibitor genes are inactivated due to mutations, and some proto-oncogenes are up-regulated, show the increase of copy number in the genome, and are activate some genes at affected sites. Some mutations can also lead to an increase of the dose of intracellular related genes and the continuous division of cells. If there is L-MYC in a small cell lung cancer cell line, the number of copies of N-MYC and C-MYC genes is amplified, especially the amplification of C-MYC is more obvious, and the number of copies is increased by dozens or even 200 times. This dose effect is very commonly observed in the oncogenes.

(4) Mutation of growth factor receptor genes leads to ligand-independent activation of the receptor

Epidermic growth factor receptor belongs to the tyrosine receptor family, which is involved in many cell signal transduction processes, such as RAS-RAF-MEK-ERK-MAPK pathway, PI3K-PDK pathway, JAK-STAT pathway and so on, and regulates the normal growth, proliferation and differentiation of cells. EGFR mutations can often be detected in non-small cell lung cancer individuals. Many mutations, such as the deletion of exon 19, can cause EGFR molecules to produce ligand-independent tyrosine kinase activity, resulting in so-called activating mutations. Abnormal EGFR eventually promotes tumor cell proliferation, migration, invasion, and differentiation.

(5) SNP could create a new promoter of genes

Among the causes of α-thalassemia, a new promoter is produced due to SNP in gene regulatory region. The new promoter competes with the original endogenous α-globin promoter, which interferes with the activity of the original endogenous promoter, and the expression of α-globin gene downstream of the promoter is significantly down-regulated, which leads to the occurrence of α-thalassemia. From the point of view of the direct biological effect of SNP mutation, it belongs to functional acquired abnormality, which increases the transcription factor binding function of the related site DNA sequence, but if you look at it from another point of view, the α-globin gene actually loses its function.

(6) Acquired RNA accumulation results in developmental disease or muscular atrophy

Ankylosing muscular dystrophy and proximal limb muscular dystrophy are related to the abnormal accumulation of RNA in cells. The pathogenicity abnormalities were caused by the amplification of different short tandem repeats in the 3'- UTR region of their respective genes. These different tandem repeats lead to the accumulation of a large number of abnormal RNA transcripts, which can inhibit the differentiation or damage of muscle cells. In addition, they can also interfere with the splicing process of normal gene mRNA transcripts by combining with antisense factors, resulting in abnormal splicing, which ultimately leads to developmental disorder and muscle atrophy.

In a word, the abnormal structure of protein or the abnormal expression of protein has various types in the molecular mechanism, which are almost very different. However, the molecular outcome caused by the abnormality of these genes and genomes, whether the loss of gene function or acquisition, can affect the normal physiological function and lead to the occurrence of genetic diseases.

15.4 Clinical outcomes of gene and genomic abnormalities

Loss or gain of gene function caused by genetic abnormalities manifests as various diseases happened at the individual level. Understanding the genetic abnormalities that cause-related diseases is the basis for revealing their pathogenic mechanisms. The genetic diseases can be classified into chromosomal diseases, somatic cell genetic diseases and Mendelian diseases also called monogenetic diseases including autosomal dominant, recessive genetic diseases and sex chromosomal genetic diseases, mitochondrial genetic diseases and complex diseases.

15.4.1 The genes in autosomes can be mutated and lead to the genetic diseases

Genetic diseases caused by gene mutation in autosomes show the dominant or recessive phenotypes. Different pathogenesis and genetic pattern are involved in dominant and recessive genetic diseases.

(1) Autosomal dominant genetic diseases are caused by the mutation of one allele in autosomes with heterozygote

Diseased individual has a normal allele and a mutant allele called the heterozygote of the gene locus. The offspring of the patient has a 50% possibility to attach the disease with the same tendency between girls and sons. It deserves to mention that genetic heterogeneity means no disease in some family members with a mutant allele, and penetrance is used to describe the phenomena that how many percentages of people carried mutant allele show the disease phenotype. The phenotype can be influenced by sex or only shows up in single-sex, for example, alopecia areata occurs only in males.

Huntington's disease (HD), as a serious genetic disease, mainly occurs in Europe, and the age of onset is concentrated around 30–50 years old. HD is characterized by progressive loss of behavioral control and the appearance of mental illness such as dementia and affective disorders. As a dominant genetic disease, the clinical phenotype and progression of mutant heterozygotes and mutant homozygotes are basically the same. The patients of 95% with genetic abnormalities are directly inherited from the deceased parents. The HD gene is located in the short arm of chromosome 4. HD protein itself is related to the production of vesicles during cell secretion. This protein is critical for the normal production of brain-derived nerve growth factors. Sequencing analysis revealed excessive CAG repeats in the coding region of the gene. Normal individuals have a repeat count of about 10–26, while individuals with 27–35 repeats have no diseased phenotype, but have the potential to pass these highly repeated fragments to their offspring. Individuals with more than 36 repetitions can exhibit the disease phenotype, and the increase in the number of repetitions is related to the advancement of the age of onset. CAG repeats can lead to excessive accumulation of glutamic acid residues at the end of the amino acid, which ultimately leads to loss of protein function.

(2) Autosomal recessive diseases are caused by mutation of both alleles in the same gene locus

The patients have a mutation of both alleles called a homozygote. Parents of the patient do not show the disease phenotype but carry a mutant allele. The patients' siblings have a 25% possibility to suffer from the disease without sex difference. The consanguineous marriage increases the disease frequency because of the enhancement of autozygosity meaning the pair of alleles derived from the common ancestor.

Cystic fibrosis (CF) is a common single-gene genetic disease in North America. About 85% of CF patients show pancreatic insufficiency, leading to chronic malnutrition. The most serious problem of this disease is the obstruction of the lungs by mucus, which makes individuals highly susceptible to Staphylococcus aureus and Pseudomonas aeruginosa infections, which in turn leads to repeated chronic obstruction and infection of the lungs and eventually damages them, this is often the cause of death for the vast majority of patients. The CF-related gene is located on the long arm of chromosome 7, which is very large, spans 250 kb and contains 27 exons. The protein product of this gene is called cystic fibrosis transmembrane transduction regulator (CFTR). CFTR protein forms cAMP-regulated chloride channels on the surface of epithelial cell membranes and also participates in the transport of sodium ions. Defects in ion transport can cause airway dehydration and mucus obstruction. Sequencing results revealed that there were more than 1000 different mutation types in the CF gene locus. The patients of 70% had a deletion of a three-base fragment encoding phenylalanine at position 508 of the CFTR protein.

15.4.2 Diseases by genetic abnormalities in sex chromosomes and mitochondria are related to gender

The genetic abnormality in sex chromosomes and mitochondrial DNA results in diseases that can be very severe and present the sex-dependent or maternal inheritance.

(1) Sex chromosome abnormalities lead to the diseases dependent on gender

Sex chromosome inheritance also shows dominant or recessive phenotypes. The patients with X chromosome dominant inheritance have the same possibility for both males and females to attach the diseases. And in the males, the disease severity is similar each other, but females show very different severity owing to the inactivation of X chromosomes. Another feature about this kind of disease is similar to the autosomal dominant diseases in many aspects. But it can be distinguished by the fact that the disease never transmits from father to son rather than to daughter. In the X chromosome recessive diseases, the males have more chance to suffer from the disease and also get more severe disease in terms of the hemizygote in their genomes. For women who have a pair of X chromosomes, the disease can occur only if a mutation takes place in both alleles of the same locus.

Duchenne muscular dystrophy (DMD) is the most serious and common muscular dystrophy, which is very common in various ethnic groups and affects about 1/3500 men. Symptoms can appear before the patient is 5 years old, the patient's calf has pseudo-hyperplasia, and the muscles are infiltrated with fat and connective tissue until the skeletal muscles are completely degraded. Both myocardial and respiratory muscles of patients are damaged, eventually leading to their death from heart or respiratory failure.

The DMD gene is located in the short arm of the X chromosome and is recessive. Female DMD mutants are usually disease-free phenotypes, with only 8%-10% showing a certain degree of muscle weakness. Dystrophin is involved in maintaining the integrity of the muscle cytoskeleton. The gene covers a 2.5Mb sequence, making it the longest known gene in humans, containing 79 exons, an mRNA transcript up to 14kb, and a mature protein consisting of 3585 amino acids. The amazing length of the DMD gene makes mutations unavoidable. One mutation site can exist for every 10 000 sites.

(2) Mitochondrial DNA mutations can lead to diseases

A distinctive feature of the mitochondrial genome is that it is passed through egg cells during the formation of fertilized eggs, that is, it shows maternal inheritance. Mitochondrial DNA (mtDNA) has high spontaneous mutations and poor repair systems. Many genetic diseases are related to abnormal mitochondrial genes. The most famous of these is Leber hereditary optic neuropathy (LHON) caused by missense mutations in protein-encoded mitochondrial DNA genes. Due to the death of the optic nerve in this disease, patients can quickly lose vision. The disease usually occurs after the age of 20 and is irreversible. Until now, no male patients with the disease have been able to pass the disease to their offspring, all of which are transmitted vertically through women. Current research suggests that the disease is mainly caused by mutations in the mitochondrial DNA at sites 11 778, 14 484, or 3460.

15.4.3 Common and complex diseases are the results of the interaction between multiple genes and environmental factors

In many common diseases, the molecular events that may be involved in the occurrence and development of the disease are extremely complicated, and various environmental factors have a certain relationship with the occurrence of the disease. The genetic features include that multigene are involved in the disease occurrence and each gene plays minor effect, and inheritance pattern does not follow Mendelian law, but prevailing in the patients' family show a significantly high compared with common population. For example, the offspring of the hypertension patients manifest a higher opportunity to get hypertension. Most common diseases such as cardiovascular disease, degenerative diseases, cancers and autoimmune diseases belong to the complicated diseases with complicated interaction between a number of genes and diverse environmental factors.

Chronic inflammatory immune disease represented by rheumatoid arthritis (RA) is a complex disease. Certain viral and bacterial infections, such as Epstein-Barr virus, parvovirus B19, influenza virus, and Mycobacterium tuberculosis, may serve as the initiating factor, triggering immune responses in individuals carrying susceptible genes, which can lead to the onset of rheumatoid arthritis. Smoking, cold, trauma, and mental irritation may also induce rheumatoid arthritis. Even focusing only on the genetic factors involved in RA can find that too many abnormalities are related to the occurrence and development of the disease. According to the results of the GWAS study in August 2012, many gene loci on multiple chromosomes are associated with RA pathogenicity. Among them, there are as many as 40 RA-related genes with p values less than 5.3×10^{-8}. See Table 15.1.

Table 15.1 **Rheumatoid arthritis related gene loci and their distribution on chromosomes**

Gene name	Chromosome number
PADI4	1
PUS10, REL, TMEM17, SPRED2, AFF3	2
RBPJ, ANXA3	4
ANKRD5, IL6ST, C5orf30, SLCO6A1, PAM, IL3, IRF1, IL13	5
CD83, HLA, TNFAIP3, OLIG3, IL22RA2, RNASET2, FGFR10P, CCR6	6
IRF5, TNPO3	7
BLK, C8orf13	8
CCL21, TRAF1, C5	9
IL2RA	10
ARID5B, PDE2A, ARAP1	11
PLD4	14
HNF4A	20
UBASH3A, AIRE, PFKL	21

15.5 Principles of identification and positional cloning of disease genes

According to the role of genetic variation in the occurrence and development of diseases, disease-related genes can be divided into disease-causing genes and disease-susceptible genes. If an abnormality in the structure or expression of the gene is the direct cause of the disease, the gene is a disease-causing gene. Such diseases are mainly single-gene diseases, that is, classic hereditary diseases, which have been less influenced by environmental factors. In complex diseases, such as tumors, cardiovascular diseases, metabolic diseases, autoimmune diseases, etc., environmental factors and genetic factors play a certain role and manifest as the addition of the "minor effect" of two or more genes with interaction each other. Therefore, this kind of disease is also called polygenic disease, and since a single gene mutation only increases the susceptibility to diseases, the genes can be called a disease susceptibility gene. We can refer to genes that affect the development of diseases in general terms as disease-related genes, or even directly as disease genes. It is important to pay attention not to be misled by gene names such as cystic fibrosis genes, diabetes genes, and so on. Many human genes were first discovered by studying the disease caused by mutations in the gene, so they are named after the disease. In fact, when these genes are normal, they all have important biological functions.

Identification of disease-related genes can not only understand the etiology and pathogenesis of diseases in detail, develop new diagnostic and intervention technologies, but also help to understand the function of genes, so it has always been the focus of biomedical research. Identification of new genes or cloning of new genes is also of great commercial value, and major biopharmaceutical companies have joined the battle for the gene discovery. The completion of the Human Genome Project provides a favorable opportunity for the discovery of disease-related genes. One of the main tasks of the biomedical field in the post-genomic era is to interpret the function of genes, and the identification of disease-related genes and the determination of gene functions are inextricably linked and can promote each other.

Determining disease-related genes is a difficult and complex systemic project that is time-consuming and expensive. The final identification of some disease genes takes decades. Although the completion of the Human Genome Project has provided many conveniences for the identification of disease-related genes, it is still not easy to clearly analyze the relationship between diseases and certain genes. Mastering the principles of identification and cloning of disease-related genes will undoubtedly play a role in doing more with less, helping to identify disease-related genes in an efficient and orderly manner. First of all, it is the key to determine the disease phenotype and the substantial relationship between genes. Second, it is a means to identify cloned disease-related genes by multiple methods and methods. Finally, identify candidate genes and clarify the relationship between changes in gene sequence and disease phenotype. The nature of the genetic disease is the core of identifying and cloning disease-related genes.

15.5.1 The key to identify and clone disease-related genes is to determine the disease phenotype and the substantial relationship between genes and phenotypes

As a genetic trait, firstly the disease must ensure its specificity and homogeneity. Proper disease diagnosis is very important. In some complex diseases, it is necessary to further classify the disease phenotype to ensure the homogeneity of the disease to reduce clinical heterogeneity. Secondly, it is necessary to determine the genetic factors of the disease, that is, to determine whether the genetic factors play a role in the pathogenesis of the disease and the degree (heritability) of the disease through pedigree analysis, twin analysis, adoption analysis, and sibling prevailing rate analysis. In diseases where genetic factors are less effective, the chances of success in identifying and eventually cloning disease-related genes are low. Once the important role of genetic factors in the disease is determined, the genes that determine the phenotype of the disease can be determined, and the location of the gene in the genome as well as the connection between the locus and other loci in the genome can be determined.

15.5.2 Identification and cloning of disease-related genes requires a comprehensive strategy of multi-discipline and multi-approach

The identification of disease-related genes is an arduous system project which requires the close cooperation of many disciplines and adopts different strategies for different diseases. As shown in Figure 15.1, these different strategies and methods complement each other in order to achieve the ultimate goal of cloning disease genes. First of all, through the linkage analysis of different disease families, the disease gene can be roughly located on a certain chromosome. These loci are called the disease locus. The exact gene is not known at this time. With the understanding of the pathogenesis of some diseases, we can further clarify the biochemical and cellular biological characteristics of disease-related proteins and take this as a starting point to find the abnormality of gene structure, determine the mutation of gene DNA base sequence and the molecular mechanism that

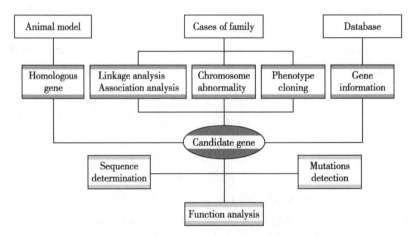

Figure 15.1 **Schematic diagram of cloning strategy for identification of disease-related genes.**

leads to abnormal protein structure or expression, and finally clone the gene of the disease. The animal model of the disease is very helpful for the cloning of disease-related genes. The study of different animal models of diseases can identify the genes that cause abnormal phenotypes in laboratory animals, and then identify the role of human homologous genes in diseases. With the help of the information about undetermined genes in the bioinformatics database, it can also greatly promote the efficiency of disease-related gene cloning.

15.5.3 The determination of candidate genes is the intersection of a variety of methods for cloning disease-related genes

There are many ways to finally identify disease-related genes, and these methods will eventually converge on the candidate genes. Once the candidate gene has been identified, mutations in the gene can be screened in patients. Candidate genes can be identified independent of their position on the chromosome, but the common strategy is to first identify the candidate chromosome region and then identify the candidate genes in this region. The completion of the Human Genome Project provides information on all human genes. Although it is still a time-consuming task to find gene mutations from many candidate genes, it is relatively easy to identify candidate genes. This is due to the reduction of location information from 20, 000 genes to 10–30 genes in the candidate region. It is not easy to predict possible candidate genes in candidate regions, and there is still a lack of such ability. What is needed is a large number of repetitions, excluded one by one, and finally, determine the mutation of the candidate gene and the link between the mutation and the disease.

15.6 Strategies and methods for identification and cloning of disease-related genes

The strategies and methods for identifying and cloning disease-related genes mainly include the cloning strategy of disease-related genes independent of chromosome mapping, positional cloning, whole-genome association analysis, and whole-exon sequencing of genes for common diseases. and bioinformatics methods in terms of the database stored a wealth of disease-related gene information.

15.6.1 The identification and cloning of disease-related genes can adopt a strategy that does not depend on chromosome location

Cloning strategies of disease-related genes independent of chromosome location include functional cloning, phenotypic cloning, and identification and cloning of disease genes using site-independent DNA sequence information and animal models.

(1) Disease genes are cloned from the function and structure of known proteins

On the basis of mastering or partially understanding the proteins of gene functional products, the method of identifying protein-coding genes is called functional cloning. This is relative to the positional cloning of genes by using gene position. This method adopts the research route from protein to DNA, aiming at some diseases that have a certain understanding of the functional proteins that affect the disease, such as hemoglobinopathy, phenylketonuria, and other molecular diseases caused by birth defects. This method can be used to locate and clone disease genes.

(2) Identification and cloning of disease-related genes are dependent on amino acid sequence information of proteins

If disease-related proteins are rich in expression *in vivo* and can be separated and purified to obtain sufficient proteins with a certain purity, mass spectrometry or chemical methods can be used for amino acid sequence analysis to obtain all or part of the amino acid sequence information. On this basis, oligonucleotide probes are designed to screen the cDNA library, and the target genes can be screened. When using this strategy, the degeneracy of codons must be taken into account, that is, except methionine and tryptophan, all amino acids have two or more codons. When designing probes, we should try to avoid the areas with degenerate codons, but in practice, it is often difficult to do so. For this reason, we can design a set of oligonucleotide probes that may contain all degenerate codon information, and use this mixed probe to screen the cDNA library and "fish" the target gene clone. In addition to cDNA library screening technology, some degenerate mixed oligonucleotides can be used as PCR primers and a variety of PCR primer combinations can be used to obtain PCR products of candidate genes.

The above methods have been successfully applied to the gene cloning of sickle cell anemia. First of all, immune electrophoresis and other methods have shown globin abnormality in patients with sickle cell anemia. After obtaining part of the amino acid sequence, degenerate oligonucleotide probes were designed to screen the cDNA library of nucleated red blood cell lines. The cDNA, of the α-globin gene, was compared with that of normal people, and the variation of the α-globin gene was found. Furthermore, the homologous complementary relationship between the cDNA probe and chromosome DNA sequence was found, and the human α-globin gene was located on chromosome 16. On this basis, the concept of molecular disease was put forward.

(3) Identification of disease genes is performed by using protein-specific antibodies

The content of some disease-related proteins in the body is so low that it is difficult to purify proteins with sufficient purity for amino acid sequencing. However, a small amount of low-purity protein can still be used to immunize animals to obtain specific antibodies to identify genes. On the one hand, the obtained antibody can be used to directly bind to the new peptide chain in the process of translation, at this time, the mRNA molecule binding to the ribosome will be obtained, and finally, unknown genes can be cloned; in addition, the specific antibody can also be used to screen the expressible cDNA library, screen the positive clones of expressed proteins that can react with the antibody, and then obtain candidate genes.

(4) Functional cloning is still a common strategy for gene cloning of single-gene diseases

Its disadvantage is that it is very difficult to identify and purify specific functional proteins, and microexpressed gene products are difficult to obtain in research, so they can hardly be used for gene isolation of polygenic diseases.

- The discovery of disease-related genes is based on the phenotypic differences of the disease. Phenotypic cloning is a new strategy in the cloning of disease-related genes. The principle of this strategy is to isolate and identify disease-related genes based on the understanding of the characteristic relationship between disease phenotype and gene structure or gene expression.
- Depending on the relationship between changes in DNA or mRNA and disease phenotype, there are several strategies.

The first strategy is to compare the difference of genomic DNA between patients and normal individuals based on the phenotype of the disease, and directly clone the mutated DNA fragments without the need for the chromosome location of the gene or other information of the gene product. For example, in some hereditary neurological diseases, the copy number of the triple repeat sequence contained in the patient's genome can change and expand with the transmission from generation to generation, which is called dynamic mutation of the gene. At this time, the use of genomic mismatch screening, representational difference analysis (RDA) and other techniques can detect whether the patient's DNA has an increase in the number of copies of the triple repeat, so as to determine the cause of the disease.

The second strategy is to target known genes. If it is highly suspected that a disease is caused by a particularly known gene, it can be determined whether the gene is a disease-related gene by comparing the difference in gene expression between patients and normal controls. The commonly used analytical methods include Northern blotting, RNA enzyme protection test, RT-PCR and real-time quantitative RT-PCR, etc.

The third strategy is aimed at unknown genes, and disease-related genes can be cloned by comparing the differences in the expression types and contents of all mRNA between disease and normal tissues. This difference may result from changes in gene structure or from changes in the mechanism of expression regulation. The commonly used techniques include mRNA differential display (mRNA-DD), suppression subtractive hybridization (SSH), serial gene expression analysis (SAGE), cDNA microarray and gene identification integrated. Only RDA and mRNA-DD technologies are introduced here.

- RDA is a PCR technology based on differential hybridization of nucleic acids. RDA is a technology that detects and captures cDNA differential fragments (representative fragments) of normal and diseased tissues by amplifying them. The basic principle is: first, sufficient DNA or cDNA fragments were obtained from the disease and normal tissues by PCR method; then differential hybridization was performed, and the hybridization was followed by a second PCR reaction with different primers; in the second PCR reaction, only the DNA fragments with different structure or expression in the two samples could be amplified (Figure 15.2).

The basic steps are as follows: (a) DNA fragment preparation: extract normal human genomic DNA (detection DNA) and patient genomic DNA (driving DNA) respectively, digest DNA with a restriction endonuclease, and obtain fragments with a length of 150–1000 bp; (b) Obtaining amplicons: add adapters to all DNA fragments of the two groups, and use the complementary sequences of the adapters as primers to perform the first step of PCR amplification; (c) Replace the adapter: cut off the adapters of all amplicons, and only add new adapters to the detection amplicons; (d) Screening of amplification products: mix the detection amplicon and driving amplicon according to the ratio of 1:100, and carry out liquid phase hybridization. Taking a small number of hybrid reactants as templates, using new splices as primers and PCR amplification for the second time, we can screen out the different fragments between the two groups of DNA samples.

The detection of DNA and driver DNA fragments in the second PCR reaction is based on

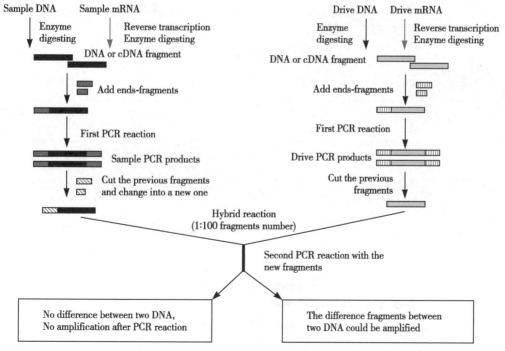

Figure 15.2 **Schematic diagram of RDA technology principle and basic process mRNA-DD.**

whether there is a difference between the two. There are two main situations: (a) The same DNA fragments between the two groups will not be amplified a lot. This is because in the hybridization reaction, the number of driving DNA fragments is much larger than the detection DNA, and the detection DNA will be preferentially bound so that there is almost no chance for the detection DNA molecule to form a homo-renaturation double strand. Therefore, there will be no amplification products in the second PCR reaction process using the new adapter; (b) The differential fragments of the two can be amplified. If a certain fragment in the detection DNA is deleted in the driving DNA, or the complementary binding ability is lost due to mutation, there is no competition from the homologous fragments in the driving DNA in the hybridization reaction. The detection DNA itself can be renatured, and since the renatured double-stranded DNA has new adapters at both ends, a large amount of PCR can be achieved. This fragment is a candidate disease-related DNA sequence. Although the non-differential fragments in the reaction may still be amplified, the amount of products is small and can be excluded.

RDA can also be used for the cloning of mRNA differentially expressed genes, only the mRNA needs to be reverse transcribed into cDNA fragments. RDA technology has a strong ability to distinguish normal and abnormal DNA fragments, high enrichment efficiency, and low requirements on starting materials. Using RDA technology, people have discovered several new disease-related genes.

- mRNA-DD is a combination of RT-PCR technology and polyacrylamide gel electrophoresis technology. mRNA-DD is also called differential display reverse transcription PCR (mRNA differential display reverse transcription PCR, DDRT-PCR). This method utilizes the mRNA that can amplify all mammals and uses a combination of several 5′ random primers and several 3′ anchor primers to amplify the cDNA of the corresponding tissues of normal and diseased individuals by PCR. The amplified products were separated by polyacrylamide gel electrophoresis, and the differences in the products between the two groups were compared (Figure 15.3). According to calculations, theoretically, the combined primers designed by this method can match the poly(A) tail of all mRNAs, so for cDNA samples of the same type and content, the types and distribution

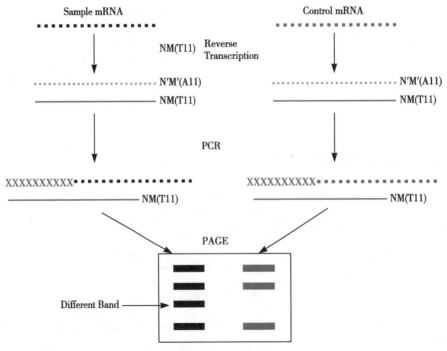

Figure 15.3 **Principle and the basic process of mRNA-DD technology.**

of PCR products should be exactly the same. If cDNA fragments of different lengths are amplified in normal and patient cDNA samples, the cDNA they represent may be related to the disease state. The advantage of this method is that it requires less mRNA, is faster, and can simultaneously show differences in multiple biological traits, and can simultaneously obtain high- and low-expressing genes. This method also has many serious defects, such as the false positive rate is as high as 70%, the obtained fragments are too short, etc. It is difficult to directly judge its function and significance. In spite of the above-mentioned shortcomings, because of its simple steps, a large amount of information can be obtained, and the method is still applied more in practical work.

15.6.2 Using animal models to identify cloned disease-related genes

Some human diseases already have corresponding animal models. If a mutant gene of a certain phenotype of an animal is located in a certain part of a chromosome, a gene with a phenotype similar to human disease is likely to exist in a homologous part of a human chromosome. In addition, when disease genes have been identified on animal models, fluorescence *in situ* hybridization can also be used to locate and isolate human homologous genes. A successful example is the cloning of the obesity-related leptin gene. Using a mutant obese inbred mouse, the leptin gene located on mouse chromosome 6 was isolated by localized cloning. According to the mouse leptin gene flanking marker, the human leptin gene was mapped to the 7q31 region of the human chromosome. Mouse and human leptin genes share 84% homology. Leptin, a secreted protein-encoding 167 amino acid residues, mainly controls food intake and promotes energy consumption. Both obese mice and some patients with hereditary obesity have a defect in this gene, resulting in loss of gene function.

15.6.3 Positional cloning is a classical method for identifying disease

Based on the general position of the disease gene on the chromosome, identifying and cloning the disease-related genes is called positional cloning. The starting point of location cloning is gene mapping, that is, determining the position of disease-related genes on the chromosome. Candidate genes are then screened by contigs of related genomic regions. Finally, the differences between these genes in patients and normal people were compared to determine the relationship between genes and disease. The candidate clones after the Human Genome Project are to locate disease-related sites in a certain chromosomal region. Then, based on the gene, EST, or the known genes of the homologous region corresponding to the model organism, the mutation screening is performed directly. Through multiple iterations, the disease-related genes were finally identified.

Gene location is the basis of gene isolation and cloning. The purpose is to determine the position of genes on the chromosome and the linear arrangement order and distance of genes on the chromosome. Gene mapping can be performed at several levels such as family analysis, cell, chromosome, and molecular levels. Due to the different methods used, multiple methods can be derived, and different methods can be used in combination to complement each other.

(1) Somatic cell hybridization locates genes through the screening of fused cells

Somatic cell hybridization, also called cell fusion, is the fusion of two cells of different origins into a new cell. Most somatic cell hybridizations use human cells to hybridize with mouse, rat or hamster somatic cells. This newly created fusion cell is called a hybrid cell and contains different chromosomes from both parents. An important feature of hybrid cells is that rodent chromosomes are retained and human chromosomes are gradually lost during reproduction and passage. Only one or a few are left in the end, and the reason is unknown. Miller et al. used somatic hybridization, combining the characteristics of hybrid cells, to prove that thymidine kinase (TK) is required for the survival of hybrid cells. Hybrid cells containing human chromosome 17 all survived due to TK activity in a special medium, but otherwise died, thereby inferring that the TK gene is located on chromosome 17. Many people's genes were mapped using this method. Tumor suppressor genes have also been discovered using somatic cell hybridization techniques.

(2) *In Situ* Hybridization is a common method for locating genes at the cell level

In Situ Hybridization is the application of nucleic acid molecular hybridization technology in gene mapping, and it is also a method of direct gene mapping. The main steps are to obtain metaphase cells in tissue culture, degenerate chromosomal DNA, and hybridize with labeled complementary DNA probes. After development, the gene can be located on a certain chromosome and a certain segment of the chromosome. If the probe is labeled with a fluorescent dye, it is fluorescence *in situ* hybridization (FISH). The cDNAs of α and β globin genes were used as probes for the first time in 1978 to hybridize with various human/mouse hybrid cells to locate human α and β globin genes on chromosomes 16 and 11 respectively. This chromosome *in situ* hybridization technique is particularly suitable for those non-transcribed repeats, which are difficult to locate by other methods. For example, *in situ* hybridization technology is used to locate satellite DNA near the centromeres and telomeres of chromosomes.

(3) Chromosomal abnormalities can sometimes provide an alternative method for disease gene location

From the perspective of gene location cloning, for any disease that is known to be directly related to chromosomal abnormalities, chromosomal abnormalities become a piece of excellent location information for disease gene location. Chromosomal abnormalities can sometimes replace linkage analysis to locate disease genes. In some sporadic and serious dominant genetic diseases, chromosome variation analysis is the only way to obtain candidate genes. Sometimes, the correct position of a gene can be obtained directly without linkage analysis, such as balanced translocation and inversion of the chromosome. The location of disease genes such as polycystic kidney disease, mega-intestine disease and Duchenne/Becker muscular dystrophy (DMD/BMD) depends largely on the abnormal karyotype of chromosomes.

If the abnormal chromosome is observed by cytology and the abnormal expression of a gene occurs at the same time, the gene can be located in the abnormal region of this chromosome. For example, in the analysis of a family with chromosome 6 arm inversion, every family with this inversion also has the expression of an HLA allele; and in the family without this inversion, there is no expression of this allele, so the HLA gene is located in the distal region of the short arm of chromosome 6.

In chromosome aneuploidy analysis, gene location can be done by the gene dose method. The activity of SOD-1 was 1.5 times higher in the patients with congenital foolishness (karyotype 47, + 21) than that in the normal people, so the gene of SOD-1 was located on chromosome 21. However, not all genes copy numbers have a significant dose effect.

(4) Linkage analysis is a common method to locate unknown genes of diseases

Linkage analysis of gene location is based on the principle that genes are arranged in a straight line on the chromosome and different genes are linked to each other to form a linkage group. In other words, it uses the characteristics of linkage between the located gene and another gene or genetic marker on the same chromosome to locate gene. If the undetermined gene and the marker gene are linked, it can be inferred that the undetermined gene and the marker gene are on the same chromosome. And according to the degree of linkage with multiple marker genes (measured by the recombination rate between them), the sequence of the undetermined gene on the chromosome and the genetic distance between the undetermined gene and the marker gene can be determined (expressed by cM). For example, it is known that the blood- group gene XS is located on the X chromosome, and the common ichthyosis and ocular albinism genes are linked with them. Therefore, it is determined that these genes are also located on the X chromosome. The relative distance between these genes can be determined by calculating the recombination rate of the patient's offspring.

15.6.4 The process of disease-related gene location and cloning includes three steps

Disease-related gene location and cloning is the main method to identify genetic disease genes, which plays an irreplaceable role in the early work of disease gene identification and has achieved great success. With the completion of the human genome project, it is easier to clone disease genes by location. The main process includes the following three steps.

(1) Reducing the candidate regions on the chromosome as much as possible

The difficulty of locating and cloning disease genes depends on the width of the candidate regions. Therefore, it is necessary to reduce the candidate regions of disease-related genes on chromosomes as much as possible. In the genetic mapping of single gene disease genes, it is necessary to select more genetic markers, find out the nearest marker, increase more families, establish the haploid type of all individuals, etc., so as to increase the chance of finding recombination, combine to find more linkage disequilibrium and accurate candidate regions of disease-related genes.

(2) Constructing gene list of target region

Due to the completion of human genome project and the establishment of a physical map at various DNA molecular levels, it has become easier to clone disease-related genes. Now, there is no need to establish DNA overlap groups. The database of human genome, such as the genome reader ensembleor the Santa Cruz reader, can directly display the confirmed or possible genes of candidate regions, but it can not rely on this information completely. It is necessary to carefully check whether the overlapping assembly is correct. Of course, more gene information of candidate regions should be obtained by combining the result of encoding program, noncoding sequences, selective transcripts and other expression profiles.

(3) Selecting gene and detecting mutation in candidate region as priority

In order to identify mutations, DNA sequencing should be carried out for unrelated patients. It can be used to detect all exons in the candidate region, as well as those with priority genes, which is depending on the research strategy, human and financial investment. The gene can be considered as the priority gene according to the following conditions: (a) Appropriate expression: the expression mode of a good candidate gene should be consistent with the disease phenotype, and the gene may not be expressed in the diseased tissue, but at least before or when the disease occurs, the diseased tissue expresses the gene, such as the gene of neural tube defect should be before neural tube atresia, that is, human embryo 3–4 weeks of development. (b) Appropriate function: the gene function of a candidate region, if known, is easy to make a decision. Such as the relationship between fibrillin and Marfan syndrome. The analysis of a new gene sequence suggests that there is a certain function, such as transmembrane motif or arginine kinase motif, which can be related to the pathogenesis of the disease and make a judgment. (c) Homology and functional relationship: if a gene in the candidate region is homologous with a known gene, whether it is an indirect homology with human or a direct homology with other species (ortholog), and it is also known that the homologous gene mutation causes a similar phenotype, the gene may be a disease gene. Candidate genes can also be determined based on close functional relationships, such as the relationship between receptors and ligands, the composition of the same metabolic or developmental pathway, etc. In recent years, more and more homologous genes have been identified, which greatly promoted the identification and cloning of human pathogenic genes.

Cloning of DMD gene is a successful example of location-based cloning. The first disease-related gene identified by location-based cloning strategy is

X-linked chronic granulomatosis gene. Moreover, the successful cloning of Duchenne muscular dystrophy (DMD) gene shows the advantage of gene location. This work is mainly divided into two stages. First of all, according to the translocation of X chromosome and No. 21 autosomal, and the small loss of Xp21.2 in male children with three other X-linked recessive genetic diseases, the DMD gene was located in Xp21 by RFLP linkage analysis. Then, two different fragments of the gene were cloned and named XJ probe and pert87 probe respectively. According to the comparison of the two fragments, DMD gene is about 2300kb, accounting for more than 1% of X chromosome. This gene encodes dystrophin, which affects the structure and contraction function of striated muscle and myocardium.

15.6.5 Identifying genes for common diseases require genome-wide association analysis and whole exon sequencing

Although we have used the method such as affected sib pair (ASP) to identify the susceptible genes of complex diseases, we have also achieved some successful examples, but generally speaking, it is not ideal. Since 2005, genome-wide association study (GWAS), based on the theory of linkage disequilibrium, has played an important role in gene location and cloning of complex diseases. GWAS method is a research method that can scan the whole genome to observe the relationship between gene and disease phenotype without hypothesis-driven. In the specific operation, we usually collect thousands of DNA samples of patients and controls, use a high-throughput chip to conduct SNP gene typing, and further determine the relationship between molecular SNP locus and disease phenotype through statistical analysis. This method has successfully identified multiple gene loci of common frequently occurring diseases, which not only simplifies the identification process of genes related to common diseases but also provides valuable information for the study of pathogenesis and intervention targets of diseases. However, the technology has high requirements for the economic strength, cooperation, bioinformatics level and the ability to identify large false-positive data of the research team, and only involves the variation of common alleles.

The whole exon sequencing technology can enrich the whole exon region of the genome to conduct high-throughput sequencing. It can selectively detect the protein coding sequence and realize location cloning. Moreover, it has high sensitivity to common and rare gene variations. Only sequencing about 1% of the genome segments can cover most disease-related gene variations of the exon. Therefore, its high-performance price ratio makes it highly respected in the research of complex disease susceptibility genes.

15.6.6 Bioinformatics database stores abundant disease-related gene information

With the completion of human genome project and multiple models of biogenomic sequencing, the development of bioinformatics, the development and application of computer software and the popularization of the Internet, People compare the homology of the obtained sequence with the nucleic acid sequence and protein sequence in the database, or conduct sequence comparison analysis and splicing among different species in the database, and then predict the new full-length gene, finally, it can be confirmed by experiments that the gene can be cloned from the tissue cells, which is called "*in silico* cloning".

Most human new gene cloning started from the homologous EST analysis. In human EST database, The methods of identifying and splicing new human genes which is highly homologous with known genes by homology comparison include (a) searching and analyzing EST database with known gene cDNA sequence (blast), finding out EST with high homology with known gene cDNA sequence; (b) constructing overlapping group with fragment assembly software of squab, and finding out the identical sequence of each overlapping group; (c) compare the relationship between the identical sequence of each overlapping group and the known gene; (d) the protein sequence of the coding region is compared with the protein functional domain of the known gene, so as to speculate the function of the new gene; (e) the sequence-tagged site (STS) database is analyzed with the new gene sequence or EST sequence, if an e If st (nonrepetitive sequence) overlaps with a certain STS, then the location of STS determines the location of the new gene. Making full use of network resources, e-cloning can greatly improve the speed and efficiency of cloning new genes. Because of the imperfection of the database, the existence of error information and the defect of analysis software, electronic cloning is often difficult to clone genes, but an electronic assisted cloning.

For the identification and cloning of disease-related genes, two strategies can be adopted, i.e. non-chromosomal function identification and location cloning. The former includes functional cloning, phenotypic cloning, location independent DNA sequence information and animal models to identify and clone disease genes; the latter is to map the gene location, determine the location of disease-related genes on the chromosome, and then find and clone genes from the region, including somatic hybridization, *in situ* hybridization, linkage analysis and chromosome abnormal location to clone disease-related genes. The cloning of the Duchenne muscular dystrophy gene is a successful example of the application of a targeted cloning strategy.

Key Concepts

- Abnormalities of gene or genome show different types, such as the increase or decrease of chromosome number, point mutation, nucleotide insertion and deletion variation, gene rearrangement, abnormal expression of genes.
- The genetic abnormalities can be derived from various reasons, including random mutation under natural conditions, pathogenic mutation induced by physical, chemical, and biological factors, natural selection, genetic drift, homologous recombination, and integration of foreign invaded genes.
- The effect of genetic abnormality on gene function can be divided into two molecular outcomes: gain of function (GOF) or loss of function of gene (LOF). Various mechanisms are involved in the LOF, which includes dose effect of the gene and down-regulation of gene expressions, such as haplotype insufficiency, dominant-negative effect, loss of heterozygosity, transcriptional RNA position effects and epigenetic modifications affect, gene variation of transcription factors regulates gene expression, the stability of mRNA. Several genetic abnormalities through various molecular mechanisms lead to the GOF, including transcriptional enhancement of the transcription regulation sequence, enhancers regulating gene overexpression, an increase of copy number in the genome, mutation of growth factor receptor genes, SNP, and acquired RNA accumulation.
- Loss or gain of gene function caused by genetic abnormalities manifests as various diseases happened at the individual level.
- The principles of identification and cloning of disease-related genes are as follow: Firstly, to determine the disease phenotype and the substantial relationship between genes. Second, it is a means to identify cloned disease-related genes by multiple methods and methods. Finally, identify candidate genes and clarify the relationship between changes in gene sequence and disease phenotype.
- The strategies and methods for identifying and cloning disease-related genes mainly include the cloning strategy of disease-related genes independent of chromosome mapping, positional cloning, whole-genome association analysis, and whole-exon sequencing of genes for common diseases. and bioinformatics methods.

Key Words

abnormalities of gene and genome(基因和基因组异常)
centMorgen, cM(厘摩, 1% 的重组率为 1 个厘摩)
copy number variance, CNV(拷贝数变异)
disease related genes(疾病相关基因)
DNA markers(DNA 标记)
gain of function(功能获得)

genome wide association study, GWAS(全基因组关联分析)
InDel(插入/缺失突变)
linkage study(连锁分析)
loss of function(功能缺失)
point mutation(点突变)
polymorphism(多态性)
positional cloning(定位克隆)
single nucleotide polymorphisms, SNPs(单核苷酸多态性)
short tandem repeats, STRs(短串联重复)

Questions

1. What is the type and result of point mutations of genes?
2. What factors may induce genetic mutations in the process of tumor development?
3. Please, give some examples how genetic variation affects proteins.
4. Why do we think that common and complex diseases are the results of the interaction between multiple genes and environmental factors?
5. Please, try to describe the principles of Identification and positional cloning of disease genes.

References

[1] Browning SR, Browning BL (2011) Haplotype phasing: existing methods and new developments. *Nat Rev Genet*, 12:703–714.

[2] Boyle EA, Li YI, Pritchard JK (2017) An expanded view of complex traits: from pogenic to omnigenic. *Cell*, 169: 1177–1186.

[3] DiMasi JA, Grabowski HG, Hansen RW (2016) Innovation in the pharmaceutical industry: New estimates of R&D costs. *J Health Econ*,47:20–33.

[4] Hunt KA, Mistry V, Bockett NA, et al. (2013) Negligible impact of rare autoimmune-loclyus coding-region variants on missing heritability. *Nature*,498:232–235.

[5] Krumholz HM, Ross JS, Presler AH, et al. (2007) What have we learned from Vioxx? *Br Med J*,334:120–123.

[6] MacArthur DG, Manolio TA, Dimmock DP, et al. (2014) Guidelines for investigating causality of sequence variants in human disease. *Nature*,508:469–476.

[7] Rober J, Vogelstein JT, Parmigiani G, et al. (2012) The predictive capacity of personal genome sequencing. *Sci Transl Med*,4:133ra58.

[8] Steinberg J, Honti F, Meader S, et al. (2015) Haploinsufficiency predictions without study bias. *Nucleic Acids Res*,43:e101.

[9] Yang J, Benyamin B, McEvoy BP, et al. (2010) Common SNPs explain a large proportion of the heritability for human height. *Nat Genet*,42:565–569.

Shemin Lyu

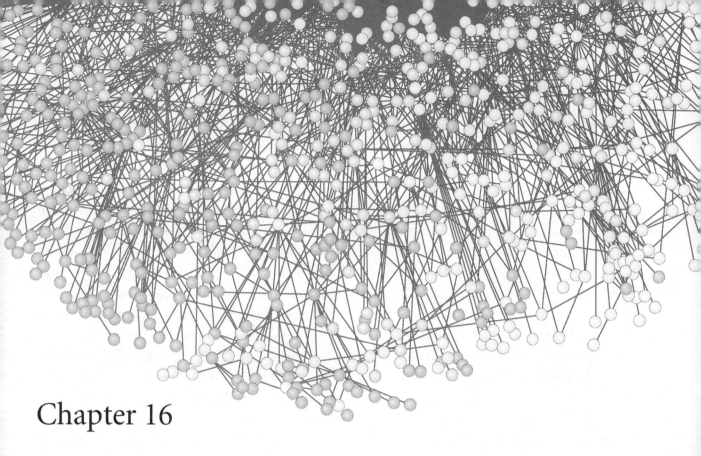

Chapter 16

Gene Diagnosis and Gene Therapy

16.1 Introduction on gene diagnosis

The occurrences of diseases are the results of abnormality of biochemical processes involving the variation of structure and function of bio-substances in human body. Development and incorporation of molecular biology into the field of medicine create a new actualization that doctors can diagnose, treat and prevent diseases at the levels of DNA or RNA, the fundamental substances of life. Any abnormality of genes including the variation of quality and quantity may cause disorders. Besides the abnormality of endogenous genes, human diseases may also be caused by the invasion of foreign organisms such as bacteria, viruses, mycoplasma, etc., which contains the genetic substances different from human genome.

Gene diagnosis is such a process to judge diseases of human body through detecting the abnormalities of DNA or RNA materials. The purpose of gene diagnosis is to find the reasons of diseases much earlier, much faster with high specificity and sensitivity. The techniques and methods used for gene diagnosis can be widely practiced in the fields of clinic medicine, preventive medicine as well as medical jurisprudence.

16.2 Gene diagnose of inheritance disease

Gene diagnosis is a very essential tool for the diagnosis of inheritance diseases because the molecular mechanism of inherited diseases involves the changes of genotypes caused by various mutations.

16.2.1 Hemoglobinopathy

Hemoglobinopathy is a group of hereditary blood diseases caused by abnormal molecular structure of globin in hemoglobin or abnormal synthesis of globin peptide chain, including abnormal hemoglobin disease and thalassemia. The abnormal hemoglobin disease is caused by a change in the gene encoding α- or β-globin chain, resulting in a change in amino acid sequence and resulting in a molecule structural defect in the hemoglobin. A certain globin chain synthesis disorder caused by the deletion or mutation of the globin gene causes hemolytic anemia, called thalassemia. Thalassemia can be divided into two types, α and β thalassemia, depending on the type of peptide chain that is abnormally synthesized. In addition, there are rare δβ and γβ thalassemias.

The clinical laboratory is essential for the diagnosis of patients with hemoglobinopathy. Although there are many methods for detecting hemoglobinopathy, molecular diagnosis plays an important role in the diagnosis of heritage diseases, genetic counseling and prenatal screening. The DNA from peripheral blood leukocytes or amniocytes can be used to diagnose various α- and β-globin chain abnormalities. For known globin chain mutations/deletions, such as Hb S, C, and E, and several thalassemia, multiplexed-amplification refractory mutation system (ARMS), PCR techniques using allele-specific probes after globin gene amplification, allele-specific primers or deletion-dependent amplification with flanking primers are used. Large deletions of the α- and β-globin gene can be identified by gap-PCR directly from dried blood spots. For the rare or unknown mutations, several PCR-based methods can be used, including single-strand conformation polymorphism analysis (SSCP) and sequencing of amplified globin gene DNA. In addition, unknown deletional mutations are also diagnosed using traditional Southern blot hybridization. Microarray approaches, such as biochip technology, are expected to be used for future high-throughput analysis.

16.2.2 Hemophilia

Hemophilia is an X-linked recessive hereditary disorder. Hemophilia can be divided into hemophilia A and hemophilia B, which are caused by a deficiency or qualitative abnormality of the coagulation factor VIII (FVIII) and coagulation factor IX (FIX), respectively. The genes encoding FVIII and FIX are the FVIII gene and the FIX gene, respectively. The FVIII gene is more complex than the FIX gene. There are many types of hemophilia gene mutations, including gene point mutation, inversion, deletion, duplication, insertion and so on. For example, inversion of FVIII gene intron 22 is the most common pathogenic mutation that causes hemophilia A, which can lead to a serious deficiency of FVIII.

The current gene diagnosis for hemophilia is divided into direct and indirect gene diagnosis. Direct gene diagnosis is mainly to detect gene defects that directly lead to disease. Direct gene diagnosis methods for hemophilia detection include Southern blot, long-distance polymerase chain reaction (LD-PCR), inverse polymerase chain reaction (I-PCR), direct sequencing (SSCP), etc. Indirect gene diagnosis is to determine the genetic status of hemophilia genes through the linkage relationship among family members, and to carry out gene diagnosis of DNA polymorphism analysis. These methods mainly include restriction fragment length polymorphisms (RFLP), variable number of tandem repeats (VNTR), short tandem repeats (STR). In clinical practice, direct gene diagnosis is generally preferred for hemophilia A. Indirect gene diagnosis is often used for hemophilia B because of these patients often have a clear family history. Performing gene diagnosis of hemophilia patients can accurately detect carriers of gene mutations in the family of patients with hemorrhagic disease and reduce the birth rate of children with hemophilia.

16.3 Gene diagnose of cancer

The development of cancer involves multiple factors including physicochemical stimulation, viral infection, function dysregulation of oncogenes and antioncogenes, aging, DNA methylation, etc. Development of cancer has the characteristics of multiple genes, multiple steps, multiple stages, so that early gene diagnosis of cancer by identification of abnormality at DNA and RNA levels is particular important.

16.3.1 Oncogene in gene diagnose

Oncogenes are those genes that have the potential to induce malignant transformation of cells, often due to abnormal activation of proto-oncogenes. Proto-oncogenes are "normal" genes in cells that promote cell growth and proliferation. Under normal circumstances, proto-oncogenes do not exhibit carcinogenic activity, and cells can only undergo malignant transformation after mutation or abnormal activation. The main reasons for the activation of proto-oncogenes as oncogenes include DNA rearrangement, gene amplification, point mutations, and so on. At present, the detection of oncogenes mainly depends on methods of gene diagnosis, including Southern blotting, Northern blotting, dot blotting, hybridization *in situ*, direct sequencing, and various PCR techniques and PCR-based RFLP and SSCP. In addition, DNA biochip is also used as a new technology for oncogene detection, which is widely used because of its high efficiency, rapidity, sensitivity and automation.

16.3.2 Gene mutation in gene diagnose

Gene mutations are closely related to the occurrence of tumors. Point mutations, small fragment deletions and insertion of tumor-related genes can cause abnormalities such as synonymous, missense, termination and frameshift of codons, resulting in changes in the protein sequence of gene expression.

(1) BRCA mutation and breast cancer

Breast cancer susceptibility gene (*BRCA*) is an important tumor suppressor gene, including *BRCA1* and *BRCA2*. The *BRCA1/2* gene plays an important role in regulating the cell cycle and participating in DNA damage repair and cell growth. However, if the *BRCA1/2* gene is mutated, its function of inhibiting tumorigenesis is affected. Current researches indicate that a significant proportion of breast cancer patients have *BRCA1/2* genetic mutations, known as hereditary breast cancer. Women with *BRCA1/2* mutations not only have an increased risk of breast cancer, but also have other risks, such as ovarian cancer, fallopian tube cancer, and pancreatic cancer. For the *BRCA* gene, it is difficult to detect *BRCA* gene mutations because the variation is distributed throughout the length of the gene and lacks hotspot mutations. At present, *BRCA* gene diagnosis generally adopts the next generation sequencing (NGS) method, but there is no uniform detection process and standard.

(2) APC and Colon Cancer

The adenomatous polyposis coli (*APC*) gene is another tumor suppressor gene. APC protein is an important part of the Wnt signaling pathway, which plays an important role in cell adhesion, cell cycle regulation, cell proliferation and cytoskeletal stability. If a point mutation and/or frameshift mutation occurs in the *APC* gene, it will result in a change in the encoded APC protein, thereby losing its protein activity and causing the cell to become cancerous. The *APC* gene is deleted or inactivated in most colon cancers, and defects in this gene are directly related to the genetic susceptibility of colon cancer. At present, the detection of *APC* gene germline mutations generally adopts denaturing gradient gel electrophoresis, RFLP, SSCP, and direct sequencing.

16.3.3 Virus gene detection in cancer

Virus is closely related to the occurrence of tumors, and many studies have found that infection of certain viruses can induce tumors. The virus that can cause tumors is called a tumor virus, and the tumor viruses mainly include DNA viruses and RNA viruses. After the DNA tumor virus infects the cells, the viral genome can be integrated into the host DNA. The RNA tumor virus contains viral oncogenes or promoters and enhancers that promote transcription of genes, and is inserted into proto-oncogenes of host cell after reverse transcription, causing proto-oncogene activation or overactivation. At present, there are mainly retroviruses, human papillomavirus (HPV), Epstein-Barr virus (EBV), and hepatitis B virus (HBV), which are closely related to human tumorigenesis.

Among them, HPV is an DNA double-stranded virus. HPV is widely spread in humans and can cause many benign papillomas in human skin and mucous membranes. In addition, a large number of studies have shown that HPV infection rate in cervical cancer cells and tissues is above 90%. At present, more than one hundred HPV subtypes have been found, and HPV subtypes are associated with the degree of malignancy of cervical cancer. HPV-DNA can be detected by dot blotting or *in situ* hybridization, and specific primers with different types can be designed for PCR amplification for HPV subtype detection. In addition, gene chip diagnostics will gradually be used for HPV subtype detection.

16.4 Basic techniques used for gene diagnosis

The basic techniques used for gene diagnosis of diseases are usually to identify the abnormalities genetic substances.

16.4.1 Molecular hybridization of nucleic acids

Molecular hybridization of nucleic acids can be used to identify the existence of abnormality of human genetic materials or foreign genetic materials so to give the evidences for the diagnosis of diseases.

Southern blotting has important values in genetic disease diagnosis. The basic principle of Southern blotting is that two nucleic acid single strands with certain homology can specifically hybridize to form a double strand according to the principle of base complementarity under certain conditions. This hybridization process is highly specific. Northern blotting is mainly used to detect whether a sample contains a gene transcription product (mRNA) and its content. Dot and slot blotting refers to that after the denaturing treatment of RNA or DNA, and then spotting it to a nitrocellulose membrane (or nylon membrane), hybridizing with a labeled probe to determine whether there is hybridization or its hybridization intensity. This method is mainly used for the detection of gene deletion or copy number alteration. Dot and slot blotting does not require electrophoresis and transfer, and the hybridization process is faster than Southern blotting and Northern blotting, which is suitable for simultaneous analysis of multiple samples.

In situ hybridization is the hybridization of nucleic acids in cells or tissues with labeled nucleic acid probes. *In situ* hybridization has been applied to basic research such as gene mapping, transgenic detection, localization of gene expression. Clinical studies are used in cytogenetics, prenatal diagnosis, diagnosis of tumors and infectious diseases, and pathogenic diagnosis of viruses. With the development of nucleic acid probes and the emergence of new labeling methods, we believe that *in situ* hybridization will be more widely used in various disciplines in the future.

16.4.2 Polymerase chain reaction

Under the precondition that an abnormal DNA sequence related to a disease has been identified, polymerase chain reaction (PCR) can be used to confirm the existence of some mutation in the suspicious DNA sequence. PCR can also be used to detect the existence of genetic substances of foreign pathogenic organisms in human body. In addition, PCR technique is a very potent tool used to amplify a segment of DNA or RNA (by RT-PCR) for further analytical research when combining with other techniques such as RT-PCR, PCR-Sequencing, PCR-Cloning, PCR-RFLP.

16.4.3 DNA Sequencing

DNA sequencing is the process of determining the nucleotide order of a given DNA fragment. On the basis of PCR amplification, DNA sequencing can be carried out successfully due to acquirement of enough amounts of DNA fragment. In addition, DNA sequencing is very helpful for the identification of property of variation or mutation in an abnormal and reason-unknown DNA fragment. DNA Sequencing will benefit to establish the connection between information about nucleotides in DNA sequence and a given disease, which will be very useful in the future diagnosis of the disease.

16.4.4 Zymogram analysis

If a DNA fragment contains recognition sites of restriction enzymes (RE), a fine zymogram of restriction enzymes can be obtained after the enzyme-digested DNA fragments are separated by gel electrophoresis and visualized in the gel (Figure 16.1). This kind of analysis can be used for gene diagnosis when some mutations have been involved in a restriction site of RE and related to a given disease. For example, a point mutation will result in disappearance of the original hydrolyzing site, (e.g., GAATTC→CAATTC, the former is the recognition and cleavage site of *Eco*R Ⅰ, a type of restriction endonucleases), or in the formation of a new restriction site (e.g., AAGGTT→AAGCTT, *Hin*d Ⅲ). Of course, deletion or insertion mutation can also cause the

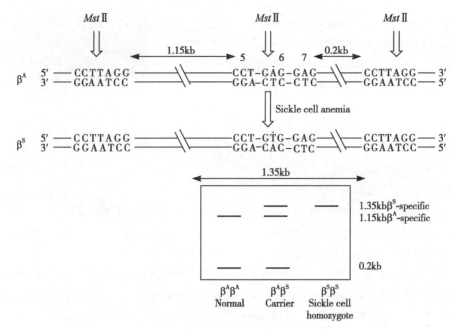

Figure 16.1 **The restriction mapping analysis of sickle cell anemia.**

change in length of enzyme-digested fragment. In this situation, this DNA fragment containing the mutation presents a different type of electrophoresis bands compared with normal zymogram after restriction enzyme digestion. Zymogram analysis can proceed following PCR amplification of DNA specimen.

16.4.5 Analysis of single strand conformation polymorphism

When double helix DNA is denatured into two single-stranded DNA molecules, each of them folds to form different spatial conformation. In other words, two single-stranded DNA molecules with the same length and complementary sequence are able to form different conformation, even if there exists the difference of only one base in their sequences. These kinds of conformational differences are known as "single-strand conformation polymorphism (SSCP)".

Analysis of SSCP is the simplest and most frequently used method of mutation detection. SSCP analysis is usually used by means of combination with PCR to form a new analytical method termed PCR-SSCP, in which PCR is used to amplify the interest DNA fragment and the resultant DNA is denatured, and separated as single-stranded molecules by a non-denaturing polyacrylamide gel electrophoresis (PAGE). One single-stranded DNA folds

differently from the other one even if they differ by a single base, therefore the mutation-induced changes of spatial conformation of DNA fragments result in different mobility of single-stranded DNA molecules in PAGE. The comparison of the electrophoresis pattern (Figure 16.2) of single-stranded PCR products from a patient with that from a normal control will reveal existence of the mutation in DNA sequence, which can be further identified by DNA sequencing.

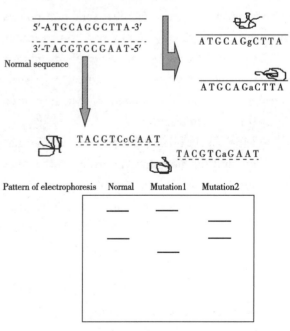

Figure 16.2 **The schematic isolation principle of single strand conformation polymorphism (SSCP).**

Because the changes of spatial conformation of single-stranded DNA occur under different conditions, e.g., temperature and ionic environment, the sensitivity of SSCP depends on these environmental conditions. The sensitivity of detection of mutation by PCR-SSCP is generally very high, up to >80% of correct identification, if the DNA fragments to be analyzed are within the length of 300 bp. The sensitivity of PCR-SSCP decreases along with the increase of the length of fragments. Now many modifications of SSCP analysis have been created to improve the detecting efficiency of mutation, including gel running conditions, different gel matrixes, high throughput SSCP, RNA-SSCP (rSSCP), restriction endonuclease fingerprinting-SSCP (REF-SSCP), multiple fluorescence-based PCR-SSCP (MF-PCR-SSCP) etc.

16.4.6 Biochip/DNA microarray

Biochip is also known as bio microarray due to the characteristic of array distributing biological molecules. Sometimes biochip may be known as DNA microarray because the material fixed on the chip is DNA molecule. In fact, the basic principle and method detecting genetic substances with biochip is molecular hybridization. However, the miniaturized biochip format represents a fundamental revolution in biological analytical techniques. In addition, biochip assays (Figure 16.3) utilize solid non-porous surface such as glass, procedures of fluorescence labelling and detecting instead of flexible membrane such as nitrocellulose and nylon, radioactivity labelling and detection of autoradiography which are used in traditional hybridization.

Biochips can be divided into two scopes: information chip and function chip. According to the types of biological materials fixed on solid surface of silicon or glass matrix, information chips can be again classified into (a) gene chips, including DNA chip, cDNA chip and oligonucleotides chip, (b) protein chips, and (c) tissue chips. Function chips include micro-fluidics chip and lab-no-a-chip. The latter will integrate many functions such as isolation, purification, extraction, mixing, chemical reaction, detection of results into a single chip, hence such a chip is just

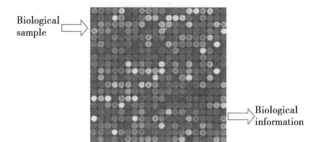

Figure 16.3 **Biochip assays.**

like a miniaturized laboratory (lab on a chip), it will represent the highest level of micro-fluidics bioanalytical techniques.

Biochips have a multitude of applications. The first application is to monitor gene expression. The changes in gene expression may correlate with the onset of a given human disease, and chip-based analysis of pathological tissues should work out the identification of abnormality at gene level whose expression will be altered in a given disease state. In addition to analysis of gene expression, biochip assay may also be used for mutation detection, polymorphism analysis, gene mapping, sequencing of genes, studies of molecular evolution and analysis of functional genome.

16.4.7 Restriction fragment length polymorphism

Restriction fragment length polymorphism (RFLP) indicates the variation in length and number of restriction enzyme-cleaved DNA fragments due to disappearance of original restriction sites or formation of new restriction sites caused by base mutations. When a restriction enzyme cuts DNA at its restriction sites, a collection of DNA fragments with precisely defined length and number are produced. The comparison of the electrophoresis pattern (Figure 16.4) of an enzyme-cleaved DNA specimen from a patient with that from a normal control will reveal the disappearance or the formation of a restriction site that is the evidence of the existence of a point mutation. This kind of comparison can also reveal the existence of the deletion or insertion in a DNA segment.

A fairly extensive RFLP mapping of human genome has been established. Sometimes, these DNA fragments are closely linked with a mutated gene of a

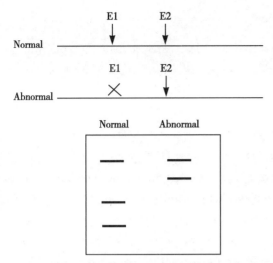

Figure 16.4 **The schematic working principle of restriction fragment length polymorphism (RFLP).**

given disease or the disease itself, thus these DNA fragments may supply useful information and a genetic marker of polymorphism for diagnosis of the disease.

16.5 Introduction on gene therapy

Gene therapy is a novel approach to treat human diseases at molecular level, besides operative therapy, chemotherapy and radiotherapy, named as the forth kind of therapeutic way for diseases. Gene therapy becomes one of the most prosperous fields of modern medicine.

Basically, gene therapy can be described as the intracellular delivery of genetic material to generate a therapeutic effect by correcting an existing abnormality and providing cells with a new function. Of course, inherited disorders are the main focus, but now a wide range of diseases including cancer, vascular disease, neurodegenerative disorders and other acquired diseases are being considered as the objects of gene therapy.

Gene therapy can be divided into germline gene therapy and somatic gene therapy. Somatic gene therapy involves the insertion of genes into diploid cells of an individual where the genetic material is not passed onto its progeny. The transfer is currently achieved by use of lipids and virus-mediated system. To date, somatic gene therapy can be accepted by majority of scientists, even ethicists. Germline gene therapy is generally not acceptable because any suddenness or oversight occurring in gene manipulation could be passed down from generation to generation, and artificially add a new inherited variation into the human gene pool. So to date all strategies and techniques of gene therapy have been only applied to somatic gene manipulation.

16.6 Main strategies of gene therapy

16.6.1 Gene replacement

The purpose of gene replacement is to replace a defective gene with the corresponding normal gene introduced into the target cells and integrated *in situ* into the chromosomes by the means of homologous recombination. But the frequency of homologous recombination is very low natively, only 10^{-6}, so it is difficult to operate at present.

16.6.2 Gene augmentation

Gene augmentation is also called as gene complementation. In this situation, the abnormal gene is not removed, and the wild target gene will be introduced to the defective cells or other cells. Thus the function of the originally affected gene is complemented or even strengthened owing to the expression of wild gene transfected in cells. This is the most common approach to gene therapy.

16.6.3 Gene interference

Gene interference is the use of a specific vector to carry small interfering RNA (siRNA) into target cells, selectively silencing the target gene, thereby inhibiting protein synthesis and achieving disease treatment. Efficient and highly specific delivery of siRNA to target cells is the key to achieve gene interference therapy. Gene interference technology has important application prospects in the treatment of human diseases.

16.6.4 Suicide gene

A suicide gene refers to a gene fragment of a part of a virus or a bacterium that is transferred to the target cell. The specific enzyme expressed by the gene converts a non-toxic or less toxic precursor drug into a substance which is toxic to the cell, thereby the recipient cells carrying the gene are killed. Such genes are called suicide genes, also known as drug-sensitive genes. The commonly used suicide gene systems include *E. coli* cytosine deaminase gene (EC-CD), TK-GCV system, and CD-5-FC system. Suicide gene therapy is often used to treat tumors and infectious diseases.

16.6.5 Gene immunotherapy for cancer

Gene immunotherapy for cancer is the use of gene transfer technology to transfer major histocompatibility complexes, costimulatory molecules, cytokines and receptors, tumor antigens, tumor suicide genes, tumor suppressor genes and other foreign genes into the human body, which induces or enhances the body's anti-tumor immune response, and achieves the purpose of inhibiting and killing tumor cells. At present, tumor immunogene therapy is clinically tested in cancer patients such as prostate cancer, non-small cell lung cancer, leukemia and melanoma.

16.7 Main strategies of gene delivery

16.7.1 Liposome vectors

Cationic lipids have been developed over the past decade as an extensively used delivery system of nuclear acids. Artificial liposome has many advantages as gene transfer carrier. After mixed together, negatively charged DNA and positively charged lipids can form a complex spontaneously. Now there are many commercially available liposomes such as "lipofectin" or "lipofectamine", which have different formulations and properties, and are capable of effectively delivering genetic material to cells. But the gene transfer with liposome *in vivo* occurs at low efficiency, although some success has been achieved in delivering genetic material to lung and liver.

16.7.2 Retrovirus vectors

Retroviruses release their genetic material into host cells in the process of infection. Their RNA genome is reversely transcribed into cDNA that is then integrated into the genome of host cells as a previrus. Previrus contains several basic elements: 5′-LTR-ψ-*gag*-*pol*-*env*-3′-LTR (LTR: long terminal repeated), where ψ is a packaging signal, *gag* gene encodes proteins which form the viral core, *pol* gene encodes reverse transcriptase (RT), *env* gene encodes the glycosylated envelope proteins which determine the tropism of the virus, LTR sequences contain the *cis*-acting sequences required to regulate viral genome replication, transcription and mediate integration into the host genome (Figure 16.5).

The vectors can be constructed from virus cDNA by deleting *gag*, *pol*, and *env* genes, leaving only the basic elements needed for gene expression, packaging, and integration of cDNA into host genome. The original coding genes of virus are replaced by antibiotics-resistant gene that is used as a selectable marker. The remained DNA is carried by a plasmid so that it is convenient to insert the therapeutic gene (cDNA) into the vector (Figure 16.6).

The advantages of retrovirus delivery system include: (a) they are relatively nonpathogenic with a few exceptions; (b) they have the ability to infect a wide variety of cell types; (c) they have the potential of integrating into host genomes. Generally, integration of the genetic material can be accomplished in dividing cells using, so that they are very useful for therapy of tumors with ability of very active division.

The main obstacles of the current retroviral system include: (a) they are unable to produce high titers; (b) they can induce immune response of human body against the viral proteins and activate complements system; (c) the lack of precise specificity for host cells allows the pseudovirus to enter those

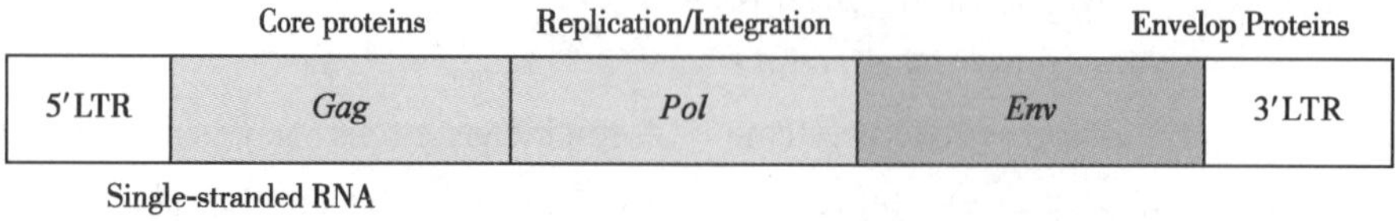

Figure 16.5 **Schematic representation of the basic retrovirus genome.**

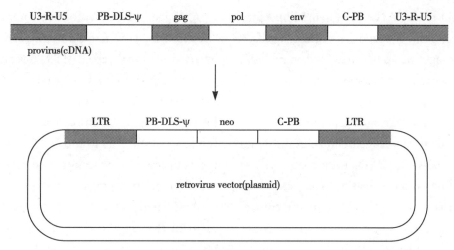

Figure 16.6　**The schematic construction elements of retrovirus vector.**

cells that are not intended to receive the foreign gene so as to reduce the amount of pseudovirus entering the targeted population, even to cause unwanted physiological effects; (d) they might disrupt an essential gene or alter genes in the way that favors cancer development, depending on where the inserted genes land, because retroviruses splice their DNA into host chromosomes randomly, instead of into predictable sites.

16.7.3　Adenovirus vectors

The ubiquitous adenoviruses (AdV) have gained the most popularity as alternatives to retroviruses in part because of their safety. The wild forms of the viruses infect readily a wide range of tissues and, more importantly, have the tropism for the lung, but cause nothing more serious than chest colds in otherwise to healthy people.

The genome of AdV consists of a single double-stranded linear DNA molecule of approximately 36 kb in length, conventionally divided into 100 map units (mu). The genome is functionally divided into two noncontiguous overlapping regions: early and late region, defined by the onset time of transcription after infection. There are five distinct early regions (E1A, E1B, E2, E3, E4) and one major late region (MLR) with five principal coding units (L1 to L5) in genome sequence (Figure 16.7). At the extremities of the viral chromosome are the inverted terminal repeats (ITRs), which constitute the origin of replication of the adenoviral genomes. The encapsidation

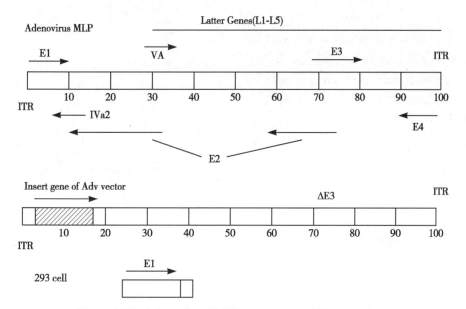

Figure 16.7　**Adenovirus (AdV) vector gene delivery system.**

signals adjacent to the 5′ ITR are essential for entry of viral DNA into empty adenovirus virion capsids.

The advantages of adenovirus vector include: (a) Adenovirus can infect a large number of different human cell types including non-dividing cells. (b) Adenovirus vectors dispatch genes to the nucleus but apparently do not insert them into chromosomes so to avoid the possibility of disturbing vital cellular genes or abetting cancer formation. This is a reason why adenovirus vectors attract the major clinical interest. (c) Adenovirus vectors have much larger cloning capacity compared with retrovirus vectors.

Because the viruses do not incorporate into genome of host's cells and the expression of inserted genes occurs only over a short period, so the effect of therapy is obvious but only temporary. Thus the drawback of this adenovirus system is that (a) it is not suitable for the therapy of chronic diseases, such as cystic fibrosis, where periodically repeated administration of viral particles containing foreign gene are needed; (b) the disadvantage, shared with retroviruses vectors, is lack of specificity for target cells; (c) the more serious stumbling block to use adenovirus vectors in patients is the body's strong immune response against them, resulting in the death of treated cells and inactivity of introduced new genes when adenoviruses vectors are delivered once again.

16.7.4 Adeno-associated virus vectors

Human adeno-associated viruses-2 (hAAV-2) is a simple, non-pathogenic, widespread, single stranded DNA virus. The replication of AAV genome in host cells needs additional genes that are provided by a helper virus (usually adenovirus or herpes similar virus). AAV DNA integrates preferentially into human chromosome 19 at site q13.4 with no noticeable effects. The AAV vectors have been shown to be weak immunogens upon delivery to some tissues.

AAV genome consists of two coding regions, *cap* and *rep*, which are flanked by the major regulatory inverted terminal repeat (ITR) at each end of the genome, and a packaging signal (Figure 16.8). The *cap* gene encodes the coat protein and *rep* encodes for proteins involved in replication and integration function. Both 3′ ITR and 5′ ITR play important regulatory role in replication, integration, scission of genome, and packaging of virion.

AAV-based vectors are constructed by the replacement of the *rep* and *cap* genes with a transgenic fragment of interest, retaining the terminal repeats and packaging sequence essential to direct replication and packaging of the genome. These vectors are therefore double helper-dependent, requiring *trans*-complementation of the *rep* and *cap* function, as well as the natural helper adenovirus functions.

The application of AAV vector is quite safe because 96% of the parental AAV genome has been deleted so that the recombinant AAV vectors contain only the gene of interest. The major drawbacks of these vector systems include: (a) the complicated process of vector production (most systems necessitate the use of adenovirus to supply helper functions), (b) the vector is small and can only accommodate the genes < 4.8 kb.

16.8 Application of gene therapy

16.8.1 Gene therapy for inherited immunodeficiency syndromes

(1) ADA-SCID

Adenosine deaminase (ADA) catalyzes the irreversible deamination of adenosine to produce inosine and 2′ deoxyinosine as part of the salvage pathway of biosynthesis of purine nucleotides. ADA is expressed in all of tissues and its action can be observed as a detoxification process. ADA deficiency results in the accumulation of the degraded products of adenos-

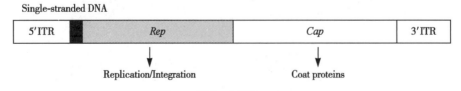

Figure 16.8 **AAV genome.**

ine, which are toxic to T lymphocytes and to a lesser extent to B-lymphocytes, rendering T and B cells lose their normal function. In clinic, this type of deficiency represents as syndrome of severe combined immunodeficiency (SCID). the Affected children generally die from severe infection for 1 to 2 years after birth. The better choice for ADA-SCID treatment is bone marrow transplantation (BMT) and an alternative pathway of treatment is extracellular enzyme replacement therapy. However, this kind of treatment fails to stabilize the immune parameters in the long term.

Since the ADA gene has been cloned, ADA-SCID was regarded as the perfect model disease for gene therapy because of well-defined target tissue and cells, easily accessible cells and their precursors, and appropriate size (1.1kb) of ADA-cDNA for delivery vectors, particularly for retroviruses vector.

Based on the above consideration and pre-clinic experiments, ADA gene therapy was approved in 1990 for the treatment in a four years old girl suffering from ADA-SCID. After that the patient's symptoms were greatly improved.

(2) SCID-X1

SCID is a primary immunodeficiency disease. There are many types of SCIDs, and a common gene defect caused by X-linked inheritance is called SCID-X1. Children with SCID-X1 are at high risk of serious infection and have a short life expectancy in the absence of a matched bone marrow donor. The pathogenesis of the disease is due to a mutation in the gene encoding the interleukin-2 receptor γ chain (IL-2Rγ), resulting in a deficiency of functional lymphocytes, so SCID-X1 patients have serious immune system defects. The traditional method of treating SCID-X1 is to perform hematopoietic stem cell transplantation, but the shortcoming of this method is that the donor's human leucocyte antigen (HLA) is usually different from the recipient's HLA, so this method will bring high mortality. Gene therapy may bring some hope to the treatment of SCID-X1. *In vitro* gene therapy generally replaces the patient's cellular IL-2Rγ unit with a normal gene by means of a viral vector that prevents cancer lesions. However, we still need more clinical trials to confirm the validity and safety of gene therapy for SCID-X1.

16.8.2 Cystic fibrosis gene therapy

Cystic fibrosis (CF) is a recessive disease that the main clinical problems are lung damage and respiratory failure. In 1989, the gene responsible for CF was cloned and designated as CFTR (cystic fibrosis transmembrane conductance regulator). The CFTR protein is a cAMP-mediated chloride channel being nonfunctional in CF patients. In CF, the treated targets are post-mitotic lung epithelial cells in which differentiation has been terminated. A delivery system and protocols suitable for gene therapy of CF have been confirmed. Two main system used are adenovirus and liposomes respectively. Among them, adeno-associated virus (AAV) is the most promising viral vector system for the treatment of CF, because it lacks the pathogenicity of the airway and does not cause any human disease. The constructed AAV-CFTR gene is transferred into airway epithelial cells, which can significantly restore the expression of CFTR protein.

16.8.3 HIV gene therapy

Human immunodeficiency virus (HIV) is an RNA retrovirus that causes acquired immunodeficiency syndrome (AIDS), and its genes can be transferred into the genes of host cells for replication.

Ribozymes are a class of catalytically active small RNAs that specifically bind to target RNA molecules and cleave to reduce target gene activity. There are many types of ribozymes, and the ribozymes currently used to treat HIV infection are mainly hairpin nuclease and hammerhead ribozyme. Because they have a small catalytic domain, they can be used as transgenic expression products or can be transported *in vivo* directly in the form of artificially synthesized oligonucleotides. Ribozymes can cleave RNA at different stages of the HIV life cycle. At present, there are many ribozyme-based strategies for the treatment of HIV infection, such as ribozymes designed for HIV long terminal repeat (LTR), *tat* gene, *pol* gene, *env* gene, and *rev* gene. They can inhibit HIV replication to a certain extent.

16.9 The future of gene therapy

16.9.1 IMLYGIC™ in cancer

Imlygic™ (Talimogene laherparepvec, T-VEC) is a new type of immunotherapy approved by the FDA in 2015, called oncolytic virus immunotherapy. T-VEC is a gene modified type 1 herpes simplex virus. This virus replicates in large amounts in tumor cells and expresses the immunostimulatory protein granulocyte-macrophage colony-stimulating factor (GM-CSF). Direct injection of T-VEC into the tumor site can result in lysis of tumor cells or recruitment of immune cells to attack the tumor. T-VEC is mainly used to treat advanced melanoma, which can infect and kill melanoma cells that are dividing, and stimulate the immune system to attack cancer cells. T-VEC is the first FDA-approved oncolytic virus treatment.

16.9.2 Precision medicine initiative

Precision medicine is an emerging method for disease diagnosis and treatment based on the population's genes, environment and lifestyle, and individual differences. It mainly relies on cutting-edge technologies such as genome sequencing technology, biological information and big data. First, genome sequencing of large sample populations is carried out to establish a huge medical data repository. Then, by analyzing the gene information of different individuals, we can further understand the common causes and special causes of various diseases. Ultimately, researchers develop therapeutic drugs and treatments for specific individuals and specific diseases. Gene detection technology, sequencing technology, gene editing technology and other new technologies represented by molecular biology technology make gene therapy more precise. In particular, gene editing technology has laid the foundation for gene therapy diseases. Among the various gene editing technologies, the new and improved CRISPR technology can target more gene sites and reduce the risk of misediting, thus making gene editing more accurate. As an accurate gene therapy technology, gene editing will gradually enter the clinical trial stage in the future.

Key Concepts

- Gene diagnosis is such a process to judge diseases of human body through detecting the abnormalities of DNA or RNA materials. Gene diagnosis is a very essential tool for the diagnosis of clinic medicine because the molecular mechanism of inherited diseases involves the changes of genotypes caused by various mutations.
- Gene therapy is a novel approach to treat human diseases at molecular level. Basically, gene therapy can be described as the intracellular delivery of genetic material to generate a therapeutic effect by correcting an existing abnormality and providing cells with a new function. A wide range of diseases including cancer, vascular disease, neurodegenerative disorders and other acquired diseases are being considered as the objects of gene therapy.

Key Words

gene diagnose(基因诊断)
gene immunotherapy(基因免疫治疗)
gene interference(基因干扰)
gene replacement(基因置换)
Gene therapy(基因治疗)
suicide gene(自杀基因)

Questions

1. What are the basic techniques used for gene diagnosis?
2. What are the strategies for gene therapy?

References

[1] Baum C (2011) Gene therapy for SCID-X1: focus on clinical data. *Mol Ther*, 19(12):2103–2104.
[2] Bhardwaj U, Zhang YH, McCabe ER (2003) Neonatal hemoglobinopathy screening: molecular genetic technologies. *Mol Genet Metab*, 80(1–2):129–137.
[3] Cavazzana-Calvo M, Hacein-Bey S, Yates F, et al. (2001) Gene therapy of severe combined immunodeficiencies. *J Gene Med*, 3(3):201–206.
[4] Franchini M, Mannucci PM (2014) The history of hemophilia. *Semin Thromb Hemost*, 40(5):571–576.
[5] Southern E (2006) Southern blotting. *Nat Protoc*, 1(2):518–525.

Juan Chen